2021 International Conference on Electrical Drives & Power Electronics (EDPE 2021)

Dubrovnik, Croatia
22 – 24 September 2021

IEEE Catalog Number: CFP21EDQ-POD
ISBN: 978-1-6654-3236-8

**Copyright © 2021 by the Institute of Electrical and Electronics Engineers, Inc.
All Rights Reserved**

Copyright and Reprint Permissions: Abstracting is permitted with credit to the source. Libraries are permitted to photocopy beyond the limit of U.S. copyright law for private use of patrons those articles in this volume that carry a code at the bottom of the first page, provided the per-copy fee indicated in the code is paid through Copyright Clearance Center, 222 Rosewood Drive, Danvers, MA 01923.

For other copying, reprint or republication permission, write to IEEE Copyrights Manager, IEEE Service Center, 445 Hoes Lane, Piscataway, NJ 08854. All rights reserved.

*** This is a print representation of what appears in the IEEE Digital Library. Some format issues inherent in the e-media version may also appear in this print version.

IEEE Catalog Number: CFP21EDQ-POD
ISBN (Print-On-Demand):   978-1-6654-3236-8
ISBN (Online):   978-1-6654-3235-1

Additional Copies of This Publication Are Available From:

Curran Associates, Inc
57 Morehouse Lane
Red Hook, NY 12571 USA
Phone: (845) 758-0400
Fax: (845) 758-2633
E-mail: curran@proceedings.com
Web: www.proceedings.com

Table of Contents

Day 1: September 22, 2021

Power Electronics and Applications I
Violet Room
11:00-12:30, Wednesday

1   Simulation verification of topologies of balancing systems with flyback converters
Marek Šimčák

7   Power Quality and Electromagnetic Compatibility - Syncretism or Dichotomy?
Mircea Buzdugan

13   Review on high power WPT coil system design
Jakub Skorvaga, Miroslav Pavelek

19   Control structure and Converter losses of a 3L-ANPC with GaN HEMTs switching at 100 kHz
Martin Geppert, Günter Schröder

25   Semi-Flexible Power Control for Grid-Connected Converter
Petr Šimek, Martin Bejvl, Viktor Valouch

Electrical Drives and Machines I
Green Room,
11:00-12:30, Wednesday

30   Qualitative comparison of the behavior for a five-phase induction motor in error states for different stator windings connections
Jakub Kellner

37   Calculation Methodology of Common Capability Diagram for Paralelly Connected Generator Group
Boris Glavan, Zlatko Hanić, Mario Vražić, Marinko Kovačić

43   Encapsulated Air Cooling System for Scalable Axial Flux Motors
Sebastian Berndl

50   Mechanical Analysis of Different Rotor Topologies for High Speed PMSM in Automotive Application
Michal Kováčik, Pavol Rafajdus, Ronald Bastovanský

56   Analysis of Temperature Change in a Permanent Magnet Synchronous Generator Under Load due to Stator Winding Inter-Turn Short Circuit
Michael Barrett

Power Electronics and Applications II
Violet Room
14:00-15:30, Wednesday

62   Simulation design of output-stage for residential smart-grid
Kristián Takács, Peter Drgona

67   Evaluation of high-voltage GAN transistor in 3-phase PFC circuit
Michal Frivaldsky, Michal Pipiska

72   A Floating Double Buck-Boost Converter as Driver for a Permanent Exited DC Machine
Felix A Himmelstoss, Helmut Votzi

78   High Step-down Converters for Battery Charging in Photovoltaic Applications
Zvonimir Malović, Nikola Vuger, Viktor Šunde, Željko Ban

88   UPS with PFC input stage for railway applications with improved immunity on input overvoltage and energy strikes
Ivan Šolc, Ante Hećimović, Viktor Šunde, Željko Ban

Electrical Drives and Machines I
Green Room,
14:00-15:30, Wednesday

97   Observability Conditions for Speed Sensorless Induction Motor Models with Neglected or Included Iron Loss Representation
Krisztián Horváth

102   A Comparative Study of Different SMO Switching Functions for Sensorless PMSM Control
Viktor Petro, Karol Kyslan

108   Contactless Energy Transmission using a Transformer with Movable Secondary
Mariusz Stepien

112   General approach of radial active magnetic bearings design and optimization
Cristina Adascalitei, Martis Radu, Claudia Martis

118   The Application of Neural Network Metamodels for Interior Permanent Magnet Machine Performance Prediction
Zlatko Hanić, Ana Hanić, Marinko Kovačić

Day 2: September 23, 2021

Power Electronics and Applications III
Violet Room
10:15-11:45, Thursday

123   Investigation of the High Voltage GaN transistor module
Richard Zelnik

129   Unbalanced Load Modeling and Control in Microgrid with Isolation Transformer
Hao Jiang, Shuyu Cao, Chew Beng Soh, Feng Wei

136   Modeling and Predictive Control of LLC Resonant Converter for Solar Powered E-Bicycle Charging Station
Karla Draženović, Ante Perić, Željko Jakopović, Viktor Šunde

144   Quality Evaluation of Jointly Used Modular Multilevel Converters and Battery Energy Storages
Alexander Bubovich, Maksims Vorobjovs, Ilya Galkin, Tenuun Dovdon

152   Harmonic content of the input current of the boost converter in quasiperiodicity
Željko Stojanović, Denis Pelin

Motion Control and Mechatronics
Green Room
10:15-11:45, Thursday

159   Determination of material parameters using video extensometry during tensile testing
Jaroslav Bulava, Libor Hargaš, Dušan Koniar, Silvia Štefunová

164   ICE Vehicle Energy Consumption Measurement and Calculation Methodology for the Purpose of EV Battery Pack Design
Antonio Peršić, Hrvoje Kristek, Mario Vražić, Vladimir Peršić

169 Sensorless speed control of brushed DC machine
Lukas Gorel, Michal Vidlák, Vladimir Vavrúš, Pavol Makyš

177 Testing predictive vehicle dynamics control algorithms using a scaled remote controlled car and a roadway simulator
Petar Makarun, Marko Švec, Goran Josipović, Šandor Ileš

183 Kalman Filter Based Sensor Fusion for Omnidirectional Mechatronic System
Blaž Korotaj, Branimir Novoselnik, Mato Baotić

189 Analysis of FPGA Implementation of Set-based Predictive Control algorithm for Grid-tied Inverters
Bruno Vilić-Belina, Renato Babojelić, Šandor Ileš, Jadranko Matuško

Day 3: September 24, 2021

Power Electronics and Applications IV
Violet Room
10:15-11:45, Friday

196 Grid-connected and Islanded Control of Energy Storage Converter
Božo Terzić, Ozren Bego, Marin Despalatović, Goran Majić, Ante Kriletić, Mislav Blajić

201 LCL Filter Design with Amorphous Core Inductor for 100 kVA Energy Storage Converter
Božo Terzić, Ozren Bego, Marin Despalatović, Goran Majić, Ante Kriletić, Mislav Blajić

206 Power Loss Analysis of Multi-converter System with Single Wire and Wireless Energy Transfer
Marcin A Zygmanowski, Marcin Kasprzak, Kamil Kierepka, Jarosław Michalak, Grzegorz Jarek, Krzysztof Przybyła

212 Analysis of Regenerative Cycles and Energy Efficiency of Regenerative Elevators
Dora Erica, Damjan, Martina Kutija, Luka Pravica, Ivana Pavlić

220 DC/DC Converter Topologies for Elevator Energy Storage Systems Based on Supercapacitors
Martin Makar, Martina Kutija, Luka Pravica, Filip Jukić

Electrical Drives and Machines III
Green Room
10:15-11:45, Friday

228 Loss Minimization Control and Energy Consumption Improvements in PMSM drive
Martin Novak

234 State of Health and Aging Estimation Using Kalman Filter in Combination with ARX Model for Prediction of Lifetime Period of Li-Ion Batteries
Lukáš Krčmář, Pavel Rydlo, Ales Richter, Jakub Eichler, Pavel Jandura

238 Online Optimization of Firing Angles for Switched Reluctance Motor Control
Peter Bober, Želmíra Ferková

243 Influence of Rotor Slot Number on Magnetic Noise in a Squirrel-cage Induction Motor for Traction Applications
Ivan Milažar, Damir Žarko

249 Application of a Simplified Inverse Fuzzy Model for an Induction Motor Drive Control
Daniela Perdukova, Pavol Fedor, Marek Fedor, Viliam Fedak

254 **Index of Authors**

EDPE 2021

35th International Conference on Power Electronics and Electrical Drives (EDPE)

10th Joint Croatia-Slovakia Conference

Dubrovnik, Croatia

September 22 - 24, 2021.

Impressum

TITLE
2021 IEEE 35th International Conference on Electrical Drives and Power Electronics (EDPE)

EDITORS

Željko Jakopović
Jadranko Matuško

CONFERENCE ORGANIZERS
KoREMA– Croatian Society for Communications, Computing, Electronics, Measurement and Control
FER – Faculty of Electrical Engineering and Computing, Zagreb, Croatia
FEEI – Faculty of Electrical Engineering and Informatics, Technical University of Košice, Slovakia
SES – Slovakian Electrotechnical Society

TECHNICAL SPONSORS
IEEE Croatia Section

Foreword

On behalf of EDPE 2021 National Organizing Committee (NOC) and International Steering Committee (ISC) it is our great pleasure to welcome you to the 35th International Conference on Electrical Drives and Power Electronics (EDPE). The Conference is the 10th joint Croatian-Slovak conference organized jointly by KoREMA (Croatian Society for Communication, Electronics, Measurement and Control), FER (Faculty of Electrical Engineering and Computing, University of Zagreb), TU (Technical University of Kosice) and SES (Slovak Electrotechnical Society) and sponsored by IEEE Croatia Section, Končar Electrical Industry Inc. and Typhoon HIL Inc.. EDPE2021 is organized in specific situation of COVID crises. Until the last moment we did not know if live event will be possible. We do hope that good scientific program and social events will enable you to forget hard pandemic times, isolation and work from home. This time we have decided that conference has only oral presentations, with three excellent keynote lecturers, one on each conference day. Keynotes are covering different fields of EDPE conferences, power electronics and electrical drives. Except scientific program, EDPE conferences are known for interesting social events. The Welcome reception is organized in the evening of the first conference day and the conference trip to the Konavle region on the second conference day. We wish you all a successful presentation of your papers, interesting and fruitful discussions and pure enjoyment during conference trip. We hope that beautiful surroundings and squares of Dubrovnik will make your free time unforgettable. At the end, we would like to express our deep thanks to the members of NOC, ISC and reviewers for their praiseworthy work done in the evaluation of the submitted papers. We want to thank all the authors and sponsors for their contribution, which is essential to make EDPE 2021 successful.

Yours sincerely,

EDPE 2021 General Chair,
Željko Jakopović
Viliam Fedák

Conference Organizing Committee

Conference General Chair:
Željko Jakopović, University of Zagreb, Croatia

Conference Co-Chairs:
Viliam Fedak, Technical University, Košice, Slovakia

National Organizing Committee Members:

Željko Jakopović, University of Zagreb, Croatia
Jadranko Matuško, University of Zagreb, Croatia
Viktor Šunde, University of Zagreb, Croatia
Stjepan Stipetić, University of Zagreb, Croatia

Martina Kutija, University of Zagreb, Croatia
Šandor Ileš, University of Zagreb, Croatia

International Program Committee Chairs
Jadranko Matuško(Croatia)
Karol Kyslan(Slovakia)

Members of the International Program Committee

Vanja Ambrožič (Slovenia)
Vassilios G. Agelidis (Denmark)
Gheorghe-Daniel Andreescu (Romania)
Goce Arsov (Macedonia)
Drago Ban (Croatia)
Zvonko Benčić (Croatia)
Jan Bauer (Czech Republic)
Pavol Bauer (Netherlands)
Ion Boldea (Romania)
Silverio Bolognani (Italy)
Pavel Brandstetter (Czech Republic)
Stefan Brock (Poland)
Josef Cernohorsky (Czech Republic)
Zdenek Cerovsky (Czech Republic)
Mihai Cernat (Romania)
Jaeho Choi (Republic of Korea)
Goga Cvetkovski (North Macedonia)
Branislav Dobrucky (Slovakia)
Tomislav Dragičević (Denmark)
Jaroslav Dudrik (Slovakia)
Frantisek Durovsky (Slovakia)
Jawad Faiz (Iran)
Viliam Fedak (Slovakia)
Zelmira Ferkova (Slovakia)
Michal Frivaldsky (Slovakia)
Zdenek Hadas (Czech Republic)
Sandor Halasz (Hungary)
Jens Bo Holm-Nielsen (Denmark)
Mikuláš Huba (Slovakia)
Šandor Ileš (Croatia)
Željko Jakopović (Croatia)
Gojko Joksimović (Montenegro)
Zbigniew Kaczmarczyk (Poland)
Vladimir Katić (Serbia)

Teresa Orlowska-Kowalska (Poland)
Marian P. Kazmierkowski (Poland)
Péter Korondi (Hungary)
Karol Kyslan (Slovakia)
Martina Kutija (Croatia)
Dan Lascu (Romania)
Elena Lomonova (Netherlands)
Dusan Maga (Czech Republic)
Miro Milanovic (Slovenia)
Sanjeevikumar Padmanaban (Denmark)
Marek Pastor (Slovakia)
John K. Pedersen (Denmark)
Denis Pelin (Croatia)
Daniela Perdukova (Slovakia)
Nedjeljko Peric (Croatia)
Ales Richter (Czech Republic)
Alex Ruderman (Kazachstan)
Asif Sabanovic (Turkey)
Helga Silaghi (Romania)
Vladislav Singule (Czech Republic)
Stjepan Stipetić (Croatia)
Viktor Sunde (Croatia)
Lorand Szabo (Romania)
Božo Terzić (Croatia)
Daniel Nistor Trip (Romania)
Viktor Valouch (Czech Republic)
Mario Vašak (Croatia)
Ján Vittek (Slovakia)
Dinko Vukadinović (Croatia)
Helmut Weiss (Austria)
Damir Žarko (Croatia)
Pavel Zaskalicky (Slovakia)
Genadij Stepanovic Zinoviev (Russia)

Simulation verification of topologies of balancing systems with flyback converters

Marek Simcak, Pavol Spanik
Department of mechatronics and electronics, Faculty of electrical engineering and information technologies
University of Zilina
Zilina, Slovakia
marek.simcakfeit.uniza.sk

Abstract— This article describes the problem of balancing systems with flyback converters. These balancing systems are divided into three types and mutually compared. The first type of balancing system is with a flyback converter for each cell separately. The second type of balancing system is a flyback converter with a multi-winding transformer with a primary winding separately for each cell and secondary side connected to whole pack. The third type of balancing system consists of a boost converter and a flyback converter for each cell separately. Simulation analyses have been focused on the evaluation of the efficiency and balancing speed for discussed topologies. The parameters of the batteries and flyback converters are given as well withing proposed paper. The experimental results received from simulation analyses could be useful as an inspiration for a specific implementation for different types of battery packs.

Keywords— *battery management system, battery, traction batteries, flyback converter, balancing system, flyback topology, multi-winding flyback topology, boost-flyback topology,*

I. INTRODUCTION

Batteries as a source of energy are living in a golden age. The market is pushing development to maximize battery capacity, life and safety. High voltage battery packs consist of electrical cells that are connected in series (higher voltage) and in parallel (higher capacity). Various battery packs consist of battery modules and the modules consist of electrochemical cells. These electrochemical cells are the basic pillar of battery pack quality. The variety of electrochemical cells reduces the capacity, life and safety of battery packs. This difference is mainly in the different internal resistance of the battery, the smaller capacity of the battery or a different degree of self-discharge [1] – [5]. This ultimately represents the difference in cell voltage or the difference in SOC. These properties of electrochemical cells must be as small as possible. Therefore, battery manufacturers give considerable resources to finance these features as little as possible. The task of the balancing system is to minimize these differences. The balancing system has the task of balancing the voltage or SOC levels. This approach can be passive or active. The passive balancing system balances the energy in the cells by discharging cells with higher energy state [6] – [8]. An active balancing system pours active energies from strong cells into weak cells. In terms of efficiency or energy savings, an active balancing system is far more efficient than a passive one. With low-capacity cells, this is not such a big impact, but with high-capacity cells or batteries, this difference is striking [9] – [11]. Therefore, in this article we will find 3 different ways and approaches to balancing electrochemical cells. These topologies are topology with 4x flyback converter, topology with multi-wind flyback converter and topology with 4x boost converter and 4x flyback converter.

II. TOPOLOGIES OF BALANCING SYSTEMS

In balancing systems, it is necessary to use converter topologies with galvanic isolation of input from output. This is a very important feature of the converter in terms of implementation [12] – [13]. Therefore, 3 different topologies were selected via a flyback converter. This drive has a simple design and is suitable for battery applications. These balancing systems were applied to a battery pack with 4 LiFePO4 traction cells Winston LFP040AHA (fig. 1). The battery simulation model is used from [14] –[15].

Fig. 1. Winston LFP040AHA cell 3,6V, 40Ah

A. Topology with 4x flyback converter

This topology consists of 4 flyback converters, which are connected by the primary side to a single cell and the secondary side is connected to the whole battery pack. With this method, one flyback converter is always started separately. The control system consists in finding the highest SOC value. If this system detects the highest SOC value, then the given flyback converter connected by the primary side to the cell with the highest SOC value starts [16] – [17]. In this way, it draws excess energy from the strongest cell and pushes it into the entire battery pack via the secondary side of the flyback converter. This method has several disadvantages. Each time the flyback converter is started, energy must be fed into the core of the flyback converter transformer. This causes relatively large losses in low voltage applications. These losses radiate to the surroundings or to cooling in the form of heating the transformer core. This reduces the overall efficiency of the balancing system. This also indirectly affects

978-1-6654-3236-8/21 $31.00 © 2021 IEEE

the cost of manufacturing a flyback converter, as these high-frequency transformers need to be oversized. The proposed system can be seen in Figure 2.[7]

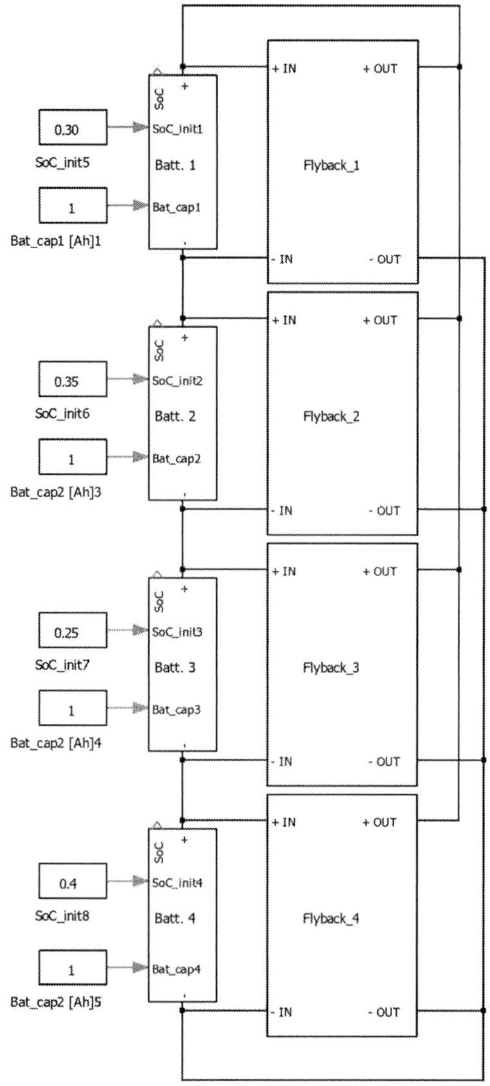

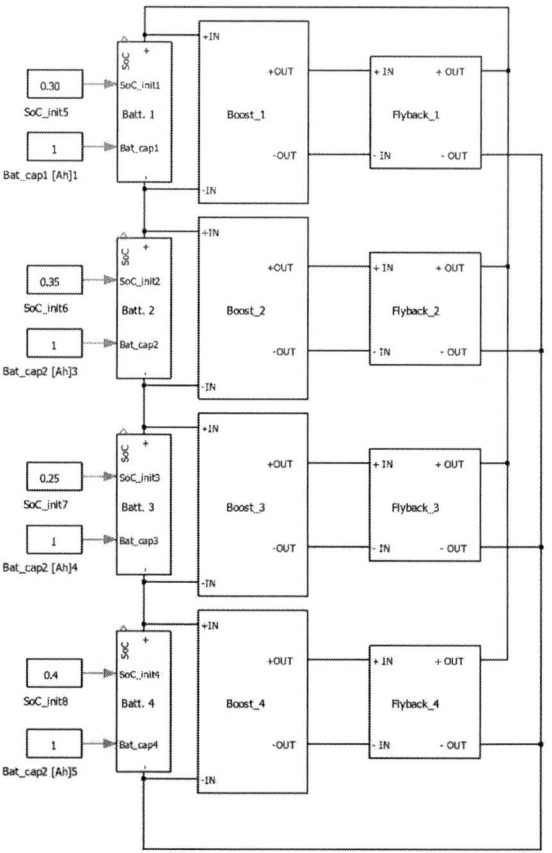

Fig. 2. Block diagram of balancing system with 4 flyback converters

B. *Topology with multiwinding flyback converter*

This topology represents a simpler topology compared to the 4x flyback converter topology. There is one high frequency transformer in this topology. This transformer consists of 4 primary sides, which are connected to each cell separately. Furthermore, this transformer contains a core and a secondary side, which is connected to the entire battery pack [18]. This balancing system is controlled similarly to the previous one. The balancing system detects the highest SOC value and always switches on the primary side that is connected to the given cell with the highest SOC value. Potential disadvantages include mainly high demands on the transformer, its core and the symmetry of the primary windings. The symmetry of the winding represents the similarity of the primary windings, the inductances of the primary winding must be the same. The main advantage is that we only need one secondary side. This can save insights and

increase the efficiency of the balancing system. The proposed system can be seen in Figure 3.

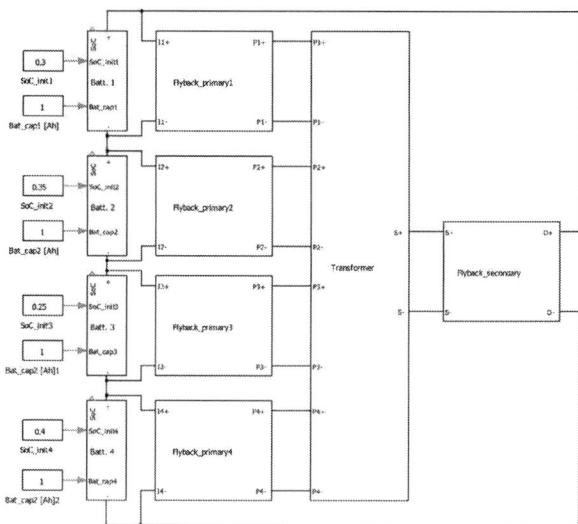

Fig. 3. Block diagram of balancing system with multiwinding flyback converter

C. *Topology with boost and flyback converter*

Fig. 4. Block diagram of balancing system with 4 boost converters and 4 flyback converters

978-1-6654-3236-8/21 $31.00 © 2021 IEEE

This topology was investigated due to high losses in the core of the high-frequency transformer. In low-voltage applications, relatively high losses occur in the flyback converter. The boost converter is a simple way to increase the efficiency of this low-voltage application. While the flyback converter achieves efficiencies of approximately 50-70% for applications up to 20V, the boost converter achieves 90% and more [19]. The main idea of this topology is to increase the cell voltage (3-4V) to the battery voltage (12V-16V). Then this higher voltage is supplied by the flyback converter with input and output voltage 16V. In this way we can reduce losses in the flyback converter. This flyback converter provides galvanic isolation. For summary, the boost converter 4V-16V is powered from a single cell and the output of the boost converter supplies a specific flyback converter 16V-16V, the output from the flyback converter is connected to the entire battery. Thus, the excess energy from the cell passes through the boost converter to the flyback converter and from the flyback converter to the entire battery [20]. The control of the balancing system is the same as in the previous cases. The SOC value is determined and the battery with the highest SOC triggers its particular boost and flyback converter. This topology is shown in Figure 4.[7]

III. SIMULATION RESULTS

This simulation was simulated with a reduced value of the capacity of the battery used. Target value of the capacity was set to 2.5% of nominal capacity (40 Ah), i.e. to 1 Ah. This was set up due to fact that simulation models requires high demands on the computational power. This is caused by requirements on high accuracy of simulation results, thus small step size is required for long time-dependent simulations, while time of the simulation is related to the capacity of the used battery. One simulation experiment using PLECS circuit simulator lasts for about 3 working days if 1Ah of capacity is defined. If 40Ah of battery capacity should be considered, then simulation times will be several times higher, putting also high demands on storage capacity.

For selected balancing topologies, the waveforms shown in Figures 5, 6 and 7 were simulated. The initial conditions were the same for each simulation. CELL1 = 30% SOC, CELL2 = 35% SOC, CELL3 = 25% SOC and CELL4 = 40% SOC. The voltage shown in these graphs represents the OCV (open circuit voltage) of the individual cells. Therefore, when starting the inverter, we do not see a faster drop in voltage and, conversely, when the inverter is switched off, the voltage rises. In a real application, when the cell is retracted, there is a drop in voltage due to the internal resistance of the battery. This simulation model simulates the output voltage of the battery only for OCV (open circuit voltage). This voltage represents the voltage of the unloaded cell.

Analyzing current waveforms is seen that the rise of the current to a constant value requires certain time. This causes a longer balance with another cell. The current from the strongest cell is directed by its converter to the entire battery pack and charges the remaining three cells. This can be seen

in the initial part of the flow chart. The blue color represents the current which charges individual cells. Behind the blue waveform there are red and green current waveforms. These waveforms represent the magnitude of the current by which weaker cells are charged. Control references were set to input current value about I_{RMS} = 35A.

We can see that the waveform of SOC changes in the last part of the graphs. This value was monitored by the control system. Based on the magnitude of this value, the given flyback converters were operated. From figures is seen that proposed control system worked accurately as required values have been balanced.

The main monitored parameters of the balancing system are balancing time and balancing efficiency. Balancing time represents how long it takes for all cells to be balanced. This parameter is directly proportional to the magnitude of the current with which the system is balanced. If the current is higher, the time is shortened. At the same time, however, increasing the current decreases the efficiency of the balancing system. This is very critical in low voltage applications. The higher the current, the greater the losses in the semiconductor components or in the magnetic components of the converters. Therefore, it is important for the designer to find a compromise between balancing time and balancing efficiency. We find out the balancing time from the graphs by finding out the time when it started to balance, and we find out the time when all cells were balanced to a similar SOC value. We determine the efficiency of the balancing system by comparing the values of the average SOC at the beginning of the simulation and comparing it with the value of the SOC after the end of balancing procedure. This difference represents the decrease in SOC by the energy required for the balancing process. The larger the decrease in SOC compared to the average SOC value at the beginning, the lower the efficiency of the balancing system. The average value at the beginning of each balancing was 32.5% SOC.

A. Topology with flyback converter

With a topology with 4 flyback converters, a few things were found. The waveforms are shown in Figure 5. During current waveforms, we see high current peaks. These peaks represent potentially higher losses. These current peaks are caused by charging the energy into the core of the high-frequency transformer. This energy must be removed (demagnetized) each time the flyback converter is started again. In real applications, this can mean high stress on the semiconductor or magnetic components of the converter. Therefore, the designer of the potential balancing system must take these current peaks into account. The fact that current peaks occur there also means that the balancing system loses energy and time by recharging energy into the core of the high-frequency transformer. We adjust the efficiency of the balancing system SOC_{INIT} (32.5) - SOC_{END} (30.75) = 1.75 SOC decrease. This system was balanced at time t = 36.5 s.

978-1-6654-3236-8/21 $31.00 © 2021 IEEE

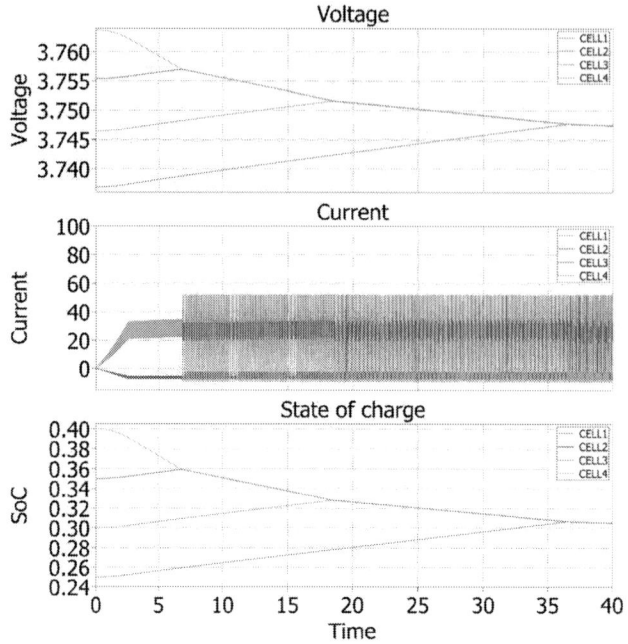

Fig. 5. Simulation results of balancing system with 4 flyback converters. (Voltage [V], Current [A], State of charge [-])

B. Topology with multiwinding flyback converter

In another simulation, a balancing system with a multi-winding high-frequency transformer was simulated. The simulation waveforms are shown in Figure 6. If we compare this system with the topology of 4 flyback converters, then at first glance we see that the current peaks there are no longer so critical. This ensures the continuous use of only one high-frequency transformer. With this method, it is necessary to provide control so that only one primary side of the transformer is used at a time. If 2 or more primary sides are switched, overvoltage of the battery pack may occur. A big problem in potential real applications can occur when designing a transformer. The symmetry and equality of each primary winding is important. This can potentially be a big problem for the transformer designer. Finally, the core of this transformer will have to be over-dimensioned. An interesting result was that at approximately the same current value, this battery pack was soon balanced as in the case of 4 flyback converters. In this case, the battery pack was balanced within 34.5s. The efficiency of the balancing system is very similar to the previous case SOC_{INIT} (32.5) - SOC_{END} (30.75) = 1.75 SOC decrease.

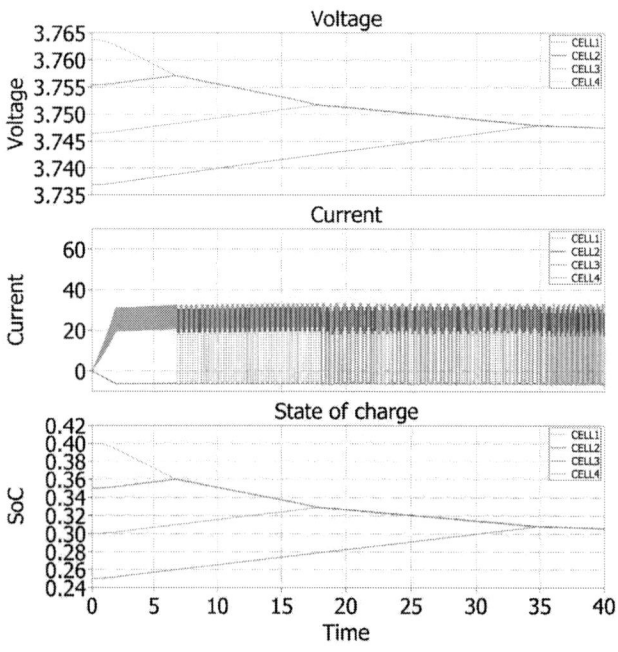

Fig. 6. Simulation results of balancing system with multiwinding flyback converter. (Voltage [V], Current [A], State of charge [-])

C. Topology with boost and flyback converter

In another case, a topology with a boost converter and a flyback converter was simulated. The results of this simulation can be found in Figure 7. We can notice that in this simulation there is the largest current ripple. This ripple potentially could increase the losses, while it represents the current entering the boost converter. Because this topology contains more semiconductor elements, it is more complex and at the same time more expensive. But with the proper circuit design, it can be the best way to balance. For the parameters used in this simulation and the real components used, the simulation came out as follows. This topology is the most complex. When implementing a boost converter, it is necessary to take into account the primary inductance of the flyback converter. This parameter was the biggest problem when debugging the simulation. Therefore, in a real application, it is necessary to consider the inductance of the primary winding of the flyback converter with the inductance of the boost converter. As can be seen in the waveforms in Figure 7, the battery pack was balanced at 37s. This time is little bit distorted by pauses after each balancing action. This pause is caused by the response of the regulators. When using other controllers, this time can be shortened, which means shortening the balancing time. In this simulation, the efficiency of the balancing system SOC_{INIT} (32.5) - SOC_{END} (30) = 2.5 SOC decrease. It is due to the higher number of semiconductor elements and the higher ripple current.

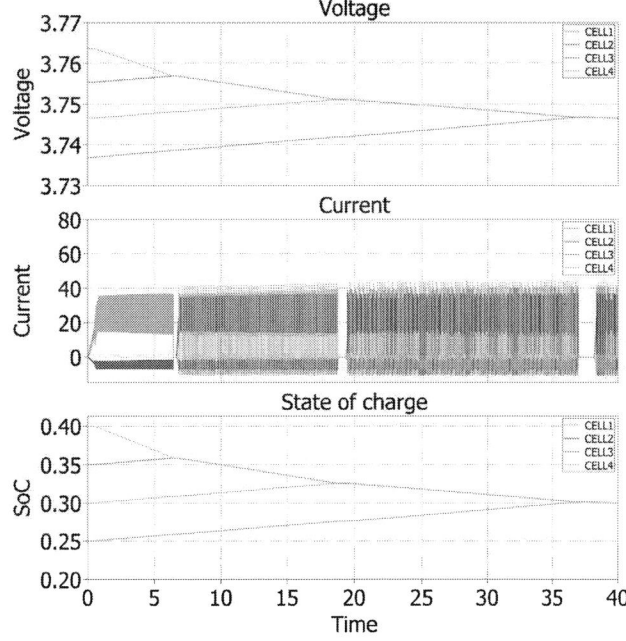

Fig. 7. Simulation results of balancing system with 4 boost and 4 flyback converters. (Voltage [V], Current [A], State of charge [-])

IV. CONCLUSIONS

This article serves as inspiration for real applications of balancing systems. From these simulations it is clear what the OCV, current and SOC waveforms should look like in different topologies of balancing systems with flyback converters. The simulations show which topology is good in different parameters. During real verifications, there will be many other things that are not very clear from the simulations. Therefore, this article serves only as a first step and to help you choose the topology for your application.

The summary of the first topology with 4 flyback converters is that this topology represents a simple implementation. The problem is that when alternately discharging two or more cells, there is a problem with the peak currents. These currents slow down the balancing time and reduce efficiency.

The summary of the second topology with a multi-winding high-frequency transformer consists in the fact that in real implementation we place high demands on the transformer. This concerns its saturation and symmetry of the primary windings of this multi-winding transformer. Management is very important here. Under no circumstances may this control of this balancing system trigger 2 or more windings at once. This would destroy or damage the battery pack. The advantage of this topology also lies in the fact that we use a single magnetic circuit.

To summarize the latest topology with boost and flyback converters, we must note that there may be great potential for implementation. In this topology, however, we must note the complexity of the production of the boost converter and the complexity of the flyback converter. This can radically increase the cost of manufacturing and developing this balancing system. The potential, however, is that the boost converter has far higher efficiencies than the flyback converter. However, this boost converter does not have galvanic isolation, so it is good to apply a higher voltage from the boost to the primary side of the flyback. Flyback does not increase or decrease the voltage only through the transformer it galvanically separates the output voltage, which recharges the entire battery pack.

ACKNOWLEDGMENT

This publication was realized with support of Operational Program Integrated Infrastructure 2014 - 2020 of the project: Innovative Solutions for Propulsion, Power and Safety

978-1-6654-3236-8/21 $31.00 © 2021 IEEE

Components of Transport Vehicles, code ITMS 313011V334, co-financed by the European Regional Development Fund". Authors would like to thank also to national grant agency Vega for project funding Vega-1/0063/21.

REFERENCES

[1] Battery university [online]. Available from: https://batteryuniversity.com/.

[2] Frivaldsky, M.; Pavelek, M. In Loop Design of the Coils and the Electromagnetic Shielding Elements for the Wireless Charging Systems. Energies 2020, 13,6661. https://doi.org/10.3390/en13246661

[3] H. Nazi and E. Babaei, "A Modularized Bidirectional Charge Equalizer for Series-Connected Cell Strings," in *IEEE Transactions on Industrial Electronics*, vol. 68, no. 8, pp. 6739-6749, Aug. 2021, doi: 10.1109/TIE.2020.3003661.

[4] P. Wen, S. Xie, H. Liu, Q. Qian, H. Feng and B. Zhang, "A Voltage-Competing Flyback-Based Balancer Combined with Auxiliary Power Supplier," 2020 IEEE Applied Power Electronics Conference and Exposition (APEC), 2020, pp. 2225-2229, doi: 10.1109/APEC39645.2020.9124446.

[5] M. Cacciato, A. Consoli and V. Crisafulli, "A high voltage gain DC/DC converter for energy harvesting in single module photovoltaic applications," *2010 IEEE International Symposium on Industrial Electronics*, 2010, pp. 550-555, doi: 10.1109/ISIE.2010.5637663.

[6] Frivaldsky, M., Sedo, J., Pipiska, M. et al. Design of measuring and evaluation unit for multi-cell traction battery system of industrial AGV. Electr Eng 102, 1579–1591 (2020). https://doi.org/10.1007/s00202-020-00982-z

[7] B. -A. Enache et al., "Flyback Converter for a Multi-Chemistry Battery Balancer," 2021 12th International Symposium on Advanced Topics in Electrical Engineering (ATEE), 2021, pp. 1-6, doi: 10.1109/ATEE52255.2021.9425205.

[8] D. D. Stefanov, T. P. Todorova and V. C. Valchev, "A Flyback Converter Based System for an Active Charge Balancing of Li-Ion Battery Packs," 2018 IEEE XXVII International Scientific Conference Electronics - ET, 2018, pp. 1-4, doi: 10.1109/ET.2018.8549663.

[9] A. M. Imtiaz and F. H. Khan, ""Time Shared Flyback Converter" Based Regenerative Cell Balancing Technique for Series Connected Li-Ion Battery Strings," in IEEE Transactions on Power Electronics, vol. 28, no. 12, pp. 5960-5975, Dec. 2013, doi: 10.1109/TPEL.2013.2257861.

[10] M. Cacciato, G. Nobile, G. Scarcella and G. Scelba, "Real-time model-based estimation of SOC and SOH for energy storage systems," *2015 IEEE 6th International Symposium on Power Electronics for Distributed Generation Systems (PEDG)*, 2015, pp. 1-8, doi: 10.1109/PEDG.2015.7223028.

[11] Frivaldsky, M.; Hanko, B.; Prazenica, M.; Morgos, J. High Gain Boost Interleaved Converters with Coupled Inductors and with Demagnetizing Circuits. Energies 2018, 11, 130. https://doi.org/10.3390/en11010130

[12] M. Cacciato, G. Nobile, G. Scarcella, G. Scelba and A. G. Sciacca, "Energy management optimization in stand-alone power supplies using online estimation of battery SOC," *2016 18th European Conference on Power Electronics and Applications (EPE'16 ECCE Europe)*, 2016, pp. 1-10, doi: 10.1109/EPE.2016.7695559.

[13] J. Koscelnik, M. Frivaldsky, M. Prazenica and R. Mazgut, "A review of multi-elements resonant converters topologies," 2014 ELEKTRO, 2014, pp. 312-317, doi: 10.1109/ELEKTRO.2014.6848909.

[14] PAVELEK, M., SPANIK, P., FRIVALDSKY, M. Voltage stress reduction on compensation capacitors of wireless charging systems for transport and industrial infrastructure. Communications - Scientific Letters of the University of Zilina [online]. 2019, 21(2), p. 50-57. ISSN 1335-4205, eISSN 2585-7878. Available from: https://doi.org/10.26552/com.C.2019.2.50-57

[15] DAOWD, M., OMAR, N., VAN DEN BOSSCHE, P., VAN MIERLO, J. Passive and active battery balancing comparison based on MATLAB simulation. In: 2011 IEEE Vehicle Power and Propulsion Conference: proceedings [online]. IEEE, 2011. ISSN 1938-8756. ISBN 978-1-61284-247-9. Available from: https://doi.org/10.1109/VPPC.2011.6043010

[16] Spanik, P.; Frivaldsky, M.; Adamec, J.; Danko, M. Battery Charging Procedure Proposal Including Regeneration of Short-Circuited and Deeply Discharged LiFePO4 Traction Batteries. Electronics 2020, 9, 929. https://doi.org/10.3390/electronics9060929

[17] M. Frivaldsky, J. Adamec, M. Danko and P. Drgona, "Traction battery (40 Ah LiFePO4) recovery after long-term short circuit," 2020 International Symposium on Power Electronics, Electrical Drives, Automation and Motion (SPEEDAM), 2020, pp. 299-304, doi: 10.1109/SPEEDAM48782.2020.9161868.

[18] SIMCAK M., SPANIK P. Comparison of control methods for power stage of battery management systems with 4 cells Marek Šimčák, Pavol Špánik. In: EPE 2020 [print, electronic]: 21st International scientific conference on electric power engineering. Prague, Czech Republic. October 19-21, 2020. - ISSN 2376-5631.

[19] SIMCAK M., FRIVALDSKY M. Design of balancing system for traction batteries [electronic] In: ELEKTRO 2020 [electronic]: conference proceedings. - 1st ed. - Danvers: Institute of Electrical and Electronics Engineers, 2020. - ISBN 978-1-7281-7541-6.

[20] D'ANGELO, V., CANNAVACCIUOLO, S., LECCE, S., BENDOTTI, V., PENNISI, O. Enhanced hotplug protection in BMS applications. Part I: Theoretical aspects and practical issues. In: 2019 AEIT International Conference of Electrical and Electronic Technologies for Automotive AEIT Automotive 2019: proceedings [online]. 2019. p. 1-5. Available from: https://doi.org/10.23919/EETA.2019.8804513

Power Quality and Electromagnetic Compatibility- Syncretism or Dichotomy?

Mircea Buzdugan
Dept. of Building Services Engineering
Technical University of Cluj-Napoca
Cluj-Napoca, Romania.
email: mircea.buzdugan@insta.utcluj.ro

Abstract—This approach intends to highlight both the dichotomy and the syncretism between two fundamental concepts in electrical systems, namely power quality, and electromagnetic compatibility. The author presents a new sort of syncretism, that is uniting parts of the two concepts in a new manner, supported by experiments. That is why the syncretism the author is seeking for does not mean mainly the unity in terms used in different standards which is clearly a shallow one, but in the physical phenomena involved and in the strong interconnection linking the two concepts. The terminology used in the paper is that of the standard IEC 60050-161 International Electrotechnical Vocabulary and of the IEEE Std 1100™-2005, IEEE Recommended Practice for Powering and Grounding Electronic Equipment.

Keywords—syncretism/dichotomy, nonsinusoidal regime, resolution of powers, harmonic limits, conducted electromagnetic emissions, passive EMI filtering.

I. INTRODUCTION

This section is devoted to a brief presentation of the concepts that represent the two pillars on which the paper rests.

According to the Cambridge dictionary (https://dictionary.cambridge.org/dictionary/english/), dichotomy is the difference between two completely opposite ideas or things, while conversely the syncretism is the combining of different religions, cultures, or ideas.

The dichotomy distinguishing between the two terms, namely electric power quality and electromagnetic compatibility is present particularly at the level of two major standardization units, namely IEEE (Institute of Electrical and Electronics Engineers), may be the world's largest technical professional organization for advancement of technology on one hand and IEC (The International Electrotechnical Commission), covering a vast range of technologies which involve electrical energy on the other hand.

The term of electrical power quality is pregnant only in IEEE vocabulary and standards, while the term adopted by IEC is electromagnetic compatibility.

On a closer inspection, one could conclude that the dichotomy resides only in the terms involved, while the physical phenomena involved are quite syncretic.

However, in the language spoken by two groups of scientists involved in the study of these phenomena, the concept of power quality is dedicated mainly to lower frequency phenomena (e.g., power quality analyzers measure frequency perturbations up to the 100-order harmonic, to five or six kHz, depending on the standard of electrical power generation), while the specific phenomena of electromagnetic compatibility are higher frequency ones. Electromagnetic compatibility generally studies specific phenomena starting

from 100 kHz. In this respect the difference (let us use the term of dichotomy), is more than significant.

The question that arises from this paper is whether power quality and electromagnetic compatibility are two syncretic or two dichotomous concepts.

The answer is at least as ambiguous as the question as the next sections will reveal.

A. Power Quality

Terms as "power quality", "safety", "reliable service", or "voltage quality" have been used over the years as some of the main objectives of electrical systems design, mainly due to the following reasons [1], [2]:

- Electronic and electric power equipment have become in the recent decades much more sensitive to voltage disturbances. Manufacturing processes have become less tolerant to possible equipment malfunctions. Consequently, production engineers have become less tolerant as well to virtual unexpected outages.

- Modern electronic equipment is sensitive mainly to voltage disturbances, causing unfortunately in turn perturbations directed towards the adjacent end users supplied from the same grid. The increasing use of equipment containing electronic converters, of lower or higher power, draws a strongly non-sinusoidal current containing an extended range of harmonic components. The rich content of current harmonics consequently determines in their turn harmonic distortion of the supply voltage. Even if each individual piece of equipment does not generate an important number of current harmonics, together they can cause serious distortions of the supply voltage. The pursuit for energy-efficient equipment is on the other hand a substantial source of power quality perturbations. For example, adjustable speed drives are an important source of current waveform distortion, being on the other hand highly sensitive to other different power quality disturbances.

- The increased need for standardization and setting performance criteria was determined mainly by the deregulation of the electricity market, which have led to the necessity of establishing relevant quality indices. Customers in their turn have become increasingly demanding, requesting more and more information about the quality of the voltage they will use.

- Last but not least, a crucial factor is that power quality can be measured in a more accurate manner. In the past, measurements were limited only at a few power quality parameters, namely RMS voltage, frequency, and long-term interruptions. Nowadays, many phenomena, negatively affecting power quality have

978-1-6654-3236-8/21 $31.00 © 2021 IEEE

been identified and became measurable, for instance variations and events. Variations are disturbances from the steady-state or quasi-stationary state that require (or allow) continuous measurements, while events are sudden disturbances characterized by start and stop moments and mainly by a very short lifetime.

A short power failure or a transient impulsive phenomenon are the most common examples of power quality.

In this respect, a transient overvoltage can be characterized in two ways: by magnitude which is either the maximum voltage or the maximum deviation of the voltage from the normal sine wave and by duration. However, there is a difficulty in defining the impulsive phenomena due to the impossibility of detecting accurately the start and end moments of events.

A unique manner to define events is by triggering them at their start moments. Unlike variations which do not require a trigger, events need necessarily a trigger. The difference between a voltage dip (sag) and a voltage variation (magnitude) consists mainly in triggering. A voltage dip has a specific start and end moment, although not always uniquely defined. Both voltage dips and voltage variations use the square mean voltage root (rms) as the basic measurement quantity. However, for further processing of voltage variations all values are important, while for further processing of voltage dips only RMS values below a certain threshold are considered [3].

About definitions, various literature sources offer different and sometimes contradictory formulations related to power quality.

The Dictionary of the Institute of Electrical and Electronics Engineers (IEEE) says that "power quality is the concept of powering and grounding sensitive equipment in a manner appropriate to the operation of equipment" [4]. From this definition it could be derived that harmonic current distortion is a problem of power quality only if it affects sensitive equipment. Another limitation of this definition is that the concept can be applied only to assess equipment performance.

On the other hand, the definition of the International Electrotechnical Commission (IEC) for power quality, given by IEC 61000-4-30, is: "The characteristics of electrical power at a given point of an electrical system, evaluated in relation to a set of technical benchmarks." This definition of power quality is not related to the performance of the equipment but to the possibility of measuring and quantifying the performance of the power supply system.

The definition that best covers the concept is that electrical power quality is a combination of voltage and current quality. A simple and direct solution resides in defining the ideal voltage as a sinusoidal voltage waveform with constant amplitude and frequency, where both amplitude and frequency are equal to their nominal value.

The ideal current is also of constant amplitude and frequency, but in addition the frequency and the phase of the current must be in phase with those of the voltage. Any deviation of the voltage or current from the ideal is a disturbance of the quality of the electric power. A disturbance can be a voltage or a current disturbance, but in practice it is often quite difficult to distinguish between the two of them.

This difficulty in distinguishing between voltage and current disturbances is one of the reasons which consecrated the use of the aggregate term of power quality. The term of voltage quality is reserved for cases where only the voltage at a certain location is considered. The term current quality is used to describe the performance of electric/electronic equipment connected to the grid.

In the specific literature many alternative definitions of power quality are used. Some of them are worth mentioning either because they express the opinion of an influential organization or because they represent an inedited approach.

Interestingly, "current quality" is often not explicitly mentioned. The quality of the current is implicitly considered only if it affects the quality of the voltage. This is also the viewpoint of the present approach the author considering that "current quality" may excite electromagnetic conducted emissions.

B. Electromagnetic Compatibility

Electromagnetic compatibility, EMC, means the ability of an equipment or system to function satisfactorily in its electromagnetic environment without introducing intolerable electromagnetic disturbances in any of those environments [5].

Lack of electromagnetic compatibility or non-compliant solutions can have the most diverse consequences: minor irritations caused by poor audio/video reception or questionable quality, unpredictable or unreliable operation of equipment in residential, commercial, or industrial environments, culminating at limit even in ecosystem risks.

The standard also defines the notion of electromagnetic disturbance (an electromagnetic noise, an unwanted signal, or a change in the propagation medium itself) as any electromagnetic phenomenon that may degrade the performance of a device, equipment, or system, or adversely affect living or inert matter. Electromagnetic interference represents an actual degradation of the performance of an equipment, transmission channel or system caused by an electromagnetic disturbance.

The terms "electromagnetic disturbance" and "electromagnetic interference" refer to cause and effect, respectively, but are often used without discrimination.

In EMC regulations, the mechanisms of electromagnetic interference are divided into two categories, conducted, and radiated electromagnetic interference.

Conducted interference consists of the energy of unwanted signals leaving a particular product through its ports (power or signal ports).

On the other hand, radiated interference (a term sometimes used to cover induction phenomena) is the electromagnetic disturbance for which energy is transferred through space in the form of electromagnetic waves. High frequency signals that "escape" from a product via a port can also radiate quite strongly.

Speaking about radiated interference, the standards and unfortunately in many situations even in the literature, do not distinguish between near field radiation and far field radiation, both being shipped in the same category of radiated emissions. While specialists in electromagnetism are familiar with the difference between near-field coupling and far-field

978-1-6654-3236-8/21 $31.00 © 2021 IEEE

emissions, the vague terminology of EMC regulations tends to lead to confusion.

An alternative term often used for interference is electromagnetic emission, the phenomenon by which electromagnetic energy emanates from a source through conducting mechanisms or through radiation.

Conducted emissions are generally controlled by filtering, radiated emissions are controlled mainly by shielding.

In terms of immunity, the opposite of emissions is the ability of a device, equipment, or system to operate without degradation in the presence of an electromagnetic disturbance. It analyzes the resistance of equipment and systems to the two emission categories mentioned above.

A complementary notion of immunity, which is simply the lack of immunity, is electromagnetic susceptibility, the inability of a device, equipment, or system to function without degradation in the presence of an electromagnetic disturbance.

II. Spectral Analysis of Signals and Powers Resolution

This section deals with a particularly low voltage disturbance, namely harmonic disturbances. This type of disturbance is measurable and provides essential indices in the analysis of electrical systems.

The spectrum, or frequency content of signals is the most important indicator in analyzing the ability of a system, not only to comply to the regulated limits, but also to ensure electromagnetic compatibility with other electrical systems.

Periodic signals $y(t \pm kT) = y(t)$; $k = 1, 2, 3, ...$ are the most common waveforms in electrical systems.

Any periodic function can be represented as an infinite sum of sinusoidal components. Each sinusoidal component represents a multiple of the fundamental frequency, $f_0 = 1/T$, respectively of the fundamental angular frequency, $\omega_0 = 2\pi f_0$.

Two, of the most common versions of Fourier discrete series are [6], [7]:

- the trigonometric series whose base is $\Psi_0 = 1$ and $\Psi_n = (\cos 2\pi n f_0 t, \sin 2\pi n f_0 t)$; $n = 1, 2, 3, ..., \infty$; the multiples of the fundamental frequency, nf_0, are the harmonics of the fundamental frequency.

- complex-exponential series, more useful due to the simplicity of the calculations, which uses as a basis the complex exponential function

$$\Psi_n = e^{jn\omega_0 t} = \cos n\omega_0 t + j \sin n\omega_0 t; \ n \in (-\infty, \infty) \quad (1)$$

A periodic signal is developed into:

$$y(t) = \sum_{-\infty}^{\infty} c_n e^{jn\omega_0 t} = ... + c_{-1} e^{-j\omega_0 t} + c_0 + c_1 e^{j\omega_0 t} + ... \quad (2)$$

The spectrum analysis of the signals, respectively their characterization as a frequency dependent function has represented and continues to represent the core of the resolution of powers in non-sinusoidal regimes.

The resolution of the power in nonsinusoidal regime has animated the scientific community in a debate of almost a century which for sure will continue as a very hot topic.

Despite all attempts to overthrow the scaffolding of harmonics in the resolution of powers, the latest standard issued by IEEE, namely IEEE Std. 1459-2010 is fundamentally based on the decomposition of non-sinusoidal signals into their harmonic components.

At the early beginning, all started from the inequality P ≤ S, observed at the induction furnaces, which led to the idea that $S^2 = P^2 + N^2$, where N is a total nonactive or fictitious power, which together with the true (real) power P contributes to line power losses. The resolution of $N \geq 0$ and its separation into subcomponents, or models, gave birth to never ending academic and scientific debates, continuing nowadays as well [8] - [15].

In 1927 C. I. Budeanu developed its theory about electric powers, based on the decomposition of the voltage and the current in harmonic components.

The equation $S^2 = P^2 + N^2$ became in Budeanu's theory: $S^2 = P^2 + Q^2 + D^2$, the power triangle becoming a tetrahedron. The first term is the total active (true or real) power P. The second one was named by Budeanu reactive power Q, analogue with the reactive power from the sinusoidal regime. The third term was named distortion power D.

Unfortunately, the reactive and distortion power introduced by Budeanu represent elegant mathematic formulae, but without physical meaning, his approach treating electric circuits in nonsinusoidal conditions as a sum of independent circuits at different frequencies which does not offer a base for the design of passive filters and the control of active filters.

Five years later, in 1933, Fryze developed his theory in the time domain, based on the physical decomposition of the current into two components, the active, respectively the residual one $i(t) = i_a(t) + i_r(t)$. The advantage of this theory is that it does not introduce new components of the power.

The biggest error of interpretation of Fryze consists in the rejection of the concept of harmonics, a very important one in spectrum analysis. However, Fryze's theory has the merit that it represented the base of further developments, especially Czarnecki's theory, who introduced the so-called currents' physical components (CPC).

In literature one can find also, several definitions belonging to Sheperd and Zakikhani, Sharon, Depenbrock, etc. [13].

In 1995, one of Budeanu's former students, A. E. Emanuel, starts from the idea that harmonic power P_H is considered electromagnetic pollution—a by-product of the energy conversion, process that takes place within nonlinear loads [8]. Emanuel's method represents the basis of the present IEEE standard resolution of the powers in nonsinusoidal conditions [14].

It is obvious that a distribution system cannot operate correctly without reactive power. The useful, fundamental magnetic flux in transformers and ac motors is supported by the fundamental reactive current. Thus, it makes good sense to separate P_1 and Q_1 from the rest of the powers.

978-1-6654-3236-8/21 $31.00 © 2021 IEEE

Consequently, one can write: $V^2 = V_1^2 + V_H^2$, $V_H^2 = \sum_{h \neq 1} V_h^2$

respectively $I^2 = I_1^2 + I_H^2$, $I_H^2 = \sum_{h \neq 1} I_h^2$.

The active power (watt) is divided also into two components, $P = P_1 + P_H$, respectively the fundamental and the harmonic (nonfundamental) active power, in which $P_1 = V_1 I_1 \cos\theta_1$ and $P_H = V_0 I_0 + \sum_{h \neq 1} V_h I_h \cos\theta_h = P - P_1$ where P_H so defined also contains components for which h is not an integer.

The fundamental reactive (var) power is $Q_1 = V_1 I_1 \sin\theta_1$

and the apparent power (VA): $S = VI$. The apparent power is divided also into the fundamental apparent power $S_1 = V_1 I_1$ and the nonfundamental apparent power S_N.

The fundamental apparent power S_1 and its components P_1 and Q_1 are the actual quantities that help to define the rate of flow of the electromagnetic field energy associated with the fundamental voltage and current $S_1^2 = P_1^2 + Q_1^2$.

The separation of the rms current and voltage into fundamental and harmonic terms resolves the apparent power in the following manner:

$$S^2 = \left(V_1^2 + V_H^2\right)\left(I_1^2 + I_H^2\right) =$$
$$\left(V_1 I_1\right)^2 + \left(V_1 I_H\right)^2 + \left(V_H I_1\right)^2 + \left(V_H I_H\right)^2 = S_1^2 + S_N^2 \quad (3)$$

S_N, the nonfundamental apparent power is resolved in the following three distinctive terms: $S_N^2 = D_I^2 + D_V^2 + S_H^2$, D_I - the current distortion power (var), D_V - the voltage distortion power (var) and S_H - the harmonic apparent power (VA).

The nonactive power lumps together both fundamental and nonfundamental nonactive components and it shall not be confused with the reactive power. Only when the waveforms are perfectly sinusoidal, results $N = Q_1 = Q$.

III. EXPERIMENTS AND DISCUSSIONS

Air handling units, abbreviated AHUs, condition and distribute air inside buildings, taking and processing the filtered fresh air from outside, while extracting the heat from the vitiated indoor air which is evacuated. The operation is essentially performed by means of two fans, BLDC driven, and a heat exchanger. In addition, an AHU may be provided with heating and/or cooling subunits as well as humidifiers for conditioning the indoor air.

Such a single-phase residential unit was tested in accordance with the standard IEC 61000-3-2: 2018 / AMD1: 2020 - Limits for harmonic current emissions Class A, using a general-purpose programmable power source. Furthermore, a spectrum analyzer has also registered the electromagnetic conducted emissions in the frequency range from 100 kHz to 30 MHz according to the IEC 61000-6-3: 2020 Emission standard for equipment in residential environments.

Because the input stage of the electric circuit of the AHU is provided with a rectifier bridge, even if the unit was provided with a common mode choke mains filter, both the harmonic content and the conducted emissions exceeded the standard limits.

	Frequency	Actual	Limit	% of Limit	Compare
1	50.000	1.391			
2	100.000	0.006	1.080	0.556	Pass
3	150.000	1.309	2.300	56.913	Pass
4	200.000	0.002	0.430	0.465	Pass
5	250.000	1.184	1.140	103.860	Fail
6	300.000	0.003	0.300	1.000	Pass
7	350.000	1.016	0.770	131.948	Fail
8	400.000	0.001	0.230	0.435	Pass
9	450.000	0.825	0.400	206.250	Fail
10	500.000	0.004	0.184	2.174	Pass
11	550.000	0.625	0.330	189.394	Fail
12	600.000	0.009	0.153	5.882	Pass
13	650.000	0.441	0.210	210.000	Fail
14	700.000	0.004	0.131	3.053	Pass
15	750.000	0.285	0.150	190.000	Fail
16	800.000	0.002	0.115	1.739	Pass
17	850.000	0.178	0.132	134.849	Fail
18	900.000	0.004	0.102	3.922	Pass
19	950.000	0.123	0.118	104.237	Fail
20	1000.000	0.001	0.092	1.087	Pass
21	1050.000	0.109	0.107	101.869	Fail
22	1100.000	0.001	0.084	1.190	Pass
23	1150.000	0.100	0.098	102.041	Fail
24	1200.000	0.002	0.077	2.597	Pass
25	1250.000	0.082	0.090	91.111	Pass

Fig. 1. Harmonic limits test report of the air handling unit

This was due mostly to the current spikes specific to single-phase bridge rectifiers. Fig. 1 displays a part of the harmonic content of the AHU. One may observe the overcome of the standard limits starting with the fifth order harmonic, up to the 23rd order harmonic.

At the same time, Fig. 2 displays the conducted electromagnetic emissions generated by the AHU (green line) in the frequency spectrum from 100 kHz to 30 MHz. In Fig. 2 the standard average limit value is displayed in blue, and the quasi-peak standard limit value is displayed in red. With the same colors are displayed the corresponding measured values. The average measured values are situated mostly within the standard limits, except in the first half of the spectrum. Instead, there is a drastic overcome of the quasi-peak values measured in almost the entire frequencies range of interest.

Obvious, a retrofitting filtering method was compulsory. An EMI filtering method consisting of a passive mains filter based essentially on a common mode choke was applied. It is presented in Fig. 3 and has the following values of the components: the differential mode capacitor $C_X = 0.033$ μF, the common mode capacitors $C_Y = 2 \times 2200$ pF and the inductances $L = 2 \times 0.5$ mH of the common mode choke [15].

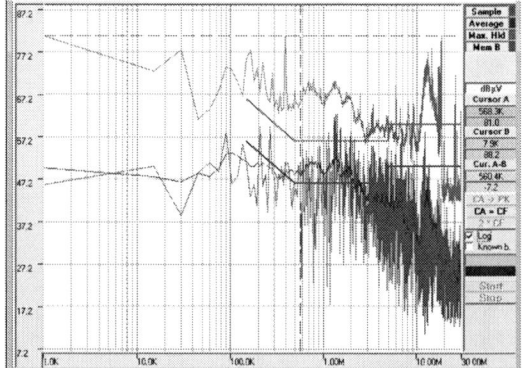

Fig. 2. Conducted interferences of the AHU in logarithmic scale

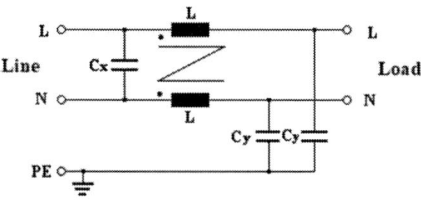

Fig. 3. Passive EMI filtering cell

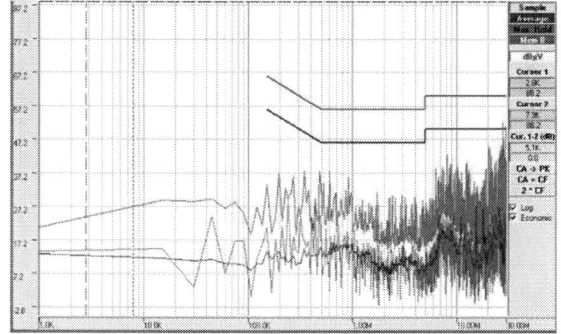

Fig. 4. Conducted interferences of the AHU in logarithmic scale after retrofitting the EMI filter

From Fig. 4 one can observe the efficiency of the applied method, the new quasi peak measured values being far below the standard limits. A second measurement of the harmonic content of the AHU no longer exceeded the limits of the IEC 61000-3-2: 2018 / AMD1: 2020 standard as well.

This was a first example of syncretism between power quality and electromagnetic interference.

Fig. 5 shows the waveform of the voltage and current drawn by a three-phase adjustable electric drive (ASD) controlled by a static converter. One can observe the highly distorted waveform of the current drawn. The current shape of the waveform is a typical one for nowadays ac adjustable-speed drives. Generally, the harmonic spectrum of the current contains mainly the 5th, 7th, 11th, and 13th harmonic components.

Fig. 6 presents the discrete spectrum and the total harmonic distortion THDi of the current drawn by the ASD. Since solely the current is distorted, the voltage may remain almost sinusoidal.

The waveform and the spectrum presented in Figs. 5 and 6 ware measured with a high-performance power analyzer. However, the particularly elegant resolution of the powers provided by the IEEE Std 1459-2010 standard is not, until now the subject of measuring equipment used in practice. Unfortunately, the results provided by the accessible equipment continue to be close to Budeanu's resolution, which was the main IEEE standard until 2010 on this matter.

The syncretism mentioned in the introductory section of this paper, is strongly manifested in this practical situation. Thus, the relatively low frequency harmonic distortions of the discrete spectrum presented above, determine an "echo" in the range of electromagnetic emissions, this time in a continuous spectrum affecting the equipment in the range of conducted interferences between 100 kHz and 30 MHz (Fig. 7).

The discrete spectrum of harmonics in Fig. 6 is the main cause of the continuous spectrum of conducted emissions.

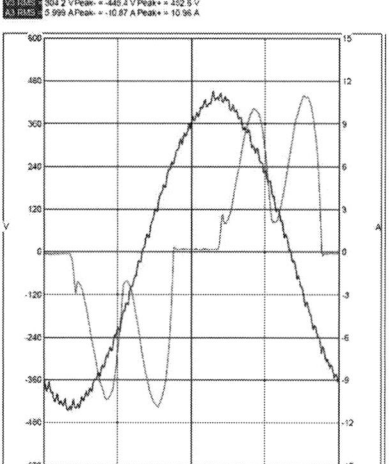

Fig. 5. Voltage supply and current drawn by a line of the three-phase ASD

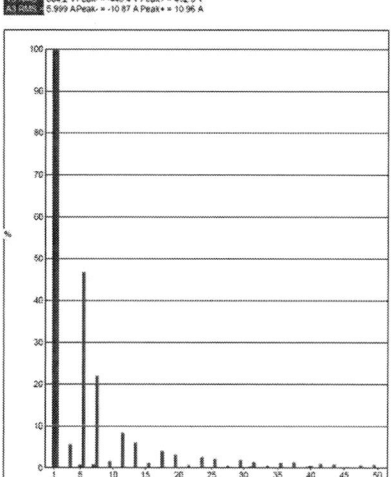

Fig. 6. THDi of the waveform depicted in Fig. 5

To mitigate these disturbances a passive filter having two filtering cells was used (Fig. 8) where the inductances are of $L_a = 3.3$ mH, $L_b = 10$ mH, and the common mode capacitors between the lines and the ground have the values of $C_{Y1} = 47$ nF, $C_{Y2} = 68$ nF [16].

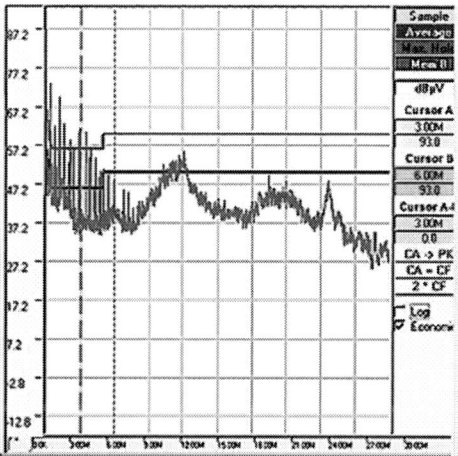

Fig. 7. Conducted emissions generated by the ASD

978-1-6654-3236-8/21 $31.00 © 2021 IEEE

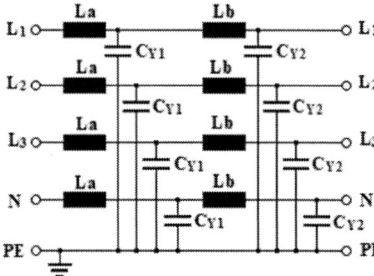

Fig. 8. Three-pahse power passive filter

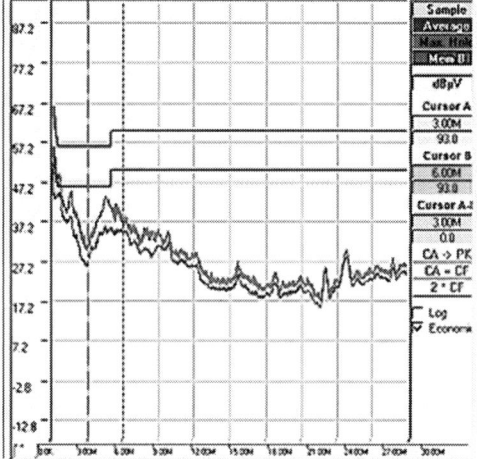

Fig. 9. Conducted interference generated by the ASD after retrofitting the passive filter

After retrofitting a passive filter at the input of the system both the harmonic content, but especially electromagnetic emissions decreased considerably (Fig. 9).

One can see that the maximum value (the red line) and the average value of standardized electromagnetic emissions according to EN 50081-1: 1992 are far below the standardized limit values, being closely related, therefore indicating the efficiency of the filtering method.

The syncretism mentioned above in the two examples is at the same time a dichotomy since harmonics are low frequency disturbance phenomena and electromagnetic interference is a relatively higher frequency perturbance (usually a radio frequency perturbance). So, the two phenomena, somehow dichotomous, are also syncretic being determined by the same causes.

IV. CONCLUSIONS

In the body text of the paper the author has presented a so-called "bijection" between dichotomy and syncretism of electromagnetic perturbation of low and higher frequency, affecting low voltage electrical equipment and systems.

The above-mentioned bijection is supported by experiments which have been depicted in Section III.

Syncretism consisted in the fact that two distinct phenomena, apparently dichotomous, harmonic limits and conducted electromagnetic emissions, measured with two

different specific equipment, were however mitigated using a single retrofitting measure, i.e., a passive filtering method applied to the mains port of the electromagnetic affected equipment.

As further work, the author intends to study the dichotomy and syncretism, between the frequencies dedicated to conducted emissions, and those of radiated emissions, the spectrum of the last ones being expanded to 3 GHz and beyond.

In this paper, no deeper analysis was performed between conducted electromagnetic emissions in differential and common mode, but as it is inferred in [17], the radiation capacity of conducted emissions is much higher in the case of common mode conducted electromagnetic emissions compared to differential mode conducted electromagnetic emissions. So, a deeper quantitative comparative analysis could lead to interesting results.

On the other hand, due to the unity of physical electromagnetic phenomena, the author considers that the interaction cause - effect on this matter presented in this paper, may evolve from bijection to multijection, leading to new results in an expanded spectrum covering at the same time the range of radiated electromagnetic emissions.

REFERENCES

[1] M. H. J. Bollen and I. Yu-Hua Gu, Signal Processing of Power Quality Disturbances, John Wiley & Sons, Inc., 2006

[2] A. Kusko and M. T.Thompson, Power Quality in Electrical Systems, McGraw-Hill Companies, Inc., 2007

[3] M.I. Buzdugan, "Voltage Dips in Power Quality - a Brief Review", AEIT International Annual Conference, Florence, Italy, 2019

[4] IEEE Std 1100-2005 IEEE Recommended Practice for Powering and Grounding Electronic Equipment

[5] IEC 60050-161 International Electrotechnical Vocabulary,

[6] A.M. Grigoryan, M.M. Grigoryan, Brief Notes in Advanced DSP Fourier analysis with MATLAB, Taylor & Francis Group LLC, 2009.

[7] R. L. Easton, Jr., Fourier Methods in Imaging, John Wiley & Sons, Ltd., 2010.

[8] A.E. Emanuel, Power Definitions and the Physical Mechanism of Power Flow, John Wiley & Sons, Ltd, 2010.

[9] Y.Beck, N. Calamaro, D. Shmilovitz, "A review study of instantaneous electric energy transport theories and their novel implementations", Renewable and Sustainable Energy Reviews 57 pp.1428–1439, 2016

[10] D. Jeltsema, "Budeanu's Concept of Reactive and Distortion Power Revisited", Przeglad Elektrotechniczny, R. 92 No. 4, 2016, pp. 68-73.

[11] L.S. Czarnecki, "Considerations on the Reactive Power in Nonsinusoidal Situations", IEEE Trans. on Instrumentation and Measurement, Vol. 34, No. 3, 1984, pp. 399–404.

[12] L.S. Czarnecki, "What is wrong with the Budeanu concept of reactive and distortion power and why it should be abandoned", IEEE Transactions on Instrumentation and Measurement, Vol. IM-36, No. 3, 1987, pp. 834-837.

[13] W. Shepherd, P. Zakikhani, D. Sharon, "Reactive", Proceedings of the Institution of Electrical Engineers, Vol. 121, Issue: 5, 1974, pp.390-392

[14] IEEE Std. 1459TM-2010, IEEE Standard Definitions for the Measurement of Electric Power Quantities under Sinusoidal, Nonsinusoidal, Balanced, or Unbalanced Conditions.

[15] M. I. Buzdugan, H. Bălan, T. I. Buzdugan, "Some Procedures in Mitigating Conducted Electromagnetic Interference", ICREPQ'11, Gran Canaria, Spain, 2011

[16] M. I. Buzdugan, H. Bălan, "Power Quality versus Electromagnetic Compatibility in Adjustable Speed Drives", ICREPQ '13, Bilbao, Spain, 2013

[17] C. R. Paul, Introduction to electromagnetic compatibility, John Wiley & Sons, Inc., 2006

Review on high power WPT coil system design

Jakub Skorvaga, Miroslav Pavelek

Department of mechatronics and electronics, Faculty of electrical engineering and information technologies
University of Zilina
Zilina, Slovakia
kristian.takacs@feit.uniza.sk

Abstract—**This article deals with an overview study of design coupling coils intended for wireless high performance transmission. Gradually represents various geometric shapes of coupling coils, their properties and comparison in terms of criteria such as the size of the power transmitted, the size of the binding coefficient, distribution of EM field to the surroundings, misalignment tolerance, complexity of control.**

Keywords—coupling coil, WPT, High power

I. Introduction

Due to increased energy demand, global fuel oil usage has increased, with the majority of fuel oil consumption going to electric energy production and internal combustion engine vehicles (ICEVs). With a way of obtaining this energy, the concentration of greenhouse gas in the atmosphere continuously increased. As a result, electric vehicles (EVs) have become increasingly common in recent years, owing to their low carbon footprint and reduced dependence on oil. EVs are expected to number more than 35 million worldwide by 2022 [1]. To ensure a sufficient amount of energy for these electric vehicles will need to build charging stations infrastructure. There are currently two battery charging methods that are conductive and inductive method. Wireless charging devices are known as inductive chargers. WCS can provide some benefits in terms of aesthetic quality, reliability, longevity, and user friendliness in general. Regardless, inductive chargers are not as widely used as conductive chargers due to issues such as electromagnetic compatibility (EMC), restricted power transfer, bulky and costly designs, shorter range, and lower performance. Higher-power wireless charging for electric vehicles is needed to reduce charging time [2], [18]. For this reason, it is necessary to propose a structure of coupling coil with optimal properties.

Several criteria must be taken into account when designing optimum binding coil structure for electric power transmission with high performance. The key criteria includes the size of the power transmitted, efficiency, the size of the distributed EM field to the environment of the system and also misalignment tolerance of the coils. In the past, a circular coil was the most discussed structure. However, in relation to the disadvantages of this structure, finding further possibilities of the shape and geometric arrangement of coils suitable for wireless charging the electric vehicle battery is necessary. Currently, most research in this area is focused on design binding coils with power level no more than 10 kW [3], [4], [5], [10], [12], [26], [25]. However, recently we can see an increased interest in designing and investigating features of double D coils for 80kW power levels [17], [18]. Managing large coil-voltages and currents in the device, as well as restricting the leakage magnetic fields to within the safety limits set by the International Commission on Non-Ionizing Radiation Protection (ICNIRP) guidelines, are the main bottlenecks for high-power inductive wireless power

transfer (WPT) systems in light-duty vehicles. In most cases, proposals are conducted by simulated 2D and 3D final elements models.

II. Coupling coils and their features

A. Circular coil

The most commonly discussed type of coil is currently a single-coil structure that is realized as a circle or as a square in terms of shape. Fig. 1 shows a structure of a circular coil with shielding. Interested in this structure is mainly due to simplicity and low price. Unipolar coils in which one pair of magnetic polar is produced when the coil is excited, provide a relatively higher coupling coefficient and generate a vertical magnetic flux. To achieve the desired coefficient of coupling when the air gap size 150-250 mm, it is necessary to increase the coil dimensions. Given the nature of the generated circular coil magnetic field do not allow large misalignment or large vertical separation [3]. In [4] is shown the possibility of optimized design of coupling coils for WPT using in-loop design consisting of a script / user interface (UI) in the MATLAB environment that is based on the models obtained by the final element method. The proposal is based on the calculation procedure for geometric shapes and coil dimensions. In [4] is also described shielding optimization using discussed user interface. The tool described in article allows to optimize shielding parameters to achieve safety limits defined by international standards. The authors achieved high accuracy between the coil parameters obtained by simulation and experimental measurement on a physical model 95%-98%. After comparing simulated and experimental results, it was confirmed that designed in the loop design and simulation models provide very accurate results for EM field distribution. [5] presented a solution of circular coil optimization using a 3-D method of finite elements. Key parameters have been investigated here and their impact on transmitted power and operation. The coil with nominal power 2 kW and with outer diameter 700 mm was designed. The tests were carried out with a horizontal radial tolerance of 130 mm (equivalent of a circular charging zone of 260 mm diameter) with an air gap 200 mm The differences between simulated and measured results are not more than 10 % confirming that the method is applicable. Circular coil has double-sided magnetic flux distribution. The height of the flux path is directly proportional to one quarter of the coil diameter. The diameter must be at least 4 times larger than the size of the air gap. In [6] general guidelines for high power IPT Systems have been described. Based on these procedures, a scaled 5 kW prototype with DC/DC efficiency more than 96,5 % was designed. The power density reached was $1,47kW/dm^2$. The outer diameter of the prototype coil was 210mm and the size of the air gap was 52 mm. [7] discusses challenges associated with the design of circular coils WPT systems such as thermal management and

978-1-6654-3236-8/21 $31.00 © 2021 IEEE

combined efficiency, volume and price with respect to vehicle dimensions and recommended standards.

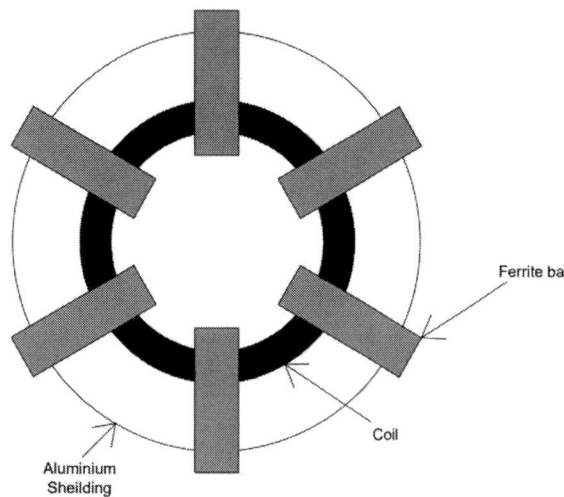

Fig. 1. Circular coil

B. Flux Pipe

This topology is composed of a ferrite bar along which the coil is wound. Benefit is that the flux height is half the length of the receiving pad. This pad is characterized by high misalignment tolerance in the horizontal direction, where the coupling coefficient is comparable to the circular coil. The disadvantage of this coil is its solenoid shape that produces a two-sided flow due to what occurs when using aluminum shielding to reduce coil quality factor. It features relatively large losses which is inappropriate for the needs of charging EV. [5] [8] describe the issue of the proposal.

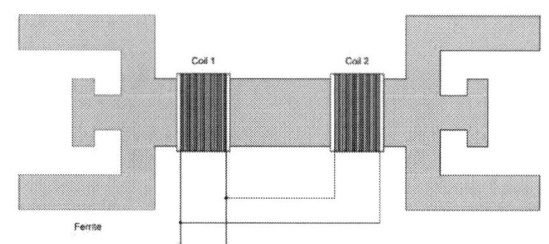

Fig. 2. Flux Pipe

C. DD Coil

This is a structure that includes the advantages of circular coil and fluxpipe. The DD structure consists of two coils of magnetically connected in the series and electrically parallel allows for a magnetic flux flow along the closed pathway. There is also the possibility of implementing in a serial connection from one litz-wire. The central part of DD Coil is realized as Flux-Pipe. Thanks to the one-side flux distribution, the size of the leakage flux is significantly reduced while increasing the coupling coefficient. The generated flux height is directly proportional to half of the DD of the coil lenght. The coil length needs to be optimized for the purpose of saving copper. The coil windings are located on the ferrite shielding through what the coils can be placed on aluminum sheet without loss of quality factor. As a result of such a structure arrangement, lower losses are in

electrical shielding, a higher coupling coefficient and a low leakage flux [10].

[10] demonstrate the process of optimizing coupling DD coils. There is an influence of the different physical parameters of DD coils by the finite element method investigated. The authors by the simulations have obtained specific DD coil parameters for power level 7,7 kW capable of working with a specific misalignment tolerance. The authors describe the procedure for optimizing the ratio between length and width and also middle occupation ratio for the primary and secondary coil. [8] shows the development of DD coils with five times larger charging zone as possible with a circular coil at a similar price and also with smaller dimensions. The authors represent 540mm x 800mm DD coils with 2 kW power transmission at 200mm. However, to achieve these properties, quadrature coil needs to be added to improve performance transmission when the horizontal misalignment in the X-axis direction occurs.

D. DDQ Pad

It is a structure derived from DD coil by adding quadrature coils as shown in Fig. 3. This coil is placed in the middle of DD Coil. The structure is discussed in [8], [11], [12]. Coils are located so that DD coil captures d-axis flow and Q coil captures q-axis flow, allowing a large range of misalignment system coils.

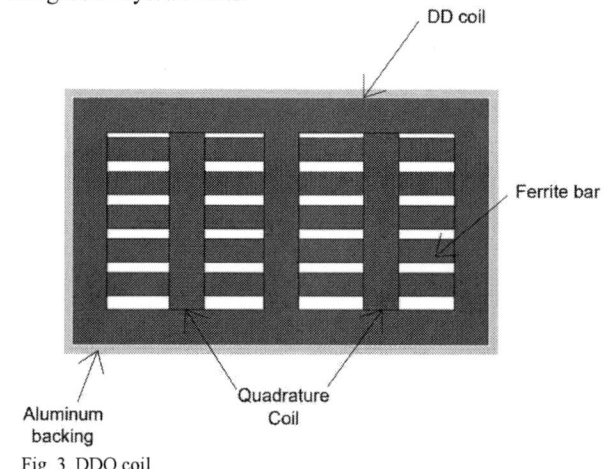

Fig. 3. DDQ coil

The disadvantage of this structure is that coil size increased approximately three times compared to the use of circular coils [11]. Both coils are mutually magnetically decoupled and therefore can be controlled independently [2].

E. Bipolar coil

It is a structure with two coils located on the ferrite shielding with a mutual overlap. Pad consists of two identical coils with mutually magnetically decoupled. Bipolar coil has high misalignment tolerance and coupling coefficient similar to DDQ. The main advantage of bipolar coil is that compared to DDQ is used by 25 % to 30 % less copper.

[3] represents a new bipolar topology of largely coplanar coupling coils, partially covered so that there was zero inductance between them. Since this arrangement prevents two coils from interacting, current flowing through them may

978-1-6654-3236-8/21 $31.00 © 2021 IEEE   14

have different amplitudes and phases. By controlling the phase and amplitude currents of both coils, the magnetic field may be shaped to help the power transmission to the receiver placed on EV. This allows tolerance of misalignment in stationary charging. Although the bipolar coil produces a lower transmitted maximum power when the phase shift between coil currents 90 ° as when both coils are in phase, the size of the power transmitted decreases slower when the misalignment occurs because the discussed system has greater misalignment tolerance.

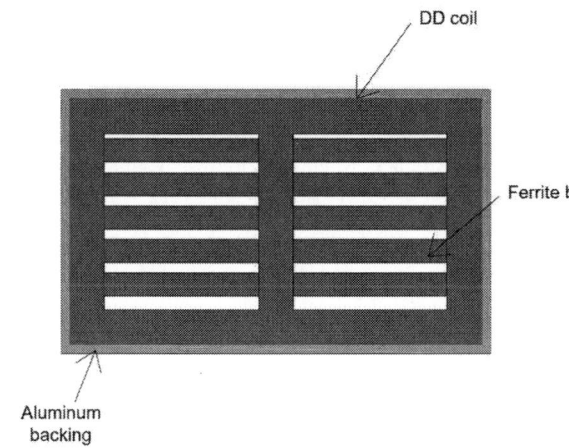

Fig. 4. DD Coil

F. Tripolar coil

Represents three-coil topology. These three coils are arranged so that they are mutual decoupled. Three coils are controlled independently to achieve the maximum coupling coefficient. The authors in [13] and [14] proved that the coupling coefficient is increased when using this configuration with bipolar or circular coil on secondary side. When using this structure, the reduction of apparent performance was achieved and also reducing the EM field radiated into the surroundings. The disadvantage of using said structure is the need for three separate inverters, which increases the price and complexity of the control algorithm [11].

G. Comparison of coupling coils properties

[11] simulates and compares the advantages of various geometric coil structures in ANSYS. Table I summarizes and compares coupling coils for different parameters. Bipolar coil and circular coil are the most common choice. Circular coil has a high coupling coefficient with single sided flux and simple implementation. Bipolar coil on the other hand achieves high misalignment tolerance. EM field radiating is also less [11].

[15] related 11kW unipolar and bipolar coil dependent WCSs, the spatial distribution and vector -pattern of the leakage magnetic field was investigated and compared. The findings reveal that the vertical leakage magnetic field in the unipolar coil-based WCS is dominant, while the horizontal leakage magnetic field in the bipolar pads coil-based WCS is dominant. The aluminum plate is found to be extremely effective for unipolar pads and almost useless for bipolar pads when it comes to shielding.

TABLE I.

COMPARISON OF COUPLING COILS PROPERTIES

Coil Structure	Misalignment tolerant	Coefficient of coupling	EMF Exposure	Magne tic flux
Circular coil	Poor	Low	High	Single sided
Flux Pipe	Medium	Medium	Low	Double sided
DD Coil	Poor	High	Low	Double sided
DDQ Coil	High	High	Low	Double sided
Bipolar pad	Medium	High	Low	Double sided
Tripolar pad	High	High	Low	Single sided

[16] compares the properties of magnetic coupling coils of various shapes (Square coil, Circular coil, Rectangular coil, DD coil) in terms of coupling coefficient and misalignment tolerance when using series-series compensation topology. Figure 5 shows the dependence of the coupling coefficient from the air gap for various types of coils [16]. The simulation has been found that the square coil has a small tolerance of misalignment in the x-axis direction and the good tolerance in the y-axis direction. For these reasons, the authors in the article investigated the impact of the coil parameters to its coupling coefficient. The diameter of litz wire has been found to have no effect on the coupling coefficient at different air gaps. It was also found that the number of threads has only minimal impact on the binding coefficient. The coupling coefficient decreases when the internal coil length increases in the range of 180 to 280 mm. On the contrary, the coupling coefficient is growing together with the outer length of the rectangular coil.

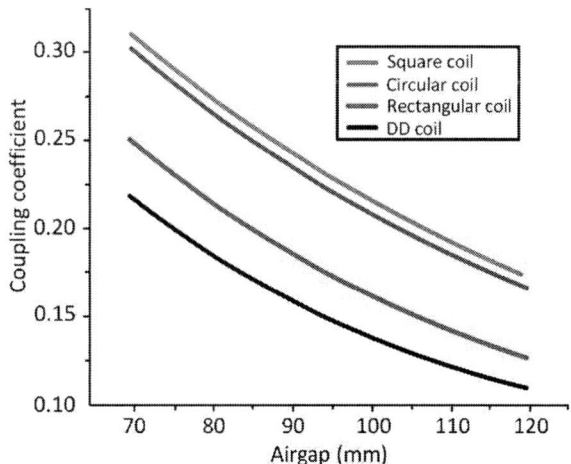

Fig. 5. Coupling coefficient vs. Airgap [16]

III. COUPLING PADS FOR HIGH POWER

An interesting option for WPT high performance transmission represents DD2Q shown in Fig. 6. Pad consists of two orthogonal DD coils to achieve a higher power density via quadrupole operation. The multi-layer structure consists of two perpendicular DD coils and one square quadrature coil. The structure thus arranged generates three magnetic poles that are perpendicular to each other. These three coils are decoupled because their magnetic fields are perpendicular to each other. At the same time, the task of DD2Q topology is simultaneously achieving a high coupling coefficient as well as a square coil and also a low leakage field as well as DD Coil when transmitting high performance. The basic philosophy of the multilayer pad topology is to divide large inductance into smaller inductances and reduce the coil voltages. Thanks to the properties that it is magnetically perpendicular coils, decoupled can be controlled with three separate inverters either in phase or out of phase with unbalanced impedance. The discussed structure is capable of taking the magnetic field in all three perpendicular axes therefore is assumed to be compatible with square and circular coils oriented in any direction. Another potential advantage of the three-phase IPT system is lower device voltage and also lower current stress.

[17] present an analysis of new multilayer topology and its properties compared to DD Coil and square coil. The authors presented a vector analysis of magnetic fields in coils and also compared DD2Q properties with square and circular coil. Thanks to its symmetry, DD2Q coil produced the same EM field. DD2Q coil due to its symmetry at misalignment produced the same EM field in the X and Y axis direction. Simulations have been shown that in terms of performance, DD2Q coil worked better than square coil but did not reach the performance that can be transferred at optimally designed DD Coil. Quadrature Coil is less than individual DD Coils to minimize leakage magnetic fields. DD Coil has better tolerance when misalignment occurs in perpendicular direction to the main flux. On the contrary, when DD Coil is misaligned in a parallel direction due to the main flux path the scaled power falls below 10 kVA. They found that optimized DDQ2 coil has a minimum change of scaled power when misalignment from the aligned position occurs.

In [18] the authors submit bilayer quadrature Double-D coil for electric vehicles using double-sided LCC compensation network. The article describes the simulation procedure for optimizing the design of coupling members WPT system for transmission of 80 kW for distance 150 mm. The proposed system consisted of a pair of bipolar coils on both primary and secondary side. Primary coils were excited separately from two inverters. Based on simulations in Simulink and COMSOL, the efficiency of the system parameters design method has been verified. After comparing it was found that BQDD coil has a higher performance density than conventional coupling coils.

More approaches to high performance charging are listed in [19]- [24].

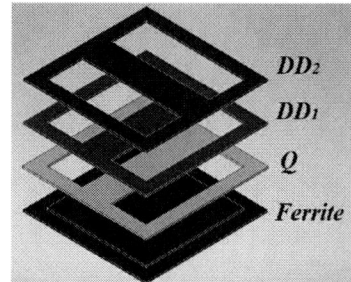

Fig. 6 DD2Q coil

IV. LCC COMPENSATION TOPOLOGY

LCC Compensation Topology shown in Fig. 7 includes the benefits of SS topology and provides more flexibility to optimize the efficiency of the system and to reach zero voltage switching. The disadvantage is that two compensation coil increases the total volume of the system. To address this problem, [25] came with an integrated compensation coil in the main coupling coil. This approach can increase the compactness of the system but the disadvantage is that 3 other types of coupling coefficients are generated. Coupling coefficient between coils on the same side k_{1f1} and k_{2f2}. There is also a cross coupling coefficient k_{1f2} and k_{2f1} and extra cross coupling coefficient k_{f1f2}. [25] reached DC / DC efficiency 95,3 % with 6 kW power. This system makes it highly effective and at the same time compact even if this method complicates the design of a wireless charging system.

This method is gradually possible to reduce 5 extra binding effects to a negligible level which simplifies proposal and analysis. L_{f1}, C_{f1}, C_1 and L_1 represent a primary resonance container that is designed and the same resonance frequency such as the switching frequency of the inverter. The secondary side is a resonance container formed by L_{f2}, C_{f2}, C_2, L_2 elements. The resonance frequency of the secondary side of the system is the same as on the primary side. For the best functioning of the WPT system is the key main coefficient of coupling between L_1 and L_2. The aim of the proposal is to maximize this main coefficient and minimize 5 extra coefficients to a negligible level.

In [26] the proposed coupler is first modeled and investigated using a 3-D finite-element analysis method to ensure its feasibility. The compensated coils location, the extra couplings range, and the misalignment tolerance are all investigated. The proposed WPT topology's circuit modeling and characteristic analysis are then performed using the fundamental harmonic approximation. Finally, a 600 mm x600 mm wireless charger prototype with a nominal 150 mm gap has been built and tested, operating at a resonant frequency of 95 kHz and rated power of 5,6 kW. At operation, a peak efficiency of 95,36 % is achieved from the dc power source to the battery load.

In [25] proposes a wireless power transfer network with a double-sided LCC compensation network and tuning system (WPT). The resonant frequency of the proposed topology and tuning method is independent of the load condition and of the coupling coefficient between the two coils, implying that the device will operate at a constant switching frequency. The proposed method's characteristics

978-1-6654-3236-8/21 $31.00 © 2021 IEEE

are demonstrated through a frequency domain analysis. Simulation and experimental findings validated the proposed compensation network and tuning method's interpretation and validity. A wireless charging system for electric vehicles was designed with an output power of up to 7,7kW and a 96 % efficiency from DC power source to battery load.

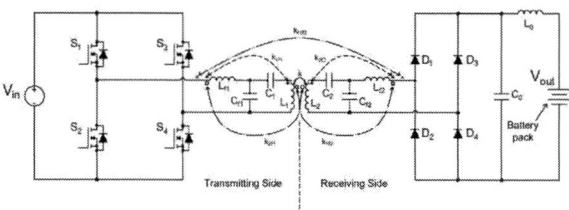

Fig. 7. WChS with LCC compensation topology

Conclusion

The article deals with an overview of the design of coupling coils for high power transmission. In the introductory part, the article presents the various types of coupling coils, their properties, advantages, disadvantages and current trends in their design. After describing the principles and properties of the coils, their properties were compared in terms of the achieved coefficient of coupling and sensitivity when misalignment occurs, distribution of the EM field to the environment. The main advantage of the other topologies versus circular coils is greater tolerance of misalignment of the primary and secondary coil. Another authors have examined the possibilities of using square coils and investigate its optimization. In the next part, DD2Q coil structures for 80 kW power transmission were described. The final part describes the LCC compensation topology, its integrated variant and the benefits that this topology brings.

Acknowledgment

This publication was realized with support of Operational Program Integrated Infrastructure 2014 - 2020 of the project: Innovative Solutions for Propulsion, Power and Safety Components of Transport Vehicles, code ITMS 313011V334, co-financed by the European Regional Development Fund". Authors would like to thank also to national grant agency Vega for project funding Vega-1/0063/21.

References

[1] Brenna, M., Foiadelli, F., Leone, C. et al. Electric Vehicles Charging Technology Review and Optimal Size Estimation. J. Electr. Eng. Technol. 15, 2539–2552 (2020).

[2] Kindl, V.; Frivaldsky, M.; Zavrel, M.; Pavelek, M. Generalized Design Approach on Industrial Wireless Chargers. *Energies* 2020, *13*, 2697. https://doi.org/10.3390/en13112697

[3] G. A. Covic, M. L. G. Kissin, D. Kacprzak, N. Clausen and H. Hao, "A bipolar primary pad topology for EV stationary charging and highway power by inductive coupling," *2011 IEEE Energy Conversion Congress and Exposition*, 2011, pp. 1832-1838, doi: 10.1109/ECCE.2011.6064008.

[4] Michal Frivaldsky & Miroslav Pavelek, 2020. "In Loop Design of the Coils and the Electromagnetic Shielding Elements for the Wireless Charging Systems," Energies, MDPI, Open Access Journal, vol. 13(24), pages 1-18, December.

[5] M. Budhia, G. A. Covic and J. T. Boys, "Design and Optimization of Circular Magnetic Structures for Lumped Inductive Power Transfer Systems," in *IEEE Transactions on Power Electronics*, vol. 26, no. 11, pp. 3096-3108, Nov. 2011, doi: 10.1109/TPEL.2011.2143730.

[6] R. Bosshard, J. W. Kolar, J. Mühlethaler, I. Stevanoviʹc, B. Wunsch, and F. Canales, "Modeling and η-α-Pareto optimization of inductive power transfer coils for electric vehicles," IEEE J. Emerg. Sel. Topics Power Electron., vol. 3, no. 1, pp. 50–64, Mar. 2015.

[7] R. Bosshard and J. W. Kolar, "Inductive power transfer for electric vehicle charging: Technical challenges and tradeoffs," IEEE Power Electron. Mag., vol. 3, no. 3, pp. 22–30, Sep. 2016.

[8] M. Budhia, J. T. Boys, G. A. Covic, and C.-Y. Huang, "Development of a single-sided flux magnetic coupler for electric vehicle IPT charging systems," IEEE Trans. Ind. Electron., vol. 60, no. 1, pp. 318–328, Jan. 2013.

[9] S. Al-Chlaihawi et al., "Inductive Power Transfer for Charging the Electric Vehicle Batteries," EEA, no. 4, 2018.

[10] S. Dong *et al.*, "Optimal Design of DD Coupling Coil for Wireless Charging System of Electric Vehicle," *2019 22nd International Conference on Electrical Machines and Systems (ICEMS)*, 2019, pp. 1-4, doi: 10.1109/ICEMS.2019.8921726.

[11] Patil, D.; McDonough, M.K.; Miller, J.M.; Fahimi, B.; Balsara, P.T. Wireless power transfer for vehicular applications: Overview and challenges. IEEE Trans. Transp. Electrif. 2018, 4, 3–37. [Google Scholar] [CrossRef]

[12] Ahmad, A.; Alam, M.S.; Chabaan, R. A comprehensive review of wireless charging technologies for electric vehicles. IEEE Trans. Transp. Electrif. 2018, 4, 38–63. [Google Scholar] [CrossRef]

[13] S. Kim, A. Zaheer, G. Covic, and J. Boys, "Tripolar pad for inductive power transfer systems," in Proc. 40th Annu. Conf. IEEE Ind. Electron. Soc. (IECON), Dallas, TX, USA, Oct./Nov. 2014, pp. 3066–3072.

[14] S. Kim, G. A. Covic, and J. T. Boys, "Tripolar pad for inductive power transfer systems for EV charging," IEEE Trans. Power Electron., vol. 32, no. 7, pp. 5045–5057, Jul. 2017.

[15] M. Mohammad, J. L. Pries, O. C. Onar, V. P. Galigekere, G. -J. Su and J. Wilkins, "Comparison of Magnetic Field Emission from Unipolar and Bipolar Coil-Based Wireless Charging Systems," 2020 IEEE Transportation Electrification Conference & Expo (ITEC), 2020, pp. 1201-1207, doi: 10.1109/ITEC48692.2020.9161726.

[16] Frivaldsky, M.; Spanik, P.; Morgos, J.; Pridala, M. Control Strategy Proposal for Modular Architecture of Power Supply Utilizing LCCT Converter. *Energies* 2018, *11*, 3327. https://doi.org/10.3390/en11123327

[17] B. J. Varghese, A. Kamineni and R. A. Zane, "Investigation of a DD2Q Pad Structure for High Power Inductive Power Transfer," 2019 IEEE PELS Workshop on Emerging Technologies: Wireless Power Transfer (WoW), 2019, pp. 129-133, doi: 10.1109/WoW45936.2019.9030644.

[18] T. Feng, C. Jiang and S. Wang, "Design and Optimization of a Bilayer Quadrature Double-D Coil for Electric Vehicle Wireless Charging System," 2020 8th International Conference on Power Electronics Systems and Applications (PESA), 2020, pp. 1-6, doi: 10.1109/PESA50370.2020.9343979.

[19] C. -H. Chung and C. -L. Yang, "Novel Parallel Serpentine Coils with High Lateral Misalignment Tolerance and Power Transfer Efficiency Optimization," *2020 IEEE Wireless Power Transfer Conference (WPTC)*, 2020, pp. 480-483, doi: 10.1109/WPTC48563.2020.9295531.

[20] M. E. Bima, I. Bhattacharya and C. W. V. Neste, "Experimental Evaluation of Layered DD Coil Structure in a Wireless Power Transfer System," in *IEEE Transactions on Electromagnetic Compatibility*, vol. 62, no. 4, pp. 1477-1484, Aug. 2020, doi: 10.1109/TEMC.2020.3002694.

[21] Frivaldsky, M.; Hanko, B.; Prazenica, M.; Morgos, J. High Gain Boost Interleaved Converters with Coupled Inductors and with Demagnetizing Circuits. *Energies* 2018, *11*, 130. https://doi.org/10.3390/en11010130

[22] X. Lv, X. Dai, C. Jiang and S. Wang, "A Cross Double-D Coil for Electric Vehicles Dynamic Wireless Transfer System to Reduce Output Power Pulsation," *2020 8th International Conference on Power Electronics Systems and Applications (PESA)*, 2020, pp. 1-5, doi: 10.1109/PESA50370.2020.9344038.

[23] Radvan, R., Dobrucky, B., Frivaldsky, M., & Rafajdus, P. (2011). Modelling and Design of HF 200 kHz Transformers for Hard- and Soft-

978-1-6654-3236-8/21 $31.00 © 2021 IEEE

Switching Application. Elektronika Ir Elektrotechnika, 110(4), 7-12. https://doi.org/10.5755/j01.eee.110.4.276

[24] G. Ke, Q. Chen, W. Gao, S. Wong, C. K. Tse and Z. Zhang, "Research on IPT Resonant Converters With High Misalignment Tolerance Using Multicoil Receiver Set," in *IEEE Transactions on Power Electronics*, vol. 35, no. 4, pp. 3697-3712, April 2020, doi: 10.1109/TPEL.2019.2936325.

[25] J. Koscelnik, M. Frivaldsky, M. Prazenica and R. Mazgut, "A review of multi-elements resonant converters topologies," 2014 ELEKTRO, 2014, pp. 312-317, doi: 10.1109/ELEKTRO.2014.6848909.

[26] Frivaldsky, M.; Morgos, J.; Prazenica, M.; Takacs, K. System Level Simulation of Microgrid Power Electronic Systems. Electronics 2021, 10, 644. https://doi.org/10.3390/electronics10060644.

Control structure and Converter losses of a 3L-ANPC with GaN HEMTs switching at 100 kHz

Martin Geppert
Chair for Electrical Machines, Drives and Controls
University of Siegen
Siegen, Germany
martin.geppert@uni-siegen.de

Günter Schröder
Chair for Electrical Machines, Drives and Controls
University of Siegen
Siegen, Germany
guenter.schroeder@uni-siegen.de

Abstract—**This paper presents a model for estimating the converter losses of a drive system consisting of a Gallium Nitride (GaN) 3-Level Active Neutral Point Clamped Converter (ANPC) and a 4.55 kW synchronous reluctance machine (SynRM). First an experimental setup of the ANPC using GaN HEMTs and the control system, which is implemented on a Zedboard (FPGA), is presented. As a second step the turn-on and -off losses of a single GaN HEMT are measured for different current, voltage and temperature values, using a double pulse test fixture. These evaluated losses are used to create a loss model of a single switch and finally a model of the 3L-ANPC. Subsequently the impact of different switching frequencies and modulation strategies on the efficiency of the inverter is investigated. Finally the results of simulations will be compared with experimental measurements of a speed controlled SynRM drive.**

Keywords—Wide Bandgap, Gallium Nitride (GaN), Voltage Source Converter, Converter Losses, Field Programmable Gate Array (FPGA)

I. INTRODUCTION

Nowadays wide bandgap materials like gallium nitride (GaN) play an important role in power electronics. GaN power devices like High Electron Mobility Transistors (HEMTs) have low conduction losses and excellent switching characteristics [1]. These devices enable the setup of efficient power converters in motor drive applications. Due to the possible high switching frequency the motor current ripple, noise and motor losses are reduced compared to conventional devices. Nevertheless the application of GaN HEMTs brings a series of challenges. On the one hand the high switching speed will cause a high di/dt and dv/dt which could lead to high EMI [2], [3]. Also the available GaN HEMTs only have a rated V_{DS} of 650 V which is normally not enough for a 600V DC link. Especially the high voltage slew rate can cause stress in the winding insulation of the machine [4]. To overcome this the 3-Level Active Neutral Point Clamped topology is chosen in this paper. On the other hand the control system, particularly the controllers and the space vector modulation, needs to be very fast. For the 3-L ANPC there exist many different modulation techniques and switching states. Beginning with the oldest and well-known switching states for the NPC [5], with its known drawback of the unequal loss distribution. Followed by the first approach on the ANPC to overcome the disadvantages of the NPC [6]. Further developments haven been investigated in particular the use of Silicon Carbide MOSFETs instead of Silicon IGBTs [7]. Lastly it is also possible to use the 3-Level inverter as a 2-Level one. In case of modulation techniques there are two main different ones. On the one hand there is the Sinusoidal Pulse Width Modulation (SPWM), which is even easy to implement for high switching frequencies. On the other hand

there is the Space Vector Modulation (SVM) which is more complex and needs more computation time. But the SVM can be reduced to mostly logic decisions [8], what makes it even usable for the mentioned high switching frequencies. As load for the inverter a 4.55 kW synchronous reluctance motor is chosen. This kind of machine is an attractive alternative to conventional alternating current machines. In the past few years it has been rediscovered since it was first described by Kostko in 1923 [9]. Greatest advantage of this machine type is the transversally laminated rotor, which is free of rare earth materials and also simple to manufacture. The rotor is designed in such way that there is an inductance saliency which generates the reluctance torque. The paper deals on one side with the development of a loss model of a GaN HEMT based 3L-ANPC converter feeding a SynRM. For this the switching losses of a single GaN HEMT are evaluated using a double pulse text fixture. Then a MATLAB Simulink Model is created and fed with these losses and the conduction losses. The SPWM and SVM is implemented including different switching states for both techniques. With the objective to find an optimal modulation strategy, even with optimal changing of the techniques online. On the other side the design of a GaN based 3-L ANPC converter operating at 100 kHz using a Xilinx Zedboard for control is shown. Finally the simulations results for driving a 4.55 kW synchronous reluctance motor are shown and verified by the described experimental hardware setup.

II. HARDWARE DESIGN OF THEN GAN BASED DRIVE SYSTEM

The hardware of the drive system can be divided into the power part and the control part. In this paper the control part is realized using a Xilinx Zedboard. The power part consists of three 3L-ANPC half bridges, including auxiliary power

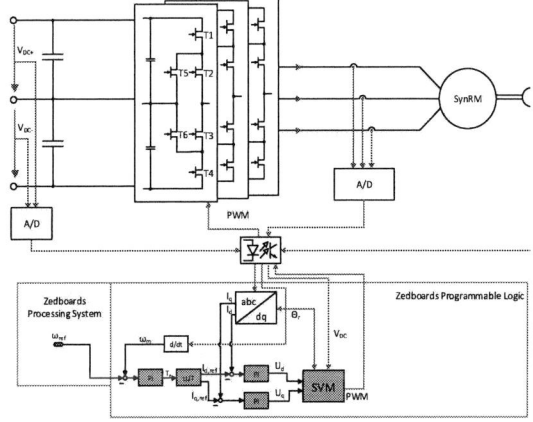

Fig. 1: Hardware setup of the motor drive system

978-1-6654-3236-8/21 $31.00 © 2021 IEEE

supplies, gate drive circuits, phase current measurement and a rotary encoder. The three-level ANPC topology is chosen despite the fact that the control of the ANPC is more difficult than the Neutral Point Clamped Converter (NPC) or a two-level topology. Due to the fact that every semiconductor needs to withstand only half of the DC-Link Voltage, it is possible to use the standard DC-Link voltage of about 600 V and standard industrial machines. Furthermore the ANPC topology is known for the additional degree of freedom for the loss distribution among the semiconductors and the midpoint balance. The entire motor drive System described in this paper is shown in Fig. 1.

A. ANPC Hardware Design

Using gallium nitride power devices the PCB layout plays a very important role. Due to the high dv/dt capability of these transistors, a very low inductance gate drive is required. Each driver cells main part is formed by the Silicon Labs Si8271 isolated gate driver IC. The separated turn-on and turn-off paths allow an optimized switching behavior. The most common way for reducing the gate loop inductance is to minimize the overall footprint. This is done by using low power surface mount packages like 0603 for the passive components. Due to the chosen topology and the negative biased voltage for a safe turn-off, each gate driver is supplied by a dc/dc converter with low isolation capacitance. Followed by a resistive voltage divider for generating the negative voltage. When using GaN HEMTs aside from the gate drive loop the power or commutation loop needs to be focused. The design of the ANPC half bridges used in this investigation is based on [11, 12]. The major idea is to have two symmetrical switching cells per phase leg on bottom and top layer of the PCB. One of these cells is formed by T1, T2 and T5 while the other one consists of T3, T4 and T6 (see Fig. 1). In addition to the basic idea there are two mid layers added between DC+ top and DC- bottom layers with the result that six layers at all are used. The used layer arrangement is shown in Fig. 2 a). These two additional layers prevent possible cross-talking between the two switching cells and inhibit the capacitive coupling between the gate drive loops on top and bottom layer. Beside the main DC-Link every phase leg has two 1 µF and 20 µF Ceralink Capacitors as separated DC-Link in its layout

for keeping the commutation loop as small as possible. Additionally the gate drive supply units are rearranged for purpose of having the same physical length of all PWM carrying traces. Hence all gate drive loop footprints including the power supplies are the same, all PWM paths are the same, so for all GaN HEMTs the PCB layout is exactly identical. As a result the parasitic inductance and resistance caused by the PCB is approximately the same for all switching devices. Fig. 2 b) shows the top layer of the final half bridge design introduced in this paper including the heatsink.

B. FPGA based Control System

When using a fast switching GaN-based inverter the use of very fast control hardware is essential. Therefore in this investigation the control of the inverter is implemented on a Xilinx Zedboard. It combines an ARM processing system with a series 7 programmable logic (FPGA). The inverter control including the different modulators is totally implemented on the programmable logic, while the ARM is used for the communication between user and FPGA and visualization of the different parameters. All peripheral connections of the converter, including AD-converters, rotary encoder and gate-signals, are carried out as fiber optic connections. This is done to protect the control hardware and prevent signals from cross-talking or any EMI-related problems. For the implemented control the phase currents and DC voltages are required. As a result there are five 25 MHz Serial Peripheral Interfaces (SPI) working in parallel for the AD-converters. The rotor angle is needed, wherefore a 2 MHz synchronous serial interface (SSI) master is utilized. As basic idea of the implemented control system in this paper the field oriented control with its two separated PI controllers for i_d and i_q is chosen. Those are put together with a PI controller for the speed, into a cascaded closed loop control system (see Fig. 1). The voltage model of the SynRM in the rotor reference frame with neglected iron losses and saturation effects is shown in (1). Where R_S means the stator resistance, while L_d and L_q describe the inductances of the rotor.

$$
\begin{aligned}
v_d &= R_s + \frac{d}{dt}\lambda_d - \omega_{me}\lambda_q, &&\text{with } \lambda_d = L_d i_d \\
v_q &= R_s + \frac{d}{dt}\lambda_q - \omega_{me}\lambda_d, &&\text{with } \lambda_q = L_q i_q
\end{aligned}
\tag{1}
$$

In this case the electromagnetic torque produced can be expressed by (2), where p means the number of pole pairs.

$$
T_e = \frac{3}{2}p(L_d - L_q)i_d i_q
\tag{2}
$$

A look-up table was created, where the values for i_d are determined empirically and via simulations for the given values of torque. Finally with a given torque, a given value for i_d and (2) it is possible to figure out i_q.

C. ANPC Space Vector modulation

Beside the well-known modulation techniques and switching states mentioned above the modified space vector modulation for high switching frequencies should be briefly discussed. These states have been introduced in [7] and are composed of three main switching states. Furthermore there are two transition states introduced, all five switching states are shown in TABLE I. If Transistors T1 and T2 are switched on, the output of the converter is +VDC/2 what is called

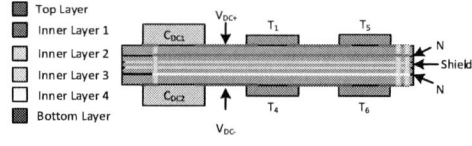

Fig. 2: Three-level ANPC half bridge design a) Used layer setup b) Mechanical setup of the half bridge

978-1-6654-3236-8/21 $31.00 © 2021 IEEE   20

positive state. During this state T6 needs to be turned on, not conducting any current, but an even voltage distribution across T3 and T4 of VDC/2 is ensured. For the transition from the positive to the zero state initially T1 need to be turned off what corresponds to the transition state 0^P from TABLE I. The converter remains in this condition for a dead time of 150 ns, meanwhile the current can commutate to the reverse channel of T5 in case of positive current. The dead time is chosen very large in case of having a safe working system. Finally transistors T5 and T3 are turned on and two parallel current paths are available for the zero voltage state. The same steps have to be fulfilled for the other commutation path. This means the switching sequences always have to be positive-0^{0-P}-0 or 0-0^{0-N}-negative both ways. Hence it is not possible to switch directly from positive to negative state.

TABLE I   ANPC Switching States

Output	T1	T2	T3	T4	T5	T6
Positive	1	1	0	0	0	1
0^{0-P}	0	1	0	0	0	1
0	0	1	1	1	1	0
0^{0-N}	0	0	1	0	1	0
Negative	0	0	1	1	1	0

The modified space vector modulation used in this paper is based on [8] and includes neutral point balance. Instead of complex and tedious mathematical operations this modulation scheme is reduced to mostly logic decisions. Due to this it is well suited for high switching frequencies and implementation in a FPGA. Starting point for this modulation is to compress the well-known two-level inverter switching state hexagon to have 45 ° diagonals instead of 60 ° ones. This is done by transforming all incoming voltage space vectors like shown in (3).

$$v_i = \text{Re}\{V_i\} + j\frac{IM\{V_i\}}{\sqrt{3}} \qquad or$$
$$v_i = \begin{bmatrix} 1 & 0 \\ 0 & \frac{1}{\sqrt{3}} \end{bmatrix} V_i \qquad (3)$$

This means a scale of the imaginary part of V_i by $1/\sqrt{3}$. Additionally he required switching times are normalized by the cycle time (T_0) of the SVM algorithm. Applying those

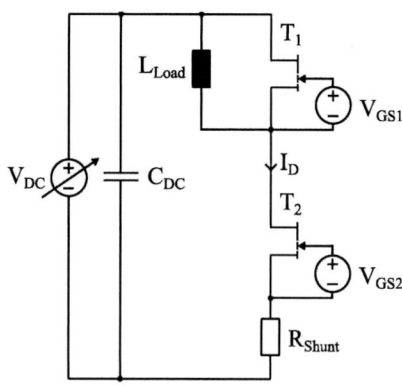

Fig. 3: Double pulse text fixture

simplifications the sector of the hexagon and the switching times (t_a and t_b) can be determined by a series of logical decisions. Finally the missing switching time t_c can be computed like shown in (4).

$$t_c = 1 - (t_a + t_b) \qquad (4)$$

For using this idea on an ANPC, its 27 switching state hexagon is divided into eight two level hexagons. Doing so results in redundant switching states, which can be used to balance the neutral point without any separated control algorithm.

III. IMPLEMENTATION OF THE LOSS MODEL

A. GaN HEMT loss evaluation

For evaluating the switching losses a half bridge double pulse test setup using an inductive load is applied. This kind of setup is a reliable verification method for identifying the switching losses independent of the semiconductor self-heating. The inductive double pulse test fixture is shown in Fig. 3. The investigated device is the GaN Systems GS66508T, which is a 650 V 35 A enhancement mode High Electron Mobility Transistor. The upper GaN HEMT T_1 is continuously switched off by V_{GS1} = -3 V, while the lower one T_2 is pulsed with +6 V and -3 V, like in the real inverter. Also the power loop design of the double pulse setup is based on the real inverter design with the aim of having an ultra-low inductive design [10]. The switching event ranges are defined from 2 % I_D to 2 % V_{DS} for the turn-on and for the turn-off

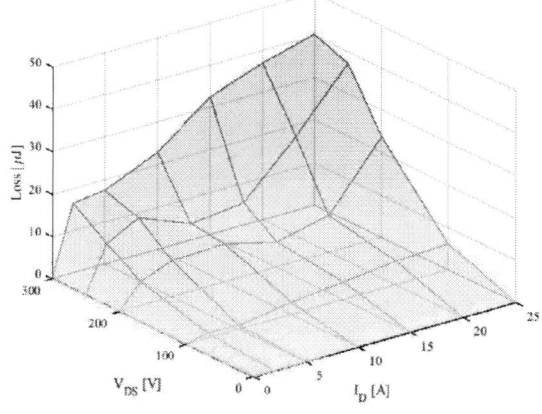

a)

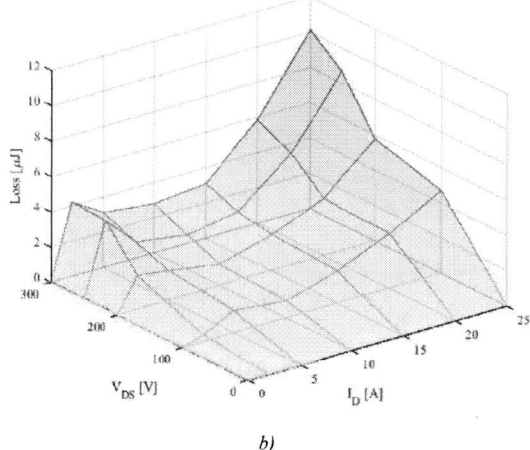

b)

Fig. 4: Measured switching loss a) turn-on losses b) turn-off losses

978-1-6654-3236-8/21 $31.00 © 2021 IEEE

respectively. Like this the results could be more expressive than measuring with the standard 10 % threshold. The current is measured with a coaxial shunt and the voltage by a passive high voltage probe. Both give a sufficient bandwidth. In addition a heat sink with an external power resistor for heating the system is mounted. So it is possible to measure the losses at room temperature, 80 °C and 100 °C. The results for measuring the switching losses at 80 °C junction temperature are shown in Fig. 4 a) for the turn-on, while the turn-off is shown in Fig. 4 b). The conduction losses of the GaN HEMT are modeled by means of a curve fitting of the output characteristics of the device given in the datasheet. As well as the switching losses, the conduction losses are modeled at a junction temperature of 80 °C. Like this the voltage drop of on-state resistance depending on the Drain-Source current is modeled.

B. GaN HEMT loss Model outline

A model of a single GaN HEMT is created. For this the measured switching losses are fed into a two-dimensional look up table. Doing so at every switching event the Drain-Source current and voltage is taken into account and the related switching loss value is picked. The chosen look up table can interpolate between the given data points, in this way for every current and voltage value a loss value is available. The same is done for the curve fitting data points of the output characteristics of the GaN HEMT. In that way the total losses of the single device can be obtained and simulated with the model.

C. 3-L ANPC loss Model outline

Finally a model of the 3-Level Active Neutral Point Clamped converter is created. Every half-bridge of the ANPC contains six of the above outlined GaN HEMT loss models. In addition to this the model includes a 4.55 kW synchronous reluctance machine, a field oriented control and five different Modulators. The implemented modulators are a SPWM and for the use of the ANPC a 2- and 3-level inverter. In the case of the 3-level SPWM there are different switching states implemented, one for the use of the inverter as a NPC and one for the use as ANPC. Also there is a SVM realized, which is done as a 2-level version and a 3-level variant with ANPC switching states, which is modified for the use of high switching frequencies [8]. Various simulations were performed for different steady-state load and speed conditions of the SynRM. Furthermore the inverters switching frequency was varied between the different operation points of the machine. All these simulations were done with respect to the inverters total losses.

Fig. 5: Simulated converter losses for 750rpm of the machine a) switching frequency of 100 kHz b) switching frequency of 200 kHz

The top of Fig. 5 show the results for a switching frequency of 100 kHz. It needs to be mentioned that the negative values of I_{rms} are representing the generator and positive values the motor operation. Whereby the 3-level NPC modulation technique is not shown because the losses are pretty much higher. This is owed to the fact that the diode behavior of the GaN HEMT when turned off with negative V_{GS} is pretty pour. So the Diode losses of the switches in the NPC are very high. In the bottom part of the figure the same modulation techniques are shown in use of a switching frequency of 200 kHz. It can been stated that the inverter losses all in all are higher than with a frequency of 100 kHz as expected due to the higher switching losses.

IV. EXPERIMENTAL RESULTS FOR THE ANPC DRIVE SYSTEM

The GaN HEMT based three-level ANPC drive system is applied with a speed step from 200 rpm to 400 rpm. The step response of the simulated system is shown in Fig. 6 a), as it can be seen the rotational speed response is very fast, also the torque is. The same step is given to the real system, the results

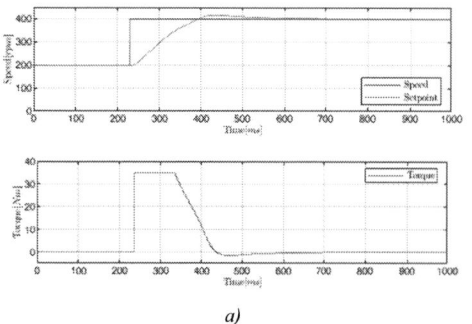
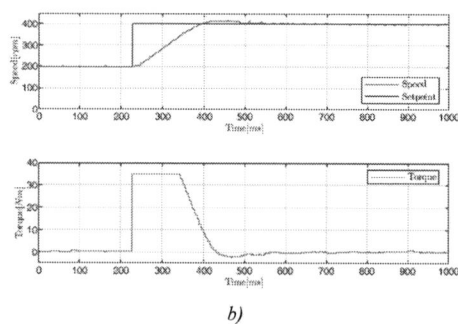

Fig. 6: Three phase ANPC drive system with a SynRM a) Simulation results b) Experimental results

978-1-6654-3236-8/21 $31.00 © 2021 IEEE

are shown in Fig 6 b). As can be seen in the figures the maximum allowed torque is 35 Nm. The response of the speed and torque in simulation and experiment is almost the same, what means the simulation model is validated and the system is working well. Also it can be seen that the dynamic behavior of the system is pretty good for the introduced simplifications of the SynRM control.

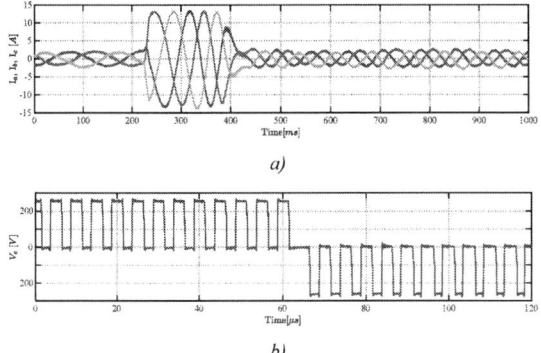

Fig. 7: Converter measurements a) Phase currents b) Line voltage phase C

The converter output is probed also, in Fig 7 a) the phase currents for the given speed step are shown. As it can be seen those are proper without any kind of output filters. Also the phase to neutral point voltage is observed and shown in Fig. 7 b). It can be assessed the voltage overshoot is very small and the switching waveform is pretty clean.

Additionally the real losses of the inverter have been measured with a power analyzer. The results for a switching frequency of 100 kHz and a speed of 750 rpm are shown in Fig. 8.

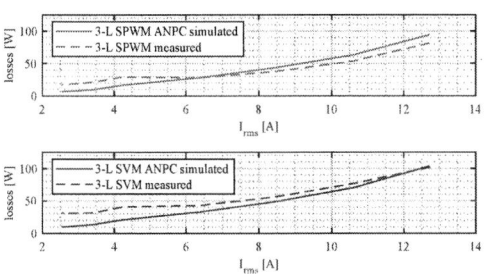

Fig. 8: Measured and simulated inverter losses for switching frequency of 100 kHz and 750 rpm

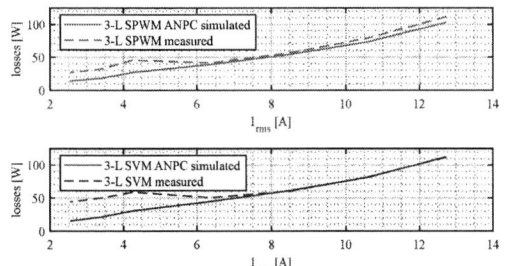

Fig. 9: Measured inverter losses for switching frequency of 200 kHz and 750 rpm

As can be seen in the figure the measured and simulated losses coincide fairly well. The slight differences could be explained by the fact that in the simulation the junction

temperature of each device is set to 80 °C. The same results could be observed in Fig. 9 where the same investigation is done for a switching frequency of 200 kHz. It can be stated that the implemented simulation model provides sufficiently accurate results regarding the inverter losses.

V. SUMMARY

The GaN HEMT based 3-Level ANPC drive System is implemented. This includes an adequate PCB layout for the special demands of high switching frequencies. Key parts of this design is the six layer PCB with two shielding layers, the symmetric gate drive design and setup of the commutation loops. A closed loop speed control including different modulators is implemented on a Xilinx Zedboard. In a second step the switching losses of a single GaN HEMT are measured. A model for the loss evaluation of a GaN HEMT just as a model of a 3-L ANPC based drive System is created. This model is used to perform simulations regarding the inverter losses dependent on the performed modulation techniques and switching states. Finally, simulation and experimental test validated on the one hand the field oriented motor control and on the other hand the correct functionality of the simulation model. It can be concluded that 3L-ANPC based GaN drive system, including a synchronous reluctance motor, under the experimental conditions of 500 V bus voltage and a switching frequency of 100 kHz is realized and working well. In addition it can be stated that the implemented loss simulation model provides sufficiently accurate results.

REFERENCES

[1] M. Guz et al., „IEEE ITRW Working Group Position Paper-System Integration and Application: Gallium Nitride: Identifying and Addressing Challenges to Realize the Full Potential of GaN in Power Conversion Applications", IEEE Power Electron. Mag., Jg. 5, Nr. 2, S. 34–39, 2018, doi: 10.1109/MPEL.2018.2822898.

[2] J. Millan, P. Godignon, X. Perpina, A. Perez-Tomas und J. Rebollo, „A Survey of Wide Bandgap Power Semiconductor Devices", *IEEE Trans. Power Electron.*, Jg. 29, Nr. 5, S. 2155–2163, 2014.

[3] X. Ding, Y. Zhou und J. Cheng, „A review of gallium nitride power device and its applications in motor drive", *CES Transactions on Electrical Machines and Systems*, Jg. 3, Nr. 1, S. 54–64, 2019.

[4] K. Hameyer, A. Ruf und F. Pauli, „Influence of fast switching semiconductors on the winding insulation system of electrical machines" in *2018 International Power Electronics Conference (IPEC-Niigata 2018-ECCE Asia)*, Niigata, 20.05.2018 - 24.05.2018, S. 740–745, doi: 10.23919/IPEC.2018.8507972.

[5] A. Nabae, I. Takahashi und H. Akagi, „A New Neutral-Point-Clamped PWM Inverter", *IEEE Trans. on Ind. Applicat.*, IA-17, Nr. 5, S. 518–523, 1981, doi: 10.1109/TIA.1981.4503992.

[6] T. Bruckner und S. Bernet, „Loss balancing in three-level voltage source inverters applying active NPC switches" in *2001 IEEE 32nd Annual Power Electronics Specialists Conference*, Vancouver, BC, Canada, 17-21 June 2001, S. 1135–1140, doi: 10.1109/PESC.2001.954272.

[7] E. Gurpinar, D. De, A. Castellazzi, D. Barater, G. Buticchi und G. Francheschini, „Performance analysis of SiC MOSFET based 3-level ANPC grid-connected inverter with novel modulation scheme" in *2014 IEEE 15th Workshop on Control and Modeling for Power Electronics (COMPEL)*, Santander, Spain, 22.06.2014 - 25.06.2014, S. 1–7, doi: 10.1109/COMPEL.2014.6877124.

[8] J. Holtz, M. Holtgen und J. O. Krah, „A Space Vector Modulator for the High-Switching Frequency Control of Three-Level SiC Inverters", *IEEE Trans. Power Electron.*, Jg. 29, Nr. 5, S. 2618–2626, 2014, doi: 10.1109/TPEL.2013.2280768.

978-1-6654-3236-8/21 $31.00 © 2021 IEEE

[9]   J. K. Kostko, „Polyphase reaction synchronous motors", *J. Am. Inst. Electr. Eng.*, Jg. 42, Nr. 11, S. 1162–1168, 1923, doi: 10.1109/JoAIEE.1923.6591529.

[10]   D. Reusch und J. Strydom, „Understanding the Effect of PCB Layout on Circuit Performance in a High-Frequency Gallium-Nitride-Based Point of Load Converter", *IEEE Trans. Power Electron.*, Jg. 29, Nr. 4, S. 2008–2015, 2014, doi: 10.1109/TPEL.2013.2266103.

[11]   E. Gurpinar, A. Castellazzi, F. Iannuzzo, Y. Yang und F. Blaabjerg, „Ultra-low inductance design for a GaN HEMT based 3L-ANPC inverter" in *2016 IEEE Energy Conversion Congress and Exposition (ECCE)*, 2016, S. 1–8, doi: 10.1109/ECCE.2016.7855540.

[12]   E. Gurpinar, F. Iannuzzo, Y. Yang, A. Castellazzi und F. Blaabjerg, „Design of Low-Inductance Switching Power Cell for GaN HEMT Based Inverter", *IEEE Trans. on Ind. Applicat.*, Jg. 54, Nr. 2, S. 1592–1601, 2018, doi: 10.1109/TIA.2017.2777417.

Semi-Flexible Power Control for Grid-Connected Converter

Petr Šimek[1]
simek@it.cas.cz

Martin Bejvl[1]
bejvl@it.cas.cz

Viktor Valouch[1, 2]
valouc@fel.cvut.cz

[1]Institute of Thermomechanics, Czech Academy of Sciences
[2]Department of Electrical Power Energy, Faculty of Electrical Engineering, Czech Technical University in Prague

Abstract— The formulas for the calculation of reference grid current components for the grid-connected converter are presented. The selection of the contribution of a negative voltage sequence to the formation of the reference current vector is possible. The current control is based on the GPCC (Generalized Predictive Current Control). The power responses to changes in power references and to grid failures are shown and discussed. The results of simulation and experimental validation are presented. The strategy is able to ensure that limit values of grid current magnitudes are not exceeded.

Keywords— *unbalanced grid voltage; positive and negative sequence; instantaneous active and reactive power; grid disturbances; power ripple; generalized predictive current control*

I. INTRODUCTION

In [1], [2] several possible current control strategies were investigated and compared. Among main problems and control strategies belong injecting negative-sequence reactive current helping to mitigate PCC voltage unbalance. Some solutions with fluctuations in either only active or reactive power were proposed. It was expended into the Flexible Positive and Negative Sequence Control (FPNSC) strategy with the maximum current peaks limitation. Flexible k_1, k_2 coefficients are either calculated using on-line measurements or may be generated through PI controllers as well [3].

A semi-flexible power control strategy of the grid-connected converter is presented here, which is based on the modified concept of the instantaneous active and reactive powers. The term the semi-flexible power control is used because the rates of the positive and negative current and power components do no depend only on the introduced flexible coefficients, but also on the voltage unbalance factor.

In the paper, the GPCC (Generalized Predictive Current Control) based power control is tested for use in voltage converters connected to the grid. The results of simulation and experimental tests are presented in the paper. As an important feature of the presented strategy is the fact that limit acceptable values for converter current magnitudes can be set and controlled even for unbalanced grid voltages.

The strategies of reference current calculation are presented in the part II, the GPCC method is reviewed in the next part III, the developed way to hold the maximal phase currents undel control is presented in the part IV. In the part V the control block diagram of the system is described and the selected results of simulation and experiments are shown in the part VI. The part VII reviews the main features of the developed methods.

II. REFERENCE CURRENT CALCULATION FOR INSTANTANEOUS POWER CONTROL

A. Converter reference current

The definition of the instantaneous active p and reactive q power is stated as follows

$$s = p + jq = 3/2\left[\operatorname{Re}\{vi^*\} + j\operatorname{Re}\{v(t-T/4)i^*\}\right] \quad (1)$$

where s is called the complex power, v, i stand for the voltage and current vectors in the SRF rotating by the synchronous speed $\omega = 2\pi/T$ in the positive direction, and T is the fundamental period.

We will present here a control strategy of the current vector i providing the precise instantaneous active and reactive power control without pulsating power components and harmonic free converter currents at the same time.

If we denote $v' = v(t-T/4)$, we can write

$$pv' - qv = \frac{3}{4}\left(v^*v' - v'^*v\right)i = -\frac{3}{2}j\operatorname{Im}\{v'^*v\}i \quad (2)$$

After some manipulations the final formula for the converter reference current (3) may be formulated on the basis of (2)

$$i = \frac{2}{3}\frac{j}{\operatorname{Im}\{v'^*v\}}(pv' - qv) = \frac{2}{3}\frac{j}{v^{p^2} - v^{n^2}}(pv' - qv) =$$
$$= \frac{2}{3}\frac{j}{v^{p^2}(1-n^2)}(pv' - qv) \quad (3)$$

where the superscripts p and n refer to the positive and negative sequences of the voltage and current vectors that are constants in steady states in the SRF^p and SRF^n, respectively, and $n = |v^n|/|v^p|$ is the unbalance factor.

B. Flexible and semi-flexible power control

The strategy of the Flexible Positive- and Negative-Sequence Control (FPNSC) has been presented in [4]. The principle consists in the selection of the measure in which the positive and negative voltage sequences contribute to the formation of the reference current vector. The reference current vector i may be viewed as the sum of two current components $i_{(p)}$, $i_{(q)}$ resulted from the demanded active p and reactive q power, respectively

$$i_{(p)} = 2/3\left[\frac{k_1 p}{v^{p^2}}v^p + \frac{(1-k_1)p}{v^{n^2}}v^n e^{-j2\omega t}\right] \quad (4)$$

$$i_{(q)} = 2/3\left[\frac{-jk_2 q}{v^{p^2}}v^p + \frac{j(1-k_2)q}{v^{n^2}}v^n e^{-j2\omega t}\right] \quad (5)$$

978-1-6654-3236-8/21 $31.00 © 2021 IEEE

where the coefficient k_1, k_2, being usually in the range <0,1>, determine the rate between the used positive and negative voltage sequence.

We can separate the current vector $\boldsymbol{i}$ (3) into similar two components

$$\boldsymbol{i}_{(p)} = \frac{2/3}{v^{p^2} - v^{n^2}} p\left(v^p - v^n e^{-j2\omega t}\right) \tag{6}$$

$$\boldsymbol{i}_{(q)} = \frac{-2/3j}{v^{p^2} - v^{n^2}} q\left(v^p + v^n e^{-j2\omega t}\right) \tag{7}$$

Similarly like the concepts presented in [4], we can also generalize the formula (6), (7)

$$\boldsymbol{i}_{\alpha\beta(p)} = \frac{2/3}{k_p v^{p^2} - (1-k_p)v^{n^2}} p_c^r\left(k_p v^p e^{j\omega t} - (1-k_p)v^n e^{-j\omega t}\right) \tag{8}$$

$$\boldsymbol{i}_{\alpha\beta(q)} = \frac{-2/3j}{k_q v^{p^2} - (1-k_q)v^{n^2}} q_c^r\left(k_q v^p e^{j\omega t} + (1-k_q)v^n e^{-j\omega t}\right) \tag{9}$$

where p_c^r, q_c^r are the final corrected power references, as it will be explained in the following paragraphs.

We are using the term the semi-flexible power control because the rates of the positive and negative components do no depend only on the coefficients k_p, k_q, but also on the unbalance factor $n = |v_n|/|v_p|$.

The coefficients k_p, k_q that can be chosen in the range <0,1> determine how much the voltage negative sequence is considered for the calculation of the active and reactive reference current vector. For $k_p = k_q = 0$ the voltage negative sequence is neglected and the converter current vector $\boldsymbol{i}$ is balanced because it is a replica of the voltage positive sequence v^p.

III. GENERALIZED PREDICTIVE CURRENT CONTROL FOR GRID-CONNECTED CONVERTER

In the GPCC (Generalized Predictive Current Control) we work with two independent SISO (Single-Input Single-Output) linear systems described by identical discrete transfer functions (10) for two axes α,β of the static reference frame

$$H_\alpha\left(z^{-1}\right) = H_\beta\left(z^{-1}\right) = \frac{B\left(z^{-1}\right)}{A\left(z^{-1}\right)} = \tag{10}$$

$$= \left(1 - z^{-1}\right)\mathbf{Z}\left\{\frac{1}{sL(s+1/\tau)}\right\} = \frac{b_1 z^{-1}}{1 + a_1 z^{-1}}$$

$$b_1 = \frac{1}{R}\left(1 - e^{\frac{T_s}{\tau}}\right), \quad a_1 = -e^{\frac{T_s}{\tau}} \tag{11}$$

where $\tau = L/R$ is the time constant of the L grid filter.

The GPCC strategy is presented in more detail in [5]-[8]. Using commonly used terminology for the variables of a control system, we can write (10) as

$$H_\alpha\left(z^{-1}\right) = H_\beta\left(z^{-1}\right) = \frac{y\left(z^{-1}\right)}{u\left(z^{-1}\right)} = \frac{B\left(z^{-1}\right)}{A\left(z^{-1}\right)} = \tag{12}$$

$$= \frac{b_1 z^{-1}}{1 + a_1 z^{-1}}$$

where the variable $u(z^{-1})$ is the control action of the GPCC and $y(z^{-1})$ is the output of the controlled system.

Generally, the GPCC control law is

$$\Delta u = \mathbf{Kw} - \mathbf{KG'}\Delta\mathbf{u}_p - \mathbf{KFy} \tag{13}$$

where $\Delta u = u(t) - u(t-1)$, $\Delta\mathbf{u}_p$ is the vector of past differences of the control action u, $\mathbf{w}$ is the vector of the future reference grid currents in the axis α or β, and $\mathbf{y}$ is the vector of the actual and past values of the grid currents in the axis α or β, respectively.

IV. LIMIT MAGNITUDE OF GRID CURRENT

Now, we will determine how to restrict the magnitude of the grid current in order to hold the converter phase currents under their demanded limit values

The total grid current vector is

$$\boldsymbol{i}_{\alpha\beta} = \boldsymbol{i}_{\alpha\beta(p)} + \boldsymbol{i}_{\alpha\beta(q)} \tag{14}$$

Substituting (8), (9) into (14) we get

$$I_{a\max} = I_{\alpha\max} = 2/3\sqrt{\begin{array}{l}\dfrac{p^2}{D_p^2}\left[(v_d^P)^2 + (v_q^P)^2\right] + \\ + \dfrac{q^2}{D_q^2}\left[(v_d^Q)^2 + (v_q^Q)^2\right] + \\ + \dfrac{2pq}{D_p D_q}(v_d^P v_q^Q - v_q^P v_d^Q)\end{array}} \tag{15}$$

$$\begin{aligned} v_d^P &= \left[k_p v_d^p - (1-k_p)v_d^n\right] \\ v_q^Q &= \left[k_q v_q^p + (1-k_q)v_q^n\right] \\ v_q^P &= \left[k_p v_q^p + (1-k_p)v_q^n\right] \\ v_d^Q &= \left[k_q v_d^p - (1-k_q)v_d^n\right] \end{aligned} \tag{16}$$

where

$$D_p = k_p v^{p^2} - (1-k_p)v^{n^2} \tag{17}$$

$$D_q = k_q v^{p^2} - (1-k_q)v^{n^2} \tag{18}$$

and $v^p = (v_d^p, v_q^p)$, $v^n = (v_d^n, v_q^n)$ are expressed in the SRF^p.

To get similar conditions for the maximum currents $I_{b\max}$, $I_{c\max}$ in the phases b, c we can repeat the same procedure, but instead of v^p, v^n we must use $v^{pb} = v^p \exp(-j2\pi/3)$, $v^{nb} = v^n \exp(-j2\pi/3)$, and $v^{pc} = v^p \exp(j2\pi/3)$, $v^{nc} = v^n \exp(j2\pi/3)$, respectively.

The current value $I_{\max}$ that is the maximal one among those calculated for all the phases a, b, c is taken for testing

978-1-6654-3236-8/21 $31.00 © 2021 IEEE

of possibility to realize the reference power values p, q or not.

Let I_{lim} is a maximum allowable RMS value of the phase current. If $\sqrt{2}\,I_{lim} < I_{max}$ then the applied references are $p^r_c = p$, $q^r_c = q$. In opposite case the applied references must be decreased.

V. CONTROL BLOCK DIAGRAM

Fig.1 shows the block diagram of the grid-connected converter with the proposed control system. The instantaneous active power p and reactive power q are the powers that are being taken/delivered from/to the grid. The values p^r, q^r are demanded values of the output variables p, q. From the adjusted power references $p_c{}^r$, $q_c{}^r$ the reference currents $i_\alpha{}^r$, $i_\beta{}^r$ are calculated using (8), (9).

The outputs of the whole control system are the demanded components of the converter voltages $v^r_{CO\alpha}$, $v^r_{CO\beta}$ calculated from the control actions u_α, u_β and grid voltage components v_α, v_β. The output converter voltage is generated applying any PWM.

The experimental setup is shown in Fig. 2.

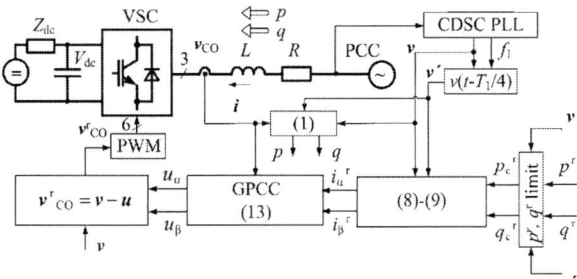

Fig. 1. Block diagram of grid-connected converter with proposed control system

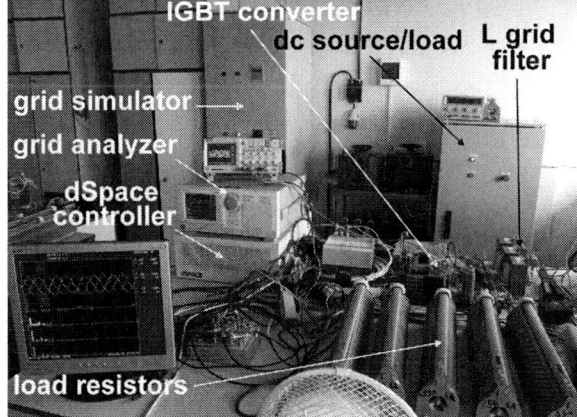

Fig. 2. Experimental setup

VI. SIMULATION AND EXPERIMENTAL RESULTS

The function of the converter connected to the grid was simulated in the Matlab™ environment and tested in the laboratory. The used parameters in simulation are summarized in TABLE I.

TABLE I. PARAMETERS OF GRID, L FILTER AND CONVERTER

Grid voltage		L filter		Sampling Frequency f_s	Switching Frequency f_{sw}	V_{dc} (V)	Reference powers	
$V^p{}_{RMS}$ (V)	$V^n{}_{RMS}$ (V)	L (mH)	R (Ohm)				p^r (W)	q^r (VA)
115 (90)	0 (25)	5	1.3	6kHz	3 kHz	600	5000 (-10000)	-5000 (-6000)

Fig. 3 shows the responses of the grid voltage and current components and both the instantaneous powers following the power reference changes for the active power p^r jumping from 0 to 5000 W and reactive power q^r jumping from 0 to -5000 VA under unbalanced grid voltage $V^p{}_{RMS}$=90 V, $V^n{}_{RMS}$=25 V. No current limit is applied. Then, the unbalanced voltage drops even to $V^p{}_{RMS}$=55 V, $V^n{}_{RMS}$=15 V in the interval 0.04-0.08 s. The stabilization of both the powers after the voltage drop by 40% of their original values is achieved during less than 10 ms.

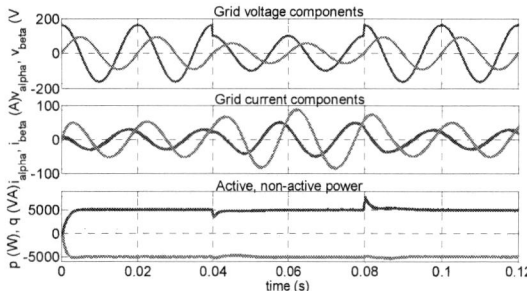

Fig. 3. Responses of grid voltage and current components and instantaneous powers after changes in references for active power and reactive power, and after voltage drop in interval 0.04-0.08 s. Grid voltage is unbalanced. None current limit is applied

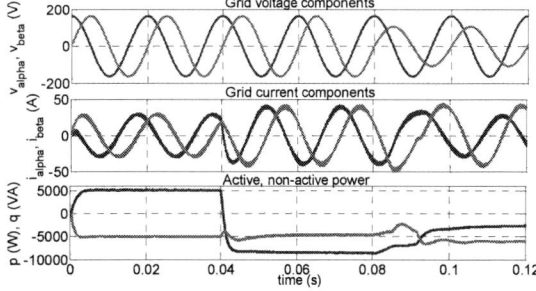

Fig. 4. Responses of grid voltage and current components and instantaneous powers after changes in references for active and reactive power and change of grid voltage. Current limit I_{maxRMS}=30 A is applied

Fig. 4 shows the responses of the grid voltage and current components and instantaneous powers after changes in the references for the active power p^r from 0 to 5000 W and the reactive power q^r from 0 to –5000 VA. Then, the change in the reference for the active power p^r from 5000 W to -10000 W is done at time 0.04 s. Finally, the grid voltage is changed to unbalanced one (V^p_{RMS}=90 V, V^n_{RMS}=25 V) and q^r from -5000 W to –6000 at time 0.08 s. The current limit I_{maxRMS}=30 A is applied.

We can see that the absolute values of both the powers are lower than those of their reference values in the time interval 0.04-0.08 s as a result of efforts not to exceed the current limit I_{maxRMS}=30 A (I_{max}=√2 30=42 A). Starting from t=0.08 s, in case of the voltage dip, the demanded reactive power –6000 VA is delivered into the grid, so the amount of the active power is further decreased.

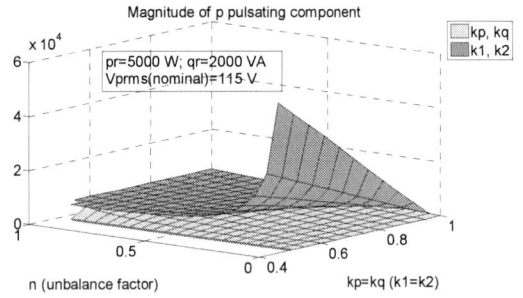

Fig. 5. Comparison of magnitudes of pulsating components of active power p for the FPNSC and developed semi-flexible method

Fig. 5 compares the magnitudes of pulsating components of the active power p for the FPNSC and developed semi-flexible method for the coefficients of both the methods changing in the range <0.5, 1> and for the unbalance factor $n \in$ <0,1>. It is evident that the magnitudes of the pulsating components are much higher for the FPNSC than those for our semi-flexible method. The red line for $k_d=k_q$=0.5 indicates that the developed semi-flexible strategy provides only pure constant power without any fluctuating component.

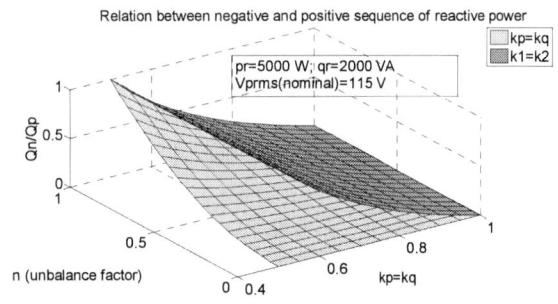

Fig. 6. Comparison of rates of the negative and positive sequence of the reactive power for the FPNSC and developed semi-flexible method

Fig. 6 compares the rates of the negative and positive sequence of the reactive power under the same conditions as those in Fig. 5. This rate can be changed using different control coefficient k_q (k_2) to decrease the negative sequence voltage component and increase the positive sequence voltage component in case of unsymmetrical grid voltage faults. We see that the rate for the developed semi-flexible method depends not only on the coefficient k_q (here $k_d=k_q$) but also on the unbalance factor n. It is due to the fact that the denominator of the equation (8), (9) depends also on this unbalanced factor $n=|v_n|/|v_p|$ which is not the case of the FPNSC (6), (7). This may be seen as a disadvantage of the developed semi-flexible method in comparison to the FPNSC, because a proper value of k_q should be continually

recalculated according to the actual value of n. So, Figs. 5, 6 elucidate the dilemma that should be solved, that is what method will be better in our current case.

The presented experimental responses were captured by the dSPACE™ control system, then transferred into PC and visualized in figures using graphical functions of the MATLAB™.

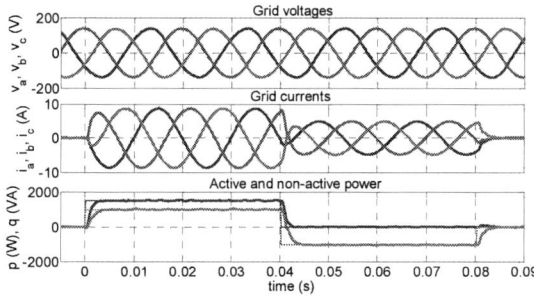

Fig. 7. Experimental captured responses of phase grid voltages and currents, and instantaneous powers after step changes in references for active p^r and non-active q^r power (symmetrical grid voltage, without current limit)

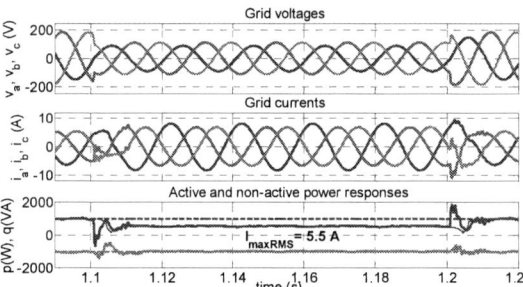

Fig. 8. Captured experimental responses of phase grid voltages and currents, and instantaneous powers after voltage dip in interval 1.1-1.2 s. Grid voltage is unbalanced (130-105-130 V_{RMS} before and after voltage dip, and 85-65-85 V_{RMS} in interval 1.1-1.2 s). Current limit I_{maxRMS}=5.5 A is applied.

Fig. 7 shows the experimental responses of the grid phase voltages and currents, and also of both the instantaneous powers after step changes in the references for the active p^r and non-active q^r power. The grid voltage is symmetrical and none current limit is applied. We can see that the generated powers follow their references almost without overshoots and the settling times are about 3-4 ms.

Fig. 8 shows the experimental responses of the grid phase voltages, currents and both the instantaneous powers.

The figure demonstrates possibility to deliver the demanded value $q = q^r$ of the reactive power to the grid in case of a voltage dip and not to exceed the phase current I_{max} at the same time. The reason for such a strategy is effort to suppress the voltage dip. Depending on the voltage dip and the values q^r, I_{max} , it may be also possible to supply even a part of the required active power p^r to the grid. In the figure the active and reactive power follow their reference values p^r=1000 W, q^r=-1000 VA for the unsymmetrical voltage (130-105-130 V_{RMS}), but after change of the voltage to (85-65-85 V_{RMS}) in the time interval 1.1-1.2 s the active power must be decreased to preserve the demanded value q^r=-1000 VA of the reactive power. The coefficients $k_d=k_q$=0.5 were applied. The transients in the grid current and

978-1-6654-3236-8/21 $31.00 © 2021 IEEE

both the powers are stabilized within the time lower than one period of the fundamental frequency.

Thus, the experimental results have confirmed the theoretical analyzes and expectations based on the results of simulations.

VII. Conclusion

The GPCC based power control for the voltage source converter connected to the grid was developed, simulated, and tested. The grid current and instantaneous power responses under different grid failures and for power reference changes were presented and evaluated.

The accuracy of achieving the required power values in steady states, minimum power fluctuations of the frequency 100 Hz together with the phase currents without low order harmonics even for a highly unbalanced grid voltage system are main advantages of the method. As an important feature of the presented strategy is the fact that limit accetable values for grid current magnitudes can be controlled even for unbalanced grid voltages. The presented algorithm is able to assure a secure fault ride-through using full converter capacity and taking into account requirements for reactive power provision during voltage dips in the grids.

Acknowledgment

The authors wish to acknowledge the financial support of the Institute of Thermomechanics, Czech Academy of Sciences (project RVO//:61388998), the research program Efficient Energy Conversion and Storage of Strategy AV21 of the Czech Academy of Sciences, and the project Advanced procedures for solving critical conditions in electrical systems (SGS20/165/OHK3/3T/13), CTU in Prague.

References

[1] Rodrigues, P., Timbus, A. V., Teodorescu, R., Liserre, M., Blaabjerg, F., "Flexible Active Power Control of Distributed Power Generation Systems During Grid Faults," *IEEE Trans. Ind. Electron.*, vol. 54, no. 5, pp. 2583–2592, Oct. 2007.

[2] Teodorescu, R., Liserre, M., Rodrigues, P., Grid Converters for Photovoltaic and Wind Power Systems. *Wiley-IEEE Press*, 2011.

[3] M. Castilla, J. Miret, A. Camacho, J. Matas, and L. G. D. Vicuna, "Voltage support control strategies for static synchronous compensators under unbalanced voltage sags,"IEEE Trans. Ind. Electron., vol. 61, no. 2, pp. 808–820, Feb. 2014.

[4] R. Kabiri, D. G. Holmes, and B. P. McGrath , "Control of active and reactive power ripple to mitigate unbalanced grid voltages,"IEEE Trans.Ind. Appl., vol. 52, no. 2, pp. 1660–1668, Mar./Apr. 2016.

[5] Vazques, S., Montero, C., Bordons, C., Franquelo, L., "Design and experimental validation of a model predictive control for a VSI with long prediction horizon," Proc. 39th annual Conf. IEEE Ind. Electr. Society (IECON'13), pp. 5788-5793, Nov. 2013.

[6] Judewicz, M. G., Gonzalez, S. A., Echeveria, N. I. Fischer, J. R., Carrica, D. O.,"Generalized predictive current control (GPCC) for grid-tie three-phase inverters," IEEE Trans. Ind. Electron., vol. 63, no. 7, pp. 4475- 4484, 2016.

[7] M. G. Judewicz, N. I. Echeverria, J. R. Fischer, S. A. Gonzalez and D. O. Carrica," Generalized Predictive Control of Three-Level Boost Rectifiers," XVII Workshop on Information Processing and Control (RPIC), Mar del Plata, Argentina, September 2017.

[8] M. G. Judewicz, S. A. González , J. R. Fischer , Juan F. Martínez, and D. O. Carrica," Inverter-Side Current Control of Grid-Connected Voltage Source Inverters With LCL Filter Based on Generalized Predictive Control," IEEE Journal of Emerging and Selected Topics in Power Electronics, Volume. 6, no. 4, December 2018, DOI: 10.1109/JESTPE.2018.282636

978-1-6654-3236-8/21 $31.00 © 2021 IEEE

Qualitative comparison of the behavior for a five-phase induction motor in error states for different stator windings connections

1st Jakub Kellner
University of Zilina
Faculty of Electrical Engineering and
Information Technology
Department of Mechatronics and
Electronics
Žilina, Slovakia
jakub.kellner@feit.uniza.sk

2nd Slavomír Kaščák
University of Zilina
Faculty of Electrical Engineering and
Information Technology
Department of Mechatronics and
Electronics
Žilina, Slovakia
slavomir.kascak@feit.uniza.sk

3rd Michal Praženica
University of Zilina
Faculty of Electrical Engineering and
Information Technology
Department of Mechatronics and
Electronics
Žilina, Slovakia
michal.prazenica@feit.uniza.sk

4th Marek Paškala
University of Zilina
Faculty of Electrical Engineering and
Information Technology
Department of Mechatronics and
Electronics
Žilina, Slovakia
marek.paskala@feit.uniza.sk

Abstract—The basis of this article is to investigate the properties of a five-phase induction motor, in various stator windings connections. The suitability of using a multi-phase machine is in applications such as EV, HEV, shipping industry, or electric aircraft. This is due to the better properties of multiphase machines, such as higher torques, lower phase loads, smoother torque, or better fault tolerance, which is ensured by further degrees of freedom of multiphase motors. We know that the stator windings of a five-phase induction motor can be connected in three configurations. It is a star connection, pentagon connection, and pentacle connection, while in all three cases the motor will have different properties. The basis of this article is to find out how the motor will behave in case of error conditions and what effect the different stator winding configurations will have on this. The result of the article is an evaluation in which of the connections the properties of the motor during the failure are the best. All results are compared with the state when the engine was running without failure. These facts are ascertained by simulation and subsequently verified on a real engine.

Keywords—*five-phase induction motor, error state, VSI, one phase fault, failure of two adjacent phases, failure of two non-adjacent phases*

I. INTRODUCTION

At present, the most frequently used motors in practice are still three-phase. This is due to the fact that their power supply is very simple. The three-phase power supply is available almost everywhere and therefore it is not a problem to use three-phase motors. But if we use an electric motor as the vehicle's propulsion, which we want to regulate in speed or torque, then we must use a converter to power the motor. In this case, the power supply is realized either using a three-phase network and a rectifier, or traction batteries are used as a DC voltage source. Subsequently, a converter with the required number of phases is designed according to the requirements of the drive. Most often, these multiphase motors are powered by Voltage Source Inverter (VSI) or matrix converters, which are still not very popular in power

electronics and practice due to their complex control and a large number of semiconductor bidirectional switches [1].

We know that the use of motors with more than 3 phases brings several advantages. Benefits include:

- Reduction of the required power per motor phase. This involves reducing the current through the motor and inverter phases without the need to increase the voltage. Mainly used in high-performance applications such as the shipping industry.

- Reduction of motor ripple.

- Reduction of engine noise. Direct connection with lower ripple.

- Reduction of torque.

- Fewer higher harmonics.

- Additional degrees of freedom.

The latter feature of the multi-phase motor ensures better fault tolerance of the machine. A single-phase failure in a multi-phase machine will result in minimal power reduction and the machine will still be able to operate smoothly. Even in the event of a two-phase failure, the motor is still able to operate, as will be described in this article.

At the same time, it should be noted that in the case of multiphase motors, it is possible to improve the motor properties by adjusting the control in fault conditions. So much so that the ripple of the machine when running without one phase will be zero. But that is not the subject of this article.

In this article, we will deal directly with the engine and study its properties in fault conditions, without adjusting the control. The basic breakdown of faults that can occur is:

- On the motor side (destruction of the motor phase winding).

- On the supply side (destruction of the inverter switch - half branch, or both switches of a given phase).

- Between power supply and motor (Power wires or connectors).

- In control (failure of PWM references of a given phase).

However, as already mentioned in the case of a multiphase induction motor, the fault condition can occur as a failure of one phase or a failure of two phases. The failure of two phases is divided into the failure of two adjacent or the failure of two non-adjacent phases. As we will see in the article.

The examination of a five-phase motor is partly mentioned in the article [2], where the result is that when the motor is connected to the star, the motor can work with a power drop of 10% in the event of a single-phase failure. When connected to a pentagon, the power drop is 20% and the use of a pentacle connection is inappropriate in the event of a single-phase failure. Articles [3,4] present simulations of the operation of a five-phase motor in operation without a single phase in the pentagon and a pentacle connection. These articles state that the engine loses its original power but can work without problems. In [5] the theory around the model of a five-phase machine in the state after failure is comprehensively solved. The article deals with obtaining a model from Clark's transformation. In recent years, research has focused mainly on the analysis of fault tolerance solutions in synchronous machines with permanent magnets [6-8].

The result of this article is a comprehensive determination of the motor properties in the event of a failure of one phase, two adjacent phases, and two non-adjacent phases in comparison with the fault-free state in different connections of stator windings. And determining which of the connections the motor has the best features.

II. FIVE-PHASE INDUCTION MOTOR

The five-phase induction motor is a type of multi-phase motor that is one of the most commonly used and researched machines in this field [9].

In a three-phase induction machine, the three phases are shifted by 120 °, while in a five-phase induction machine, five phases are shifted by 72 °. The stator winding of an n-phase machine can thus be designed so that the spatial displacement of two successive secondary phases is always:

$$\alpha = 2\pi/n \qquad (1)$$

In this way, a symmetrical multiphase machine is created. This occurs when the number of phases is an odd prime number. Figure 1 shows the spatial distribution of the windings of a five-phase induction motor.

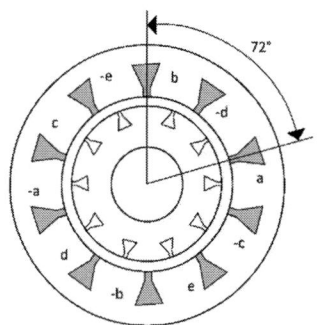

Fig. 1 Winding distribution of a five-phase induction motor

As mentioned in the introduction, a five-phase machine can be connected to three different connections. It is a star connection, pentagon connection, and pentacle connection, as shown in Figure 2, where Figure 2 a.) Represents the star connection, 2 b.) Represents the pentagon connection and 2 c.) Represents the pentacle connection [10].

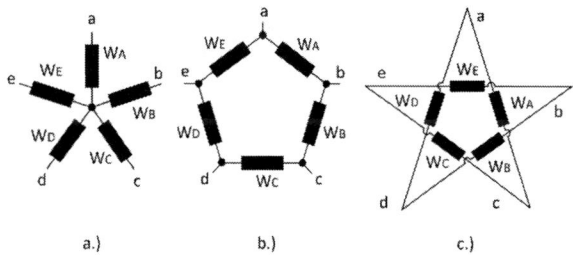

Fig. 2 Stator windings connections of a five-phase induction machine. a.) Star. b.) Pentagon. c.) Pentacle

The basic difference between the individual connections is given by the arrangement of the individual windings. It is appropriate to show this in the phase diagram, where we can better see the difference between these connections. When connected to a star, the stator voltages and thus also the currents are the same as the voltages and currents of the power supply. However, the difference occurs when connecting to the pentagon and the pentacle. The phase diagram of the pentagon connection is shown in Figure 3 a.). It is obvious that the voltage on the stator windings of the motor will be 1.17 times higher than the supply voltage. The phase diagram of the pentacle connection is shown in Figure 3 b.). It is obvious that the voltage on the stator windings of the motor will be 1.9 times higher than the supply voltage [11].

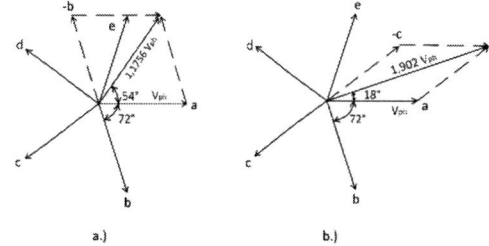

Fig. 3 Phase diagram for a five-phase motor. a.) Pentagon. b.) Pentacle

It is necessary to realize what is affected by the voltage change on the stator windings of the motor. If it is a symmetrical multiphase machine, then the input power of the motor is given:

978-1-6654-3236-8/21 $31.00 © 2021 IEEE

$$P = m \cdot U_1 \cdot I_1 \cdot cos\varphi \ [W] \qquad (2)$$

where:

m – is the number of stator winding phases [-]. U_1 – is one phase voltage [V]. I_1 is one phase current [A]. $cos\varphi$ is power factor [-].

The relation for calculation of mechanical torque is:

$$T_{mech} = \frac{m_1 \cdot p \cdot U_1^2 \frac{R_2'}{s}}{2 \cdot \pi \cdot f_1 \cdot \left[\left(R_1 + \frac{R_2'}{s}\right)^2 + (X_{r1} + X_{r20}')^2\right]} \ [Nm] \qquad (3)$$

where:

R_2'/s is the total active rotor resistance [Ω]. X_{r20}' is leakage reactance of the rotor winding converted to a stator [Ω]. X_{r1} is stator leakage reactance [Ω]. R_1 is the stator winding resistance [Ω]. R_2' is the resistance of the rotor converted to stator [Ω]. p is the number of pole pairs [-]. s is the motor slip [-].

The power calculation on the motor shaft is as follows:

$$P = T \cdot \Omega = T \frac{2\pi n}{60} \ [W] \qquad (4)$$

where:

T is the motor torque [Nm]. Ω is the angular velocity of the rotor [rad.s^{-1}]. n is the motor speed [rpm].

In this article, the simulation, and measured properties of the motor in fault states and fault-free operation will be presented [12-13].

III. SIMULATIONS

The simulations listed in this heading were performed for the machine parameters listed in Table 1. These are the parameters of a five-phase induction motor on which measurements were subsequently performed to confirm the accuracy of the simulations.

TABLE I. FIVE-PHASE INDUCTION MOTOR PARAMETERS

Signature	Name	Value
R_s	Stator resistance	15.05 [Ω]
R_r	Rotor resistance	5.926 [Ω]
L_s	Stator inductance	0.8714 [H]
L_r	Rotor inductance	0.8714 [H]
L_m	Mutual inductance	0.85 [H]
p	Number of pole pairs	2 [-]
J	Moment of inertia	0.007 [kg.m^2]
P_n	Mechanical power	1.1 [kW]

A. Simulation of motor torque in error states

We first detected the torque drop of a five-phase induction machine in error states such as the failure of one phase, the failure of two adjacent phases, and the failure of two non-adjacent phases for all three stator winding connections.

We created the engine model in the Matlab / Simulink environment. It is a mathematical model, which is given by standard equations for a multiphase induction machine. The basis of the model is the Clarke matrix, current and flow calculations, and then the feedback Clarke matrix. The model works with the basic dq components as well as with the xy and 0 components, which come into force in fault states. The motor was powered by a hard five-phase voltage source 5x230V, 50Hz. The connection type of the stator winding is taken into account in the model by connecting the beginnings and ends of the individual windings according to Figure 2.

Figure 4 shows a torque characteristic of a motor simulation in the star connection of stator windings. From the characteristic, we can see how the motor torque decreased in fault conditions. The characteristic lists all error states in which the engine is able to operate without a change in control.

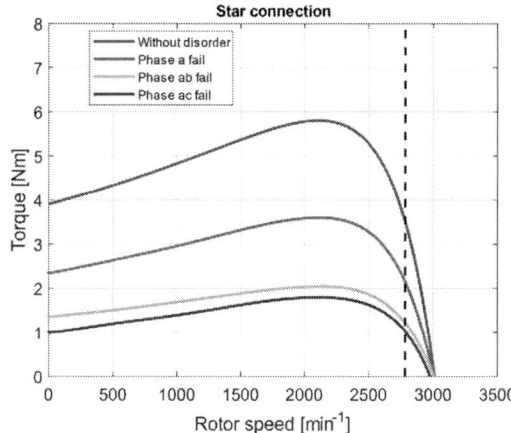

Fig. 4 Torque characteristic of five-phase induction motor in a star connection

Figure 5 represents the same simulation as in Figure 4 but for the pentagon connection. Similarly, Figure 6 represents a simulation of the motor torque, but for the pentacle connection.

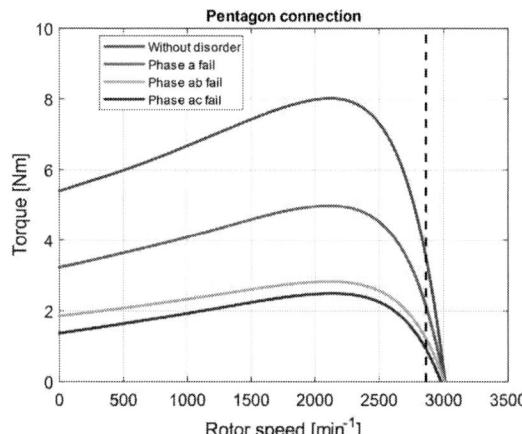

Fig. 5 Torque characteristic of five-phase induction motor in a pentagon connection

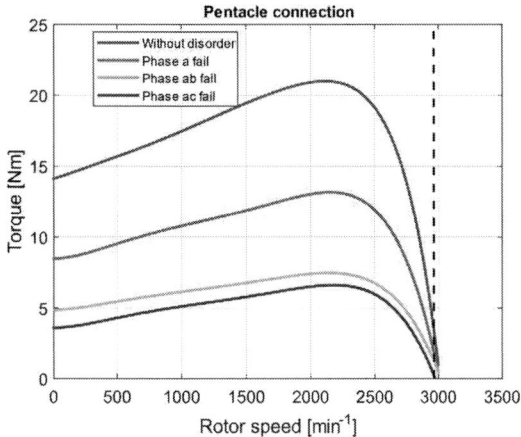

Fig. 6 Torque characteristic of five-phase induction motor in a pentacle connection

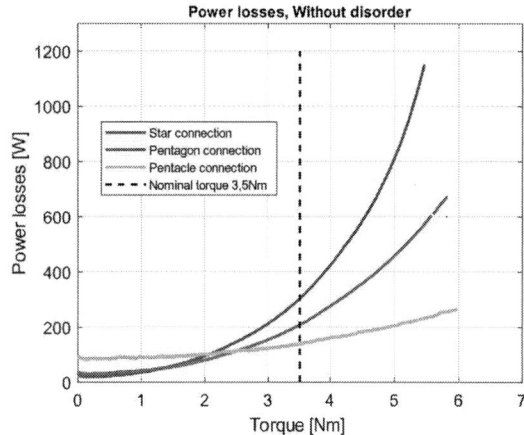

Fig. 7 Motor losses of five-phase induction motor in a free-fault condition

The results of the star torque motor simulation are as follows. In the event of a single-phase failure, the motor torque decreased by 38.62%. In the case of failure of two adjacent phases, the decrease of torque was by 65.28% and in the case of failure of two non-adjacent phases, it was up to 70.77%.

When connected to a pentagon, the torque drop was 39.23% with one phase failure, 65.20% with two adjacent phases, and 73.72% with two non-adjacent phases. The percentage reduction is essentially the same/similar, but the resulting torque is 1.37 times greater. What is given by the connection of stator windings.

When connected to the pentacle, the decrease in torque at the failure of one phase was 45.15%, at the failure of two adjacent phases it was 72.3%, and at the failure of two non-adjacent phases 38%. The resulting torque is 3.62 times greater. What is given by the connection of stator windings.

From the curves, we can see that the moments when they are connected to the pentagon are 1.369 (1.17^2) times greater and 3.61 (1.9^2) times greater at the pentacle. This means that the five-phase induction motor has the same characteristics when connected to a star as when connected to a pentagon when the supply voltage is reduced by 1.17 and for the pentacle when the supply voltage is reduced by 1.9 times.

B. Simulation of motor power losses in error states

This subheading shows engine losses in fault and free-fault conditions. Figure 7 shows the characteristics of the motor during trouble-free operation. From the waveforms we see that the pentacle connection has the smallest losses, then the pentagon connection, and the worst is the star connection.

It follows from this figure that it makes sense to operate the engine (economic operation) up to the point where the losses in the pentagon/pentacle circuit are equal to the losses in the star circuit at rated torque/power. The economical operation represents a loss of power of approximately 300 W or more, but only to the extent that the motor reaches 80% of rated torque.

All operations above 300W are unsuitable and motor power should be reduced to reduce these losses. Figure 15 shows that it does not make sense to use a pentacle connection even in the case of a single-phase outage, only if we accept larger losses.

Figure 8 shows motor losses in the event of a single-phase failure. From the waveforms, we see that the connection in the star and pentagon has similar losses, but the connection in the pentacle has higher losses.

Figures 9 and 10 show the losses of the motor in operation with the failure of two adjacent and two non-adjacent phases. From the waveforms, we see that the losses in the star connection and pentagon are similar, but the losses of the motor in the pentacle connection are very large.

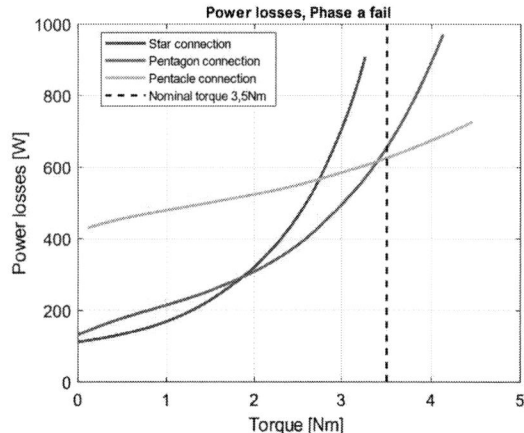

Fig. 8 Motor losses of five-phase induction motor in a single-phase error

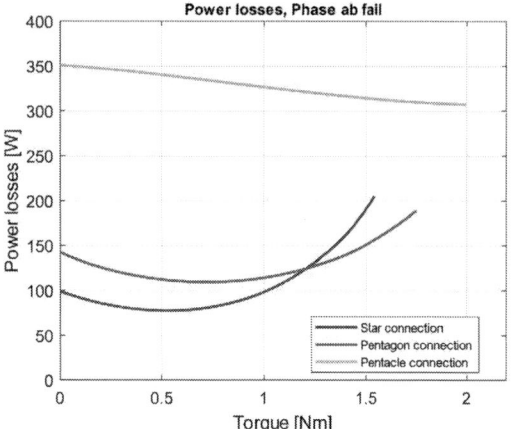

Fig. 9 Motor losses of five-phase induction motor in case of failure of two adjacent phases

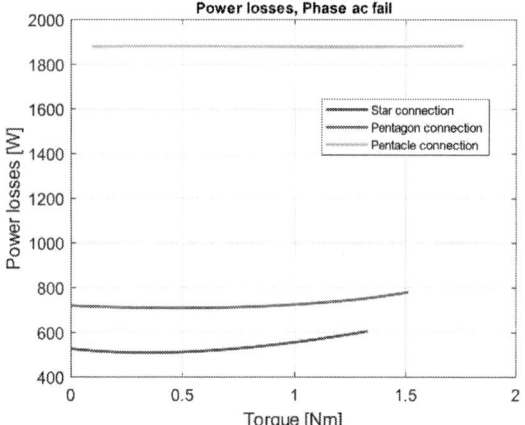

Fig. 10 Motor losses of five-phase induction motor in the event of a failure of two non-adjacent phases

From the simulation curves of the five-phase induction motor we can say that the pentacle connection has the best properties, ie. the smallest losses in the fault-free state but has the largest losses when the motor is operated with a fault (phase a, ab, or ac fault). Although the star connection has the greatest losses in the fault-free state, in the event of a phase failure, the losses are almost comparable to a pentagon. The stars connection has better properties in the first half of the characteristic, but then has fewer losses at higher loads. Therefore, the pentacle connection is inappropriate in this aspect. But a pentagon connection has significantly better properties than a star connection.

However, in the event of a failure of these two phases, the losses are very large and reach up to 50% of the rated power. Therefore, operation in this mode is not possible. Only in the case of a star connection for emergency travel.

IV. MEASUREMENT AND COMPARISON WITH SIMULATION

In this subheading, we measured a five-phase induction motor that was powered by a five-phase VSI. We loaded the engine with a dynamometer. The measurements resulted in speed-dependent torque characteristics. We measured one phase current during trouble-free operation. The others we considered the same because of the symmetry of the engine.

We verified the measured parameters by simulating the five-phase model used in heading 3 powered by a five-phase VSI. We did this to verify the correctness of the engine model.

The measurement was performed only for a certain range of loads due to overheating of the machine. Fault conditions were caused by the disconnection of the motor phases from the inverter. Figure 11 shows the basic circuit diagram of the measurement with the disconnected phase a.

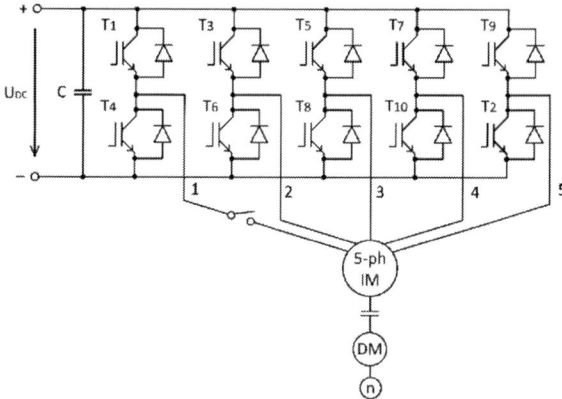

Fig. 11 Basic measurement scheme of a five-phase motor with a phase error

A. Measurement of five-phase induction motor in star connection

As we mentioned above, we measured the engine only in a certain range of speeds due to engine overheating. The supply voltage of the DC line was 500V. Figure 12 shows the characteristics from the measurement (red curve) and simulation (blue curve) of the motor torque in fault-free operation. And also, in case of failure of one phase measurement (yellow curve) and simulation (gray curve).

Figure 13 shows a comparison of the phase current by one phase from the measurement (red) and the simulation (blue). From these curves, we see that the relative error is 6.1% which means that the simulation model is almost comparable to a real engine.

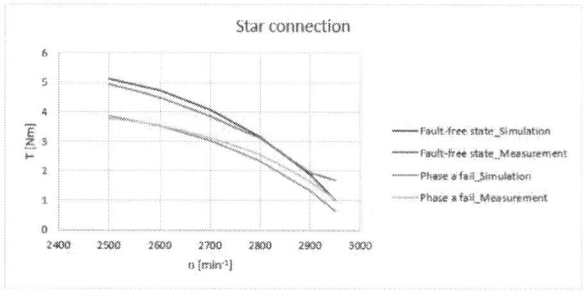

Fig. 12 Measurement of torque and comparison with simulation in free-fault condition and failure of one phase in a star connection

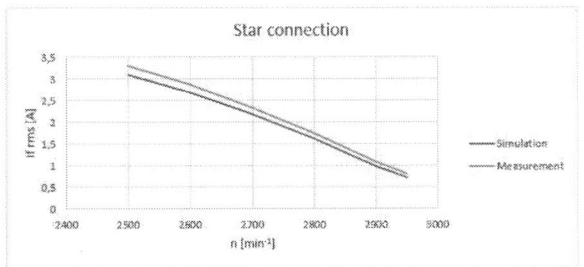

Fig. 13 Measurement of phase current and comparison with simulation in the free-fault condition in a star connection

B. Measurement of five-phase induction motor in pentagon connection

We performed the same measurements for the pentagon motor connection, which we then compared with the simulation. The supply voltage of the DC line was 420V. Figure 14 shows the measured and simulated waveforms of torque. And Figure 15 shows a comparison of measured and simulated phase currents in a pentagon circuit. And we can see from the curves that the relative error is 7%.

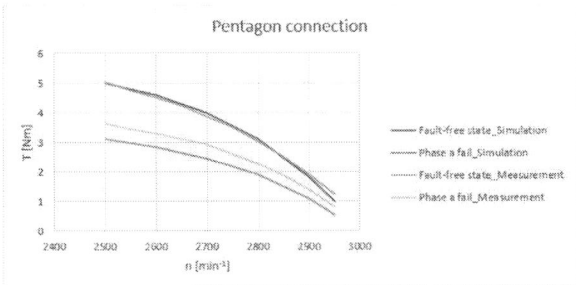

Fig. 14 Measurement of torque and comparison with simulation in free-fault condition and failure of one phase in a pentagon connection

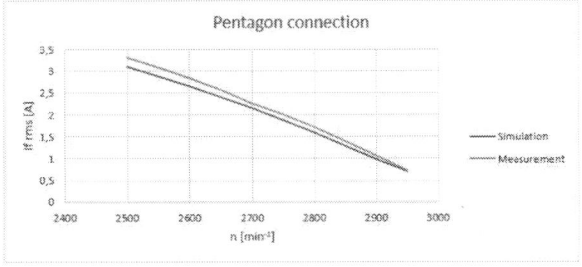

Fig. 15 Measurement of phase current and comparison with simulation in the free-fault condition in a pentagon connection

C. Measurement of five-phase induction motor in pentacle connection

At the pentacle of the motor connection, we performed the same measurements, which we then compared with the simulation. The supply voltage of the DC line was 270V. Figure 16 shows the measured and simulated waveforms of torque. And Figure 17 shows a comparison of measured and simulated phase currents in a pentagon circuit. The relative error is less than 1%.

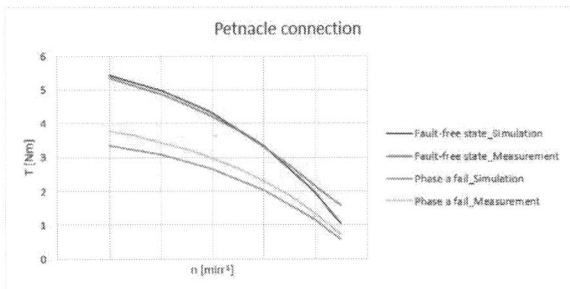

Fig. 16 Measurement of torque and comparison with simulation in free-fault condition and failure of one phase in a pentacle connection

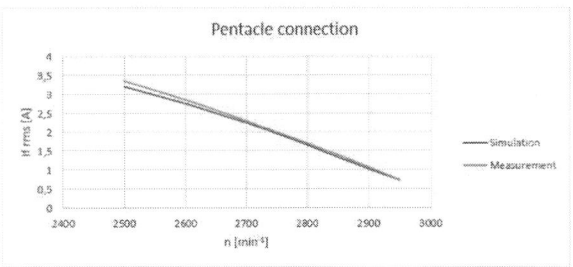

Fig. 17 Measurement of phase current and comparison with simulation in the free-fault condition in a pentacle connection

D. Measurement of five-phase induction

Figure 18 shows the measurement stand on which the measurements described in this heading were performed.

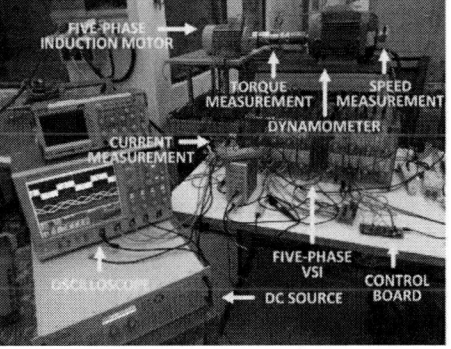

Fig. 18 Measurement of a five-phase induction motor in fault conditions

As mentioned successively, in these measurements, the input voltage of the DC line for the star connection was 500V, and the value of the first harmonic voltage was 170V. When pentagon connection, the value of the DC line was 420V and the first harmonic voltages on the motor phase were again 170V. Similarly, when connected to the pentacle, the input voltage was reduced, but the value of the first harmonic voltage of the machine winding was 170V. This ensures that the results can be comparable in all conditions.

Figure 19 – 21 shows the waveforms of phase current measurement in individual connections of stator windings. In all cases, the measurement is performed when the engine is idling. Figure 19 shows the phase currents for star connection. Figure 20 shows the phase currents for the pentagon connection. Figure 21 shows the phase currents for the pentacle connection. The fifth waveform is not shown in Figures 19-21 because only a four-channel oscilloscope was used.

978-1-6654-3236-8/21 $31.00 © 2021 IEEE

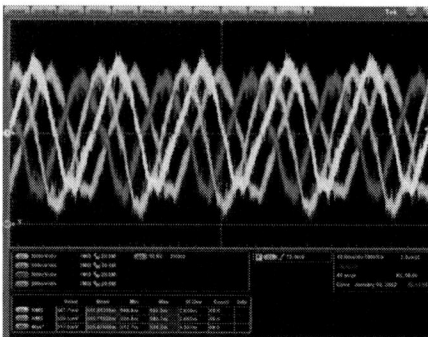

Fig. 19 Measured motor phase currents in star connection, no-load.

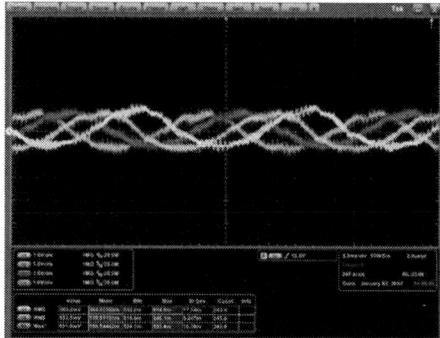

Fig. 20 Measured motor phase currents in pentagon connection, no-load.

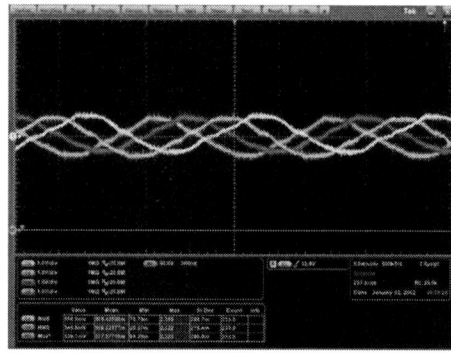

Fig. 21 Measured motor phase currents in pentacle connection, no-load.

V. CONCLUSION

The aim of this work was to find out how a five-phase induction machine will behave in fault states, in the event of a failure of one phase, two adjacent phases, and two non-adjacent phases. It was compared in which of the stator windings connection (star, pentagon, pentacle) the motor will have the best properties.

The motor properties were first determined by simulation using a model in the Matlab / Simulink environment. The results of these simulations are presented in heading 3. Subsequently, the engine model was verified on a real engine, on which we further determined the properties of the engine in fault conditions. The results of the measurements are given in heading 4. Heading 2 presents a general theory regarding five-phase induction motors and heading 1 presents an introduction to the issue as well as an overview of previous research on fault conditions in multiphase motors.

The result of the work is a comprehensive evaluation of the behavior of the motor in fault conditions, which is given from headings 3 and 4. We have found that during trouble-free operation, it is most advantageous to operate the motor in a pentacle circuit. And in the event of a failure or malfunction, switch to the pentagon connection.

ACKNOWLEDGMENT

This research was funded by VEGA 1/0085/21.

REFERENCES

[1] G. K. Singh "Multi-phase induction machine drive research-a survey," Electric Power Systems Research 61 (2002) 139 – 147, March 2002, doi: 10.1016/S0378-7796(02)00007-X.

[2] L. Schreier, J. Bendl and M. Chomat, "Analysis of fault tolerance of five-phase induction machine with various configu-rations of stator winding," 2015 International Conference on Electrical Drives and Power Electronics (EDPE), Tatranska Lomnica, 2015, pp. 196-203, doi: 10.1109/EDPE.2015.7325294

[3] P. Zaskalicky, "Pentagon Connected Five-Phase Induction Machine Working under One-Phase Fault," *2020 IEEE 29th International Symposium on Industrial Electronics (ISIE)*, Delft, Netherlands, 2020, pp. 339-344, doi: 10.1109/ISIE45063.2020.9152419.

[4] P. Zaskalicky, "Behavior of a Five-Phase Pentacle Connected IM Operated under One-Phase Fault," *2019 International Aegean Conference on Electrical Machines and Power Electronics (ACEMP) & 2019 International Conference on Optimization of Electrical and Electronic Equipment (OPTIM)*, Istanbul, Turkey, 2019, pp. 126-131, doi: 10.1109/ACEMP-OPTIM44294.2019.9007166.

[5] H. Guzmán, M. J. Durán and F. Barrero, "A comprehensive fault analysis of a five-phase induction motor drive with an open phase," 2012 15th International Power Electronics and Motion Control Conference (EPE/PEMC), Novi Sad, 2012, pp. LS5b.3-1-LS5b.3-6, doi: 10.1109/EPEPEMC.2012.6397474.

[6] L. Zhang, X. Zhu, J. Gao and Y. Mao, "Design and Analysis of New Five-Phase Flux-Intensifying Fault-Tolerant Interior-Permanent-Magnet Motor for Sensorless Operation," in IEEE Transactions on Industrial Electronics, vol. 67, no. 7, pp. 6055-6065, July 2020, doi: 10.1109/TIE.2019.2955407.

[7] S. S. R. Bonthu, M. T. B. Tarek, A. K. M. Arafat, M. Z. Islam and S. Choi, "Fault-tolerant performance comparisons between external and internal rotor PMa-SynRMs," 2018 IEEE Applied Power Electronics Conference and Exposition (APEC), San Antonio, TX, 2018, pp. 2933-2939, doi: 10.1109/APEC.2018.8341436.

[8] T. Tao, Y. He, R. Xue, W. Zhao and T. Zeng, "Fault-Tolerant Predictive Model Control for Five-Phase PM Motor With Optimal Duty Modulation Strategy," 2019 22nd International Conference on Electrical Machines and Systems (ICEMS), Harbin, China, 2019, pp. 1-5, doi: 10.1109/ICEMS.2019.8921565.

[9] P. Jandura, A. Richter and Ž. Ferková, "Flywheel energy storage system for city railway," 2016 International Symposium on Power Electronics, Electrical Drives, Automation and Motion (SPEEDAM), 2016, pp. 1155-1159, doi: 10.1109/SPEEDAM.2016.7525923.

[10] E. Levi, M. Jones, S. N. Vukosavic, A. Iqbal and H. A. Toliyat, "Modeling, Control, and Experimental Investigation of a Five-Phase Series-Connected Two-Motor Drive With Single Inverter Supply," in IEEE Transactions on Industrial Electronics, vol. 54, no. 3, pp. 1504-1516, June 2007, doi: 10.1109/TIE.2007.894694.

[11] M. I. Masoud, "Five phase induction motor: Phase transposition effect with different stator winding connections," IECON 2016 - 42nd Annual Conference of the IEEE Industrial Electronics Society, Florence, 2016, pp. 1648-1655, doi: 10.1109/IECON.2016.7793400.

[12] J. Kellner, M. Praženica, "Two five-phase induction motors used as an electronic differential," Transportation Research Procedia 55(2002):896-903, January 2021, DOI: 10.1016/j.trpro.2021.07.058.

[13] K. Kyslan, M. Lacko, Ž. Ferková and P. Záskalický, "V/f Control of Five Phase Induction Machine Implemented on DSP using Simulink Coder," 2020 ELEKTRO, 2020, pp. 1-6, doi: 10.1109/ELEKTRO49696.2020.9130358.

Calculation Methodology of Common Capability Diagram for Parallelly Connected Generator Group

Boris Glavan
HEP-Proizvodnja d.o.o.
Hydro Power Plant Vinodol
Tribalj, Croatia
boris.glavan@hep.hr

Zlatko Hanić
University of Zagreb
Faculty of Electrical Engineering and Computing
Zagreb, Croatia
zlatko.hanic@fer.hr

Mario Vražić
University of Zagreb
Faculty of Electrical Engineering and Computing
Zagreb, Croatia
mario.vrazic@fer.hr

Marinko Kovačić
University of Zagreb
Faculty of Electrical Engineering and Computing
Zagreb, Croatia
marinko.kovacic@fer.hr

Abstract—**In this paper, a new synchronous generator steady-state model is presented with the purpose of determining the limits of the capability diagram. Those limits can be expressed by computationally efficient functions that are suitable for implementation in the real-time online environment. Computationally efficient limit functions are used in the calculation methodology of a common capability diagram for a parallelly connected generator group. Finally, as a result, the common capability limits are shown, and it is visible that their shape and width depend on operating modes of active power distribution between the generators in the group.**

Keywords—synchronous generator, steady-state conditions, capability diagram, flux linkage functions

I. INTRODUCTION

Besides supplying the power system with active power, which is their primary task, electric energy producers can provide a variety of auxiliary services to the electric system, such as active power and frequency control, reactive power and voltage control, black start and isolated operation. However, the power system management's standard strategy was to use synchronous generators as an active power source. In addition, synchronous generators are the best solution for reactive power and voltage control in the power system because of their ability to continuously, precisely, and quickly control reactive power in both inductive and capacitive directions. Operators often restrain from the operation in the underexcited capacitive region due to problems with the stability and overheating of the construction elements in the end region of the generator [1], [2]. In order to control reactive power and power system's voltage more reliably, especially in the underexcited area, a model-based capability diagram is required.

The manufacturers of the generators provide operational limits, but they are often derived from constant synchronous inductances. In real-world operation, synchronous inductances are not constant, and they change with the level of magnetic saturation of the generator, which varies with the operating point. Taking this into consideration, it is possible to establish a wider allowable operating area, but it is more difficult to determine them.

There is a relatively small number of models which describe steady states of synchronous generators with the purpose of determining the limits of the capability diagram

[3], [4]. A saturation description is essential for the accurate modelling of the synchronous generators. In [5], the authors use polynomial functions for the description of the saturation in the d and q axis. However, that description does not work well outside of the identification zone and have poor accuracy. In [6], authors use equivalent magnetising current in d and q axis, and they fit the L_d and L_q inductances to that equivalent current. However, the representation of the model using flux linkages is a more general approach.

For determination of the synchronous generator allowable operating area real limits in the capability diagram, a new steady-state model is developed and presented in this paper, which takes into account magnetic saturation and the voltage change, and is based on measurements of real generator operating points. Although measured operating points may only be achieved inside allowable limits, the model describes the behaviour of the machine successfully outside the allowed operating limits.

Visualisation is a very important aspect of the utilisation of the capability diagram in practice. In [7], the authors developed the virtual environment for the simulation of the operation of the synchronous generators using the capability diagram, and they use neural networks to model the limits in the capability diagram. In [8], the authors use Labview application for the visualisation of the capability diagram and they use measured data to estimate the operational limits. In [9] a similar aproach is done in Matlab enviroment.

In practice, capability diagram limits should be suitable for implementation in power plants and power system control systems, which means that the model must be based on a computationally efficient function. It is convenient for power system operators to treat multi-generator power plants as single units, with their common capability diagrams. In [10], the authors showed an example of a coordinated reactive power-voltage control for multimachine power plants, and in this paper, we propose the P-Q capability diagram for a powerplant with parallelly connected generators. These diagrams are constructed using computationally efficient functions. The limits of the multi-generator power plant capability diagram depend on the way demanded plant total active power is distributed between the individual generators. In that process, optimisation methods with the aim of the width of the generator group capability diagram limits can be successfully used.

978-1-6654-3236-8/21 $31.00 © 2021 IEEE

II. SYNCHRONOUS GENERATORS STEADY-STATE CONDITIONS MODELING WITH CURRENT DEPENDENT FLUX LINKAGE FUNCTIONS Ψ(I)

A. The Problem of Determination of the Capability Diagram Limits Outside of the Identification Zone

The main problem in determining accurate limits of the capability diagram is the fact those limits are placed outside the allowable generator operating area. In practice, the model is derived from the measured operating points. However, that very same model is used to determine the operational limits outside of the identification zone. It is known that the mathematical functions used for the interpolation have poor accuracy outside of the identification zone. Therefore, it is required to find the functions that are also accurate outside of the identification area when applied to the identification of the synchronous generator model. Such a model should be as accurate as possible in the operating area outside the allowable operating limits, although in that area measurements for the model identification are not possible. This means that a model should be developed and identified using operating points inside the allowable operating area and also used with high accuracy along certain zones outside the allowable operating area. In the model structure development phase, the electromagnetic conditions in a certain zone outside the allowable operating area should be known for its validation. To address that problem, operating points were simulated by FEM (Finite Element Method) [11]. In order to keep consistency in the model structure development, operating points inside the allowable operating area used for model identification are also determined by FEM. It should be noted that this procedure is applied only for model structure development and validation and should not be done for each generator and model identification in future use after this model is established.

There is a model developed [5] in which the flux linkages in direct and quadrature axis are expressed as Current Dependent Flux Linkage functions (in further text Ψ(I)) in the form of a polynomial of degree 3, dependent on three variables – armature currents in the direct (I_d) and quadrature axis (I_q), and field current (I_f).

$$\Psi_d\left(I_d, I_q, I_f\right) = d_1 + d_2 I_d + d_3 I_q + d_4 I_f + d_5 I_d^2 + d_6 I_q^2$$
$$+ d_7 I_f^2 + d_8 I_d I_q + d_9 I_d I_f + d_{10} I_q I_f$$
$$+ d_{11} I_d^3 + d_{12} I_q^3 + d_{13} I_f^3 + d_{14} I_d^2 I_q \qquad (1)$$
$$+ d_{15} I_d^2 I_f + d_{16} I_q^2 I_d + d_{17} I_q^2 I_f$$

$$\Psi_q\left(I_d, I_q, I_f\right) = q_1 + q_2 I_d + q_3 I_q + q_4 I_f + q_5 I_d^2 + q_6 I_q^2$$
$$+ q_7 I_f^2 + q_8 I_d I_q + q_9 I_d I_f + q_{10} I_q I_f$$
$$+ q_{11} I_d^3 + q_{12} I_q^3 + q_{13} I_f^3 + q_{14} I_d^2 I_q \qquad (2)$$
$$+ q_{15} I_d^2 I_f + q_{16} I_q^2 I_d + q_{17} I_q^2 I_f$$
$$+ q_{18} I_f^2 I_d + q_{19} I_f^2 I_q + q_{20} I_d I_q I_f$$

The vector-phasor diagram representing the vectors relations is shown in the Fig. 1.

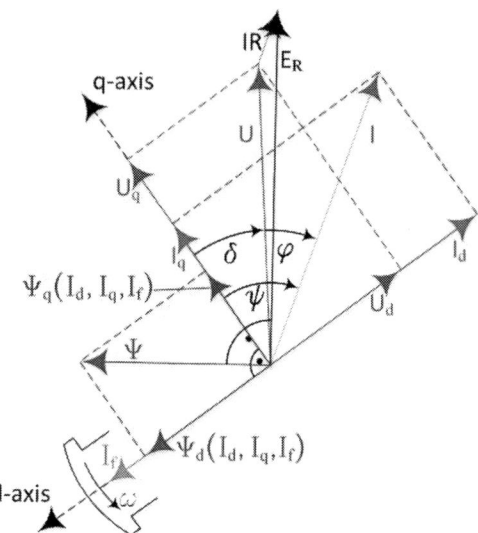

Fig. 1   Vector-phasor diagram of the synchronous generator based on the flux linkage model

The identification of the functions consists of calculating the function's coefficients using the least square quadrats optimisation method as in (3) and (4), according to known measured operating points, which for example for i_{th} operating point are quadruplets ($I_{d,i}$, $I_{q,i}$, $I_{f,i}$ and $\Psi_{d,i}$) for d axes and ($I_{d,i}$, $I_{q,i}$, $I_{f,i}$ and $\Psi_{q,i}$) for q axes. The procedure for d axis is as follows:

$$G_d = \sum_{I=1}^{N} \left[\psi_{d,i} - \psi_d\left(I_{d,i}, I_{q,i}, I_{f,i}\right)\right]^2 \to min! \qquad (3)$$

$$\frac{\partial G_d}{\partial d_k} = 0 \qquad for \quad k = 1\dots20 \qquad (4)$$

$$\frac{\partial G_d}{\partial d_1} = -2 \sum_{I=1}^{N} \left\{\left[\psi_{d,i} - \psi_d\left(I_{d,i}, I_{q,i}, I_{f,i}\right)\right]I_{d,i}\right\} = 0$$
$$\vdots \qquad\qquad (5)$$
$$\frac{\partial G_d}{\partial d_{20}} = -2 \sum_{I=1}^{N} \left\{\left[\psi_{d,i} - \psi_d\left(I_{d,i}, I_{q,i}, I_{f,i}\right)\right]I_{d,i}I_{q,i}I_{f,i}\right\} = 0$$

For the q axis, the procedure is analogue. From (5) two systems (for d and q axis) of 20 linear equations and 20 unknowns ($d_1...d_{20}$ and $q_1...q_{20}$) can be set and solved in matrix calculation. Once the coefficients $d_1...d_{20}$ and $q_1...q_{20}$ are calculated, the model is identified and can be used for the calculation of any steady-state electromagnetic quantities.

It is known that the mathematical functions in the form of a polynomial used for interpolation have poor accuracy outside of the identification zone. In this aspect it is the problem and the motive for investigating other ways for more accurate modelling of the synchronous generator steady states outside the allowable operating area.

B. Sigmoid Functions

The functions that have a similar shape to the synchronous generators' natural no-load magnetic saturation curves are modified sigmoid functions. Those functions are expressed in their basic form, in one-dimensional space as in

$$f(x) = a_1 \cdot x + a_2 \cdot \frac{x}{\left(a_3 + abs\left(\frac{x}{a_4}\right)^{a_5}\right)^{\frac{1}{a_5}}} \qquad (6)$$

Comparison of function with the no-load curve of the synchronous generator on which the function is fitted is presented in Fig. 2.

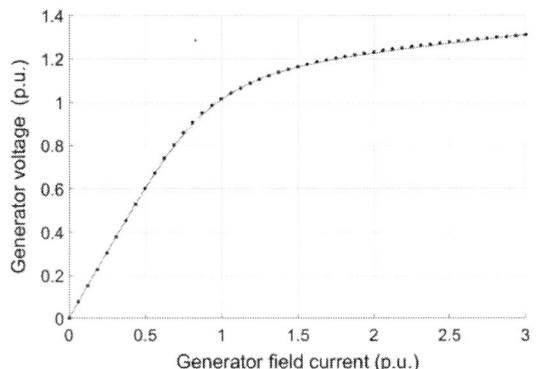

Fig. 2   No-load curve of the synchronous generator (line) and fitted Sigmoid function (dots)

Generalising into a three-dimension I_d - I_q - I_f space, the sigmoid function can be used to express $\Psi(I)$ functions, in the following form for d and q axis flux linkage, respectively:

$$\Psi_d\left(I_d, I_q, I_f\right) = d_1 I_d + d_2 I_f +$$
$$\left(\frac{d_3 I_d + d_4 I_f}{\left(d_5 + d_6 I_d^2 + d_7 I_q^2 + d_8 I_f^2 + abs(d_9 I_d + d_{10} I_f)^{d_{11}}\right)^{\frac{1}{d_{11}}}}\right) (7)$$

$$\Psi_q\left(I_d, I_q, I_f\right) = q_1 I_q + \left(\frac{q_2 I_q}{\left(q_3 + q_4 I_d^2 + q_5 I_q^2 + q_6 I_f^2\right)^{\frac{1}{q_7}}}\right) \qquad (8)$$

The identification of the sigmoid functions is not possible like in the case of the polynomial functions through matrix calculation of the linear equations systems. The unknown sigmoid function's coefficients $d_1...d_{11}$ and $q_1...q_7$ are fitted similarly, but using the least square method and heuristic optimisation algorithm, such as the Differential evolution method.

C. Comparison of the Accuracy

In Table I., the accuracy between the classic vector-phasor model, polynomial $\Psi(I)$ and sigmoid $\Psi(I)$ models is compared. Inside the identification area, the best accuracy is obtained by the polynomial $\Psi(I)$ model, slightly better than the sigmoid $\Psi(I)$ model, and the classic vector-phasor model has significantly the worst accuracy for at least two orders of magnitude. Outside the identification area, in the model verification operating area, which is about borderline and in the close zone outside the allowable operating area, the sigmoid $\Psi(I)$ model has the best accuracy, unlike the polynomial $\Psi(I)$ model, which reaches extremely large values of the error (1.3249 p.u.), as expected.

TABLE I.   ACCURACY COMPARISON

Model Error		d-axis			q-axis		
		Ident	Verif	All	Ident	Verif	All
Max	Classic	0.3174	0.5543	0.5543	0.1466	0.1773	0.1773
	Polynom.	0.00044	1.3249	1.3249	0.00095	0.3844	0.3844
	Sigmoid	0.0013	0.0266	0.0266	0.0068	0.0894	0.0894
Mean	Classic	0.1619	0.2223	0.1911	0.0469	0.059	0.0527
	Polynom.	0.00011	0.0872	0.0421	0.00032	0.0373	0.0182
	Sigmoid	0.00032	0.0063	0.0032	0.0019	0.0129	0.0072

D. Capability Diagram Limits Determination and Comparison

Presented $\Psi(I)$ models are used for capability diagram limits determination in the following ways:

1) Maximum Apparent Power Limit

Determination of the maximum apparent power limit is trivial since the value of the armature current has a maximum value (1 p.u.) and since apparent power in p.u. system S_{max} is given as the product of the armature voltage U and maximum armature current I_{max}, and this limit is represented with a circle whose radius corresponds to the p.u. value of the armature voltage.

2) Constant Field Current Limit

In the constant field current limit calculation, it is required to ensure that the field current is fixed to the constant value.

$$I_f = const. = I_{f,const} \qquad (9)$$

Capability diagram limits are calculated for the constant armature voltage, so for an arbitrary value of the I_d, the value of the I_q can be calculated from the following voltage equation.

$$U = \sqrt{\begin{array}{l}\left(\omega\Psi_d\left(I_d, I_q, I_{f,const}\right) - I_q R\right)^2 + \\ +\left(-\omega\Psi_q\left(I_d, I_q, I_{f,const}\right) - I_d R\right)^2\end{array}} =$$
$$= const. = U_{const} \qquad (10)$$

where U is the constant armature voltage value.

Equation (10) can be solved using Newton's method or any other suitable numerical method. Once the equation (10) is solved, according to the shown procedure (11) a triplet $(I_{d,arb}, I_{f,const}, I_q)$ is obtained, Ψ_d and Ψ_q calculated, which can be substituted into expressions for reactive and active power

$$\begin{array}{ccc}
\begin{array}{cc} I_{f,const} & I_{d,1,arb} \\ \vdots & \vdots \\ \vdots & \vdots \\ I_{f,const} & I_{d,n,arb} \end{array} & F_{CF}(I_q) = (U - U_{const})^2 \to min! \atop =\!=\!=\!=\!=\!=\!=\!=\!=\!=\!=\!=\!=\!=\!=\!=\Longrightarrow & \begin{array}{c} I_{q,1} \\ I_{q,2} \\ \vdots \\ I_{q,n} \end{array}
\end{array}$$
$$(11)$$

$$\begin{array}{ccc}
\Psi(I) & \begin{array}{c} \Psi_{d,1}, \Psi_{q,1} \\ \Psi_{d,2}, \Psi_{q,2} \\ \vdots \\ \Psi_{d,n}, \Psi_{q,n} \end{array} & \begin{array}{c} (12), (13) \\ =\!=\!=\!=\!=\!=\!=\!=\Longrightarrow \end{array} & \begin{array}{c} Q_{CF,1}, P_{CF,1} \\ Q_{CF,2}, P_{CF,2} \\ \vdots \\ Q_{CF,n}, P_{CF,n} \end{array}
\end{array}$$

(12), (13) in order to obtain a pair (Q_{CF}, P_{CF}) which represents of one of the points that lie on the constant field current limit. The following procedure should be conducted for various arbitrary values of the $I_{d,arb}$ in order to obtain the

978-1-6654-3236-8/21 $31.00 © 2021 IEEE   39

constant field current limit in the capability diagram. The

$$Q = \omega \begin{pmatrix} -\Psi_d(I_{d,arb}, I_q, I_{f,const})I_{d,arb} \\ -\Psi_q(I_{d,arb}, I_q, I_{f,const})I_q \end{pmatrix} \quad (12)$$

$$P = \omega \begin{pmatrix} \Psi_d(I_{d,arb}, I_q, I_{f,const})I_q - \\ \Psi_q(I_{d,arb}, I_q, I_{f,const})I_{d,arb} \end{pmatrix} - (I_{d,arb}^2 + I_q^2)R \quad (13)$$

procedure can also be applied for different values of the armature voltages.

3) Theoretical Steady-State Stability Limit

For each value of field current in the complete operating range raster, from minumum to maximum of field current, at a constant voltage, there are operating points with a maximum value of the active power for each particular field current. The main part of the theoretical steady-state stability limit determination procedure consists of the equation's (13) maximum calculation. It can be performed by the optimisation algorithm which looks for the minimum, so in that case the equation (13) should be put in the negative form, i.e. should be multiplied by -1. The constant voltage condition can be formulated in the optimisation algorithm as the constraint in the following way:

$$|U - U_{const}| =$$

$$\left| \sqrt{\begin{array}{c} (\omega\Psi_d(I_d, I_q, I_{f,arb}) - I_q R)^2 + \\ +(-\omega\Psi_q(I_d, I_q, I_{f,arb}) - I_d R)^2 \end{array}} - U_{const} \right| \le 10^{-5} \quad (14)$$

By solving the extreme of the equation (13), the values of I_d and I_q are obtained, and according to the shown procedure (15) a triplet $(I_d, I_{f,arb}, I_q)$ is obtained, Ψ_d and Ψ_q calculated, which can be substituted into expressions for reactive and active power (12), (13) in order to obtain a pair of coordinates (Q_{TS}, P_{TS}) of one of the points that is consisted in the theoretical steady-state stability limit. The following procedure should be conducted for various arbitrary values of the $I_{f,arb}$ in order to obtain the theoretical steady-state stability limit in the capability diagram. The procedure can also be applied for different values of the armature voltages.

$$\begin{array}{ccc} I_{f,1,arb} & & I_{d,1}, I_{q,1} \\ I_{f,2,arb} & F_{TS}(I_d, I_q) = -P \to min! & I_{d,2}, I_{q,2} \\ \vdots & ================\Rightarrow & \vdots \\ I_{f,n,arb} & & I_{d,n}, I_{q,n} \end{array}$$
$$(15)$$
$$\begin{array}{ccc} & \Psi_{d,1}, \Psi_{q,1} & Q_{TS,1}, P_{TS,1} \\ \Psi(I) & \Psi_{d,2}, \Psi_{q,2} & (12),(13) & Q_{TS,2}, P_{TS,2} \\ ====\Rightarrow & \vdots & ================\Rightarrow & \vdots \\ & \Psi_{d,n}, \Psi_{q,n} & Q_{TS,n}, P_{TS,n} \end{array}$$

4) Practical Steady-State Stability Limit

The practical steady-state stability limit is achieved in a way that for each arbitrary value of field current $I_{f,arb}$ in the complete operating range raster, from minimum to maximum of field current, on its curve of constant field current one operating point should be found, whose maximum active power (theoretical steady-state stability limit) is decreased by a safety margin of 0.1 p.u..

At a constant voltage, the pair of coordinates (Q_{PS}, P_{PS}) represents one of the points on the practical steady-state stability limit, which also lies on the same constant field current curve. The procedure (17) is similar to the theoretical steady-state stability limit determination, with the same constraint, and with the only difference in the aim function of the optimisation process, as follows:

$$F_{PS}(I_d, I_q) = (P - (P_{TS} - 0.1))^2 =$$

$$= (\psi_d(I_d, I_q, I_{f,arb})I_q - \psi_q(I_d, I_q, I_{f,arb})I_d \quad (16)$$

$$-(I_d^2 + I_q^2)R - (P_{TS} - 0.1))^2$$

$$\begin{array}{ccc} I_{f,1,arb} & F_{PS}(I_q, I_q) = & I_{d,1}, I_{q,1} \\ I_{f,2,arb} & (P - (P_{TS} - 0.1))^2 \to min! & I_{d,2}, I_{q,2} \\ \vdots & ================\Rightarrow & \vdots \\ I_{f,n,arb} & & I_{d,n}, I_{q,n} \end{array}$$
$$(17)$$
$$\begin{array}{ccc} & \Psi_{d,1}, \Psi_{q,1} & Q_{PS,1}, P_{PS,1} \\ \Psi(I) & \Psi_{d,2}, \Psi_{q,2} & (12),(13) & Q_{PS,2}, P_{PS,2} \\ ===\Rightarrow & \vdots & ================\Rightarrow & \vdots \\ & \Psi_{d,n}, \Psi_{q,n} & Q_{PS,n}, P_{PS,n} \end{array}$$

Furthermore, using the $\Psi(I)$ models for the determination of the capability diagram and its limits, the comparison shows the differences between the referent FEM, classic vector-phasor model, sigmoid $\Psi(I)$ and polynomial $\Psi(I)$ models.

In Fig. 3., one can see that the limits that correspond to the polynomial $\Psi(I)$ model has lost physical sense in the verification zone (zone with operating points signed by "x"), outside of the identification area. However, there is a good agreement of the limits between the sigmoid $\Psi(I)$ model with the referent FEM model, and their result of the limits in the capacitive part of the capability diagram describe a significantly wider allowable operating space than the space obtained by the classic vector-phasor model.

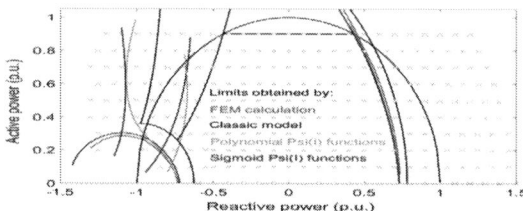

Fig. 3   Different models capability diagram limits comparison

III. CAPABILITY DIAGRAM LIMITS EXPRESSED BY COMPUTATIONALLY EFFICIENT FUNCTIONS

A. Application

Using the method described in section II it was proven that the most convenient model uses Sigmoid functions. It can be successfully used for effective calculations of operating points in any part of the three-dimensional operating space. In the case of using this model to determine the capability diagram and its limits, it involves complex calculations that entail solving the nonlinear equation systems and utilise optimisation algorithms. Such calculations are time-consuming, so that is the reason why they are conducted offline. Since the shape of the capability diagram limits depends on the actual operating point parameters, in practice, capability diagram limits should be calculated in real-time conditions of the power plant and power system control system environment containing PLCs (Programmable Logic Controllers) and the standard industrial SCADA (Supervisory Control and Data Acquisition) system.

The main idea is to offline calculate the detailed data set of the operational limits in all conditions and for the different armature voltages by the method described in parts *1)* to *4)* of section II, and then fit that data to the functions that can be evaluated quickly. A computationally efficient method should be developed for the representation of the capability limits in the real-time online environment, as it is done in the condition-monitoring system installed in the hydro power plant [3].

B. Computationally Efficient Capability Diagram Limit Functions

1) Maximum Field Current Limit and Practical Steady-State Stability Limit

For representations of both the maximum field current limit and practical steady-state stability limit, the same technique is used. A large set of the operating point triplets (Q, P, U) that lie on the limits is calculated by the method described in parts 2) and 4) of section II, for different values of the armature voltage that can be expected in real conditions. The same predefined reactive power polynomial function of degree 4, with the armature voltage and active power as independent variables (18) is fitted, identifying the coefficients l for both types of limits in two separate processes of optimisation, consisted of the least square method and heuristic optimisation algorithm, such as the Differential evolution method:

$$Q_{limit} = f(U, P) = l_{10} + l_{10}(U + l_2) + l_{20}(U + l_2)^2 +$$
$$l_{30}(U + l_2)^3 + l_{40}(U + l_2)^4 + l_{11}(U +$$
$$l_2)(P + l_3) + l_{12}(U + l_2)(P + l_3)^2 +$$
$$l_{13}(U + l_2)(P + l_3)^3 + l_{14}(U + l_2)(P +$$
$$l_3)^4 + l_{21}(U + l_2)^2(P + l_3) + l_{22}(U +$$
$$l_2)^2(P + l_3)^2 + l_{31}(U + l_2)^3(P + l_3) \quad (18)$$

Once the coefficients l are identified, function (18) can be used universally in the complete range of variables in a computationally efficient way, of course with two different sets of coefficients l for maximum field current limit and practical steady-state stability limit calculation.

2) Minimum Field Current Limit

For the minimum field current limit, a problem lies in the shape of a curve that has two different values of reactive power for each value of the active power and in the fact that the maximum value on the limit depends on the armature voltage. Because of that, the active power points cannot be predefined, as was the case in the maximum field and practical steady-state stability limits. That is the reason why the minimum current limits are approximated with ellipses in the polar coordinate system. Once the minimum field current

limits operating points are calculated for various armature voltages, they are fitted to the polar ellipse functions that have the following form in the polar coordinate system:

$$P_{limit} = f(R_p, \alpha) = R_p(U)\sin(\alpha) \quad (19)$$

$$Q_{limit} = f(R_q, \alpha) = Q_0 + R_q(U)\cos(\alpha) \quad (20)$$

where:

$$R_p(U) = a_1 U^{a_2} + a_3 U^{a_4} \quad (21)$$

$$R_q(U) = b_1 U^{b_2} + b_3 U^{b_4} \quad (22)$$

$$Q_0(U) = c_1 U^{c_2} + c_3 U^{c_4} \quad (23)$$

The ellipse functions are fitted using the least square method, and heuristic optimisation algorithm, such as the Differential evolution method, whose aim function in this case is:

$$F_{MinFC} = \sum_{I=1}^{N} [P_i - P_{MinFC}(Q_i, U_i)]^2 = 0 \quad (24)$$

where Q_i, P_i and U_i are triplets in the large set of the operating points, which represent the minimum field current limits and are calculated by the method described in part 2) of section II, for the different values of the voltage that can be expected in real conditions. Function $P_{MinFC}(Q, U)$ is derived from (19) and (20) in the following way:

$$P_{MinFC}(Q_i, U_i) = R_p(U)\sqrt{1 - \frac{|Q - Q_0(U)|^2}{R_q{}^2(U)}} \quad (25)$$

Once the coefficients *a*, *b* and *c* are identified, function (25) defining the minimum field current limit can be used universally in the complete range of variables in a computationally efficient way.

IV. COMMON CAPABILITY DIAGRAM FOR PARALLELLY CONNECTED GENERATOR GROUP

A. The Aim and the Methodology Description

The methodology assumes that all the limits in the capability diagram's limit functions are known for all the parallelly connected generators.

In a multi-generator power plant, each value of the total active power (P_{tot}) can be allocated between the individual generators in the infinite number of active power distribution combinations. Different principles for the active power allocation can be used, such as equal-parallel mode (contribution of the generators at the same time on the same parts of the active power), sequential mode (generators sequentially contribute one by one, after the previous reaches its full active power), or in an optimal mode. The optimal mode can be determined with different aims, and one of them can be the width of the generator group capability diagram limits. For example, in the three generators group, the aim function in such an optimisation process in the capacitive part of the capability diagram is:

$$F_{Q,multi}(P_1, P_2) = Q_{1,limit}(P_1, U) + Q_{2,limit}(P_2, U) +$$
$$Q_{3,limit}((P_{tot} - P_1 - P_2), U) \rightarrow min! \quad (26)$$

with the following constraints:

$$P_{tot} \geq P_1 + P_2 \qquad (27)$$

$$P_{tot} - P_1 - P_2 \leq P_{3,max} \qquad (28)$$

Determining the common capability diagram limits for parallelly connected generators groups consists of the following steps. First, the power plant common total active power is distributed between the individual generators with the chosen method. Second, for each allocated active power of an individual generator, the corresponding reactive power from its capability diagram limit is read. Third, the sum of the read reactive powers of the individual generators represents the x-coordinate of the power plant operating point, of which the corresponding y-coordinate is the common total active power. By repeating this procedure for each of the equidistant raster points in the full range of the power plant common total active power and connecting them with a line, a common capability diagram limit for the parallelly connected generator group is achieved.

B. Results

In Fig. 4 an example of common capability diagram limits in three different generators group's capacitive operating region is shown. There are three common capability diagram limits obtained from three different operating modes.

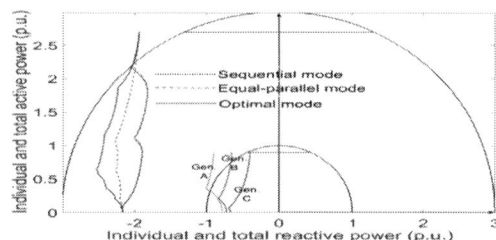

Fig. 4   *Common capability diagram for different generators group*

The widest possible power plant allowable operating area is achieved when the operating mode was determined using optimisation. From that point of view, the equal-parallel mode is second, and the sequential mode is the worst in that way, i.e. it has the narrowest limits.

V. Conclusion

A new steady-state model of the synchronous generator is presented. The most accurate model uses Sigmoid functions, and it can be successfully used for effective calculations of operating points at any part in the three-dimensional operating space and for the determination of the capability diagram and its limits.

The results of determining the limits of the capability diagram with this model have confirmed the results obtained by the FEM calculation, i.e. that the allowable operating space is wider compared to the classic vector-phasor model in the capacitive part of the capability diagram, which is of more interest. Some very useful parts of the generator

operating space could be exploited, which was not done up to date because the limit determination was based on the classic vector-phasor model. Presented computationally efficient functions enable approximation with the points on the capability diagram limits precisely, and with its simplicity, they are convenient for use in real-time conditions of the power plant and power system control system's environment containing PLC (Programmable Logic Controllers) and the standard industrial SCADA (Supervisory Control and Data Acquisition) systems.

In addition, the calculation methodology of the common capability diagram for a parallelly connected generator group is presented. In the example of the capacitive part of the capability diagram in the three different generators power plant, it is shown that limits depend on the principle of active power distribution between the generators. If the optimisation algorithm is used, the widest operating area of the capability diagram is achieved, so it can be said that a quality synergy of the generator operation is obtained.

References

[1] Choi, S. S., and X. M. Jia. "Under excitation limiter and its role in preventing excessive synchronous generator stator end-core heating." IEEE Transactions on Power Systems 15.1 (2000): 95-101.

[2] Maljković, Zlatko, and Ivan Gašparać. "Operating limits of underexcited synchronous generator." 4th International Conference on Power Engineering, Energy and Electrical Drives. IEEE, 2013.

[3] Glavan, Boris, Zlatko Hanić, Marinko Kovačić, and Mario Vražić. "Condition-monitoring system for identification and representation of the capability diagram limits for multiple synchronous generators in a hydro powerplant." Energies 13, no. 15 (2020): 3800.

[4] Ilić, I., Maljković, Z., Gašparac, I., Pavlica, M., Ilić-Zubović, D., Jarić, V., Višković, A. and Belobrajić, R., 2007. Methodology for Determining the Actual PQ Diagram of a Hydrogenerators. Journal of Energy: Energija, 56(2), pp.144-181.

[5] Hanic, Zlatko, Mario Vrazic, and Zlatko Maljkovic. "Steady-state synchronous machine model which incorporates saturation and cross-magnetisation effects." 4th International Conference on Power Engineering, Energy and Electrical Drives. IEEE, 2013.

[6] M. Despalatovic, M. Jadric and B. Terzic, "Modeling of Saturated Synchronous Generator Based on Steady-State Operating Data," in IEEE Transactions on Industry Applications, vol. 48, no. 1, pp. 62-69, Jan.-Feb. 2012, doi: 10.1109/TIA.2011.2175429.

[7] Lee, Chun-Yao, and Maickel Tuegeh. "Virtual Visualization of Generator Operation Condition through Generator Capability Curve." Energies 14.1 (2021): 185.

[8] D. Vuljaj, Z. Hanic and M. Vražić, "LabVIEW application for visualisation of synchronous machine real-time capability diagram with variable operational limits," 2015 International Conference on Electrical Drives and Power Electronics (EDPE), 2015, pp. 291-295, doi: 10.1109/EDPE.2015.7325308.

[9] Pejovski, Dejan, Bodan Velkovski, and Krste Najdenkoski. "MATLAB Model for Visualization o f PQ diagram of a Synchronous Generator." URL: https://www. researchgate. net/publication 312190893.

[10] Dragosavac, J., Janda, Ž. and Milanovic, J.V., 2012. Coordinated reactive power-voltage controller for multimachine power plant. IEEE Transactions on Power Systems, 27(3), pp.1540-1549.

[11] Hanic, Zlatko, Stjepan Stipetic, and Mario Vrazic. "Computationally efficient finite-element-based methods for the calculation of symmetrical steady-state load conditions for synchronous generators." IET Electric Power Applications 8, no. 9 (2014): 357-365.

Encapsulated Air Cooling System for Scalable Axial Flux Motors

Sebastian Berndl
Faculty of Electrical and Industrial Engineering
University of Applied Sciences Landshut
Landshut, Germany
sebastian.berndl@haw-landshut.de

Alexander Kleimaier
Faculty of Electrical and Industrial Engineering
University of Applied Sciences Landshut
Landshut, Germany
alexander.kleimaier@haw-landshut.

Abstract — A Scalable Axial Flux Motor (AxMDM) with geometrically optimized U-core laminations and a NdFeB-magnet equipped GRP disc rotor for an aviation application was developed, assembled and tested. The novel machine concept is easy to fabricate and can be scaled by the number of U-core laminations to meet different requirements for output torque and power. It shows a high torque-to-weight ratio and has an excellent utilization rate of the magnet material. In this paper, the adaption of the AxMDM concept for a propeller drive is discussed. One of the main challenges for the aviation application of electric motors is a dust robust air-cooling approach. Therefore, an encapsulated air-cooling system for the AxMDM concept was developed. The impact of the minimization of the phase elements of the motor will be introduced, followed by an investigation of the temperature behavior of the motor at variable motor speed and current injection. In the last section, the temperature behavior of the motor at nominal and maximum operating conditions will be presented.

Keywords — axial flux motor; air-cooling; cooling system characterization; aviation application

I. INTRODUCTION

For sustainable and livable cities in the future, transport infrastructure should be rethought, especially in the large cities. A good opportunity are electric air taxies and cargo drones. They are able to reach any destination within cities in a short time and will reduce congestion on the roads. One of the most important parts of such air taxis are the electric motors. They have to be lightweight and robust to operate in a wide range of climates. Therefore, the requirements for such motors are that they do not contain large amounts of cooling fluid to reduce weight and increase fail-safety. They should also be robust against sand and other particles in the air. Based on these requirements, the encapsulated air cooling system for the AxMDM concept was developed. In this paper, the improvement of the thermal behavior by downsizing the motor phase elements and the encapsulated air cooling system will be discussed. Furthermore, the thermal behavior at nominal and maximum operation condition will be introduced.

II. MOTOR CONCEPT AND MEASUREMENT SETUP

A. Motor Concept

The stator consists of two parallel aluminum plates, which are equipped with phase elements (Fig. 1). They are connected by a housing ring, ensuring stability and distance of the stator plates. A phase element consists of a U-core lamination and a plug-in coil. Compared to previous test versions of the AxMDM motor, the U-core lamination are downsized with an optimized geometry, here called "U-16", referring to "U-30"

for standard UI30 core laminations; see Fig. 3. The U-16 cores are especially laser cut for this motor. The advantage of refining the pole pitch is a significant reduction of weight [1], but one has to deal with increased pole changing frequencies and iron losses. In this paper, the thermal aspects of this strategy will be addressed in section III A.

Air channels were placed into the stator plates for an encapsulated cooling system with internal air circulation (Fig. 1). Details will be described in section II B. The disc rotor is arranged between the phase elements. It is made of glass fiber reinforced plastic (GRP), which is equipped with NdFeB magnets so that the magnet material can be utilized very efficiently from both sides (Fig. 1). A magnet element consists of three electrically isolated magnet segments. This reduces the eddy currents within the magnets and thus the magnet temperature [2]. A fan wheel is mounted on the disc rotor, which creates an airflow inside the motor to deheat stator phase elements as well as rotor magnets (see Section II B).

The motor (AxMDM prototype 3.0) consists of 24 U-core laminations (resulting in 48 stator teeth) on each stator plate. The disc rotor is equipped with 56 NdFeB-magnet-elements (p = 28), so the number of slots per pole is q = 2/7, where an optimum of torque can be generated. Hence, the torque is generated with the 7th harmonic of the stator field, with separated flux paths for each pair of U-cores in the stator [3]. In order to use plug in coils, it was necessary to use a fractional slot winding [4].

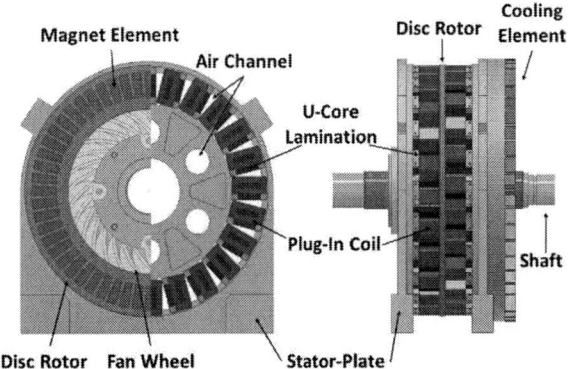

Fig.1: AxMDM concept

B. Encapsulated Air Cooling System

To ensure that the motor can operate in an environment with dust and sand, an encapsulated air cooling system was developed. Heat production in the active parts of the motor is caused by the ohmic losses in the coils (copper losses), the iron losses within the U-core laminations and eddy currents in the magnets.

The internal air-cooling is done by forced convection (Fig.2): Cold air is drawn into the stator plate and pushed by the fan wheel in the direction of the phase elements and the magnets in the disk rotor, where it can absorb their heat. The heated air is guided out of the stator plates into the cooling elements, where the heat is dissipated to the inner cooling fins. An external airflow, which will be generated by the propeller, cools the outer cooling fins.

In addition, cooling by heat conduction (Fig.2) is used: The heat generated by the coils and the U-core laminations is conducted via the stator plate into the cooling fins and the housing ring. These are also cooled by the external propeller airflow. In contrast, the magnets can be cooled only by the internal airflow, since heat conduction through GRP is negligible. This is due to the low thermal conductivity of the GRP, which is $0.2 - 0.3 \frac{W}{K \cdot m}$ [5] and thus acts like a thermal insulator.

The maximum acceptable temperatures in the motor are determined by different elements of the motor such as the fan wheel, the plastic coil bodies, the wire insulation, the 2-component adhesive for magnet pasting and the magnets themselves. For this reason, the maximum temperature of the coils is limited to 120 °C and the maximum internal air temperature to 100 °C. Exceeding these temperatures can cause severe mechanical damage to the motor, as the fan wheel, which is made of polycarbonate and therefore loses its mechanical stability [6]. If the magnets exceed a temperature of 120 °C, the magnetic flux destiny can be permanently reduced since the temperature stability of the magnets depends on the operation points on the B-H-Curve [7], given by the motor design. Therefore, a reliable cooling system is very important.

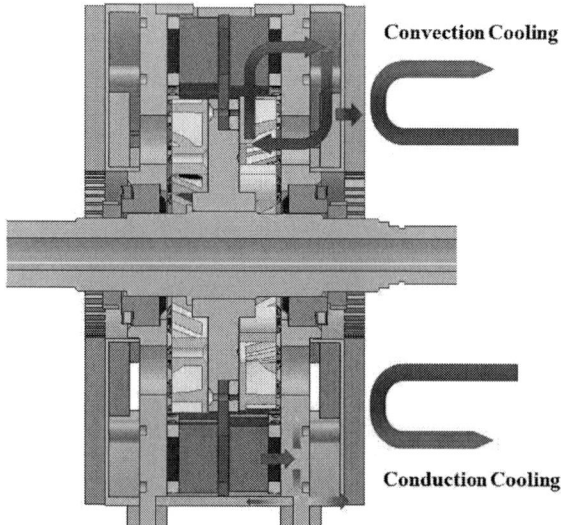

Fig. 2: Encapsulated cooling system

C. Measurement Setup

The AxMDM prototype 3.0 was measured at the test stand of the electric drives lab at the UAS Landshut. The results provided in section III are gained by the following measurement procedures and calculations methods:

- The motor speed and motor torque are measured with a gauge bar. The torque measurement rage is from 0.1 to 200 Nm and the speed measurement rage is from 0 to 15,000 rpm. The mechanical power is automatically calculated by the software with the equation $P_{mech} = \omega \cdot T$.

- The temperature is recorded with PT100 sensors and an NI-DAQ system. It records 8 temperatures simultaneously.

- The air velocity is measured with a hot-wire anemometer. It can be mounted on the inside of the cooling element and on the air duct of the stator plate behind the phase elements. The measurement range is from 0.1 to 30 $\frac{m}{s}$.

- The thermal images were taken with a calibrated Testo 882 IR-Camera. To ensure the correctness of the thermal images, black chalk spray with $\varepsilon = 0.95$ was applied to the DUTs. A comparison with PT100 sensors was performed at room temperature.

- The applied voltage is measured with a voltage probe TT-SI 9001 from Testtec. The current is measured with a current probe E3N from Chauvin Arnoux. Both are read-in with a DAQ from National Instruments, so the electrical power can be calculated.

- The magnet temperature is calculated by using the EMF constant, which corresponds to the B_r of the magnets. It can be derived when the motor is dragged at open circuit operation. The temperature coefficient for the B_r (datasheet [7]) is used to calculate the magnet temperature. The calculated magnet temperatures were verified by an IR-Camera monitoring at different motor speeds.

- The thermal resistance R_{th} and thermal conductance G_{th} can be calculated with Equation 1 and 2. The required parameters are measured at the test stand.

$$R_{th} = \frac{1}{G_{th}} = \frac{\Delta T}{P_{el}} = \frac{T_1 - T_2}{U_{rms} \cdot I_{rms}} \quad (1)$$

$$G_{th,total} = \frac{P_{el}}{\Delta T} = \frac{U_{rms} \cdot I_{rms}}{T_{coil} - T_{Cooling\ Fin}} \quad (2)$$

- The measurements for thermal characterization of the motor were performed with DC injection and with a dummy rotor without magnets. This allowed the heat generated by the copper losses to be observed independently of the motor speed.

- The iron losses were measured at open circuit operation, when the AxMDM was dragged by the test stand machine. Therefore, the GRP disk rotor equipped with magnet was used. The iron losses were calculated by $P_{Fe} = \omega \cdot T$.

978-1-6654-3236-8/21 $31.00 © 2021 IEEE

III. MEASUREMENT RESULTS

In this section, the measurement results regarding the thermal behavior of the phase elements and the motor will be presented and discussed.

A. Phase Element Characterization

Two types of phase elements are compared in their thermal behavior. The U-30 phase element is used as a reference for the U-16 phase element, which is used in the here discussed AxMDM motor. The dimensions of the phase elements can be seen in Fig.3. All dimensions are in millimeters (mm).

The U-30 phase elements (Fig. 3, left side) were used for previous prototypes, where weight optimization was not an issue. It consists of standard U-30 core laminations and a coil body made of acrylonitrile butadiene styrene (ABS) with a wall thickness of 1mm. The copper wire has a diameter of 0.9mm with 186 windings per plug-in coil; the overall nut filling factor is 39.4%.

The U-16 phase elements (Fig. 3, right side) have been shrunk to achieve a finer pole pitch and thus a higher torque-per-active-weight ratio. This enables a total weight reduction for aerospace applications. The U-16 laminations were designed to fit more copper between the limbs; therefore, the limb width was reduced. Since the application does not require high overload operation of the motor, the magnetic flux density in the U-cores does not reach saturation within the limbs; a torque-limiting saturation occurs only in the tooth edges. The coil body is made of polycarbonate (PC) with a wall thickness of 0.5mm. The copper wire has a diameter of 0.7mm with 118 windings per plug-in coil; the nut filling factor is 42.0%, despite of the miniaturization of geometry.

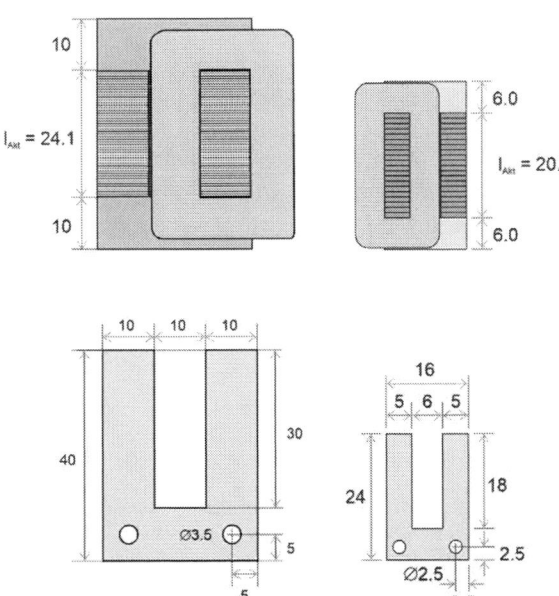

Fig. 3: Phase element geometries for "U-30" and "U-16"cores

Fig. 4 and Fig. 5 show the IR-Images of the two phase element types on the bottom of the figures. The IR-Image is taken from top view, similar to the one shown in Fig. 3. The phase elements are placed on a copper block so only conduction and free convection were present. On the top of these images, the temperature profile of the phase elements,

along the line P1 is shown. The line P1 can be seen in white color on the IR -Images. The different graphs represent different points in time of the heating (I_{on}) and cooling (I_{off}) phase. A Part of the plastic material of the coil body has been removed from the lower part of the coils. Therefore, the temperate of the copper windings can be seen in the IR-Images as well. For the measurement, a current density of $10.2 \frac{A}{mm^2}$ (U-16) and $10.5 \frac{A}{mm^2}$ (U-30) was applied. Applying equal current densities results in equal ohmic power losses per volume unit in the coils. This makes the temperature results comparable. These current densities were defined for the nominal operation of the motor. The measurements were carried out at an ambient temperate of 21 °C.

Fig. 4 shows the results from the U-30 phase element. The current density is $10.5 \frac{A}{mm^2}$, resulting in a coil current of 6.7 A_{RMS}. In the temperature over image pixel plot, from pixel ~50 to ~140 the left limb and from pixel ~210 to ~290 the right limb of the phase element; can be seen. The high temperatures from pixel ~140 to ~210 and from pixel ~290 to ~350 represent the coil body. As the heating time increases (Fig. 4; from green to red curve), the coil heats up rapidly, but the heat is transferred slower into the laminations stack, comparted with the U-16 phase element (Fig. 5). The hot spot temperature difference between the coil and the coil body is 23.2 K, because the coil body poorly conducts the heat as it can be seen in the IR-Image.

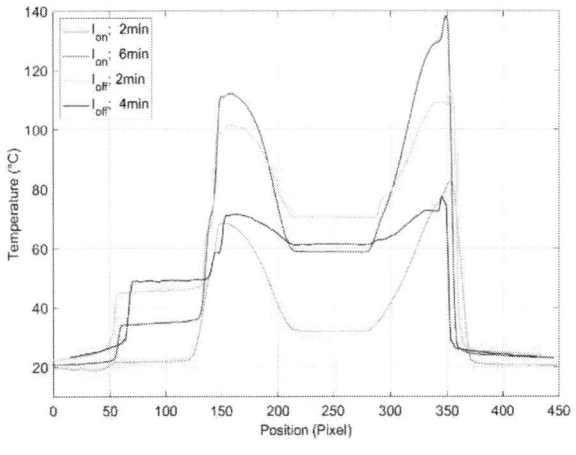

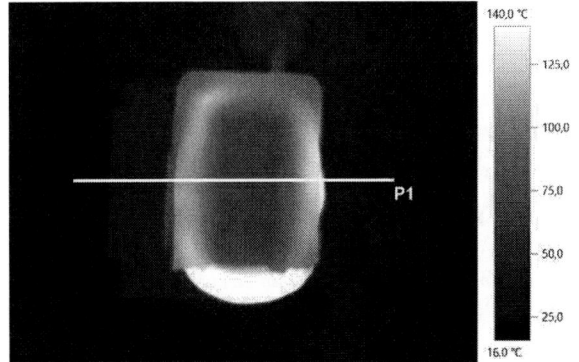

Fig. 4: U-30 temperature profiles at different times (top)
U-30 IR-Image after 6 min applied load current (bottom)

After the current was turned off, the DUT is cooling down (Fig.4; from turquoise to blue graph). The temperature equalization with the environment is slow compared to the U-16 phase element (Fig. 5).

Fig. 5 shows the results from the U-16 phase element. Here, the current density is $10.2 \frac{A}{mm^2}$, resulting in an coil current of 3.9 A_{RMS}. In the temperature over image pixel plot, from pixel ~140 to ~180 the left limb and from pixel ~240 to ~280 the right limb of the phase element, can be seen, followed by the copper block. The higher temperatures from pixel ~80 to ~140 and from pixel ~180 to ~240 are from the coil body. With increasing heating time (Fig. 5; from green to red curves), the coil temperature increases and the heat is transferred into the lamination stack. From the IR-Image, it can be seen that the temperature difference between the coil and the coil body is just 4.5 K (hot spot temperature difference). Additionally, the outer coil windings stays cooler than the windings in the center. This can be seen from the curve shape of the temperature profile as well as in the IR-Image. After the current was turned off, the DUT cools down (Fig 5; from turquoise to blue graph). The temperature equalization with the environment is fast, compared to the U-30 phase element (Fig. 4).

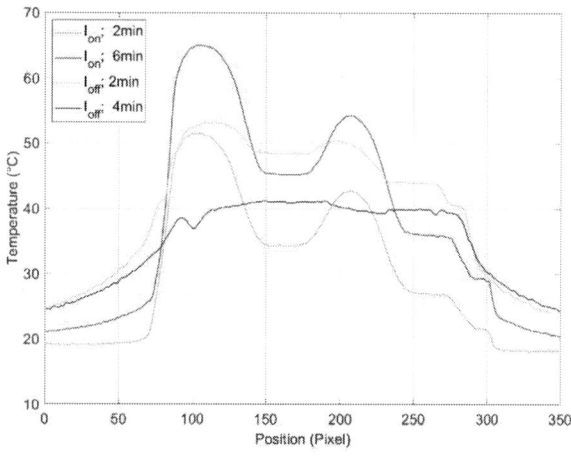

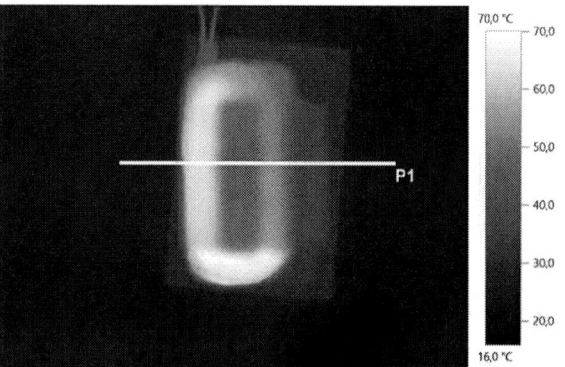

Fig. 5: U-16 temperature profile at different times (top)
U-16 IR-Image after 6 min applied load current (bottom)

From Fig. 4 and Fig. 5, it can be deduced that the U-16 phase element exhibits significantly better thermal behavior than the U-30 phase element at similar current density. This can be explained with the better thermal conductivity of the coil body and the higher surface-area-to-volume-ratio of the U-16 phase element – a clear benefit of the miniaturization.

The coil body increases the thermal conductivity for the U-16 coil mainly because of the thinner wall thinness. The thermal conductivity of ABS and PC is similar with $0.17 \frac{W}{K \cdot m}$ respectively $0.21 \frac{W}{K \cdot m}$ [8]. The surface-area-to-volume-ratio is $0.23 \frac{1}{mm}$ for the U-16 coil and $0.18 \frac{1}{mm}$ for the U-30 coil. This means the surface-area-to-volume-ratio is 28% greater for the U-16 coil. This improves heat transfer to the air and thus the forced convection cooling. Heat transfer to the phase element is also improved and thus the conduction cooling.

Further optimization for following prototypes can be done by using another coil body material and a further reduction of the wall thickness. Moreover, the surface-area-to-volume-ratio can be increased by geometry optimization.

B. Encapsulated Air Cooling System Characterization

For these measurements, DC-Current was applied, a dummy rotor without magnet was used and the AxMDM was dragged with an induction machine. Therefore, the thermal power losses of the coils and the motor speed can be investigated independently of each other. The total thermal conductivity of the motor increases with increasing motor speed and constant power injection as shown in Fig. 7.

The thermal conduction values were determined with the following experiment setup:

- A DC coil current of 3.8 A_{RMS} (phase current: 30 A_{RMS}) was injected; this leads to a thermal power injection, equal in this case equal to the power losses $P_{v,,}$ of about 330 W, depending on the temperate of the coils.

- The thermal conductivity was calculated at the thermal equilibrium of the motor. The motor was externally cooled with an air speed of 1.8 $\frac{m}{s}$. This was generated from an external fan. Additional an external airflow was crated from the cooling fan of induction machine who drags the AxMDM motor.

- At 0 rpm, the first measurement was performed and for the following measurements, the motor speed was increased by 500 rpm steps until 2000 rpm.

- From the power data and the temperature data the total thermal conductivity $G_{th,total}$ was calculated as it can be seen in equation 2.

- The value at 0 rpm was considered as heat conduction only, since in this condition the free convention can be considered negligible due to the encapsulated housing and the internal airflow measurement. The internal airflow at the inner cooling fins was measured to < 0.1 $\frac{m}{s}$; this is below the noise level of the air-flow-sensor.

- From these data, a network (Fig. 6) of two thermal conductances can be modeled. The thermal conductivity value of the heat conduction $G_{th,cond}$ is constant over the motor speed (determined at 0 rpm) [Fig. 7]. The thermal conductivity value of the forced convention $G_{th,conv}$ is dependent on the motor speed (determined at 500, 1000, 1500 and 2000 rpm) [Fig. 7].

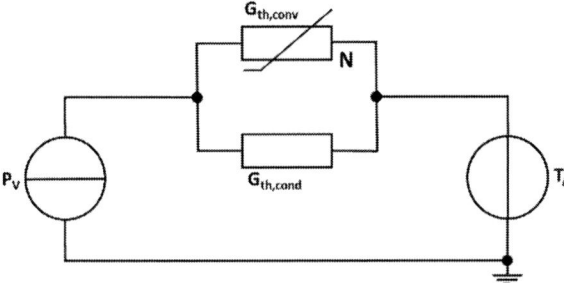

Fig. 6: Static thermal network of the AxMDM prototype 3.0

In Fig. 7 the internal thermal conductivity of the motor over the motor speed is plotted. It can be seen that above 1600 rpm the share of thermal convection is larger than of the heat conduction. The linear approximation of the thermal convection depending on the motor speed is given by equation 3.

$$G_{th,conv} = 0.0038 \frac{W/K}{rpm} * N\ [rpm] + 0.154 \frac{W}{K} \qquad (3)$$

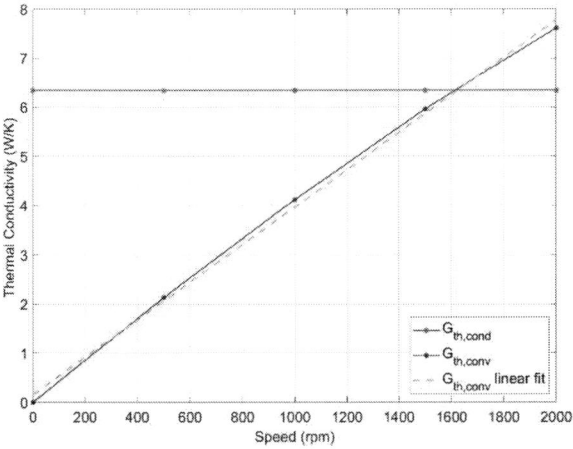

Fig. 7: Thermal conductivity over motor speed

With increasing motor speed, the cooling gets more effective, as the air velocity increases with the motor speed. The heat flow is proportional to the mass flow and thus to the air velocity [9]. Furthermore, the heat-transfer-coefficient increase with increasing air velocity [9].

The encapsulated air cooling system can be improved by optimizing the thermal properties of the phase elements (see section III A). Furthermore, it also can be enhanced by increasing the air velocity in the motor, by optimizing the fan wheel geometry. Research on these topics will be conducted at future AxMDM prototypes.

C. Temperature Behavior at Operation Conditions

For the time behavior of temperature related to copper losses at operation conditions, DC-Current was applied and the AxMDM motor was dragged with an induction machine. For these measurements, a dummy rotor without magnet was used.

Fig. 8a shows the temperature behavior at nominal current injection (Coil current: 3.9 A_{RMS}; Phase current: 31.1 A_{rms}) and nominal speed (1653 rpm). It can be seen that the coil reaches a stable temperature of 67°C. The air inside the motor reaches 48°C whereas the stator plate reached for 44°C. The connection between the motor and the environment, the outer cooling fin, reaches a temperature of 38°C at an ambient temperature of 21°C.

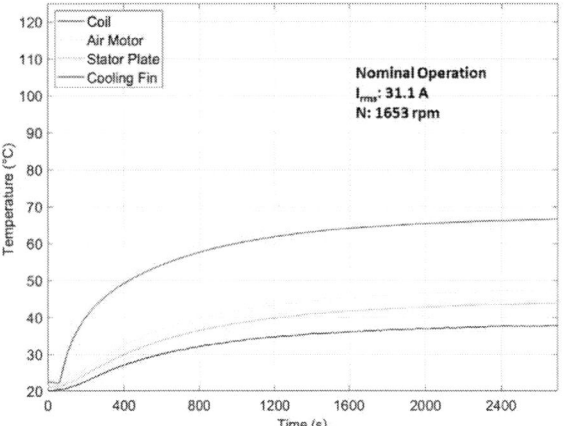

Fig. 8a: Copper losses temperature behavior of AxMDM prototype 3.0 at nominal operation

Fig. 8b shows the temperature behavior at maximum current injection (Coil current: 5.4 A_{RMS}; Phase current: 43.5 A_{rms}) and maximum speed (1908 rpm). It can be seen that the coil reaches a temperature of 120°C and still will increase slightly further. The air inside the motor reaches 78°C whereas the stator plate reached for 70°C. The connection between the motor and the environment, the outer cooling fin, reaches a temperature of 59°C at an ambient temperature of 22°C.

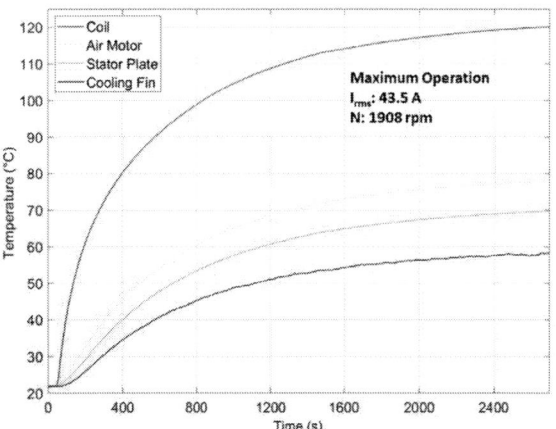

Fig. 8b: Copper losses temperature behavior of AxMDM prototype 3.0 at maximum operation

In both measurements, it can be seen that the largest temperature drop and therefore the largest thermal resistance (Eq. 1) is from the coil to the stator plate respectively the air. The thermal resistance from the stator plate respectively the air to the cooling fin is comparably small.

For the temperature behavior generated by iron losses (eddy currents in the magnets and U-core laminations) at operation conditions, the AxMDM motor was dragged with an induction machine and the temperatures were measured respectively calculated (see section II C). The current injection during the measurement was 0A.

Fig. 9a shows the temperature behavior at nominal speed (1653 rpm). It can be seen that the magnets reach a stable temperature of around 94°C. The U-core laminations reach a temperature of 64°C. The air inside the motor reaches 57°C whereas the stator plate reached for 52°C. All this temperatures where measured at an ambient temperature of 31°C.

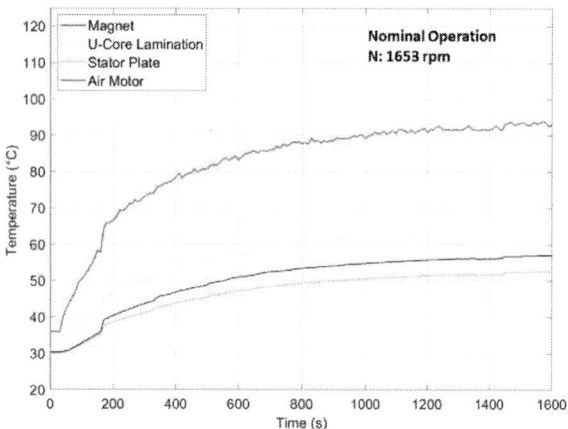

Fig. 9a: Iron losses temperature behavior of AxMDM prototype 3.0 at nominal operation

Fig. 9b shows the temperature behavior at maximum speed (1908 rpm). It can be seen that the magnets reach a stable temperature of around 107°C. The U-core laminations reach a temperature of 70°C. The air inside the motor reaches 62°C whereas the stator plate reached for 56°C. All this temperatures where measured at an ambient temperature of 31°C.

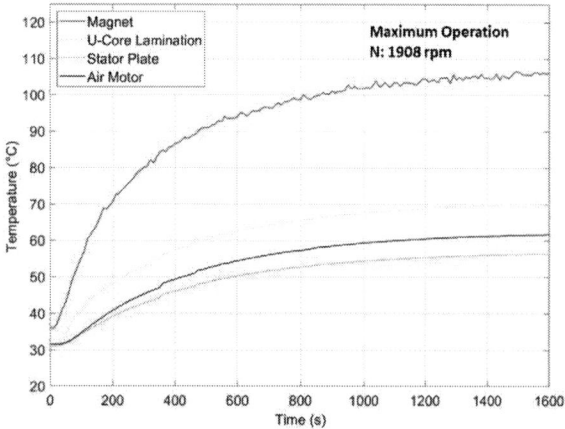

Fig. 9b: Iron losses temperature behavior of AxMDM prototype 3.0 at maximum operation

Magnet temperatures during the nominal and maximum operations are critical, as a temperature above 120 °C will permanently reduce the magnetic flux density [7]. An additional temperature increase of the magnets results from higher ambient temperatures, heating due to copper losses, and increased eddy currents during active motor operation.

Since the magnets in the GRP rotor are thermally isolated, a good way to improve the deheating of the magnets is to increase the air velocity at the magnet surfaces. This will be achieved by optimizing the fan wheel geometry and by installing small cooling fans inside the encapsulated cooling system.

Another optimization potential of the cooling system focuses on the large thermal resistance from the coil to the stator plate respectively the air. Here, a higher air velocity will be helpful as well. In addition, the thermal conduction from the coils to the stator plate can be increased, as discussed in section III A. These optimization topics are currently being investigated and will be implemented in the next prototype.

IV. CONCLUSION AND FUTURE STEPS

In this contribution, the AxMDM concept in combination with an encapsulated internal air-cooling concept was presented. The thermal behavior of two phase element types was investigated. It was shown that minimizing the phase elements has a positive influence on the thermal behavior, both by the thinner wall thickness of the coil body and by the lager surface-area-to-volume-ratio.

A G_{th}-Network was modeled based on the speed dependent thermal conductivity value of the motor. It shows that the thermal conductivity value of the complete motor increases with increasing motor speed. This can be used for further optimization of the encapsulated air-cooling system.

The temperature behavior of the motor at nominal and maximum operation was presented. It could be shown that the coils can be effectively cooled with the encapsulated air-cooling system. As the internal air velocity increases with increasing motor speed, the cooling becomes more effective at higher motor speed. This is in the favor of a propeller drive, as the torque shows a quadratic relationship with the motor speed ($T \sim N^2$). The magnet temperatures however may become critical, when the heat flow of the coils, caused by copper losses, is superposed. Therefore, the cooling system needs to be optimized in general by lowering the thermal resistance from the coil to the motor housing and increase the air velocity inside the motor.

It can be concluded that the concept of the encapsulated air-cooling system basically works, but still needs to be optimized. The next step of the testing will be the active motor operation, where the copper and iron losses occur simultaneously.

ACKNOWLEDGMENT

The authors like to thank the Bavarian Ministry of Economic Affairs, Regional Development and Energy who fund this research project. We also like to thank our industrial project partner Silver Atena for the good cooperation during this project.

REFERENCES

[1] S. Berndl, A. Kleimaier, R. Kennel "Verification of the Analytical Torque Calculation and Active Weight Optimization for a Scalable Axial Flux Motor", 23rd European Conference on Power Electronics and Applications (EPE) 2021, virtual, Sept. 2021

[2] S. Berndl, A. Kleimaier," Comparison between Different Air Gaps and Rotor Magnet Geometries of a Scalable Axial Flux Motor with Standard Core Laminations", 10. International Drive Production Conference (EDPC) 2020, Ludwigsburg, Dec. 2020

[3] A. Kleimaier," Scalable Axial Flux Motor with Standard Core Laminations", 10. Expertenforum Elektrische Fahrzeugantriebe - Internationale Konferenz für elektrische Fahrzeugantriebe und Elektromobilität, Stuttgart, Sept. 2018

[4] A. Binder, "Elektrische Maschinen und Antriebe; Grundlagen, Betriebsverhalten", Springer Vieweg, 2017, p.105

[5] https://www.pluessag.ch/de/technische-informationen/gfk-technische-daten.html

[6] PolyMax, PC Technical Data Sheet, Version 4.1; Polymaker; Utrecht; Nov. 2018

[7] Sintered Neodymium Iron Boron (NdFeB) Magents; Eclipse Magnetics Ltd; Sheffield

[8] H. Czichos, "Die Grundlagen der Ingenieurwissenschaften, D Werkstoffe, Wärmeleitfähigkeit von Werkstoffen", 31. Auflage. Springer, 2000, p. 54

[9] U. Hahn, " Physik für Ingenieure", Oldenbourg Wissenschaftsverlag GmbH, 2007, p. 335-337

Mechanical Analysis of Different Rotor Topologies for High Speed PMSM in Automotive Application

Michal Kovacik
*Department of Power Systems and
Electric Drives
University of Zilina
Slovak Republic
michal.kovacik@feit.uniza.sk*

Pavol Rafajdus
*Department of Power Systems and
Electric Drives
University of Zilina
Slovak Republic
pavol.rafajdus@feit.uniza.sk*

Ronald Bastovansky
*Department of Design and Mechanical
Elements
University of Zilina
Slovak Republic
bastovansky@uniza.sk*

Abstract—This paper describes a comparative FEA-approach analysis of 4 different rotor topologies of cylindrical type high-speed PMSM, which is planned for a turbocharger assisted with electric motor. In the first part, considered rotor topologies are modelled in the Ansys Workbench 19.0 and their mechanical performances are simulated at rated speed 100 000 rpm. This way, the maps of equivalent von-Mises stress and total deformation are obtained. Appropriate rotor topology is selected and further analysed in two constructional options. In the second part, options of selected topology are further modelled and analysed for electromagnetic performance, using the Ansys Motor-CAD. This way, the equivalent circuit parameters, losses and output performance are given. In the third part, the results provided by the simulations are compared and the options of selected rotor topology are further discussed regarding mechanical safety and desired output parameters.

Keywords— PMSM, FEA, high speed, rotor

I. INTRODUCTION

For their favourable parameters, PMSMs are widely used in automotive applications. Automotive industry brought PMSMs not only into the traction drives, but also into the vehicle on-board systems, e.g. servo drives, power steering, generators [1]. Among these are also high speed applications, such as assisted turbochargers, turbo-compressors, cooling fans etc., which interpose special requirements to electric drives. A high speed PMSM design has to deal with electromagnetic, mechanical and thermal problematics. From the electromagnetic point of view, appropriate construction materials need to be selected to provide sufficient output parameters. High speed means high operational frequencies of supply source, which leads to an increase of core losses. Therefore, steel materials with low core loss properties are selected. Reduction of core loss is also reached by using laminated stator and rotor body and sectionalizing permanent magnets. From the mechanical point of view, rotor is crucial part of a high speed PMSM. To resilient centrifugal forces, an appropriate construction of rotor and safe fixation of permanent magnets (PM) are required. From the thermal point of view, it is important to check operational temperatures in aim of avoiding damage to winding, permanent magnets and machine construction, caused by an excessive thermal overload [2], [3], [4]. Modern approaches were developed in the last few decades to deal with electrical machines design problematics. Nowadays, most prevalent methods are based on finite element analysis (FEA) [5], [6].

II. INVESTIGATED ROTOR TOPOLOGIES

In aim to satisfy the high-speed PMSM design, which is planned for 100 000 rpm assisted turbocharger, four different topologies of rotor are supposed as suitable, according to [5] and [7]. In [8], considered topologies are modelled in the Ansys Maxwell2D and investigated using FEA approach to obtain electromagnetic performance. The topologies are simulated with one the same stator arrangement. Simulation models are denoted MOTOR 1; MOTOR 2; MOTOR 3; MOTOR 4, according to Fig. 1. a); b); c); d) respectively. Output performances given by the electromagnetic analysis in [8] for each topology are concluded in Tab. I. The highest produced torque and output power are given by MOTOR 2. On the other hand, this topology shows the highest core losses. Nevertheless, due to supposed short ON-time in operation cycle in order of few seconds, the MOTOR 2 is selected as the topology with the best electromagnetic performance.

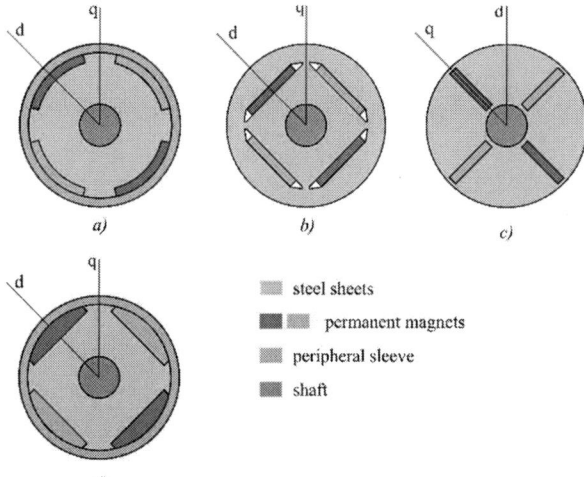

Fig. 1. Investigated rotor topologies: (a) Surface mounted PMs with peripheral sleeve; (b) Tangentially embedded PMs; (c) Radially embedded PMs; (d) Surface mounted bread-loaf PMs with peripheral sleeve. [8]

TABLE I. OUTPUT PARAMETERS AT RATED SPEED [8]

Topology	Torque [mNm]	Output Power [W]	Core Losses [W]
MOTOR 1	156	1 664	480
MOTOR 2	292	3 061	585
MOTOR 3	102	1 069	538
MOTOR 4	204	2 145	503

978-1-6654-3236-8/21 $31.00 © 2021 IEEE

III. MECHANICAL ANALYSIS OF ROTOR TOPOLOGIES

The operational speed 100 000 rpm sets limitations to the rotor construction. To avoid an excessive centrifugal forces, circumferential speed is limited to 200 m/s and thus overall rotor diameter can be no greater than 38 mm. The active magnetic length of rotor is 55 mm. Such dimensions lead to a rotor topology with simple arrangement, which will be mechanically smooth to eliminate windage losses. At first, rotor steel sheets alone, then complete structures of rotors are modelled in the Ansys Workbench 19.0. At second, FEA-based simulation is utilized, to verify their mechanical properties. With this approach, the maps of equivalent von-Mises stress and total deformation for the rotor cross-sections at rated speed 100 000 rpm are obtained.

A. Steel Sheets of Rotor Body

The Rotor body is constructed of 0.1 mm laminated steel sheets. The low core losses steel material 10JNEX900 is used, which is convenient for the planned high speed PMSM. It's yield strength is 600 MPa. As it is shown in Fig. 2 and Fig. 3, MOTOR 1, MOTOR 3 and MOTOR 4 have got appropriate values, while MOTOR 2 requires modifications to improve mechanical safety. It has to be declared that in MOTOR 2, sharp angle of air barriers was modelled in the simulation to obtain values for extreme condition.

B. Complete Rotor Structures

The complete rotor structures are modelled with regard to construction fitting of each component. The materials used in simulations are: 10JNEX900 for rotor the steel body, sintered NdFeB for the permanent magnets and Inconel for the peripheral sleeve. The shaft is left undefined for this simulation. The PMs shall be fit in slots of rotor body and additionally fixed in place. Small dimensions and low mass of the rotor components allow fixing e.g. with a heat resistant glue material. The models are first simulated with all the construction components fixed with glue. The result is given that additional fixation of PMs (for all topologies) and peripheral sleeve (for MOTOR 1 and MOTOR 4) shall reduce the mechanical stress and deformation values and thus make the rotor structures more resilient to centrifugal forces. Further simulations are obtained for several conditions without the additional glue fixing. Fig. 4 and Fig. 5 show such example. The absence of additional glue fixing shall cause increase of stress caused by the centrifugal forces. In some cases, this can lead to a mechanical damage. It can be seen, that MOTOR 3 provides lowest stress and deformation. In MOTOR 1 and MOTOR 4, stress is caused mainly by the mass of PMs. In MOTOR 2, the width of bridge between air barriers, the shape of PM slot cut and the mass of steel situated above PMs require further attention (IV).

TABLE II. MATERIALS USED FOR MECHANICAL ANALYSIS

Material	Young's modulus [GPa]	Poisson's Ratio	Solid density [g/cm³]
10JNEX900	210	0,33	7,49
NdFeB	160	0,24	7,5
Inconel	200	0,31	8,19

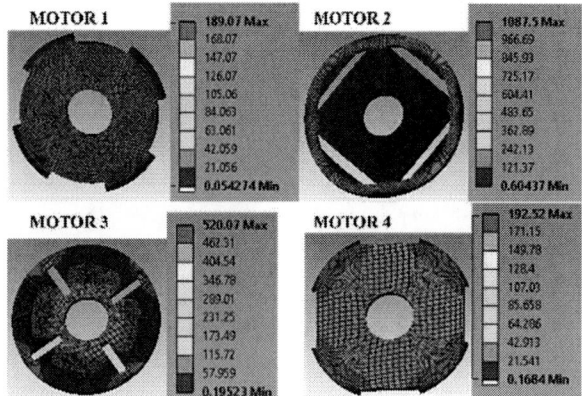

Fig. 2. Equivalent von-Mises stress of rotor steel sheets measured in MPa (Ansys Workbench).

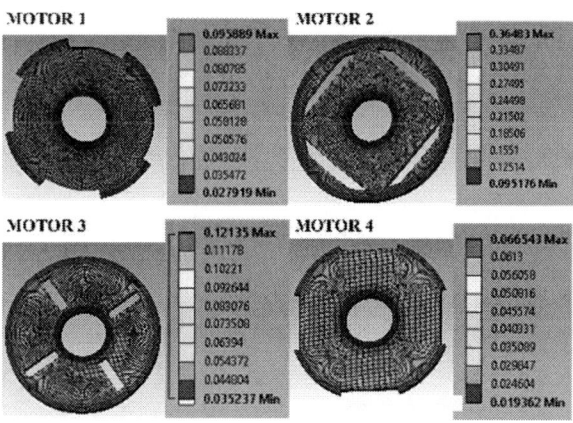

Fig. 3. Total deformation of rotor steel plates measured in mm (Ansys Workbench).

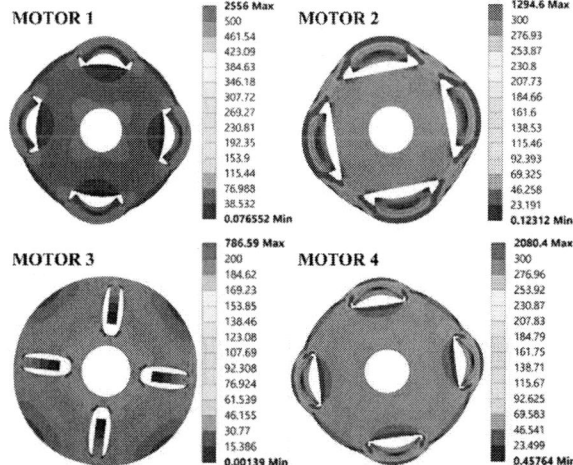

Fig. 4. Equivalent von Mises stress of complete rotors measured in MPa - example without additional fixation of components (Ansys Workbench).

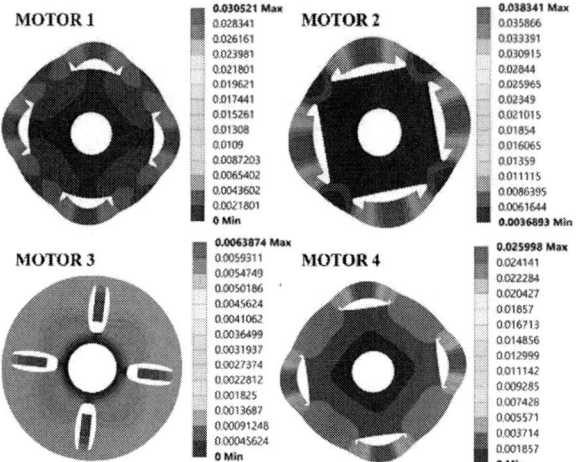

Fig. 5. Total deformation of complete rotors measured in mm - example without additional fixation of components (Ansys Workbench).

IV. OPTIONS OF SELECTED ROTOR TOPOLOGY

Considering mechanical properties of the investigated topologies, the rotor with tangentially embedded PMs (MOTOR 2) is selected as a basic concept for the planned high speed PMSM design. The bridges between air barriers and the edges surrounding slots for PMs are crucial from the mechanical stress point of view. Embedding PMs deeper into the rotor will reduce width of bridge between air barriers. On the other hand, embedding PMs closer to the air gap will increase the stress of steel mass above PMs caused by the centrifugal force. Closer position to surface leads to reduction of PM´s width and thus to deterioration of output performance. Hence this is not appropriate. To fulfil mechanical safety and satisfying desired output performance, a compromise in construction is defined. Two options of construction arrangement are simulated and compared.

A. Optimisation of Air Barrier Edges

To reduce the mechanical stress, rounding the angle of air barrier edge is possible, with respect to the width of bridge between barriers. The bridge behaves as magnetic path for undesired leakage flux of PMs. The wider is the bridge, the higher is the leakage flux. On the other hand, mechanical resilience increases with the bridge width (Fig. 6).

B. Dividing PM Blocks by Inserting Reinforcing Joists

Other option is to insert reinforcing joists to the middle of PM slots. The joists will improve mechanical resilience by reducing stress in bridges, caused by the mass of steel above PMs. As shown in Fig. 7, it was found by optimisation, that 2 mm thick joists will satisfy mechanical safety, with respect to yield strength of 10JNEX900 steel, which is 600 MPa. A high stress occurs in edge angles surrounding the joists. This can be reduced by additional glue fixation. Inserted steel joists shall reduce PM width and behave as an additional magnetic path for leakage flux, thus deteriorating desired output parameters. However, splitting PM blocks in half sections by joists shall reduce eddy current loss in magnets (Fig. 7).

Rounding the angle of air barrier, together with glue fixation of PMs will reduce mechanical stress for both options. In aim to investigate impact of inserted reinforcing joists to output performance, electromagnetic properties of both options are analysed in (V).

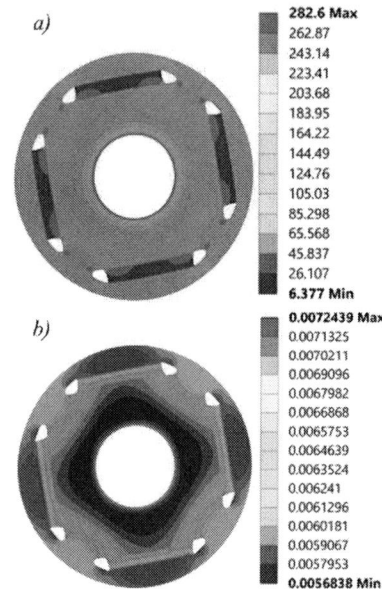

Fig. 6. Option with rounded air barrier angles: (a) equivalent von Mises stress measured in MPa; (b) total deformation measured in mm (Ansys Workbench).

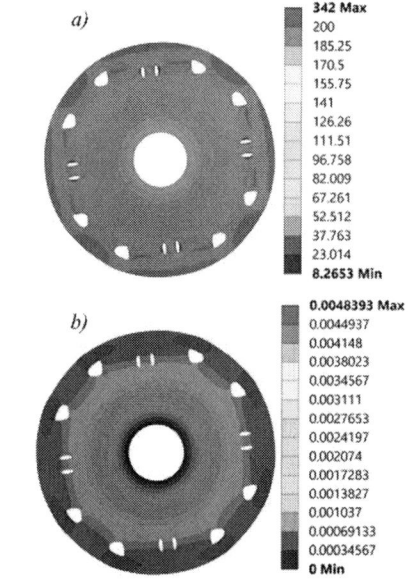

Fig. 7. Option with inserted reinforcing joists in magnet slots: (a) equivalent von Mises stress measured in MPa; (b) total deformation measured in mm (Ansys Workbench).

V. ELECTROMAGNETIC ANALYSIS OF SELECTED ROTOR TOPOLOGY

To analyse the electromagnetic performance, both options of selected topology are modelled in the Ansys Motor-CAD v14.1.6. The models are analysed with one stator arrangement and results are compared. For each option, different number of turns per phase coil is required, which is discussed bellow. In the geometry mode, body of machine is modelled with regard to dimensions. In the winding mode, parameters of winding are set (Tab. III). The winding mode enables to select winding type according to design requirements. When input parameters are set, program automatically generates specifications of pattern and definition of winding.

978-1-6654-3236-8/21 $31.00 © 2021 IEEE

TABLE III. PARAMETERS OF DESIGNED MOTOR

Design Parameters	
Max. output power	8 kW
DC-bus supply voltage	48 V
Rated speed	100.000 rpm
Max. rotor circumferential speed	200 m/s
Required Dimensions	
Active length	55 mm
Stator outer diameter	110 mm
Rotor diameter	38 mm
Air gap	0.3 mm
Winding Parameters	
Type of winding	segment, concentric
Phases	3
Phase connection	Y
Stator Slots	6
Poles	4
Parallel Paths	1
Coil Pitch	1
Turns per Phase Coil	2 or 3

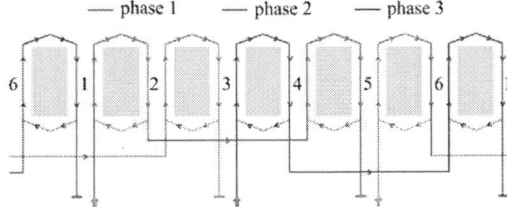

Fig. 8. Winding layout pattern of designed motor (MotorCAD).

After defining winding properties, construction materials from program library are assigned to the model according to Tab. IV. In the calculation mode, operational requirements are set according to Tab. III. Limit of peak current is set to 200 A. This is the assumed current limit of supposed power inverter, which will control the motor. Program also enables temperature settings, however, the thermal model will be calculated during final optimization of design. Here, the electromagnetic model can be solved by FEA-method. The duration of simulation is dependent on model complexity and number of parameters selected for calculation. This way, the equivalent circuit parameters and output performance are obtained. At first, both options are simulated with one stator arrangement, where number of turns per phase coil is set to 2. As expected, adding the reinforcing joists shall lead to decrease of torque and back-EMF, which leads to deterioration of desired output performance. However, the torque ripple and cogging torque will also decrease. This shall lead to more silent operation of the motor. At second, option with reinforcing joists is simulated with number of turns per phase set to 3, to compensate deterioration of output performance. Tab. V provides the comparative summary of parameters obtained with simulations. For each option, graphs

of air gap flux density, back-EMF, output torque and cogging torque are shown in Fig. 11, 12 and 13. The results show that adding one turn per phase coil for option with reinforcing joists shall increase its output torque and power, utilizing an advantage of lower torque ripple and cogging torque. On the other hand, disadvantage is an increase of the iron losses. Adding more turns is problematic, since Back-EMF at rated speed 100.000 rpm must not exceed 48 V of supply DC-bus. 3 turns per phase coil still adhere to this condition.

TABLE IV. CONSTRUCTION MATERIALS ASSIGNED TO MODEL

Construction Part	Assigned Material
Stator Body	JFE-10JNEX900
Rotor Body	JFE-10JNEX900
Permanent Magnets	N30UH
Armature Winding	Copper (Pure)
Shaft	Iron (Pure)

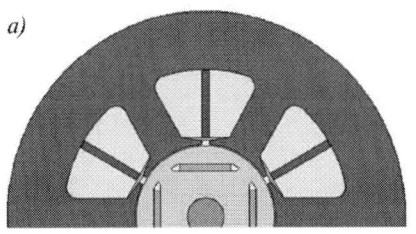

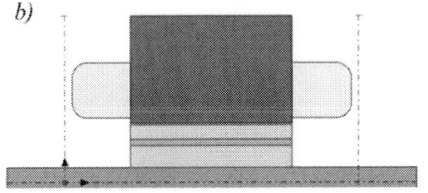

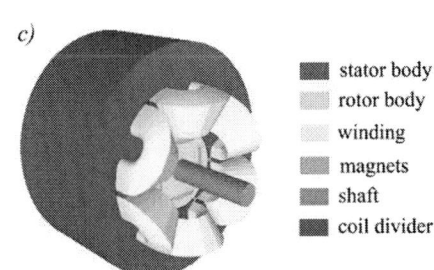

Fig. 9. Geometry- option with simple PM arrangement: (a) radial view; (b) axial view; (c) 3D view. (MotorCAD).

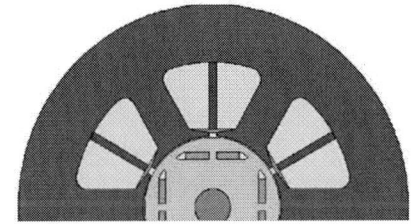

Fig. 10. Geometry- option with reinforcing joists between PM blocks-radial view; axial and 3D views are identical to Fig. 9. (MotorCAD).

TABLE V. SUMMARY OF ELECTROMAGNETIC PARAMETERS

Parameter	unit	Option		
		Simple magnet	*Reinforcing joists*	*Reinforcing joists*
Turns per phase coil	-	2	2	3
D-axis inductance	[mH]	0.008887	0.008655	0.0146
Q-axis inductance	[mH]	0.01881	0.01913	0.032
Phase resistance	[mΩ]	1.159	1.159	1.739
Back-EMF (peak)	[V]	37.06	29.05	43.57
Average torque	[mNm]	698	571	795.5
Torque ripple	[%]	26.73	14.92	19.64
Cogging torque (peak to peak)	[mNm]	51.6	37.4	37.4
Armature DC copper loss	[W]	69.57	69.57	104.4
Magnet Eddy current loss	[W]	130.7	31.21	58.2
Stator iron loss	[W]	304.3	303.1	362.7
Rotor iron loss	[W]	54.36	58.13	69.33
Total losses	[W]	559	462	594.6
Peak output power	[W]	7311.9	5983.7	8335.7
Efficiency	[%]	92.55	92.41	93.03

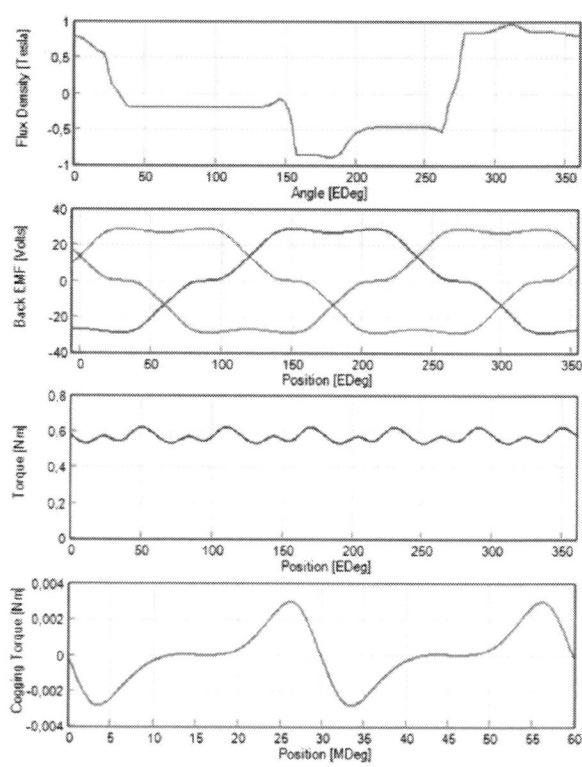

Fig. 12. Option with reinforcing joists- 2 turns per phase coil: diagrams of air gap flux density, back-EMF, torque and cogging torque (MotorCAD).

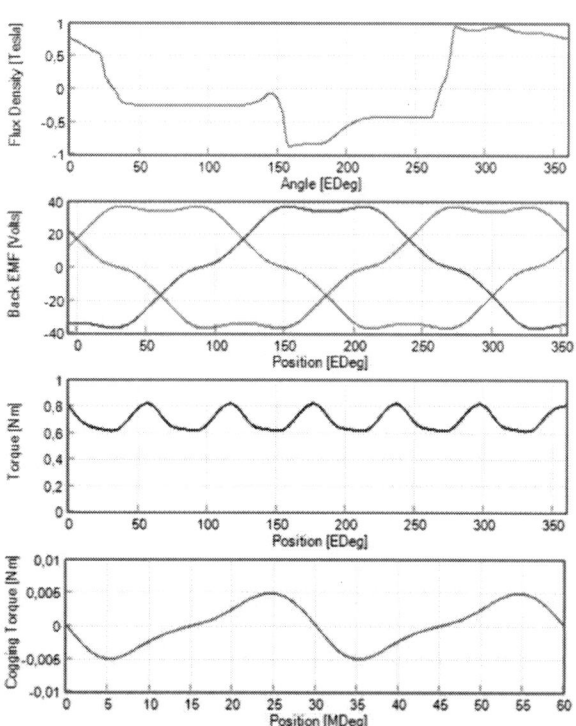

Fig. 11. Option with simple magnet arrangement- 2 turns per phase coil: diagrams of air gap flux density, back-EMF, torque and cogging torque (MotorCAD).

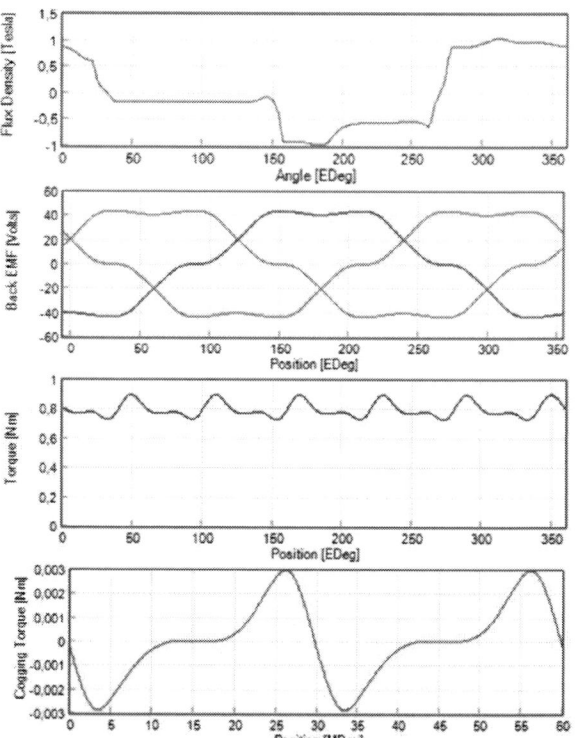

Fig. 13. Option with reinforcing joists- 3 turns per phase coil: diagrams of air gap flux density, back-EMF, torque and cogging torque (MotorCAD).

VI. Conclusion

In aim to adhere mechanical safety and desired output performance of designed motor, a feasible compromise in form of 2 constructional options of rotor topology is defined:

- The option with simple magnet arrangement is easier to be manufactured and requires stator with 2 turns per phase coil. This is advantageous in lower armature copper loss caused by flowing current and lower iron losses. Disadvantage is higher torque ripple and cogging torque.

- The option with reinforcing joists is more complex for manufacture than option with simple magnet arrangement, yet it is supposed as feasible. Using 3 turns per phase coil will compensate the deterioration of output performance caused by inserted reinforcing joists. The advantage is increase of output torque and power and decrease of torque ripple and cogging torque. On the other hand, an increase of iron loss caused by adding turns per phase coil is disadvantageous.

For both options, all components have to be additionally fixed with proper glue material in aim to increase mechanical resilience of the rotor. To reduce eddy current loss in magnets, sectionalisation of PM blocks can be also done. Both options are close to desired 8 kw output power. Subsequent work shall focus on improvements, which can be achieved by optimisation of design. Flux density saturation maps and a thermal analysis shell be provided. After completing design procedures, a prototype of motor is planned to be manufactured and its properties shell be verified by means of laboratory measurements.

Acknowledgment

This work was realized with support by Slovak Scientific Grant Agency VEGA No. 1/0615/19, by projects ITMS 26220120046, cofounded from EU sources and European Regional Development Fund, Operational Program Integrated Infrastructure 2014- 2020 of the project: Innovative Solutions for Propulsion, Power and Safety Components of Transport Vehicles, code ITMS 313011V334, co-financed by the European Regional Development Fund".

References

[1] A. Emadi, 2014. "Advanced electric drive vehicles," McMaster University, Hamilton, Ontario, Canada, CRC- Press, Taylor & Francis group, 2014. ISBN 978-1-4665-9769-3.

[2] D. Gerada, A. Mebarki, N. L. Brown, C. Gerada, A. Cavagnino, A. Boglietti, "High speed electrical machines: Technologies, trends and developments." IEEE Transactions on Industrial Electronics, Volume: 61, Issue: 6, June 2014. ISSN 1557-9948. 2014.

[3] J. Ahn, C. Han, C. Kim, J. Choi, "Rotor design of high-speed permanent magnet synchronous motors considering rotor magnet and sleeve materials." 2018.

[4] M. V. Casey, P. Krähenbühl, C. Zwissig, "The design of ultra-high-speed miniature centrifugal compressors," European Conference on Turbomachinery Fluid Dynamics and Thermodynamics ETC 10, 2013.

[5] J. Pyrhönen, T. Jokinen, V. Hrabovcova, "Design of rotating electrical machines, Second Edition." John Wiley & Sons Ltd, West Sussex, United Kingdom, ISBN 978-1-118-58157-5. 2014.

[6] V. Hrabovcova, P. Rafajdus, P. Makys, "Analysis of Electrical Machines." EDIS, IntechOpen 2020, ISBN: 978-1-83880-208-0. 2020.

[7] J. F. Gieras, M. Wing, "Permanent magnet motor technology, design and applications, Second Edition, Revised and Expanded." BT Cellnet, London, ISBN 0-8247—0739-7. 2002.

[8] M. Kovacik, P. Rafajdus, and S. Kocan, "Comparison of various PMSM rotor topologies for high-speed drives in automotive applications," Transcom 2021- 14th International Scientific Conference on Sustainable, Modern and Safe Transport., May, 26- 28, 2021.

Analysis of Temperature Change in a Permanent Magnet Synchronous Generator Under Load due to Stator Winding Inter-Turn Short Circuit

Michael Barrett, Member, IEEE
Engineers Ireland (IEI)
Dublin, Ireland
barrett.mike1@gmail.com

Abstract— The permanent magnet synchronous generator (PMSG) finds application in wind and tidal-stream generation. The reasons for its deployment may be summed up as its efficiency and the fact that it is self-sustaining, that is, it does not require external rotor electrical power. Because of the high installation and maintenance costs of both tidal and wind energy systems on-going condition monitoring (CM) is advisable. In CM many parameters are monitored, one of those being temperature. However, absolute temperature values may provide limited information about the equipment being monitored. The temperature of a PMSG under full load will be higher than an unloaded one which, of itself, is not relevant. The proposal here, explores the relative temperatures values over time for the purpose of detecting stator inter-turn short circuits. For this, an enhanced experimental setup is proposed. A previous work proposed a methodology to detect inter-turn short circuits in an unloaded generator by temperature analysis. This work is an extension of that with, in this case, a generator under different load conditions. A range of experiments verified a detectable temperature rise, in a short time, above that which would be expected in a healthy generator under load to one with a faulted stator winding. Further, this proposal automatically allows for a generator starting from cold or restarting with an elevated temperature. This proposed system, while still at the laboratory stage, has been successful in capturing temperature change over time and robust in detecting stator inter-turn short circuits within a short period of start-up. Further, while in this work a programmable controller (PLC) is used as the CM platform the methodology proposed here may be integrated into existing monitoring systems utilizing existing temperature sensors by modifying the monitoring software program.

Keywords— PMSG, inter-turn short circuits, condition monitoring

I. INTRODUCTION

The essential difference in three-phase generators is the method by which the rotor gets its magnetic field. In the separately excited type, the magnetic field is generated by a DC current flowing through a wound rotor. In the PMSG the magnetic field is from permanent magnets mounted on the rotor, see Fig.1. The three-phase stator winding may be configured in star or delta. Its generated frequency is a function of its rotor speed rotation and the number pairs of rotor poles and is described by (1)

$$F = N_s \, P \qquad (1)$$

where F is the frequency in Hz., N_s the speed in revolutions per second and P the number of pole pairs.

Fig. 1 Disassembled PMSG showing the stator windings and rotor magnets

One of the recurring issues, regarding PMSGs, identified in the literature as being a cause for detailed research is the detection of stator winding insulation breakdowns. In a fault condition, as the PMSG is self-sustaining, that is, it does not require rotor excitation, and will continue to generate power as long as the permanent magnet rotor continues to rotate, therefore has the possibility to cause further damage to itself and its associated equipment. Further, in order to reduce costs manufacturers may reduce stator winding diameter which can impact on copper losses and reduce the amount of laminated steel in the rotor or stator cores with implications for temperature handling [1]. For those, and other maintenance issues, monitoring of PMSG status is desirable. Condition Monitoring (CM) [2] is a useful tool for detecting electrical and mechanical faults in equipment generally and also has the potential to inform equipment development. The CM platform in this work is the Programmable Logic Controller (PLC).

The PLC is widely acknowledged in industry and in the literature as the preferred platform for production control as it is a tried and tested technology with a wide range of logic and mathematical functions. Its functionality may be extended by the addition of extra modules, thus extending its

978-1-6654-3236-8/21 $31.00 © 2021 IEEE

flexibility. The PLC programming languages are defined by IEC 61131-3 [3] which facilitates standardization and portability across different manufacturers' platforms. Its use in CM, however, is not widely reported in the literature. A secondary role of this paper is to further assess the feasibility of the PLC in CM. In a previous work [4] the early detection of stator inter-turn short circuits in an unloaded PMSG by temperature analysis was explored and found to be feasible. In this work the temperature rise in a PMSG under load in an inter-turn short circuit condition is examined in the case study. The PMSG has wide acceptance because it has high power density, high efficiency and requires minimum maintenance [5]. It can, however, be prone to mechanical and electrical issues. The major electrical problem is stator-winding insulation breakdown. In electrical machines generally, stator-winding faults account for 30 % - 40% of faults [6] while He et al. [7] report stator insulation faults in permanent magnet machines as accounting for 36% of problems in low voltage machines, rising to some 66% in high voltage machines. Stator-winding faults in electrical machines may be described as:

1. Phase-to-phase shorts;
2. Phase-to-ground-wall shorts;
3. Phase inter-turn shorts.

Inter-turn short circuits occur for several reasons:

1. Mechanical, electrical and environmental stress
2. The generator being over driven causing excess phase current, resulting in temperature rise and insulation degradation.

Stator winding issues may start as inter-turn short circuits, that is where wires of the same phase short together. The inter-turn short circuit may be regarded as a separate circuit within the faulted phase. The rotating permanent magnet rotor, apart from generating the phase emf resulting in phase current, generates an emf in the shorted loop. This results in a voltage, v_{sc}, across the short circuit resistance described by (2),

$$v_{sc} = r_{sc} \cdot i_{sc} + \frac{d}{dt} \Psi \qquad (2)$$

where r_{sc} and i_{sc} are the shorted loop resistance and current respectively and Ψ the loop flux generated. The shorted loop current, which can be very high [8], will generate heat, as described by Joule's Law. Mohammed et al. [9] define 4 factors determining this heat:

1. The short circuit contact point current;
2. The current through the shorted coil turns and the faulted coil structure;
3. The current through the healthy part of the coil;
4. The ratio between the faulted and healthy coil turns.

Thermal conduction occurs between the heat generated in the faulted phase to the rotor permanent magnets, the stator-core laminations and to the generator frame.

Stator inter-turn short circuits result in several issues. The faulted phase current will be reduced in proportion to the shorted to healthy turns ratio causing an output current imbalance resulting in positive (+), negative (-) and zero (0) sequence current components as described by (3)

$$\begin{matrix} IA \\ IB \\ IC \end{matrix} = \begin{matrix} IA_+ \\ IB_+ \\ IC_+ \end{matrix} + \begin{matrix} IA_- \\ IB_- \\ IC_- \end{matrix} + \begin{matrix} IA_0 \\ IB_0 \\ IC_0 \end{matrix} \qquad (3)$$

This represents power output loss and has implications for power quality [10]. Further temperature rises are likely as long as the permanent magnet rotor continues to rotate resulting in further insulation breakdown with the resultant increase in inter-turn short circuits and the possibility of phase-to-phase and phase-to-ground shorts. Grubic et al. [11] report a 50% decrease in stator insulation life being caused by a temperature rise of 10% while making the point that while normal ageing does not, in itself, cause breakdown it does make the insulation more sensitive to other stress factors. Those factors include unbalanced phase current, overloading and ambient temperature. High temperatures, also have the potential to lead to permanent demagnetization of the rotor's permanent magnets [12]. Another less reported consequence is the extra torque loading on the generator's prime mover due to the braking (Lenz) effect of the inter-turn magnetic field. This could have implications for mechanical couplings and bearings.

In a rotating machine several losses occur [13]. Losses in generators arise from several sources. Mechanical losses include core (P_{core}) and rotation losses (P_{rot}). Electrical losses are $I^2 R$ copper losses in the stator windings (P_{cu}). In the event of a stator inter-turn short circuit an extra loss, copper loss in the short circuit (P_{sc}) must be included. Those losses result in temperature rise.

$$P_{tot} = P_{core} + P_{rot} \qquad (4)$$

$$P_{tot} = P_{cu} + P_{core} + P_{rot} \qquad (5)$$

$$P_{tot} = P_{cu} + P_{core} + P_{rot} + P_{sc} \qquad (6)$$

Expression (4) describes the losses which lead to temperature rise in an unloaded rotating generator, while (5) defines a healthy generator supplying a load. The temperature rise in a stator faulted condition is expressed by (6). The common factor in (5) and (6) is current flow which determines the extra temperature rise. Fig.2, obtained from the experimental system used in the case study, illustrates the significance of (4), (5) and (6) in a generator. The starting temperature, 22°C, is the ambient or room temperature while the generator output frequency is 55Hz. The phase current is 0.8Amps. while the short-circuited loop current is 3.2Amps. Phase currents are functions of the load and the frequency. The short-circuited loop currents are predominately determined by the frequency.

978-1-6654-3236-8/21 $31.00 © 2021 IEEE

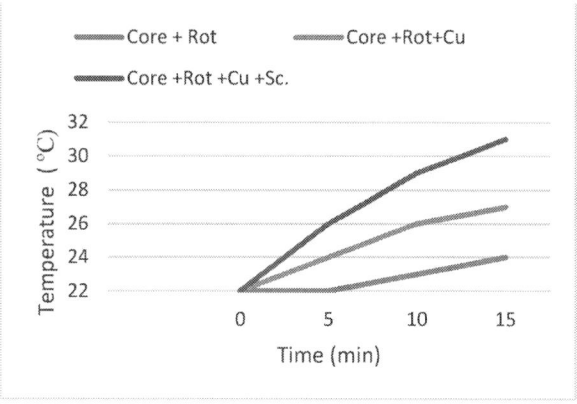

Fig.2 Temperature rise due to losses in unloaded, healthy and faulted stators

Fig.2 also reveals a distinct temperature difference between the temperature trajectories of the open-circuit, the healthy and the faulted stator. This will be explored further in the case study.

II. DETECTING INTER-TURN SHORT CIRCUITS

A. Issues

Because the temperature distribution in generators is not uniformly distributed, measurement is challenging. One method of monitoring the stator winding temperature would be to embed a range of sensors, such as Resistance Temperature Detectors (RTDs) or thermocouples, in the windings themselves. The number of such sensors would depend on the physical size of the generator and it would still be possible to miss temperature hot-spots. Further, those devices are susceptible to electro-magnetic interference. Embedding sensors would only be possible in new machines at the manufacturing stage. Such sensors would have to be protected from the high voltages being generated by the windings. Retrofitting such sensors in existing machines would probably not be feasible in machines outside of the laboratory. In many cases RTDs and thermocouples are mounted on the generator body and while they would reasonably represent an average generator temperature this would lag the winding temperature. Many researchers have proposed a range of methodologies for determining generator temperature.

B. Temperature monitoring methods in the Literature

Mohammad and Djurović [14] report the implementation of Fiber Bragg Grating (FBG) as a temperature monitoring system for a prototype induction motor stator winding. In FBG individual periodic reflective gratings are spaced along a section of optical fiber. An incident light spectrum propagates through the fiber. A specific wavelength (the Bragg wavelength (λ_b)) is reflected back. A change in the refractive index of the gratings (due to strain or temperature) causes the reflected wavelength to change. This relationship as applied to temperature change is defined by (7) [14].

$$\Delta\lambda_b = \lambda_b \ (\alpha + \xi)\Delta T \qquad (7)$$

where α is the fiber expansion coefficient, ξ the fiber thermo-optic coefficient and T the temperature. Independent sensors may be deployed along the same fiber. So, by multiplexing, an array of distributed sensors is available along a length of optical-fiber which would increase the possibilities of detecting temperature hot-spots. FBG has the added advantage of being immune to Electro Magnetic Interference (EMI) and being (electrically) nonconductive.

In a permanent magnet machine, the rotor permanent magnet temperature has a direct influence on several of the machine's parameters and on its remanent flux density [15]. This fact is exploited in temperature measurement of the permanent rotor magnets by measuring the magnetic flux and relating it to temperature. Researchers propose several approaches. Fernandez et al. [16] propose a methodology which required the embedding of three Hall-effect sensors, about the stator of a PM machine. The Hall output voltages, being proportional to the flux, were a function of the temperature. While the method presented challenges, among them being interference from the stator current and the non-linearity of the Hall sensors, this methodology was found to be sufficient for the detection of high temperatures which could result in PM damage. Sprecht et al. [17] propose a flux observer model, for an internal permanent magnet synchronous motor, based on the rotor reference frame orthogonal components v_{dq} and i_{dq} . The orthogonal components are inputs to the observer model which outputs the rotor temperature. The authors report good precision for this model.

As the resistance of copper wire changes with temperature in a known manner, determining stator temperature of-line is a simple matter of measuring its resistance. On-line measurement is complex. Lee et al. [18] detail two approaches for determining the stator temperature of an induction motor, a stator-resistance estimation method and a DC signal injection method. The stator-resistance estimation requires knowledge of the stator voltage, current, rotor speed, stator flux linkage, rotor resistance, slip and the supply frequency. The accuracy of the estimation is dependent on those parameters. Wilson et al. [19] report limited results in the use of the stator-resistance method to determine the temperature of a permanent magnet synchronous motor (PMSM). The same paper focuses on the DC signal injection method in which a high value DC pulse is injected through a combination of phases. The technique, which can produce torque ripples, is reported as having an estimation error of less that 10°C at temperatures greater than 60°C.

III. THE CASE STUDY

The focus of the experiments is a small three-phase Permanent Magnet Synchronous Generator (PMSG). The generator is driven by a three-phase drive motor (the prime mover) controlled by an inverter. This allows start/stop and speed control of the PMSG. The overall control and monitoring functions are implemented from the Human Machine Interface (HMI) via the PLC, see Fig.3.

978-1-6654-3236-8/21 $31.00 © 2021 IEEE

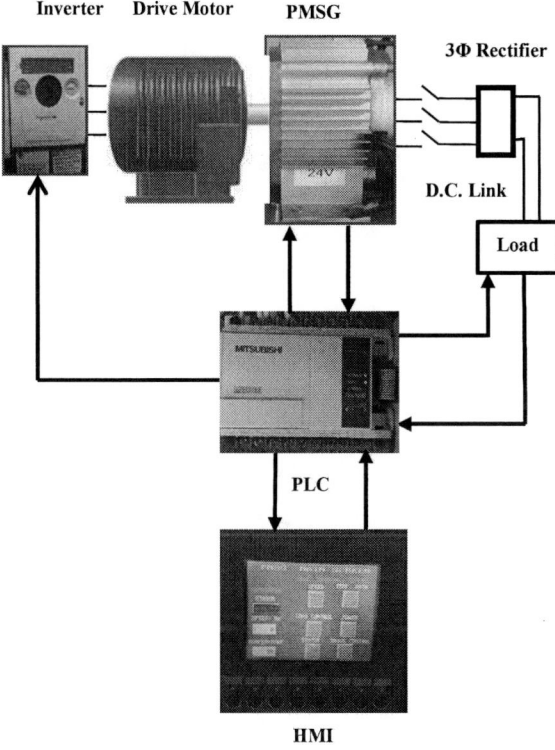

Fig.3 Over-view of PMSG monitoring and control system

The PMSG outputs to a variable non-linear load, consisting of a three-phase rectifier and two switchable resistor loads to simulate the electrical power conversion system.
HMI control functions are:
1. Drive motor start, stop and speed control;
2. Stator Inter-turn short circuit insertion;
3. Load select.

The HMI monitors:
1. PMSG temperature;
2. PMSG generated frequency;
3. DC load voltage;
4. DC load current;
5. Stator inter-turn short circuit current.

The drive-motor electrical power and the PMSG output power are monitored separately using current and voltage meters. PMSG speed is captured by a proximity sensor, interfaced to the PLC, detecting a keyway on the shaft. The rotation pulses are counted by the PLC high-speed counter facility and converted to frequency by the Ladder Logic program using (1). The HMI thus displays the PMSG output in Hz. For temperature measurement three PT100 RTDs are mounted in the louvres of the PMSG, one at the top and the other two on the right- and left-hand sides 90° displaced from the top. The challenges of correct temperature measurement of generators are already described, however, in this case the PMSG is small and the proposed methodology focusses on the relative temperature change between a healthy and faulted

stator rather than absolute values. An inter-turn short circuit of 8% of the phase winding number of turns may be switched in the "U" phase of the PMSG, see Fig.4. As the proposed methodology requires the ability to detect faults in the short-term, temperatures are captured using the algorithm [4], see Fig.5, with values time-stamped at the start of an experiment and at 5-minute intervals thereafter up to 15 minutes. The algorithm to implement this is one of the functions in the PLC program. Automating this process ensures accuracy of readings. The average value of the temperature inputs from the three RTDs are used as the temperature input. The three RTD value differences were, in fact, not significant.

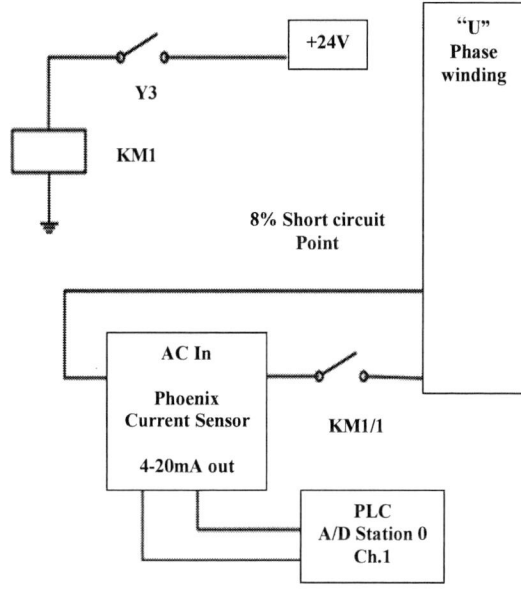

Fig.4   Fault insertion circuit

The fault insertion circuit is switched in via the touch screen facility of the HMI, activating Y3 of the PLC output. Also, the actual shorted-loop current is displayed on the HMI, verifying that the short circuit is in place. Further, the facility to measure the shorted-loop current will be a factor in a further study.

A. Experiments

A range of experiments was carried out to compare the temperature changes in a healthy PMSG under load and a faulted one at different generated frequencies and starting temperatures using the algorithm [4] in Fig.5. The output current is, of course, a function of the load and the generated frequency. The terms "low load" and "high load" used later describe a range of load currents, the high load range being significantly higher than the low range. The ambient or room temperature range for the experiments was 20° to 22°C. Starting temperatures from 25°C upwards were achieved by running the PMSG under load until the required temperature was reached. The experiments were run continuously over 5-6 hours per day for several months.

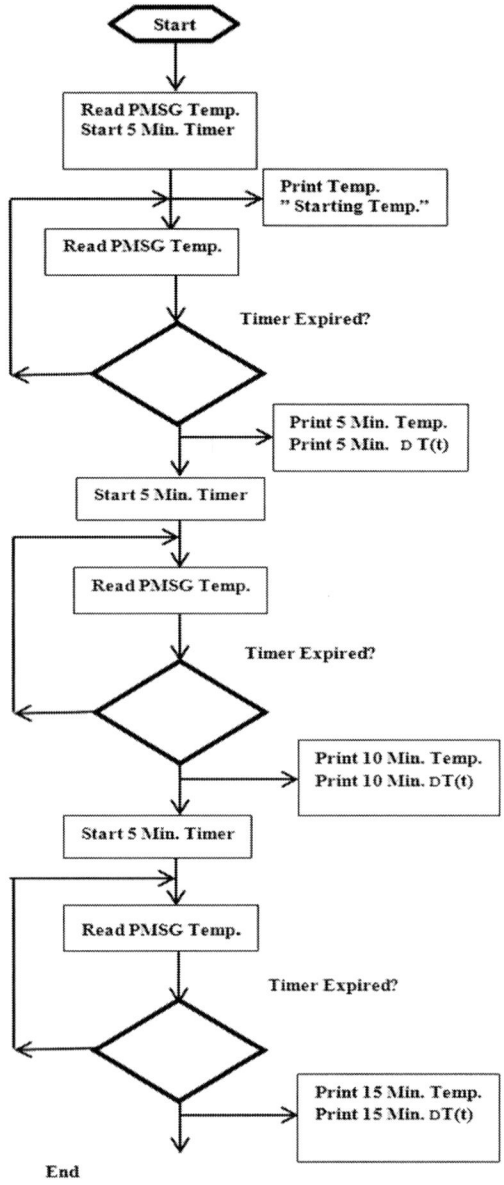

Fig.5 Data-capture algorithm [4]

The experiments were run at in the format shown in the Table 1 matrix which details the range of starting temperatures against the range of frequencies at which the PMSG was run. Experiments were run for the PMSG in a healthy stator condition, under low and high load, for each of the temperatures and frequencies shown and repeated for the PMSG with a faulted stator (with the same loads).

TABLE I.   EXPERIMENT MATRIX

Starting Temps.	Freq.	Freq.	Freq.	Freq.
22°C	45Hz.	55Hz.	65Hz.	75Hz.
25°C	45Hz.	55Hz.	65Hz.	75Hz.
30°C	45Hz.	55Hz.	65Hz.	75Hz.
35°C	45Hz.	55Hz.	65Hz.	75Hz.

B. Experimental Results

An example of the results of the temperature trajectory for a healthy and faulted stator with the generator output frequency of 55Hz. at high load starting at 25°C is detailed in Fig.6.

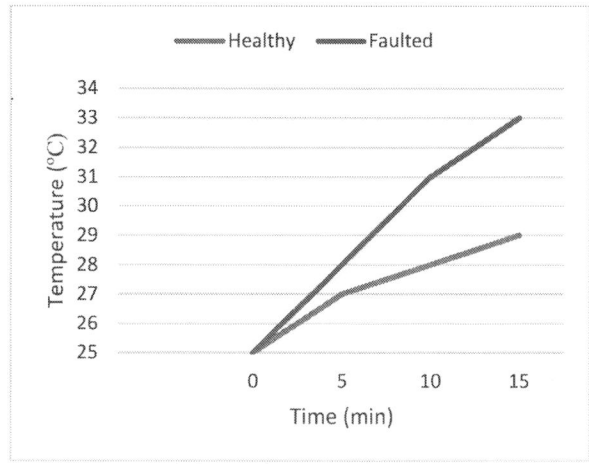

Fig.6 The temperature trajectory for healthy and faulted stator at 55 Hz. at high load condition

As may be observed from Fig.6 there is a discernable difference between the healthy and faulted stator temperature trajectories within the first 5-minutes which increases with time. A quantifiable difference was evident in all of the experiments specified in Table I. Tabulating the differences between healthy and faulted temperature trajectories has two purposes, first to show that such differences exist across the defined frequency and temperature spectra for all load conditions and secondly, to provide the data necessary to construct a healthy temperature profile for the PMSG. Further, this data will inform the design of the PLC program to detect and flag abnormal temperature rise within a 5-minute time scale. Table II presents the average-temperature slope per minute in the first 5-minute period from start for the PMSG under high load. This difference is the key to the proposed inter-turn fault detection algorithm. Similar differences were evident for the PMSG under low load conditions.

TABLE II.   TEMPERATURE RISE AT HIGH LOAD

Freq.	Temperature Range	Healthy	Faulted
45Hz	22°C -35°C	0.1°C/Min.	0.35°C/Min.
55Hz	22°C -35°C	0.2°C/Min.	0.6°C/Min.
65Hz	22°C -35°C	0.3°C/Min.	0.7°C/Min.
75Hz	22°C -35°C	0.4°C/Min.	0.93°C/Min.

C. Fault Detection

The inter-turn fault detection system is implemented in the PLC program. As may be observed from Table II, and is evident from similar data for low load condition, the temperature trajectories are functions of frequency, temperature range, whether the generator is working at low or high load and whether a stator fault is present. Those functions are inputs to the PLC program to detect abnormal temperature rises within a 5-minute period of the PMSG startup. The PLC program will compare the current temperature to healthy temperature profiles held in a lookup table constructed from the data-capture algorithm depicted in Fig.5, the frequency, the temperature and the load status. Any difference, designated, in this work as the "faulted temperature" will flag an error. The basis of the fault detection algorithm may be summarized is the pseudocode statement:

If (within a 5-minute period from start):
current temperature => starting + faulted temperature
Then:
Flag "Abnormal Temperature Rise"
Else:
Flag: "Normal Temperature"

Fig.7 Abnormal temperature detection pseudocode

The PLC based fault-detection program was found to be effective in discriminating between healthy and faulted stators within the parameters indicated in Table I (low load and high load conditions). Results were compromised, however, if the PMSG speed was changed during the 5-minute period.

IV. RESULTS AND DISCUSSION

This work was successful in developing a methodology based on temperature analyses to detect a stator inter-turn short circuit in a short time period for a PMSG under load conditions. This is significant because inter-turn shorts have the potential to cause serious damage in a relatively short time. The limitation of this work is that, while inter-turn short circuits of 8% and above of the phase are captured, values below that have not sofar been tested. A generic model of the methodology proposed would require healthy generator temperature profiles to be captured. Any deviation in a short time from start, should be flagged. A CM system, to be of greatest value, should, as well as recording all relevant process variable values, have a diagnostic function. In the diagnostic mode it may be necessary to compare data from several machine parameters, for example in this work if an abnormal temperature rise is evident and a phase output imbalance is detected then stator inter-turn short circuit is strongly indicated. This work identifies the PLC as being a suitable candidate as a CM platform because of its flexibility from a hardware and software perspective. There is however a need to identify sensors to capture parameters such as vibration and harmonics with PLC interface. Further work will address the detection of inter-turn short circuits at values below 8% and the issue of frequency change during the detection window as identified earlier. Work will continue on the development of the PLC based CM system to identify signal conditioned sensors suitable for vibration and harmonic analysis.

REFERENCES

[1] G. C. Stone, M. Sasic, D. Dunn and I. Culbert, "Recent problems experienced with motor and generator windings," 2009 Record of Conference Papers - Industry Applications Society 56th Annual Petroleum and Chemical Industry Conference, 2009, pp. 1-9, doi: 10.1109/PCICON.2009.5297173.

[2] C. Sheng, Z. Li, L. Qin, Z. Guo, Y. Zhang, "Recent progress on mechanical condition monitoring and fault diagnosis", Procedia Engineering,Volume 15, 2011.

[3] R. Ramanathan, "The IEC 61131-3 programming languages features for industrial control systems," 2014 World Automation Congress (WAC), 2014, pp. 598-603, doi: 10.1109/WAC.2014.6936062.

[4] M.Barrett, "The initial temperature rise in a permanent magnet synchronous generator as a tool in the detection of stator inter-turn short-circuits", International Conference on Fundementals of Electrical Engineering 2020 (ISFEE), Bucharest, November 2020. (In Print)

[5] Y. Sun, S. Wang, Z. Huang and S. Mu, "Research on inter-turn short circuit of armature windings in the multiphase synchronous generator–rectifier system", The Journal of Engineering, 2018: 625-630. https://doi.org/10.1049/joe.2018.0026

[6] F. Cira, M. Arkan and B. Gumus, "Detection of stator winding wnter-turn short circuit faults in permanent magnet synchronous motors and automatic classification of fault severity via a pattern recognition system",*Journal of Electrical Engineering & Technology, 2016, volume11, pages 416-424*

[7] J. He, C. Somogyi, A. Strandt and N. A. O. Demerdash, "Diagnosis of stator winding short-circuit faults in an interior permanent magnet synchronous machine," 2014 IEEE Energy Conversion Congress and Exposition (ECCE), Pittsburgh, PA, USA, 2014, pp. 3125-3130, doi: 10.1109/ECCE.2014.6953825.

[8] F. Wu, C. Tong, Y. Sui, L. Cheng and P. Zheng, "Influence of third harmonic back EMF on modeling and remediation of winding short circuit in a multiphase PM machine With FSCWs", IEEE Transactions on Industrial Electronics, vol. 63, no. 10, pp. 6031-6041, Oct. 2016, doi: 10.1109/TIE.2016.2577552.

[9] A. Mohammed, J. I. Melecio and S. Djurović, "Stator winding fault thermal signature monitoring and analysis by in situ FBG sensors," in IEEE Transactions on Industrial Electronics, vol. 66, no. 10, pp. 8082-8092, Oct. 2019, doi: 10.1109/TIE.2018.2883260.

[10] E. Hossain, M.R. Tur, S. Padmanaban, S. Ay, I. Khan, "Analysis and mitigation of power quality issues in distributed generation systems using custom power devices". IEEE Access, 6, 16816-16833.https://doi.org/10.1109/ACCESS.2018.2814981, 2018

[11] S. Grubic, J. M. Aller, B. Lu, T. G. Habetler, "A survey on testing and monitoring methods for stator insulation systems of low-voltage induction machines focussing on turn insulation problems", IEEE Transactions on Industrial Electronics, Vol. 55, No. 12, December 2008

[12] H. Guo, Q. Ding, Y. Song, H. Tang, L. Wang, and J. Zhao, "Predicting temperature of permanent magnet synchronous motor based on deep neural network," *Energies*, vol. 13, no. 18, p. 4782, Sep. 2020.

[13] P. Irasari, H.Syaeful A, M. Kasim, " Thermal analysis on radial flux permanent magnet generator (PMG) using finite element method", *IPTEK, The Journal for Technology and Science*, Vol. 22, No.2, May 2011

[14] A. Mohammed, S. Djurović, "Stator winding internal thermal monitoring and analysis using *in situ* FBG sensing technology", *IEEE Transactions on Energy Conversion*, Vol.33, No.3, September 2018

[15] O. Bilgin , F. A. Kazan, "The effect of magnet temperature on speed, current and torque in PMSMs," 2016 XXII International Conference on Electrical Machines (ICEM), 2016, pp. 2080-2085, doi: 10.1109/ICELMACH.2016.7732809.

[16] D. Fernandez, D. Hyun, Y. Park, D. D. Reigosa, S. B. Lee, D-M. Lee, F. Briz, "Permanent magnet temperature estimation in PM synchronous motors using low cost hall effect sensors," 2016 IEEE Energy Conversion Congress and Exposition (ECCE), Milwaukee, WI, 2016, pp. 1-8, doi: 10.1109/ECCE.2016.7855349.

[17] A. Specht and J. Böcker, "Observer for the rotor temperature of IPMSM," Proceedings of 14th International Power Electronics and Motion Control Conference EPE-PEMC 2010, Ohrid, Macedonia, 2010, pp. T4-12-T4-15, doi: 10.1109/EPEPEMC.2010.5606818.

[18] S. B. Lee, T. G. Habetler, R. G. Harley and D. J. Gritter, "An evaluation of model-based stator resistance estimation for induction motor stator winding temperature monitoring," IEEE Transactions on Energy Conversion, vol. 17, no. 1, pp. 7-15, March 2002, doi: 10.1109/60.986431.

[19] S.D.Wilson, P.Stewart, B.P.Taylor, "Methods of resistance estimation in permanent magnet synchronous motors for real-time thermal management", IEEE Transactions on Energy Conversion, 25(3):698-707, October 2010

Simulation design of output-stage for residential smart-grid

Kristian Takacs, Peter Drgona
Department of mechatronics and electronics, Faculty of electrical engineering and information technologies
University of Zilina
Zilina, Slovakia
kristian.takacs@feit.uniza.sk

Abstract—This paper is an overview of a preliminary attempt at the modeling procedure of a bi-directional power converter system for design an output-stage of a residential smart-grid based solid-state transformer. The aim was to develop a simulation model, which has been performed with a PLECS circuit simulator. The studies of this manuscript dealt with the analysis and design of bi-directional power converters and their control strategies based on the requirements of the residential smart-grid power flow. Moreover, the simulation model was performed for 10 kW power flow in both directions and results from the simulation such as total system efficiency, power factor, or total harmonic distortion were evaluated.

Keywords—smart-grid, control strategy, phase shift modulation, space vector PWM modulation, dual active bridge, voltage source inverter.

I. INTRODUCTION

As a result of the growing penetration of green energy sources and modern big loads, such as an electric vehicle (EV) charging stations, many operational and technical challenges in distribution grids have emerged [1] – [3]. Recent innovations, using power electronics systems, enable the control of electric power flow in a wide range of environments, allowing for the implementation of the more consider smart-grid concept [4] – [5]. Smart-grids are also appropriate for situations involving a large variety of generators (photovoltaic panels, wind generators) and energy storage units (batteries, electric vehicles). At the same time, a two-way energy transfer and control is needed.

The output stage of a smart-grid is an intermediate segment between the smart-gird and the AC grid. In this manuscript, the output-stage is in for of a three-stage solid-state transformer (SST). Due to the smart features offered, the three-stage SST topology is the most common type in research compared to the other SST topologies [6]. This SST does have a lower volume and weight, and it can also optimize the efficiency of distribution and transmission grids [7]. The two DC links in the three-stage SST topologies are capable of addressing power quality (PQ) issues as well as supplying and using any devices connected to those DC links [8]. In terms of voltage control, current limit, protection, and power factor, three-stage topology outperforms one-stage and two-stage topologies [9].

The requirements of the proposed output stage are lower volume of a transformer and galvanically isolate a DC-bus and the mains three-phase grid to avoid critical scenarios. For that purpose, this manuscript describes the application of a solid-state transformer (SST). The used three-stage topology of SST is in form of a cascaded connection of a Dual Active Bridge (DAB) and a Voltage Source Inverter (VSI) as an output-stage for the residential smart-grid.

The purpose of the DAB is to act as a bi-directional isolated buck-boost converter. A VSI is used as a three-phase power inverter or as an active rectifier for allowing a bi-directional power flow.

II. ANALYSIS OF THE OUTPUT-STAGE OF RESIDENTAL SMART-GRID

The design requirements are set from a previously designed smart-grid according to [4]. The planned smart-grid block scheme is shown in figure Fig.1. A common photovoltaic panel installation with an energy storage unit for residential use is depicted in this block diagram. The power level of the desired smart-grid is estimated approximately to 10 kW.

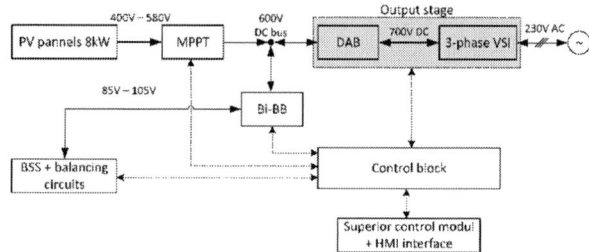

Fig. 1. Block diagram of designed residental smart-grid solution

The planned smart-grid includes several power converters, energy source, and storage elements such as; PV panels, Maximum Power Point Tracking (MPPT) converter, energy management block, battery storage system (BSS), bi-directional DC-DC converter (Bi-BB), communication and control blocks, and the output stage block, that is consisting of a DAB and 3-phase VSI converters.

A. Dual active bridge converter

A DAB converter is a galvanic isolated power converter with a high-power density and efficiency [10], [11]. It is made up of two active power switching H-bridges and a high-frequency transformer. The volume and direction of power flow are regulated by the phase shift between the two H-bridges, which operate at a set percentual duty ratio. Since it decreases the weight and volume of passive magnetic systems, a high-frequency transformer is used in conjunction with high-frequency switching devices. The high-frequency transformer has some leakage inductance in its primary and secondary windings, which together serve as an energy storage component, in addition to galvanic isolation.

978-1-6654-3236-8/21 $31.00 © 2021 IEEE

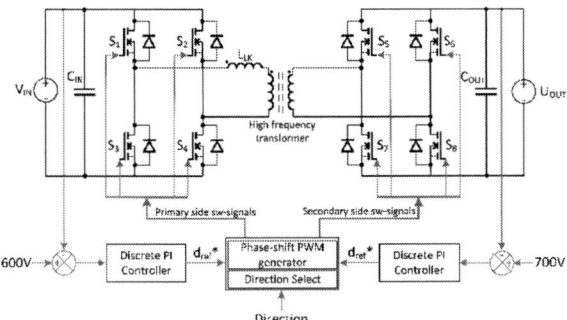

Fig. 2. Schematic configuration of the Dual Active Bridge converter with a bi-directional control structure

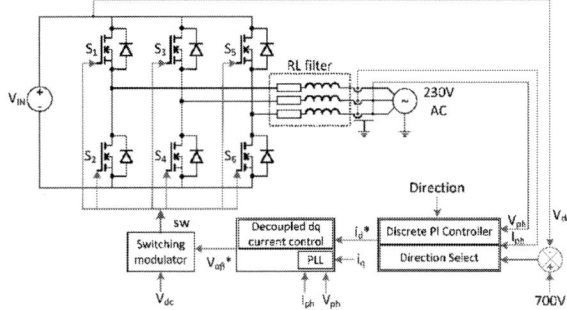

Fig. 3. Schematic configuration of the Voltage Source Inverter with bi-directional control structure

Figure Fig. 2 shows a schematic of a one-phase Dual Active Bridge dc-dc converter with the proposed control unit. Since it allows for zero-voltage switching (ZVS), bidirectional power transfer, and lower component stresses, the DAB topology is an attractive type of converter [10], [11]. The input voltage is converted to an intermediate high-frequency AC voltage by the primary side H-bridge. The high-frequency square wave AC voltage is converted back to a DC output voltage by secondary side H-bridge [8] - [10]. According to [11] the output power can be described as:

$$P = V_{out}I_{out} = \frac{(1 - |d|)dTV_{in}V_{out}}{nL_{Lk}} \quad (1)$$

With d represents the phase shift duty percentage of the primary and secondary bridges, n represents the turns ratio of the high-frequency transformer, T represents the duration of a half-cycle of the period, and L_{LK} is the leakage induction of the high-frequency transformer. The expression (1) shows that the power delivered to the output is a function of the phase shift between the two bridges, the converter's switching frequency, and the energy transfer inductance.

The DAB converter is controlled by Phase-Shift Modulation (PSM) method. Controlling the phase shift duty percentage between those H-bridges controls both the direction and the volume of power flow. Changing the secondary bridge pulses by +d instantiates power delivery from the primary bridge to the secondary bridge, taking into consideration the control pulses for switches S1,4 of the primary bridge and S5,8 of the secondary bridge. Similarly, shifting the secondary bridge by -d makes it the leading bridge, resulting in power being delivered to the primary bridge [12] – [13]. The bi-directional control is designed with discrete PI type regulators with a negative feedback loop for each power flow direction.

B. Three-phase Voltage Source Connverter

Voltage Source Inverters are widely used in many grid-connected applications, and their control systems directly regulate their input current or provide an internal current feedback loop [14], [15]. In this manuscript, a vector control strategy of the grid-tied VSI converter was chosen. The proposed control structure consists of decoupled dq-current control, three discrete PI regulators, and phase-locked loop (PLL) modules. The topology of the three-phase VSI converter with the control block diagram structure is shown on figure Fig.3.

An inner *dq*-current controller receives a current setpoint from an outer discrete voltage controller implemented as a PI regulator. The discrete voltage controller estimates the error by comparing the DC-link voltage with a reference value (700 V_DC). Separate discrete PI regulators for direct and quadrature currents produce a V_{dq} reference in the current controller. The voltages and currents in each AC phase are calculated and fed into the current controller. The reference phase angle for the *abc* to *dq* transformations is generated by a PLL.

III. PARAMETRIC SPECIFICATION OF THE SIMULATION MODEL

According to the previous section and designed smart-grid the design of the proposed output-stage SST consists of a cascade combination of DAB and VSI bi-directional converters. The designed output stage specification is given in table Tab.1. Selecting the switching devices was considered about their electrical specifications, such as peak voltage and current ratings, R_{ds-on} resistance, and the stresses they must tolerate. As a power switch device for DAB and VSI converters was used the C3M0030090K SiC-based high-frequency MOSFET transistor with fast intrinsic diode.

TABLE I.   SPECIFIACATION OF THE OUTPUT-STAGE CONVERTER SYSTEM

Parameter	Value/part
Dual active bridge	
Primary side DC voltage	600 V
Secondary side DC voltage	700 V
Transformer turn ratio	1.167
Switching frequency	250 kHz
Leakage inductance of transformer	12 µH
Output capacitance	600 µF
Input capacitance	30 µF
Three-phase Voltage Source Inverter	
Input DC voltage	700 V
3-phase AC voltage (RMS) output	3 x 230 V
Output frequency	50 +/- 0.25 Hz
Switching frequency	25 kHz
Filter inductance	1.5 mH
Equivalent resistance of the L filter	5 mΩ

For detailed simulation result a high-frequency transformer model with included saturation behavior of the DAB converter was used [16] – [17]. The transformer was modeled with the PLECS magnetic domain library. The magnetic model's parameters are directly linked to the structure and material properties of the core. TDG's TP4A ferrite material and industry standard E43/28 core are used in the transformer.

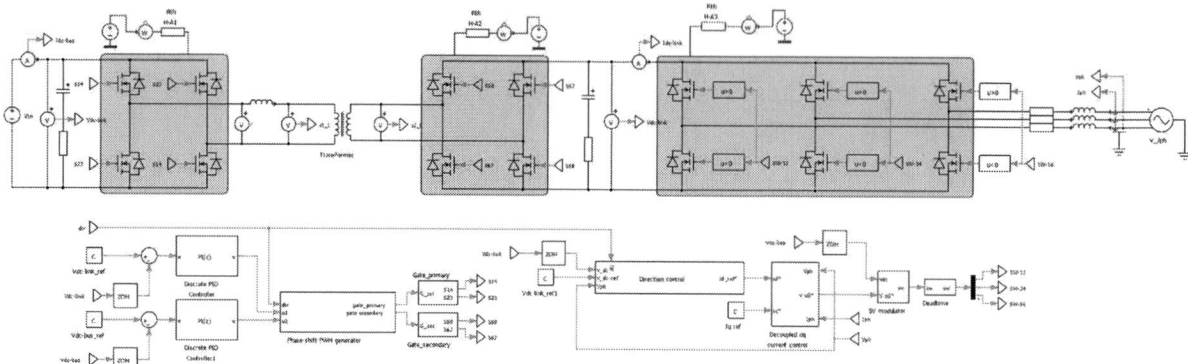

Fig. 4. Simulation model of the output-stage for resitental smart-grid

The leakage inductance of the high frequency transformer is calculated by the expression (2):

$$L_{Lk} = \frac{(1 - |d|)dV_{in}V_{out}}{2f_s n P_{max}} \qquad (2)$$

Where the maximum output power (P_{max}) was set to 10 kW. The value of the duty cycle was set to 0.5, because of the power transfer characteristic of DAB converter [10].

IV. SIMULATION RESULTS

The focus at this point is on the evaluation of the SST simulation model. Figure Fig.4 shown the simulation model of the output-stage of the desired smart-grid. The loss models of the chosen semiconductors in the converter prototypes were used to model them. It is possible to obtain the correct behavior of circuit schematics that are modeled using a linearized approach using this method [14], [15].

$$PF = \frac{P_{out}}{V_{out} \cdot I_{out}} \qquad (3)$$

$$System\ efficiency = \frac{P_{out}}{P_{in}} \cdot 100\% \qquad (4)$$

$$THD_I = \frac{\sum_{i=n}^{\infty} I_{out\ i}}{I_{out\ 1}} \qquad (5)$$

The simulation of the SST model as a cascade connection of DAB and VSI was tested in two scenarios. In bought scenarios, the value of DC bus was set to 600 V. The Global direction signal is set to bought subordinate controllers of converters. This signal evaluates the DC bus voltage and undervoltage and switches off the bought converters when any limits are exceeded. For evaluation of the simulation model, the power factor (PF), system efficiency, and total harmonic distortion (THD) were considered.

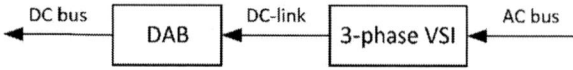

Fig. 5. Block diagram of power flow in scenario n. 1

In the first scenario, the direction of energy flow was set from the three-phase AC source to the DC bus (see Fig. 5). The three-phase AC source supplied the DC bus with a 10 kW power limit. The DC-link voltage is regulated by the VSI as an active rectifier and the DC bus voltage is regulated by the primary side's PI controller of DAB.

Figures Fig. 6-8 shows the main variables of the inspected converter system i.e., DC bus voltage, and current, DC-link voltage and current, AC grid currents, voltages, real and reactive power. For the first scenario the PF is calculated by expression (3), where the P_{out} is the output power, I_{out} is output current and V_{out} is output voltage of the primary side of DAB. The system efficiency is calculated by the ration of VSI's input power and output power of BAD's primary side (see expression (4)). The THD is calculated by the output currents of BAD's primary side (see expression (5)).

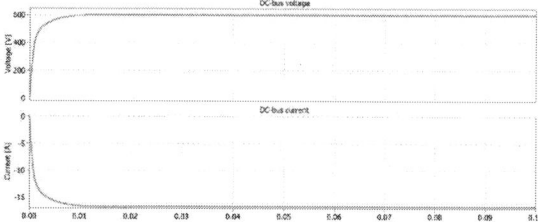

Fig. 6. DC-bus voltage and current in scenario n. 1

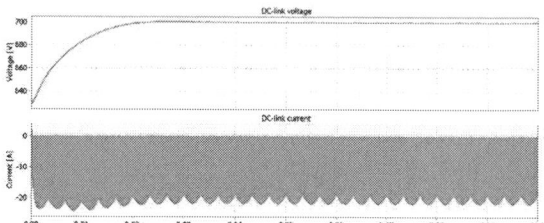

Fig. 7. DC-link voltage and current in scenario n. 1

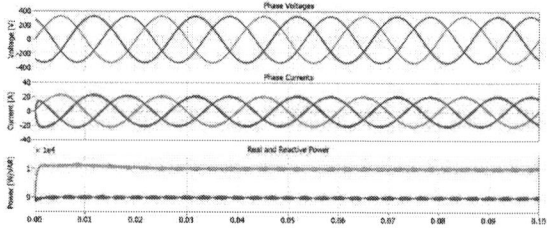

Fig. 8. Main variables of the AC grid in scenario n. 1

978-1-6654-3236-8/21 $31.00 © 2021 IEEE

Table Tab. 2 shows the results of the qualitative indicators of the concept in scenario n. 1. It can be seen, that the simulation result of total system efficiency is 79.86 % and the value of total harmonic distortion reaches 3.689.

TABLE II. EVALUATION OF THE OPERATIONAL VARIABLES OF THE SIMULATED OUTPUT-STAGE DURING SCENARIO N. 1.

Variable	Value
V_{BUS} [V]	599.992
ΔV_{BUS} [V]	0.375
I_{Bus} [A]	16.664
ΔI_{Bus} [A]	0.037
$V_{DC\text{-}link}$ [V]	699.933
$\Delta V_{DC\text{-}link}$ [V]	0.247
V_{grid} [V RMS]	230
PF [-]	0.999714
System efficiency [%]	79.86
THD [-]	3.689

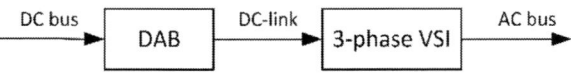

Fig. 9.   Block diagram of power flow in scenario n. 2

In the second scenario, the direction of energy flow was set from the DC bus to the three-phase AC source (see Fig. 5). The DC-bus supplied the AC source with a 10 kW power limit. The DC-link voltage is regulated by the secondary side's PI controller of DAB and the id* current value for VSI is calculated by predetermined power and voltage reference of AC bus.

Figure Fig. 10-12 shows the main variables of the inspected converter system i.e., DC bus voltage and current, DC bus voltage, and current, AC grid currents, voltages, real and reactive power. For the second scenario the PF is calculated by expression (3), where the P_{out} is the output active power, I_{out} is output current and V_{out} is output voltage of the VIS converter. The system efficiency is calculated by the ration of input power of BAD's primary side and output power of VSI (see expression (4)). The THD is calculated by the output currents of VSI converter (see expression (5)). Table Tab. 3 shows the results of the qualitative indicators of the concept in scenario n. 2.

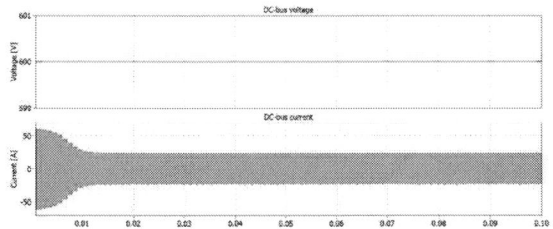

Fig. 10. DC-bus voltage and current in scenario n. 2

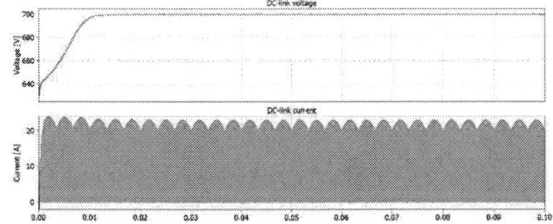

Fig. 11. DC-link voltage and current in scenario n. 2

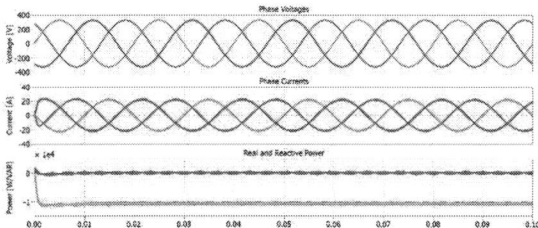

Fig. 12. Main variables of the AC grid in scenario n. 1

TABLE III. EVALUATION OF THE OPERATIONAL VARIABLES OF THE SIMULATED OUTPUT-STAGE DURING SCENARIO N. 2.

Variable	Value
V_{BUS} [V]	600
ΔV_{BUS} [V]	0
I_{Bus} [A]	19.48
ΔI_{Bus} [A]	46.452
$V_{DC\text{-}link}$ [V]	700.047
$\Delta V_{DC\text{-}link}$ [V]	0.858
V_{grid} [V RMS]	230
PF [-]	0.999736
System efficiency [%]	85.74
THD [-]	3.495

In scenario n. 2, the DC-bus of the smart-grid was modeled as a voltage source with the voltage set to 600 V. It can be seen, that the simulation result of a total system efficiency is 85.74% and the value of total harmonic distortion reaches 3.495. The relatively high efficiency in both power flow directions is partly achieved by DAB converter with high-frequency transformer.

V. CONCLUSION

This paper describes a design of a simulation model of the bi-directional output-stage for a residential smart-grid. The converter system consists of a cascade connection of a dual active bridge converter and a three-phase voltage source inverter in form of a solid-state transformer configuration. For each power converter, a control strategy for bi-directional power flow was designed. The direction of the power flow is defined and set by the superior control module. The main focus was to design an output-stage converter system for the smart-grid with the aim of 10 kW bi-directional power flow.

The system efficiency of designed output-stage reached 79.86% for the first operational scenario and 85.74% for second operational scenario. In case of sole VSI, the system efficiency achieved higher values in both scenarios (86.28% for first scenario and 91.61% for second scenario) in comparison with cascade connection of DAB with VSI. But the DAB converter in the proposed output-stage of smart grid has an advantage of galvanically isolate the DC bus from AC bus. Another advantage of a smart grid based on SST is the voltage correction and compensation of DC-link and DC bus.

It should be mentioned that the results were gained with a simulation model. The proposed control strategy could be developed further by focusing on different types of loads that occur within the AC grid or DC bus, as well as an impact of power-factor correction circuits or an impact of an EV charging.

ACKNOWLEDGMENT

This research was funded by VEGA 1/0593/20 Research of power flow control in smart grid using smart transformer.

References

[1] J. Sallan, J. L. Villa, A. Llombart and J. F. Sanz, "Optimal Design of ICPT Systems Applied to Electric Vehicle Battery Charge," in IEEE Transactions on Industrial Electronics, vol. 56, no. 6, pp. 2140-2149, June 2009. doi: 10.1109/TIE.2009.2015359.

[2] W. Zhang and C. C. Mi, "Compensation Topologies of High-Power Wireless Power Transfer Systems," in IEEE Transactions on Vehicular Technology, vol. 65, no. 6, pp. 4768-4778, June 2016. doi: 10.1109/TVT.2015.2454292.

[3] J. Koscelnik, M. Frivaldsky, M. Prazenica and R. Mazgut, "A review of multi-elements resonant converters topologies," 2014 ELEKTRO, 2014, pp. 312-317, doi: 10.1109/ELEKTRO.2014.6848909.

[4] Frivaldsky, M.; Morgos, J.; Prazenica, M.; Takacs, K. System Level Simulation of Microgrid Power Electronic Systems. Electronics 2021, 10, 644. https://doi.org/10.3390/electronics10060644.

[5] M. A. Hannan et al., "State of the Art of Solid-State Transformers: Advanced Topologies, Implementation Issues, Recent Progress and Improvements," in IEEE Access, vol. 8, pp. 19113-19132, 2020, doi: 10.1109/ACCESS.2020.2967345.

[6] L. F. Costa, G. De Carne, G. Buticchi, and M. Liserre, ''The smart transformer: A solid-state transformer tailored to provide ancillary services to the distribution grid,'' IEEE Power Electron. Mag., vol. 4, no. 2, pp. 56–67, Jun. 2017.

[7] D. Das and C. Kumar, ''Operation and control of smart transformer based distribution grid in a microgrid system,'' in Proc. Nat. Power Electron. Conf. (NPEC), Dec. 2017, pp. 135–140.

[8] R. De Doncker, D. Divan, and M. Kheraluwala, "A three-phase soft-switched high-power-density dc/dc converter for high-power applications," IEEE Transactions on Industry Application, vol. 27, pp. 63–73, Jan./Feb. 1991.

[9] Frivaldsky, M.; Hanko, B.; Prazenica, M.; Morgos, J. High Gain Boost Interleaved Converters with Coupled Inductors and with Demagnetizing Circuits. *Energies* 2018, *11*, 130. https://doi.org/10.3390/en11010130

[10] Qin, Hengsi, "Dual active bridge converters in solid state transformers" (2012). Doctoral Dissertations, https://scholarsmine.mst.edu/doctoral_dissertations/1914.

[11] M. N. Kheraluwala, R. W. Gascoigne, D. M. Divan, and E. D. Baumann, "Performance characterization of a high-power dual active bridge," IEEE Transactions on Industry Application, vol. 28, pp. 1294–1301, Jun. 1992.

[12] George, K., Design and Control of a Bidirectional Dual Active Bridge DC-DC Converter to Interface Solar, Battery Storage, and Grid-Tied Inverters, 2015, Electrical Engineering Undergraduate Honors Theses Retrieved from https://scholarworks.uark.edu/eleguht/45.

[13] Frivaldsky, M.; Spanik, P.; Morgos, J.; Pridala, M. Control Strategy Proposal for Modular Architecture of Power Supply Utilizing LCCT Converter. *Energies* 2018, *11*, 3327. https://doi.org/10.3390/en11123327

[14] Chenhao Nan; Ayyanar, R., "Dual active bridge converter with PWM control for solid state transformer application," in Energy Conversion Congress and Exposition (ECCE), 2013.

[15] M. Kazmierkowski and L. Malesani, "Current control techniques for three-phase voltage-source PWM converters: A survey," IEEE Trans. Ind. Electron., vol. 45, no. 5, pp. 691–703, Oct. 1998.

[16] Radvan, R., Dobrucky, B., Frivaldsky, M., & Rafajdus, P. (2011). Modelling and Design of HF 200 kHz Transformers for Hard- and Soft-Switching Application. Elektronika Ir Elektrotechnika, 110(4), 7-12. https://doi.org/10.5755/j01.eee.110.4.276

[17] Turzynski, M.; Chrzan, P.J. Behavioural model of IGBT transistor. Przeglad Elektrotechniczny, 2010, R. 86, nr 2, 86.

[18] B. Bahrani, A. Karimi, B. Rey and A. Rufer, "Decoupled dq-Current Control of Grid-Tied Voltage Source Converters Using Nonparametric Models," in IEEE Transactions on Industrial Electronics, vol. 60, no. 4, pp. 1356-1366, April 2013, doi: 10.1109/TIE.2012.2185017.

[19] Turzyński, M.; Banach, P.; Murawski, P.; Peplinski, R.; Chrzan, P. A predictive estimation based control strategy for a quasi-resonant dc-link inverter. Bull. Pol. Acad. Sci. Tech. Sci. 2013, 61, 757–762, doi:10.2478/bpasts-2013-0081.

Evaluation of high-voltage GAN transistor in 3-phase PFC circuit

Michal Frivaldsky, Michal Pipiska

Department of mechatronics and electronics, Faculty of electrical engineering and information technologies
University of Zilina
Zilina, Slovakia
michal.frivaldsky@feit.uniza.sk

Abstract—**The aim of this paper is evaluation of new 1200 V GaN power transistor technology within selected target application circuit. As a evaluating circuit, 3-phase PFC in dual interleaved topology was selected. Consequently, identification of commercially available 1200 V GaN structure was realized. Selected transistor was implemented withing PFC circuit and evaluated. Testing under nominal operational parameters revealed that developed 1200 V GaN power transistor has lack of performance to reliably operate at industrial application due to missed optimization and testing during development phase. Asa a result it was concluded that 1200 V GaN technology still requires additional research and development time to successfully compete robust SiC transistor structures.**

Keywords—component, formatting, style, styling, insert (key words)

I. Introduction (*Heading 1*)

Continuous development in the field of power semiconductors is heading power electronic systems to the new dimensions characterized by improved efficiency, power density and thermal performance. All these aspects are related to relationships that must be conceived in the form of a compromise during development. One of the ways how to optimize solutions is research and development of a completely new system. Another way is substitution of the main components and parts of the existing system using new technological materials and structures. Hereby we are discussing about upcoming trend, where silicon power transistors are continuously substituted with the perspective components based on SiC and GaN technology. The process should be sometimes simple, just exchange the part for part, but several solutions requires additional modification to the circuit thus making process more complex and costly. Regarding the fact that each electronic device requires power factor correction, it is worth to consider the efficiency performance of this systems at all. Currently SiC transistors are popular for modern PFC circuits, however there is also GaN technology which should undergo investigation whether it is not better solution to improve operational characteristics of PFC themselves.

Considering requirements on the voltage blocking capability of the power transistors if 3-phase system is designed, here we talk about 900 Vdc or 1200 Vdc level for V_{DS} rating of the transistor itself As was mentioned previously, SiC and GaN are currently leading technologies for optimization of the properties of power electronics systems. Regarding high-voltage blocking capability, SiC offers several verified solutions and manufacturers are providing components with voltage rating up to 1600 Vdc. On the other side GaN technologies are continuously id development while verified solutions are represented by devices with 600 Vdc blocking voltage capability. There are some manufacturers with devices declaring 900 Vdc or even

1200 Vdc. Thus, perspectives for improvements of switching dynamics even for high voltages are on the technological horizon.

Based on these facts, in this paper, the evaluation of high voltage GaN power transistor with blocking capability of 1200 Vdc is evaluated within the power circuit of 3-phase dual interleaved boost type PFC.

II. Operational Parameters of the 3-Phase PFC Circuit

A. PFC topology and nominal parameters

The physical prototype of the PFC converter, whose circuit diagram of the main circuit is shown in Figure 1, is shown in Figure 2.

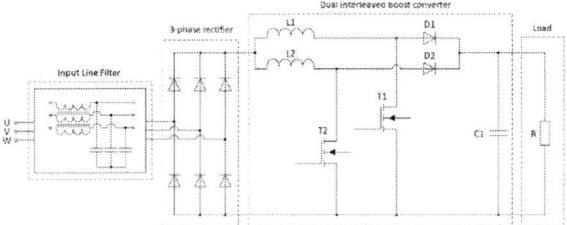

Fig. 1. Principle schematic of considered 3-phase PFC circuit

Fig. 2. Phyysical prototype of considered 3-phase PFC converter

It is clear from the parameters listed in Table 1 that the industrial use of such a system is predestined for the universal grid network. Therefore, GaN transistors (Table II), whose blocking voltage is at the level of 1200 Vdc will be analysed.

TABLE I.   PARAMETERS OF PFC CONVERTER

Input Voltage	180-528V AC (3L-PE)
Output Voltage	800V DC
Output Power	4 kW
Switching Frequency	41 kHz
Converter Topology	Double interleaved boost converter
Cooling type	Water cooled aluminum heatsink

978-1-6654-3236-8/21 $31.00 © 2021 IEEE

B. Analysis of available 1200 V transistors in GaN technology

By analyzing the semiconductor components market, we found that there are currently very few manufacturers of GaN transistors on the market with a terminal voltage greater than 650V. Our application requires a transistor terminal voltage greater than 1000V to secure the 20% voltage reserve required in practical applications. In Table III. lists the manufacturers and types of high voltage GaN transistors produced.

TABLE II.   OVERVIEW OF 1200 V GaN POWER TRANSISTORS

Manufacturer	Transistor type	Note
VisIC	VM40HB120D	Unavailable, manufacturer released all samples
IGaNPOWER	GPIHV30SB5L	Non-optimized transistor, datasheet lacks informations about limiting and SOA values
Transphorm	TP90H180PS	Low voltage margin for target application parameters.

From Table II is seen, that from the selected types of transistors the most suitable variant for testing of 1200V GaN technology is the transistor GPIHV30SB5L. However, the data sheet does not contain data such as the maximum power load of the transistor chip, the maximum pulse current flowing through the transistor under specific conditions, SOA (safe operating area) and other parameters needed to determine the possible transistor load. The transistor is supplied in a TO-263 package in a 5-pin variant, where the middle terminal is cut off at the case by the manufacturer (Fig. 3). However, this package is unsuitable for mass-produced power supplies operating at voltages greater than 200V. I tis due to the fact that the cut-off terminal from the housing slightly protrudes and is internally connected to the cooling surface, thus being connected to the source electrode of the transistor. As a result, it is located at a smaller distance from the drain electrode than the recommended distance for these voltage levels. In this way, industry standards are violated, which significantly increases the risk of shorting the terminals through arcing, which can ultimately cause the transistor to destroy. When mounting the transistor on a printed circuit board, there is a risk of tin particles evaporating between the terminals of the transistor, which could again lead to an electric arc. Basic electrical parameters of selected transistor are listed in Table III.

TABLE III.   ELECTRICAL PROPERTIES OF IGANPOWER – GPIHV30SB5L

Parameter	Value
V_{DS}	1200 V
$R_{DS(ON)}$	65 mΩ
I_{DS}	30 A
Qg	8.25 nC

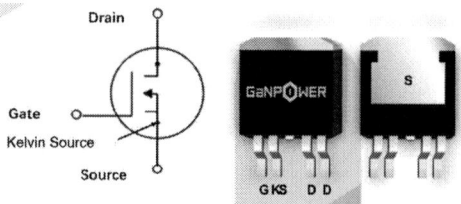

Fig. 3.   1200 V GaN transistor GPIHV30SB5L

C. Practical implementation within considered PFC circuit

The PFC converter contains the original transistors in the TO-247 package design. As we tried to avoid the need to produce a forged PCB due to the cooling of the SMD package of the GaN transistor, we decided on a simpler and more efficient variant. The GaN transistors were mounted on a tinned copper plate, adapting the terminals to the printed circuit board so that they could be mounted in the original locations for the transistors (Fig. 4). We subsequently tested the transistor with a DC high-voltage non-destructive power supply for 1200V.

Fig. 4.   Comparison of original SiC transistor in TO-247 package with modified GaN transistor

The original transistor driver, which was designed for SiC transistors, generated driving pulses in the range of 0 to 15 volts at the output. For transistors with GaN technology, which are not in cascade connection with a MOSFET transistor, the gate voltage levels should be within the range of 0V to 6V. For this reason, the original driver was modified in the way that the control pulses are in the range of 0V to 5V. We used a 5.6V Zener diode to protect against voltage surges at the transistor´s gate input. The mounted transistors in the PFC converter are shown in Figure. 5.

Fig. 5.   Interpretation of mounted 1200 V GaN transistors in PFC converter

III. EXPERIMENTAL VERIFICATION WITHIN OPERATION OF CONSIDERED PFC CIRCUIT

Initial testing of selected GaN transistor was performed only as an informative measurement at reduced input and output voltage. The value of the input combined voltage was 190V and the value of the output voltage was 400V. The test was performed up to an output power of the PFC converter of 800 W and a switching frequency from 40 kHz to 120 kHz. The measured graphs of efficiency and losses as a function of output power are shown in Figs 6 and Fig. 7. From the obtained results it is clear that in PFC converter does not exhibit any significant problems in terms of power losses if reduced operational parameters are considered within the main circuit. It is also clear from the results that the increase in frequency leads to a decrease in efficiency, what is in line with the theory of the increase of switching losses of transistors. However, it can be stated that the difference in maximum efficiency was in the range of 0.5%.

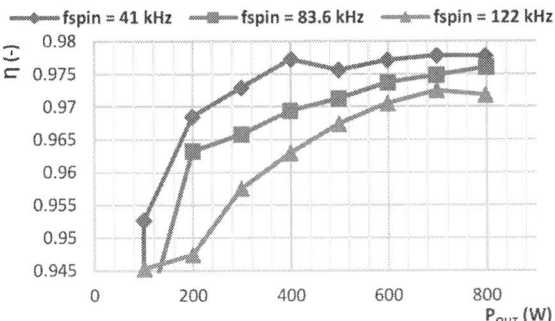

Fig. 6. Efficiency waveform at reduced operational parameters of PFC converter equiped with high-voltage GaN transistor GPIHV30SB5L

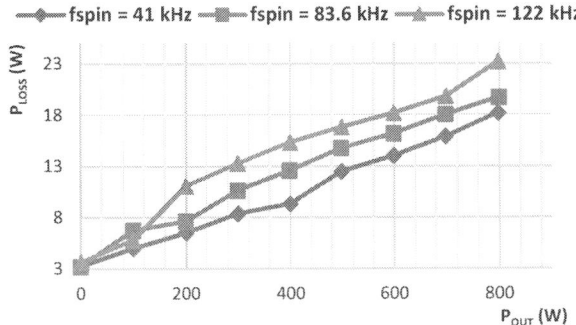

Fig. 7. Power losses of PFC converter operating at reduced operational parameters equiped with high-voltage GaN transistor GPIHV30SB5L

Another test was performed at the nominal input and output voltages. After setting the voltage feedback to the required value (800 V) and after turning on the operation of proposed PFC converter, one of the GaN power transistors was destroyed (Fig.8). In this case, it is expected that an electric arc raised between the drain and the source, which caused the that one of the transistor terminals evaporated and thus destroyed the transistor itself. This problem can be eliminated by increasing the creepage distance between these terminals. One of the possible ways is to apply liquid isolation for example in the form of application of glue Loctite HYSOL (Fig. 8 - right).

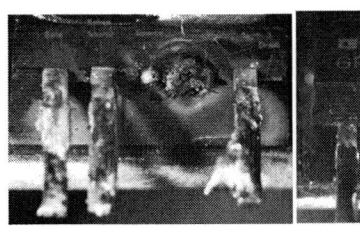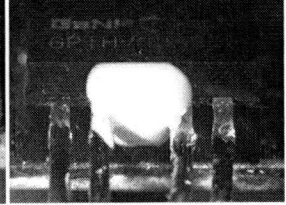

Fig. 8. Destroyed transistor GPIHV30SB5L due to electric arc between drain and source (left), possible way how to increase voltage sustainability (right)

By gluing the middle terminal, the voltage sustainability between the Drain and Source terminals was increased. After

repeated restart of the PFC converter, the converter operated up to an output voltage of 800V. During gradual increase of converter power, we reached a value of 700W. At this point one of the transistors was destroyed again due to unknown reasons. At an output power of 700W, the temperature of the package of the transistor was 35 °C, while voltage waveforms at the gate-source and drain-source have been recorded without any oscillations during turn-on / turn-off process.

After this unsuccessful testing, the manufacturer of the selected high-voltage GaN transistor was contacted. Main reason was to be informed more in detail about the limiting values of the operational parameters of transistor. By inspection of the manufacturer's information it was revealed that the selected type of high voltage GaN transistor was not subjected to any detailed testing in dynamic modes, i.e., within switching process.

Based on achieved results, it was found that the R&D process in the field of high-voltage GaN transistors in terms of their practical application remains relevant and intensively discussed problem. Therefore, for applications where blocking capability above 600 V is required, SiC or Si transistor structures are the only possibility of the utilization. Within the technology development and manufacturers effort, it is expected that GaN will be another choice for high voltage applications, however it needs more time and careful testing of developed structures to become more reliable and robust solution.

IV. PFC CONVERTER MODIFICATION FOR POSSIBILITIES OF GaN TRANSISTOR TECHNOLOGY IMPLEMENTATION

The original PFC converter was adapted considering input/output parameters to be able to utilize available GaN power transistors. Currently, 600 V power transistors based on this technology are almost well experimentally verified and tested. The changes of PFC converter mainly concern the value of the output voltage, which will be reduced to 480V and the value of input voltage, which will have a value of 210V phase-to-phase (Table IV.).

TABLE IV. MODIFICATION OF INPUT – OUTPUT PARAMETERS OF PROPOSED PFC CIRCUIT

Input voltage	210V AC (3L-PE)
Output voltage	480V DC
Output power	4 kW

From parameters listed in Table IV is clear that modified PFC will not be applicable to the European grid, but will be applicable to the US network, where grid parameters are like the modified ones. Reducing the output voltage will allow us to analyze the use of 600V to 650V GaN transistors, of which there are several types on the market from various manufacturers. For testing, two types have been selected from two different manufacturers, which will be analyzed in terms of evaluating the efficiency of the modified PFC converter.

TABLE V. PREVIEW OF THE SELECTED 600 V GaN TRANSISTORS FOR TESTING

Transistor ID	Manufacturer	Technology	$U_{DS\,max}$	Package
TP65H050WS	Transphorm	Cascode GaN	650V	TO247-3
IGT60R070D1	Infineon	CoolGaN	600V	PG-HSOF-8-3

A. Switching performance and efficiency evaluaion

The first type of transistor tested is TP65H050WS from Transphorm. Since it is a Cascode GaN type, the advantage is that it is not necessary to use a special type of driving circuit, but it is possible to use conventional driving circuits for MOSFET transistors. On the other hand, a limitation in terms of switching dynamics exists, which is influenced by the MOSFET transistor. This is considered as disadvantage.

Fig. 9 shows turn-on process of the TP65H050WS transistor, and Fig. 10 turn-off process. The figure shows the extent to which the unipolar MOSFET part of the Cascode GaN transistor has an on and off delay. The process from off to on and vice versa in this case takes approximately 13 ns - 14 ns. The waveforms also show the progressive switching of the individual parts of the Cascode connection. The delay of switching the GaN part after switching on the MOSFET part is clearly visible during turn-on and turn-off.

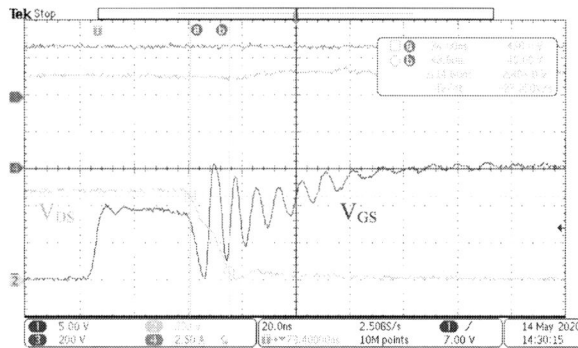

Fig. 9. Turn – on waveforms of TP65H060WS

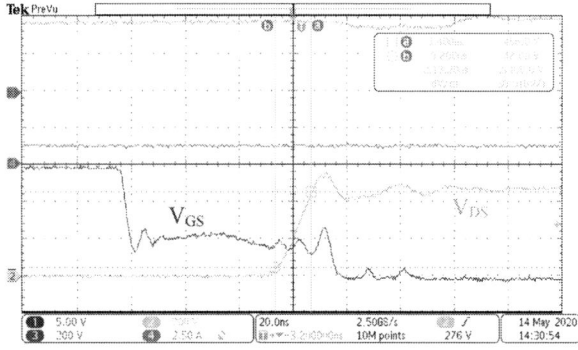

Fig. 10. Turn – off waveforms of TP65H060WS

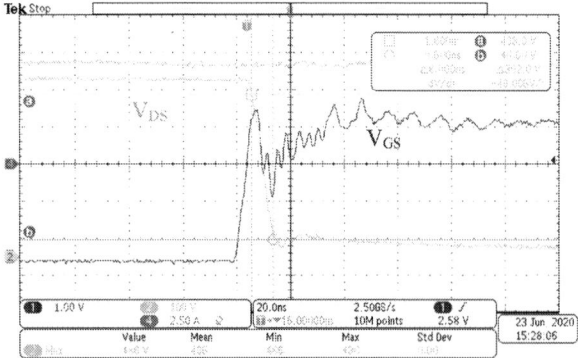

Fig. 11. Turn – on waveforms of IGT60R070D1

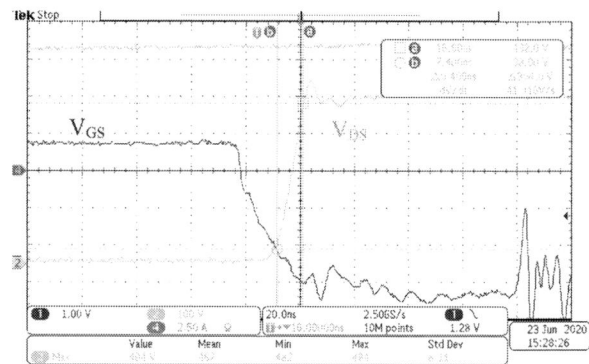

Fig. 12. Turn – off waveforms of IGT60R070D1

Naopak transistor od Infineon s označením IGT60R070D1 vyžaduje využitie špeciálne konfigurovaného budiaceho obvodu, pričom tento tranzistor je charakteristický prúdovým budiacim signálom. Pri analýze priebehov zapínacieho a vypínacieho procesu môžeme vidieť (Fig. 11 a Fig. 12) okamžité zapnutie/vypnutie bez oneskorenia budiaceho signálu. Čas prechodu z vypnutého stavu do zapnutého a naopak je na úrovni 8 ns – 9ns. Z tohto výsledku možno usúdiť vyššiu dynamiku druhého testovaného tranzistora.

In contrast, a transistor from Infineon called IGT60R070D1 requires the use of a specially configured driving circuit. This transistor is characterized by a current driving signal after transistor turns-on. When analyzing the on and off processes, it is possible to see (Fig. 11 and Fig. 12) immediate turn-on and turn-off without delay of the driving signal signal. The transition time from off to on and vice versa is within time interval of 8 ns – 9 ns. From this result, the higher dynamics of the second tested transistor can be concluded.

The following figure (Fig. 13) graphically interprets the efficiency curves for the selected 600 V GaN transistors. It can be seen from the results that the IGT60R070D1 CoolGaN transistor achieves an efficiency higher by approx. 0.3% for the entire operational power range of the modified PFC converter.

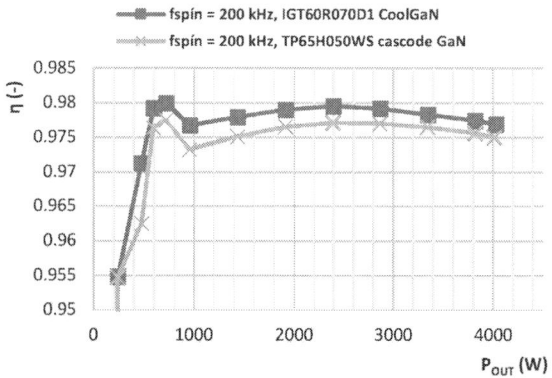

Fig. 13. Efficiency characteristic of modified PFC converter equiped with selected GaN transistors

The achieved results represent an excelent performance with respect to the operating frequency of 200 kHz and the power of the converter of 4 kW. Based on this, it can be stated that in the future GaN transistors will naturally replace SiC transistors in industrial power supplies as well. Currently, the biggest disadvantage of available GaN transistors is that they are not applicable to blocking voltages higher than 480 Vdc, making SiC technology a high priority for high voltage applications.

CONCLUSION

In this paper, the testing of the new high voltage power transistor structure was tested and evaluated. Initially target application device was specified, i.e. 3-phase PFC converter. For this application 1200 V transistors are required because the output voltage of PFC reaches 800 V. Currently there are not many transistors based on GaN technology, which are suited for such application. One of commercially available 1200 V GaN transistor was selected and implemented within proposed converter circuit. First tests have been provided under reduced input/output operational parameters to verify functionality. Later, after successful initial testing, PFC was adapted to its nominal parameters. Selected transistor must undergo simple modification related to increase of the creepage distance between its electrodes, however final tests had shown, that this technology is currently not well mannered and concrete component is not suitable for selected application even due to its datasheet values. Main problem was also related to dynamic tests of selected transistors, while they lacked during product development and release. However, in order to verify, how GaN transistors can influence operation of selected PFC, a 600 V structures have been selected for testing, while input/output operational parameters of PFC have been modified to US supply network /i.e. voltage levels have been reduced. As a main contribution of GaN transistor technology is high operational efficiency of target system app. 98%, even delivered power reached high values (4 kW) and switching frequency (200 kHz) was about standardly used value. For future purposes and requirements for high voltage applications, development of GaN must undergone relevant research to meet performances of SiC structures.

ACKNOWLEDGMENT

This research was funded by national grant agency VEGA 1/0063/21. Atuhors would like to thank to project support funded by APVV under Nr. 0396-15.

REFERENCES

[1] H.Z.Azazi, S.A.Mahmoud, S.S.Shokralla (2010): *Review of passive and active circuits for power factor correction in single phase, low power AC-DC converters.* Electrical engineering department, Faculty of engineering, Menoufiya university, Shebin El-Kom, Egypt

[2] B. Liu, R. Ren, E. Jones, F. Wang, D. Costinett and Z. Zhang, "A compensation scheme to reduce input current distortion in a GaN based 450 kHz three-phase Vienna type PFC," *2016 IEEE Energy Conversion Congress and Exposition (ECCE)*, 2016, pp. 1-7, doi: 10.1109/ECCE.2016.7854663.

[3] Hruska, Karel; Kindl, Vladimir; *Pechanek, Roman, Concept, Design and Coupled Electro-Thermal Analysis of New Hybrid Drive Vehicle for Public Transport*, In: Conference: 14th International Power

Electronics and Motion Control Conference (EPE-PEMC) Location: Ohrid, MACEDONIA Date: SEP 06-08, 2010

[4] Cacciato, Mario; Consoli, Alfio, New Regenerative Active Snubber Circuit for ZVS Phase Shift Full Bridge Converter, In: 26th Annual IEEE Applied Power Electronics Conference and Exposition (APEC) Location: Fort Worth, TX Date: MAR 06-11, 2011

[5] H. Zhu, B. Li, K. Wang and X. Yang, "Common-Mode EMI Modeling and Analysis for GaN-Based Full-Bridge CRM PFC under Unipolar PWM Scheme," *2020 IEEE Workshop on Wide Bandgap Power Devices and Applications in Asia (WiPDA Asia)*, 2020, pp. 1-5, doi: 10.1109/WiPDAAsia49671.2020.9360281.

[6] Attansio, R; Cacciato, M; Consoli, A; et al., *A novel converter system for fuel cell distributed energy generation, In:* Conference: 36th Annual IEEE Power Electronic Specialists Conference (PESC 05) Location: Recife, BRAZIL Date: JUN 12-16, 2005

[7] Do, N.-N.; Huang, B.-S.; Phan, N.-T.; Nguyen, T.-T.; Wu, J.-H.; Liu, Y.-C.; Chiu, H.-J. Design and Implementation of a Control Method for GaN-Based Totem-Pole Boost-Type PFC Rectifier in Energy Storage Systems. *Energies* 2020, *13*, 6297. https://doi.org/10.3390/en13236297

[8] Sylvain LECHAT SANJUAN (2010). Voltage oriented control of three-phase boost PWM converters. Master of science thesis in electric power engineering, 2010 Goteborg, Sweden, Chalmers university of technology.

[9] Microsemi PPG. Gallium Nitride (GaN) versus Silicon Carbide (SiC) in the high frequency (RF) and power switching applications

[10] T. Kachi : Current status of GaN power devices, TOYOTA Central R&D Labs. Japan Science and Technology Agency, November 2013

[11] J. W. Kolar, J. Muhlethaler. The essence of three-phase PFC rectifier systems, Swiss Federal Institute of Technology (ETH) Zurich, Power Electronic systems Laboratory

[12] X. Huang, W. Chen : An Improved Calculation Method of High-frequency Winding Losses for Gapped Inductors, Fujian University of Technology, July, 2017

[13] F. K. Wong, B. Eng., M. Phil. : High frequency transformer fo switching mode power supplies, School of Microelectronic Engineering, Griffith University, Brisbane, Australia, March 2004

[14] F. Zeng, J. X. An, G. Zhou, W. Li, G. Wang, T. Duan, L. Jiang, G. Yu : A comprehensive review of recent progress on GaN high electron mobility transistors : Devices, Fabrication and Reliability, Southern University of Science and Technology, December 2018

[15] Dr. John Schonberger. Space vector control of a three-phase rectifier using PLECS. Plexim GmbH Technoparkstrasse 1 8005 Zurich.

[16] H. O. Jimenez: AC resistance evaluation of foil, round and litz conductors in magnetic components, Master of Science thesis, Chalmers university of technology, Goteborg, Sweden 2013

[17] Ch. R. Sullivan, R. Y. Zhang : Simplified design method for Litz Wire, Thayer School of Engineering at Dartmouth, M.I.T. Cambridge, March 2014

[18] H. Akagi. (2006). Modern active filters and traditional passive filters. Department of electrical and electronic engineering, Tokyo Institute of Technology, Tokyo Japan

[19] L. Rossetto, G. Spiazzi, P. Tenti.. *Control Techniqes for power factor correction converters* , University of Padova, Department of electrical engineering

[20] Turzynski, Marek; Chrzan, Piotr J., Behavioural model of IGBT transistor, In: *PRZEGLAD ELEKTROTECHNICZNY*, Volume: 86 Issue: 2

[21] Hruska, Karel; Kindl, Vladimir; Pechanek, Roman, Design and FEM Analyses of an Electrically Excited Automotive Synchronous Motor, In: Conference: 15th International Power Electronics and Motion Control Conference and Exposition (EPE-PEMC ECCE Europe) Location: Novi Sad, SERBIA Date: SEP 04-06, 2012

A Floating Double Buck-Boost Converter as Driver for a Permanent Exited DC Machine

Felix A. Himmelstoss
Electronics Department
UAS Technikum Wien
Wien, Austria
felix.himmelstoss@technikum-wien.at

Helmut L. Votzi
Electronics Department
UAS Technikum Wien
Wien, Austria
helmut.votzi@technikum-wien.at

Abstract—**Every year still hundreds of millions of DC drives are built in the low voltage area because of the simplicity of the machine. But only very few papers are published concerning converters for them. Here a floating double buck-boost converter is treated with some interesting features (reduced voltage stress across the devices and higher voltage transformation ratio). The converter is explained and important connections between the voltages and the currents are derived. The design is presented, the large signal and the small signal models are given, the transfer function derived and a simple feedforward-controller is designed and verified by simulations.**

Keywords— *DC/DC converter, permanent magnet DC machine, modelling*

I. Introduction

Only few papers were published in the last years concerning converters for brushed DC-motor drives. Looking at market analyzes [1], however, one can see a still rising market for brushed DC-machines. The necessary voltage transformation ratio must have a step-down feature to start the machine with low voltage, because the source voltage of the machine is dependent on the speed. The difference of the machine source voltage and the voltage across the terminals of the machine evaluates the current through the machine. The output voltage of the converter must therefore start with about ten percent (depending on the machine) of the input voltage. Especially for lower voltage systems an additional step-up feature can be useful. In this paper we study a possible converter topology with step-up-down characteristic, which is derived from [2]. Extensive information concerning converters can be found in the textbooks [3, 4, 5]. An adaption of the modified buck-boost converter [6] can be found in [7] as an example for the adaptation of converter topologies for the control of brushed DC machines. Important topological research for converters can be found e.g. in [8-11].

Fig. 1 shows the circuit of the converter for achieving a one-quadrant drive. The converter consists of one active switch S_1, two diodes D_2, D_3 (the numbers are chosen in this way, because in the two-quadrant version all semiconductors have to be active switches), two capacitors C_1, C_2, and an inductor L_1. Either the input or the output is floating. This limits to a certain degree the application of this converter for security reasons.

The converter in the one-quadrant version is applicable for driving pumps and blowers, were no controlled braking is

necessary. When controlled braking is required, the diodes have to be replaced by active switches (Fig. 2). To avoid a short-circuit at the bridge, a dead time has to be inserted. Therefore, diodes antiparallel to the active switches are necessary (especially when IGBTs are used).

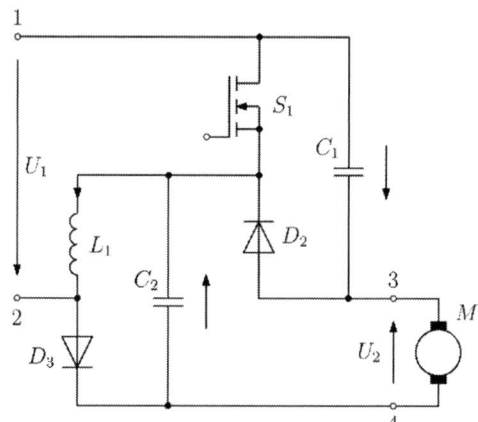

Fig. 1. 1-Q (quadrant) motor drive.

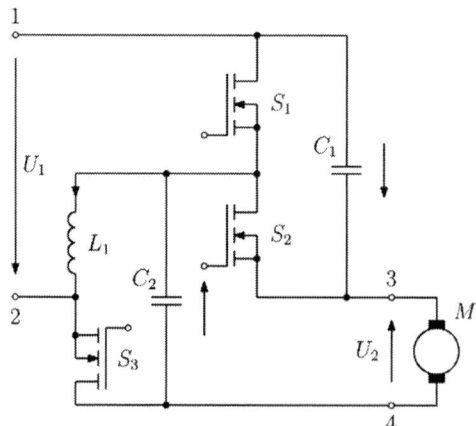

Fig. 2. 2-Q (quadrant) motor drive.

For larger power and higher voltages one would use IGBTs with antiparallel diodes. In this case S_1 would be controlled and S_2 and S_3 would be blocked for the motoring operation. So only the diodes would be used. When braking is necessary, the active switches S_2 and S_3 are used and S_1 is blocked, so only the diode of S_1 feeds back the energy. When MOSFETs are used, the forward voltage of the transistor is lower than that of a diode. Therefore, it can be advisable to use the active switches also when only one-quadrant operation is necessary (synchronous rectification).

II. BASIC FUNCTION

We study the basic function of the floating double buck-boost converter with permanent DC machine as load. Floating means that the input or the output can be grounded, but not both.

We simplify the drive with the usual assumptions:

- Ideal devices (even the machine has no armature resistor in this first step)
- The drive is in steady-state mode
- The capacitors are so large that the voltage across them is nearly constant within one switching period

Continuous inductor current mode CICM, that means that the current through the inductor of the converter L_1 and through the machine L_A does not reach zero during one switching period.

The system has two modes. In mode M1 the active switch S_1 is conducting and the diodes are blocked, and in mode M2 the diodes D_2 and D_3 are conducting and the switch S_1 is turned off.

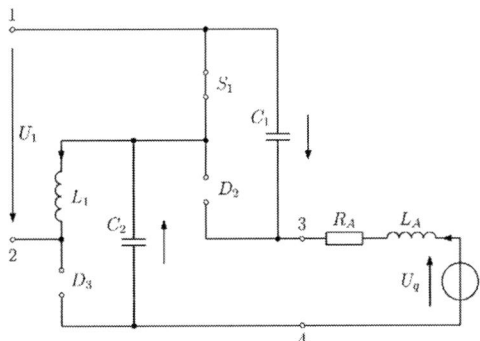

Fig. 3. Equivalent circuit for mode M1.

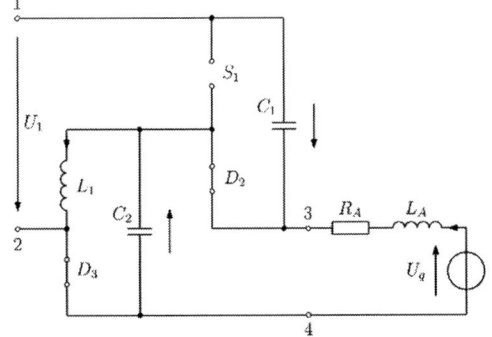

Fig. 4. Equivalent circuit for mode M2.

In the two-quadrant case (or when synchronous rectification is used) in mode M1 S_1 is conducting and in mode M2 S_2 and S_3 are conducting. Fig. 3 and Fig. 4 show the equivalent circuits for Mode M1 and M2, respectively. The armature resistor is included, because this is the dominant resistor in the systems. The machine is modelled with this resistor R_A, the source voltage U_q and the armature inductor L_A.

A. Voltage transformation ratio

For the voltage-time balance one can write for the converter inductor

$$U_1 d = U_{C2}(1-d) \tag{1}$$

and for the armature inductor with U_q as the source voltage of the machine

$$\left(-U_q + U_{C2} + U_{C1}\right)d = \left|-U_q + U_{C2}\right|(1-d) . \tag{2}$$

Furthermore, one can see that the input voltage is always

$$U_1 = U_{C1} - U_q + U_{C2} . \tag{3}$$

Therefore, one can simplify (2) to

$$U_1 d = \left|-U_q + U_{C2}\right|(1-d) \tag{4}$$

which leads with the help of (1) to

$$U_q = \frac{2d}{1-d} U_1 . \tag{5}$$

The connection between the source voltage of the machine and the input voltage can be interpreted as a double buck-boost converter.

The speed of the machine, with C_E as its armature voltage constant, is now given according to

$$n = \frac{1}{C_E} \frac{2d}{1-d} U_1 . \tag{6}$$

Fig. 5 shows the voltages across the inductors L_1, L_A and across the capacitors C_1, C_2, the input and output voltages of the converter and the source voltage which is proportional to the speed. The voltage across the terminals is pulsating.

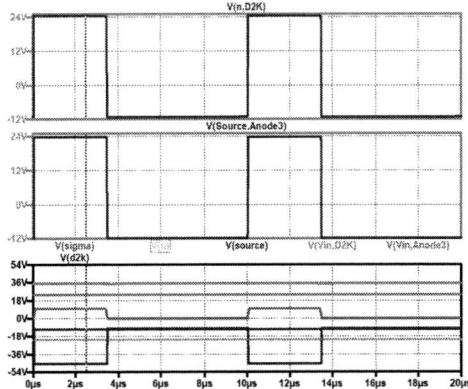

Fig. 5. Up to down: voltage across L_2, L_1; voltage across C_1 (turquoise), input voltage (violet), control signal (red), voltage across C_2 (blue), armature voltage (black), machine source voltage (green).

B. Voltage stress across the devices

For the voltages across the capacitors one gets in dependence on the input voltage and the duty cycle

$$U_{C2} = \frac{d}{1-d} U_1.$$ (7)

$$U_{C1} = \frac{1}{1-d} U_1.$$ (8)

The voltage across C_2 has a buck-boost and across C_1 a boost behavior. Now one can easily calculate the voltage stress U_{SD} across each semiconductor (S_1, S_2, S_3/S_1, D_2, D_3)

$$U_{SD} = \frac{1}{1-d} U_1.$$ (9)

With the mean voltage across the armature terminals U_2, the duty cycle d can be given as

$$d = \frac{U_2}{2U_1 + U_2}.$$ (10)

This leads to

$$U_{SD} = \frac{2U_1 + U_2}{2}.$$ (11)

One can also write for the voltages across the capacitors

$$U_{C1} = \frac{2U_1 + U_2}{2}.$$ (12)

$$U_{C2} = \frac{U_2}{2}.$$ (13)

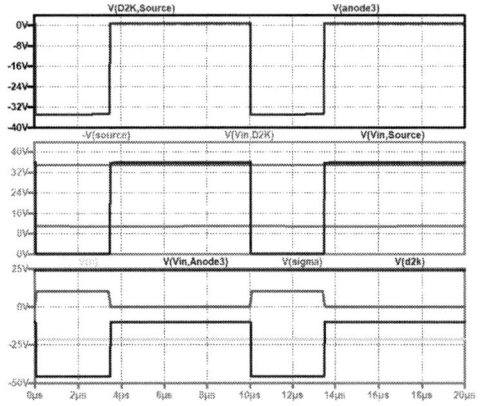

Fig. 6. Up to down: voltages across the diodes; voltage across active switch (black), C_1 (grey), C_2 (violet); input voltage (blue), control signal (red), armature voltage (black), source voltage of the machine (green).

The maximum voltage across the inductor of the converter depends on the input and output voltages according to

$$|U_{L1}| = \max\left\{U_1, \frac{U_2}{2}\right\}.$$ (14)

Fig. 6 shows the voltages across the devices of the converter and across the machine and the source voltage. The load current was chosen to 5 A and the armature resistor has the value of 0.2 Ω.

C. Connection between the currents

The current through the armature inductance is in steady state equal to the load current (which depends on the load torque, in this case equal to the torque that the machine produces, C_T is the torque constant of the machine)

$$\bar{I}_{LA} = I_{Load} = \frac{M_{Load}}{C_T}.$$ (15)

For the capacitor C_2 one can write for the charge balance

$$\left| -\bar{I}_{LA} \right| d = \frac{\bar{I}_{L1} - \bar{I}_{LA}}{2}(1-d).$$ (16)

The reason is that the two capacitors are in parallel during mode M2. Therefore, one can write for equal capacitors

$$i_C = \frac{i_{L1} - i_{LA}}{2}.$$ (17)

The current in the mean through L_1 is therefore

$$\bar{I}_{L1} = \frac{1+d}{1-d} I_{Load} = \frac{1+d}{1-d} I_{LA}.$$ (18)

Fig. 7 shows the currents through the devices. Up to down one can see the currents through C_2, C_1, D_3, D_2, S_1, and the currents through the inductors.

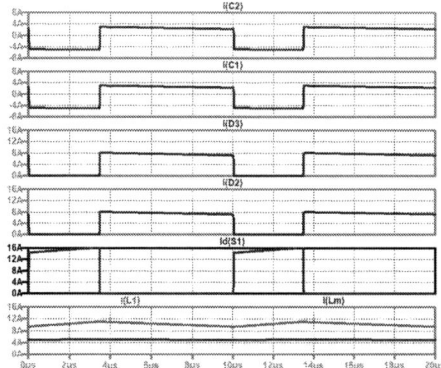

Fig. 7. Up to down: currents through capacitor C_2; capacitor C_1; diode D_3; diode D_2; active switch S_1; converter inductor L_1 (red), armature current L_M.

III. MODELLING OF A FLOATING DC MOTOR DRIVE

A. State space model

An interesting aspect of this converter is the fact, that during mode M1 the same current is flowing through both capacitors. Therefore, the model is reduced to a fourth order one. It is clever to choose for both capacitors the same value, so that the change of voltage is the same, because in mode M2 both capacitors are connected (in the AC model, the input voltage is shorted) in parallel and a balance current would occur. The voltage across C_2 can be given by

978-1-6654-3236-8/21 $31.00 © 2021 IEEE

$$u_{C2} = -u_1 + u_{C1}. \tag{19}$$

Using the state-space averaging method one gets now a fourth-order large signal model with equal capacitors, C_M as torque constant, and C_E as constant of the source voltage of the machine (emc-constant)

$$
\frac{d}{dt}\begin{pmatrix} i_{L1} \\ i_{LA} \\ u_{C1} \\ n \end{pmatrix} =
\begin{bmatrix}
0 & 0 & \dfrac{d-1}{L_1} & 0 \\
0 & -\dfrac{R_A}{L_A} & \dfrac{d+1}{L_A} & -\dfrac{C_E}{L_A} \\
\dfrac{1-d}{2C_1} & -\dfrac{d+1}{2C_1} & 0 & 0 \\
0 & \dfrac{C_M}{2\pi J} & 0 & 0
\end{bmatrix}
\begin{pmatrix} i_{L1} \\ i_{LA} \\ u_{C1} \\ n \end{pmatrix} +
$$
$$
+ \begin{bmatrix}
\dfrac{1}{L_1} & 0 \\
-\dfrac{1}{L_A} & 0 \\
0 & 0 \\
0 & -\dfrac{1}{2\pi J}
\end{bmatrix}
\begin{pmatrix} u_1 \\ m_{Load} \end{pmatrix}. \tag{20}
$$

The linearized model can be calculated to

$$
\frac{d}{dt}\begin{pmatrix} \hat{i}_{L1} \\ \hat{i}_{LA} \\ \hat{u}_{C1} \\ \hat{n} \end{pmatrix} =
\begin{bmatrix}
0 & 0 & \dfrac{D_0-1}{L_1} & 0 \\
0 & -\dfrac{R_A}{L_A} & \dfrac{D_0+1}{L_A} & -\dfrac{C_E}{L_A} \\
\dfrac{1-D_0}{2C_1} & -\dfrac{D_0+1}{2C_1} & 0 & 0 \\
0 & \dfrac{C_M}{2\pi J} & 0 & 0
\end{bmatrix}
\begin{pmatrix} \hat{i}_{L1} \\ \hat{i}_{LA} \\ \hat{u}_{C1} \\ \hat{n} \end{pmatrix} +
$$
$$
+ \begin{bmatrix}
\dfrac{1}{L_1} & 0 & \dfrac{U_{C10}}{L_1} \\
-\dfrac{1}{L_A} & 0 & \dfrac{U_{C10}}{L_A} \\
0 & 0 & -\dfrac{I_{L10}+I_{LA0}}{C_1} \\
0 & -\dfrac{1}{2\pi J} & 0
\end{bmatrix}
\begin{pmatrix} \hat{u}_1 \\ \hat{m}_{Load} \\ \hat{d} \end{pmatrix} \tag{21}
$$

For the connections between the values at the working point one gets

$$
U_{C10} = \frac{1}{1-D_0}U_{10}, \quad N_0 = \frac{1}{C_E}\left(-R_A I_{LA0} + \frac{2D_0}{1-D_0}U_{10}\right) \tag{22}
$$

$$
I_{L10} = \frac{1+D_0}{1-D_0}I_{LA0}, \quad M_{Load0} = C_M L_{LA0}. \tag{23}
$$

B. Transfer functions

With four state variables and three input variables twelve transfer functions can be calculated. The most important ones are those which describe the influence of the input variables on the speed. The denominator is calculated from the determinant of the coefficient matrix and is the same for all transfer functions.

Using abbreviations for the matrix elements in (21)

$$
\frac{d}{dt}\begin{pmatrix} \hat{i}_L \\ \hat{i}_{LM} \\ \hat{u}_C \\ \hat{n} \end{pmatrix} =
\begin{bmatrix}
0 & 0 & A_{13} & 0 \\
0 & A_{22} & A_{23} & A_{24} \\
A_{31} & A_{32} & 0 & 0 \\
0 & A_{42} & 0 & 0
\end{bmatrix}
\begin{pmatrix} \hat{i}_L \\ \hat{i}_{LA} \\ \hat{u}_C \\ \hat{n} \end{pmatrix} +
\begin{bmatrix}
B_{11} & 0 & B_{13} \\
B_{21} & 0 & B_{23} \\
0 & 0 & B_{33} \\
0 & B_{42} & 0
\end{bmatrix}
\begin{pmatrix} \hat{u}_1 \\ \hat{m}_{Load} \\ \hat{d} \end{pmatrix} , \tag{24}
$$

and using the Laplace transformation leads to

$$
\begin{bmatrix}
s & 0 & -A_{13} & 0 \\
0 & s-A_{22} & -A_{23} & -A_{24} \\
-A_{31} & -A_{32} & s & 0 \\
0 & -A_{42} & 0 & s
\end{bmatrix}
\begin{pmatrix} I_L(s) \\ I_{LA}(s) \\ U_C(s) \\ N(s) \end{pmatrix} =
\begin{bmatrix}
B_{11} & 0 & B_{13} \\
B_{21} & 0 & B_{23} \\
0 & 0 & B_{33} \\
0 & B_{42} & 0
\end{bmatrix}
\begin{pmatrix} U_1(s) \\ M_{Load}(s) \\ N(s) \end{pmatrix} \tag{25}
$$

Now one calculate the denominator according to

$$
D = s^4 - s^3 A_{22} - s^2\left(A_{24}A_{42} + A_{13}A_{31} + A_{23}A_{32}\right) + \\
+ sA_{13}A_{22}A_{31} + A_{13}A_{24}A_{31}A_{42}. \tag{26}
$$

The most important transfer function describes the connection between speed and duty cycle, this describes the system to be controlled. The numerator can be calculated with the help of Cramer's law according to

$$
N\_ND = s^2 A_{42}B_{23} + sA_{23}A_{42}B_{33} + A_{23}A_{31}A_{42}B_{13} - A_{13}A_{31}A_{42}B_{23}. \tag{27}
$$

The disturbances of the system are the input voltage and the momentum of the load. The numerators can be calculated according to

$$
N\_NU = A_{42}\left[s^2 B_{21} - A_{13}A_{31}B_{21} + A_{23}A_{31}B_{11}\right], \tag{28}
$$

$$
N\_NM = B_{42}\left(s^3 - s^2 A_{22} - s\left(A_{13}A_{31} + A_{23}A_{32}\right) + A_{13}A_{22}A_{31}\right), \tag{29}
$$

respectively.

Now one can construct Bode diagrams and can use the linear control theory to design a PI or PID controller. In this paper, however, a feedforward controller is described.

IV. FEEDFORWARD CONTROL OF THE CONVERTER

A simple control technique which avoids the measurement of the speed and also the feedback is a feedforward control. In this case changes of the input voltage can be compensated, but load changes lead to deviation of the speed. Starting from (6) one can calculate the duty cycle for a speed n_{ref} and an input voltage U_1 according to

$$
d = \frac{C_E n_{ref}}{C_E n_{ref} + 2U_1}. \tag{30}
$$

or for the source voltage of the machine $U_q = C_E n$ according to

$$d = \frac{U_{qref}}{U_{qref} + 2U_1} \quad . \tag{31}$$

This duty cycle can be easily calculated by a microcontroller or better by a signal processor. A correction factor which is dependent on the nominal load torque of the drive can reduce the speed error at the nominal point.

Fig. 8 shows in the upper picture the input voltage. 250 ms after applying the input voltage, the voltage makes a drop of about 17 % from 24 V down to 20 V. In the lower picture one can see the reference value (the blue curve) and the source voltage of the machine (green, which is proportional to the speed). The speed follows the reference. The difference is caused by the armature resistor of the machine. The drop of the input voltage has only a marginal influence on the speed due to the used control law.

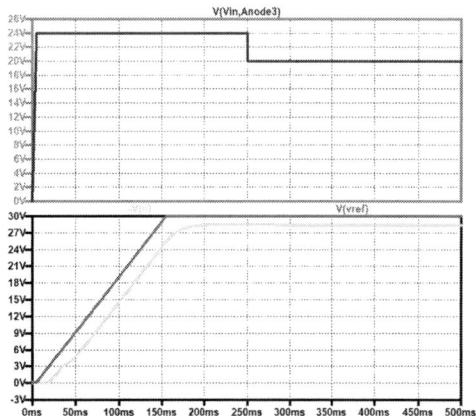

Fig. 8. up to down: input voltage (blue, step at 250 ms); reference value (red), source voltage of the machine corresponding to the speed (green).

Fig. 9 shows a load step. The load increases by 40 %. The armature current increases with the mechanical time constant according to an exponential function. The lower traces show the reference value and the speed. The speed decreases after the step of the load torque.

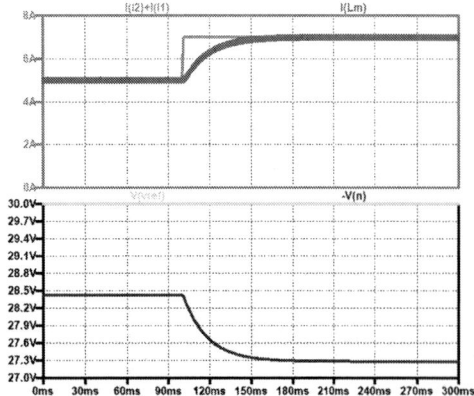

Fig. 9. Load step: load current (turquoise, which is proportional to the load torque), armature current (red); reference value (green), source voltage of the machine (blue, proportional to the speed).

Fig. 10 shows two filtered armature voltages. The voltage is filtered by simple low-pass filters with time constants of 100 µs (green) and 1 ms (red). The switching frequency is 100 kHz, therefor the time constants are 10 and 100 times larger than the switching period. The reference value of the output voltage (the source voltage) is 30 V.

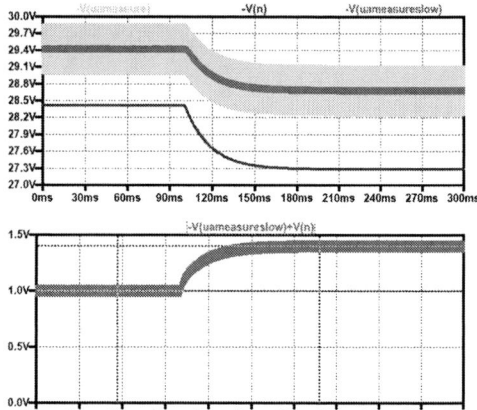

Fig. 10. Load step: filtered armature voltage (100 µs, green) (1ms, red), source voltage (blue, speed); error between filtered armature voltage and source voltage.

One can see that the mean value of the armature voltage is 0.6 V lower and after the load step 1.3 V. This can be interpreted by an output (nonlinear, not constant) resistor of the converter (with a value of 0.12 Ω or 0.19 Ω). The differences between the armature voltage and the source voltage are 1 V and 1.4 V, respectively, caused by the armature resistor of 0.2 Ω. When the first load is taken as the nominal load, one can calculate as correction resistor R_{cor} 0.32 Ω. U_q is therefore by $R_{cor} \cdot I_{Loadnom}$ lower than the reference value U_{ref}.

$$U_q = \frac{2d}{1-d} U_1 - R_{cor} I_{Loadnom} \quad . \tag{32}$$

This leads to the corrected control law

$$d = \frac{U_{ref} + R_{cor} I_{Loadnom}}{2U_1 + U_{ref} + R_{cor} I_{Loadnom}} \quad . \tag{33}$$

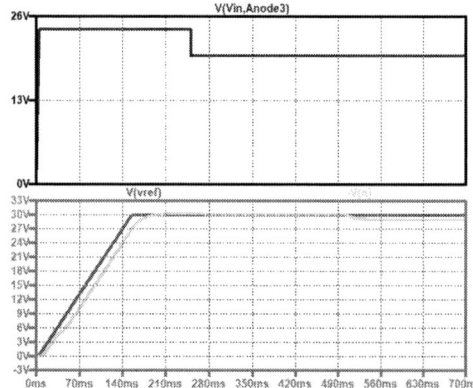

Fig. 11. Corrected control law: reaction to an input voltage step (after 250 ms) and a load step (after 500 ms) up to down: input voltage (blue); reference value (red), source voltage of the machine (green).

Fig. 11 shows the reaction to an input voltage and to a load step for the corrected control law. The reference value and the real value are equal in the stationary case. One has to keep in mind, however, that the temperature of the machine changes and therefor the value of the armature resistor changes too. The correction should be designed for an expected stationary temperature value. A change of the load from the nominal value leads naturally to an error. The load step response with the correction can be seen enlarged in Fig. 12 (same scale as Fig. 9).

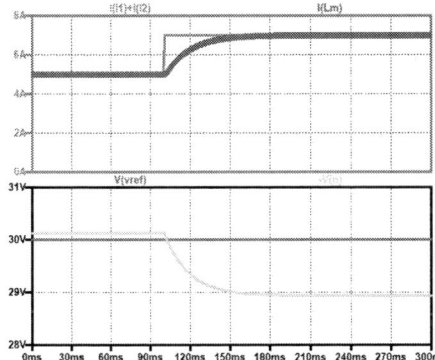

Fig. 12. Load step with correction: load current (turquoise, which is proportional to the load torque), armature current (red); reference value (green), source voltage of the machine (blue, proportional to the speed).

V. SIMULATIONS

Fig. 13 shows the feed-forward controlled converter with an input voltage step at 250 ms. The reference value is constant and the speed decreases only a little bit. The current through the converter coil and the armature current show a damped ringing.

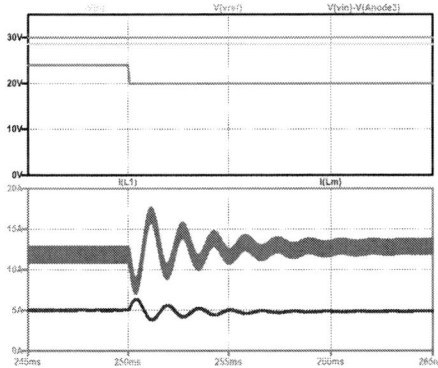

Fig. 13. Input voltage step: up to down: reference value (turquoise), speed (green), input voltage (violet); current through the converter coil (red), armature current (blue).

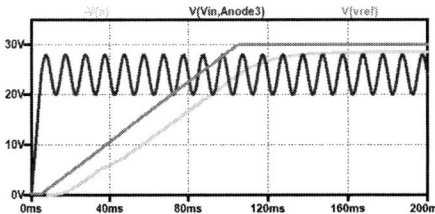

Fig. 14. Up to down: reference value (turquoise), speed (green), input voltage (blue) with 100 Hz ripple.

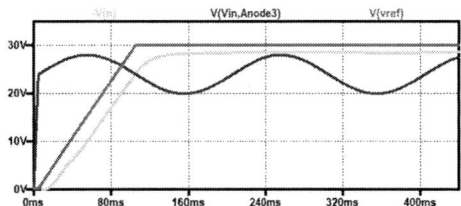

Fig. 15. Up to down: reference value (red), speed (green), input voltage (blue) with 5 Hz ripple.

Fig. 14 shows the start-up and an input voltage with a 100 Hz ripple. One can see that the feed-forward controlled converter can eliminate 100 Hz. The momentum of inertia also damps the input voltage ripple in this case, but in Fig. 15 a 20 times lower disturbance frequency is shown. The stabilization of the speed is now achieved by the controller.

VI. CONCLUSIONS

A double buck-boost with only one active switch and two diodes for driving a permanent DC motor is presented. To achieve a voltage transformation ratio of three, only a duty cycle of 0.6 is necessary, and for a voltage equal to the input voltage a duty cycle of one third must be used. The converter is a step-up-down structure, which is especially useful for high step-up rates e.g. for sources with low voltage and possible high currents like batteries. It is also possible to use three active switches and no diodes. In this case the losses can be significantly reduced because of the much lower forward losses. The derived equations can also be used in this case.

REFERENCES

[1] https://www.databridgemarketresearch.com/reports/global-brushed-dc-motor-market (visited Sept 1st, 2021).

[2] N. Elsayad, H. Moradisizkoohi and O. A. Mohammed, "Design and Implementation of a New Transformerless Bidirectional DC–DC Converter With Wide Conversion Ratios," in IEEE Transactions on Industrial Electronics, vol. 66, no. 9, pp. 7067-7077, Sept. 2019

[3] N. Mohan, T. Undeland and W. Robbins, Power Electronics, Converters, Applications and Design, 3nd ed. New York: W. P. John Wiley & Sons, 2003.

[4] F. Zach, Power Electronics, in German: Leistungselektronik, Wien: Springer, 6th ed.,

[5] Y. Rozanov, S. Ryvkin, E. Chaplygin, P. Voronin, Power Electronics Basics, CRC Press, 2016.

[6] F. A. Himmelstoss, and K. H. Edelmoser, "Modified Basic DC-DC Converters," Power Conversion and Intelligent Motion PCIM 2018, pp. 1076-1083.

[7] Himmelstoss, F. A., and Edelmoser, K. H., "Modified Buck-Boost Converter as DC Motor Driver with Continuous Input Current," WSEAS Transactions of Circuits and Systems, Vol.19, pp. 186-194 2020

[8] G. Chen, Z. Jin, Y. Liu, Y. Hu, J. Zhang and X. Qing, "Programmable Topology Derivation and Analysis of Integrated Three-Port DC–DC Converters with Reduced Switches for Low-Cost Applications," IEEE Transactions on Industrial Electronics, vol. 66, no. 9, pp. 6649-6660, Sept. 2019.

[9] S. Cuk, "General topological properties of switching structures," IEEE Power Electronics Specialists Conference, San Diego, CA, USA, 1979, pp. 109-130.

[10] D. Maksimovic and S. Cuk, "Switching converters with wide DC conversion range," IEEE Transactions on Power Electronics, Vol. 6, No. 1, pp. 151-157, Jan. 1991.

[11] B. W. Williams, "Generation and Analysis of Canonical Switching Cell DC-to-DC Converters," IEEE Transactions on Industrial Electronics, Vol. 61, No. 1, pp. 329-346, Jan. 2014.

High Step-down Converters for Battery Charging in Photovoltaic Applications

Zvonimir Malović*
zvonimir.malovic@fer.hr

Nikola Vuger*
nikola.vuger@fer.hr

Viktor Šunde*
viktor.sunde@fer.hr

Željko Ban**
zeljko.ban@fer.hr

*Department of Electric Machines, Drives and Automation, University of Zagreb, Faculty of Electrical Engineering and Computing, Zagreb, Croatia

** Department of Control and Computer Engineering, University of Zagreb, Faculty of Electrical Engineering and Computing, Zagreb, Croatia

Abstract—**This paper describes non-isolated high step-down converters used to connect photovoltaic modules to the battery storage of a bicycle charging station. Due to the high voltage difference between input and output, some problems may occur with conventional non-isolated DC-DC buck converters. The characteristics of alternative topologies of high step-down converters have been explored. Ideal converter models were created in the PLECS simulation software package based on which the voltage and current stress of the semiconductor switches were determined. Based on measured voltage and current relations, commercial switches were selected. Based on catalogue data, electro-thermal models of the switches were created. A comparative analysis of converters based on conduction and switching losses was performed. Converters were also compared based on the number of components and complexity of control.**

Keywords—photovoltaic module, batteries, step-down converters, duty cycle, conduction losses, switching losses

I. INTRODUCTION

Photovoltaic (PV) energy can be stored in a suitable storage device, used to power DC or AC loads or delivered to the power grid [1], [2], [3]. In this paper, the storage of energy generated from a photovoltaic module in the batteries of a bicycle charging station is considered. For appropriate conditioning of photovoltaic energy, different types of converters with control algorithms are used depending on the load to ensure that the photovoltaic module always operates at the optimal operating point [4], [5], [6]. In this particular case, the energy of the photovoltaic module is stored in the battery storage at a nominal voltage several times lower than the nominal voltage of the module. For this application, isolated or non-isolated DC-DC converters can be used [4]. In applications with extreme conversion ratios, conventional non-isolated step-down converters may experience several problems. Disadvantages of conventional buck converters under these conditions are described in [7] and [8].

Alternative topologies of non-isolated high step-down converters [9] - [23] are considered in this paper. A comparative analysis of the characteristics of these converters with particular reference to conduction and switching losses is performed. Section II describes the design of DC nano-grid of electric bike charging station with battery storage. Current and voltage relations of a conventional non-isolated step-down converter are presented, as well as the problems of operating

with an extremely small duty cycle. The disadvantages of isolated step-down converters are also described in this section. Section III gives an overview of non-isolated high step-down converter topologies. Section IV describes simulation models of converters and results obtained from simulations. A comparative analysis of the basic characteristics with emphasis on conduction and switching losses of the described converters is given in Section V.

II. CHARGING STATION NANO-GRID

The charging station nano-grid consists of the PV module, the battery storage, DC and AC loads and the power converters for connecting the source and loads to the DC bus and the public power grid, Fig. 1.

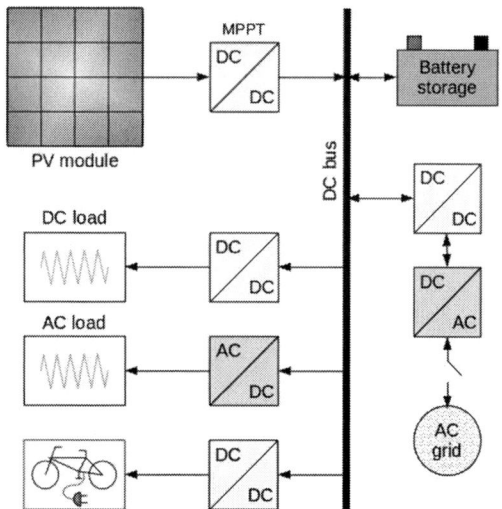

Fig. 1. DC nano-grid of the bicycle charging station

The battery storage directly connected to the DC bus, maintains the voltage level of the nano-grid. The PV module is connected to the DC bus via a step-down converter. Power is supplied to DC and AC loads via DC-DC converters and inverters. Charging of bicycle batteries is done through a buck-boost DC-DC converter. The connection of the charging station nano-grid to the AC power grid is done through a bidirectional DC-DC converter and a PWM inverter that allows bidirectional power flow. In cases where excess energy is generated from the photovoltaic module, energy can be transferred to the power grid. In cases where energy consumption is greater than production, the difference in energy can be drawn from the grid.

All the costs of publishing of this paper are co-financed by the "Development of an Advanced Electric Bicycle Charging Station for a Smart City" project co-funded under the Operational Program from the European Structural and Investment Funds

978-1-6654-3236-8/21 $31.00 © 2021 IEEE

A. Photovoltaic module

From the calculated energy balance of the charging station, it was concluded that a PV module with a power of 2.3 kW is required. The rated voltage of the PV module is 240 V. Photovoltaic modules of the manufacturer SOLVIS were chosen. According to the data sheet, the efficiency of the modules is about 19 %, the area of a module is 1.98 m² and the nominal voltage is 40 V. The required power is achieved by interconnecting 6 of these modules in a string with a rated string voltage of 240 V and the area of 11.88 m².

B. Battery storage

The battery storage stores the excess energy of generated from the photovoltaic modules and maintains the voltage level of the DC bus. The voltage of the nano-grid is 48 V. Valve Regulated Lead-Acid (VRLA) 12 V batteries were chosen for this application. The capacity of one battery is 55 Ah. Although the energy density of lead-acid batteries is not as good as that of lithium-ion batteries, VRLA batteries were chosen because of their low price and easy maintenance.

C. PWM inverter

The DC bus is connected to the AC power system through the DC-DC converter and the PWM inverter which ensures bidirectional power flow. The bidirectional DC-DC converter ensures that the voltage level of the DC part of the inverter is matched to the voltage level of the grid. Single-phase inverters are used in low power systems, while three-phase inverters are preferred in higher-power systems. The bicycle charging station is connected to the grid using a three-phase inverter with a nominal power of 4 kW and the appropriate grid synchronization algorithm.

D. DC-DC converters

In the considered nano-grid of the electric bike charging station, several DC-DC converters are used: (i) for storing the energy generated in the PV modules in the battery storage, (ii) for supplying the DC loads of the charging station from the battery storage, (iii) for connecting the DC bus of the charging station to the inverter, and (iv) for charging the electric bikes from the battery storage. The battery storage and DC loads are connected using a non-isolated DC-DC step-down converter, while an isolated buck-boost converter is used to charge the electric bicycles.

The output voltage of the photovoltaic module (240 V) is higher than the nominal voltage of the battery storage (48 V). Therefore, a step-down converter is needed to connect the module and the battery storage. The converter regulates the voltage of the PV module and ensures maximum utilization of the available power using MPPT algorithms.

The module's operating point is calculated using current and voltage measurements. If the module is not operating at MPP, a change in the control signal is calculated to ensure the transition to a new operating point. MPPT algorithms can be classified into three categories: (i) indirect algorithms, (ii) hill-climbing algorithms and (iii) intelligent algorithms. A detailed analysis of MPPT algorithms can be found in [24]. The P&O algorithm is easy to implement [25], which is why it is preferred for the charging station.

In cases with a large voltage difference between the output voltage of the PV module and the nominal voltage of the battery storage, certain problems may occur in the operation of the conventional non-isolated buck converter, Fig. 2.

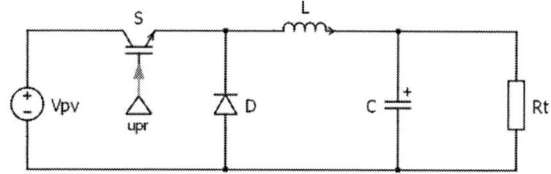

Fig. 2. Conventional non-isolated buck converter

The buck converter operates at a low duty cycle, which causes converter control problems and reduces efficiency [7]. The increase in losses is highest at the main switch. The high voltage stress on the switch results from the large difference between input and output voltage. Also, due to the short conduction time, the peak current through the switch is large, which together with the voltage stress leads to large switching losses of the switch. Due to these losses, the switching frequency is limited, which ultimately leads to larger and more expensive filter components. In addition, the short conduction time of the switch can affect the proper operation of the switch. An increase in losses can also be observed in the diode. Unlike a switch, a diode has a long conduction time, which is why the conduction losses increase. Due to the large current ripple, the stress on the inductor is higher.

A possible solution is to use a converter with a high-frequency transformer, Fig. 3. The voltage ratio of the primary and secondary windings is used to match the voltage level of the photovoltaic module to the voltage level of the battery pack. The galvanic isolation also allows easy grounding of the photovoltaic module. However, transformers increase the volume and price of the converter. Copper and core losses increase, while leakage inductance increases the voltage stress on the switches. An alternative solution for connecting PV modules to batteries with a large voltage difference between the nominal voltages of both elements is to use special non-isolated high step-down topologies.

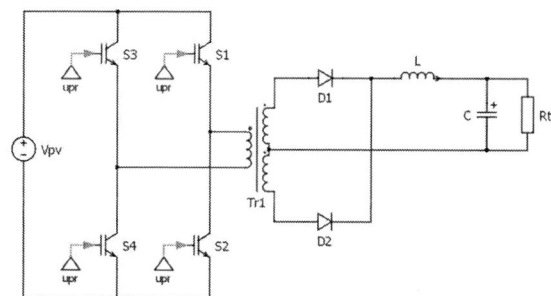

Fig. 3. Isolated DC-DC converter

III. Non-Isolated Step-Down Converters for High Voltage Ratio Conversions

The basic idea is to use the DC converter topology which increases the duty cycle of the switch. With this approach, the conduction time of the switch is increased and hence the efficiency is also increased. Due to the longer conduction time, the peak current value is lower for the same power. The switching losses are reduced for the same switching voltage.

The simplest method to increase the duty cycle is to cascade conventional buck converters [9]. The conversion ratio is shared among the converters, allowing a wider range of duty cycles to be used. However, the increased number of components reduces the reliability and efficiency of the system. Each converter requires separate control loops, which makes the converter more expensive.

978-1-6654-3236-8/21 $31.00 © 2021 IEEE

A. Quadratic step-down converter

The quadratic buck converter, [10], was created as a modification of a cascaded converter to be controlled with only one switch, Fig. 4. The duty cycle of the converter is equal to the product of the duty cycles of the individual converters. Current and voltage waveforms of the semiconductor switches and a waveform of a load current are shown in Fig. 5.

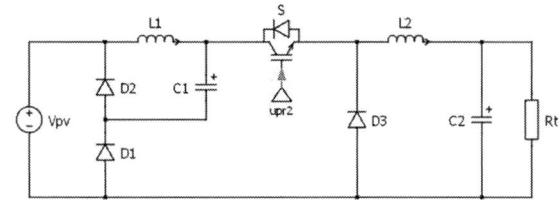

Fig. 4.   Quadratic buck converter

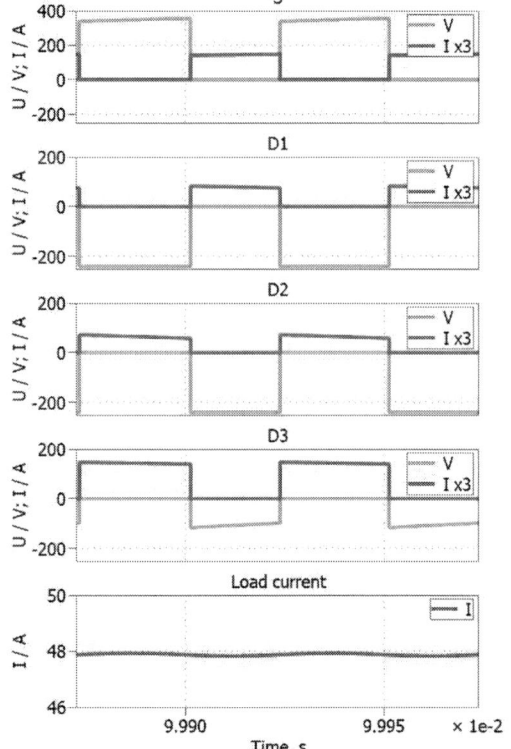

Fig. 5.   Current and voltage waveforms of the switches and a waveform of the load current of the quadratic buck converter

In the interval from 0 to DT, switches S and D1 are conducting. Inductors L1 and L2 accumulate energy through switch S. Capacitor C1 discharges through diode D1. In the interval from DT to T, the switch S is off. The inductor L1 discharges and delivers the energy to the capacitor C1 through the diode D2. The inductor L2 delivers the energy to the capacitor C2 and the load through the diode D3.

By using only one active switch, the control of the converter is simplified. However, the main switch is under high voltage stress which reduces the efficiency of the converter. The topology of the quadratic converter can be modified to a double quadratic buck converter to reduce the voltage stress of the main switch, Fig. 6. The switches S1 and S2 are controlled with the same control signals. The operating principle of the converter and the relation between a duty cycle and the conversion ratio have not changed [11]. The reduced voltage stress of the switches can be seen in the voltage waveforms of the double quadratic buck converter in Fig. 7.

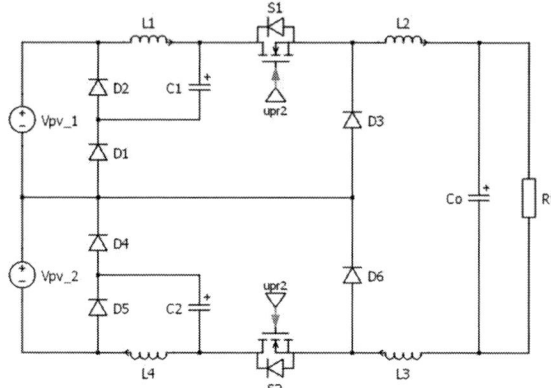

Fig. 6.   Double quadratic buck converter

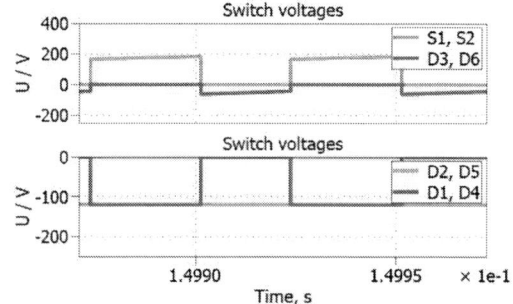

Fig. 7.   Voltage waveforms of the switches of the double quadratic buck converter

B. Series capacitor buck converter

The series capacitor buck converter, [12], was created by modifying a conventional interleaved buck converter, Fig. 8. The added capacitor in one branch of the converter acts as a DC voltage source. The value of the capacitor voltage is equal to half the input voltage which doubles the duty cycle of the converter. The normal operating range of the converter is for the duty cycles up to 50 %. Larger duty cycles increase the voltage and current stress of switches, so this operating range is avoided [13].

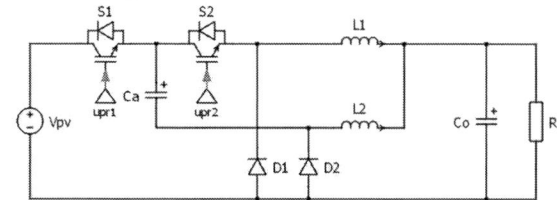

Fig. 8.   Series capacitor buck converter

The phase shift in the control signals of switches S1 and S2 is 180°. The switching period consists of 4 intervals. In the first interval, switch S1 is on. The inductor L2 and the capacitor Ca accumulates the energy from the source. The inductor L1 delivers its energy to the output capacitor Co and the load via the diode D1. In the second interval, both switches are closed. The inductor currents flow through the diodes. In the third interval, switch S2 begins to conduct. The capacitor Ca discharges through the switch S2 and the diode D2 and delivers the energy to the inductor L1. At the same time, the inductor L2 delivers the energy to the load through the diode D2. The fourth interval is the same as the second interval. The voltage and current waveforms of the switches and the waveform of the load current are shown in Fig. 9.

978-1-6654-3236-8/21 $31.00 © 2021 IEEE

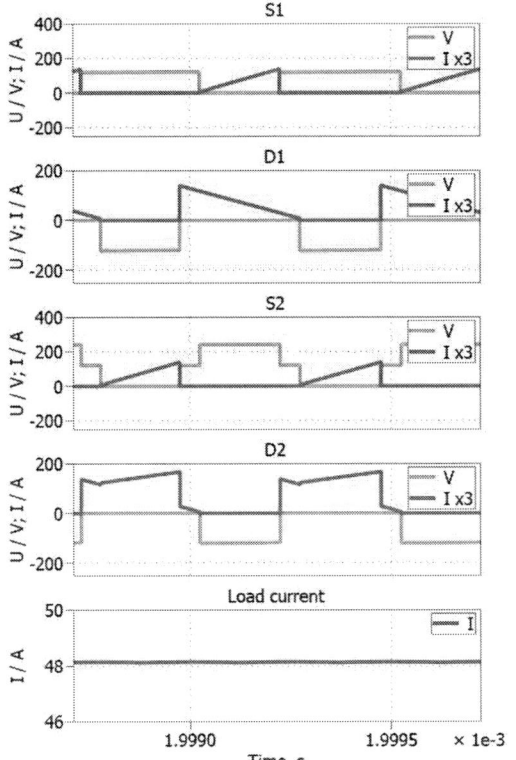

Fig. 9. Current and voltage waveforms of the switches and a waveform of the load current of the series capacitor buck converter

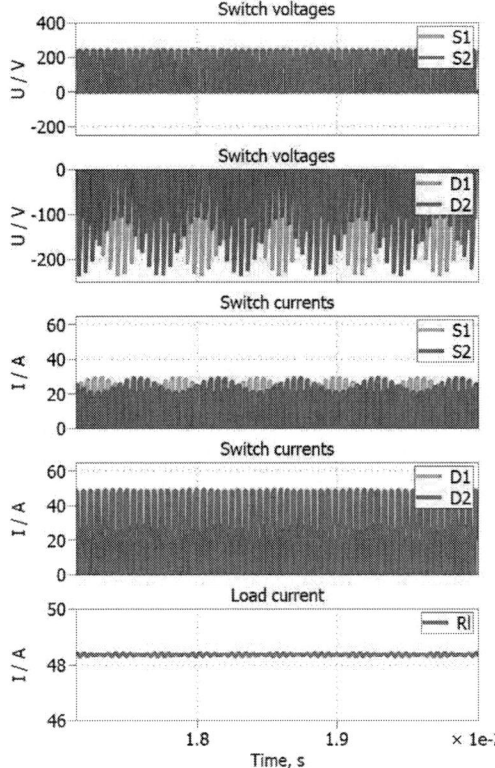

Fig. 10. Oscillations in waveforms of the series capacitor buck converter

Compared to the conventional interleaved buck converter, the control is simplified due to the automatic current balancing through the inductors. The voltage stress of active switches during switching is reduced, but the current stress of diodes is increased. The ripple of the output current is reduced as with the conventional interleaved buck converter.

In combination with certain parameters of the converter (switching frequency, inductance of the coils, capacitance of a series capacitor), oscillations in voltages and currents may occur due to the automatic current balancing mechanism [14], Fig. 10. These oscillations are in the counter phase through the branches of the converter, so the influence on the output waveforms is minimal. Moreover, the oscillations are reduced at higher switching frequencies, which is why this converter is usually used in systems with switching frequencies from a few hundred of kHz up to several MHz [12] – [15], [28], [29].

The topology can be modified to further extend the duty cycle and reduce the voltage stress of the switches. With two more switches and three more capacitors [15], the duty cycle can be expanded four times while the voltage stress can be reduced to a quarter of the input voltage. Additional switches require more complex control, which affects the size and price of the converter.

C. Switched inductor step-down converter

The switched inductor step-down converter, [16], consists of an additional inductor and an additional diode compared to a conventional buck converter, which allows the expansion of the duty cycle, Fig. 11. It operates on the principle of charging two inductors in series and discharging them in parallel, which increases the duty cycle. Voltage and current waveforms of the converter are shown in Fig. 12.

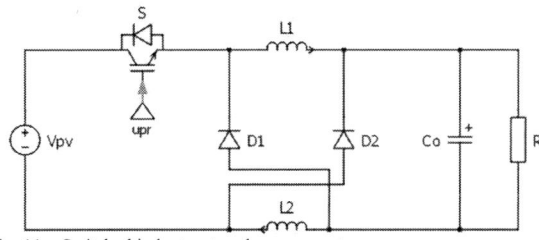

Fig. 11. Switched inductor step-down converter

When the active switch S is on, the inductors L1 and L2 are connected in series between the source and the load and are accumulating the energy from the source. When the switch S is off, the inductors discharge in parallel through the diodes D1 and D2, delivering the energy to the load.

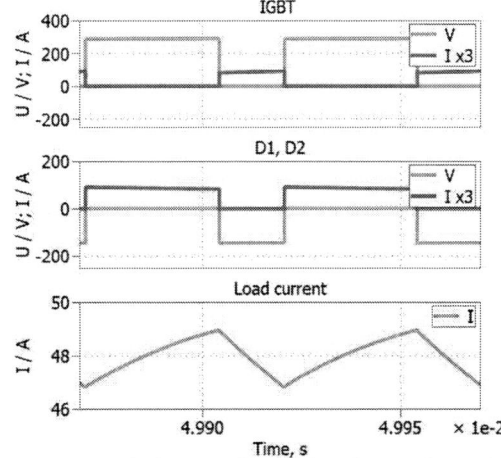

Fig. 12. Current and voltage waveforms of the switches and a waveform of the load current of the switched inductor step-down converter

978-1-6654-3236-8/21 $31.00 © 2021 IEEE

The control of the converter is the same as that of a conventional buck converter. The voltage stress of semiconductor elements is reduced, but the ripple of the output current is increased, Fig. 12.

D. Switched capacitor step-down converter

The switched capacitor step-down converter operates on a similar principle to the switched inductor step-down converter, with two capacitors charged in series and discharged in parallel. Two additional capacitors, three diodes and an inductor are added to the conventional buck converter to form a switched capacitor step-down converter, Fig. 13.

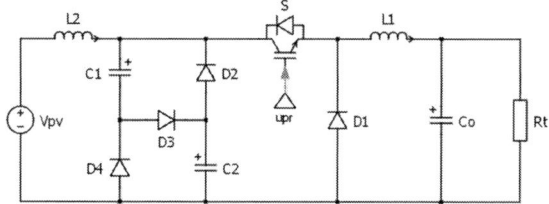

Fig. 13. Switched capacitor step-down converter

When switch S is on, capacitors C1 and C2 are discharged in parallel through the diodes D2 and D4, delivering energy through the switch S to the output filter and load. When the switch is off, the capacitors are connected in series through the diode D3 and are charged with energy from the source. Energy accumulated in inductor L1 is delivered to capacitor Co and the load through diode D1. The inductor L2 limits the current peak caused by discharged capacitors. Voltage and current waveforms of the converter are shown in Fig. 14.

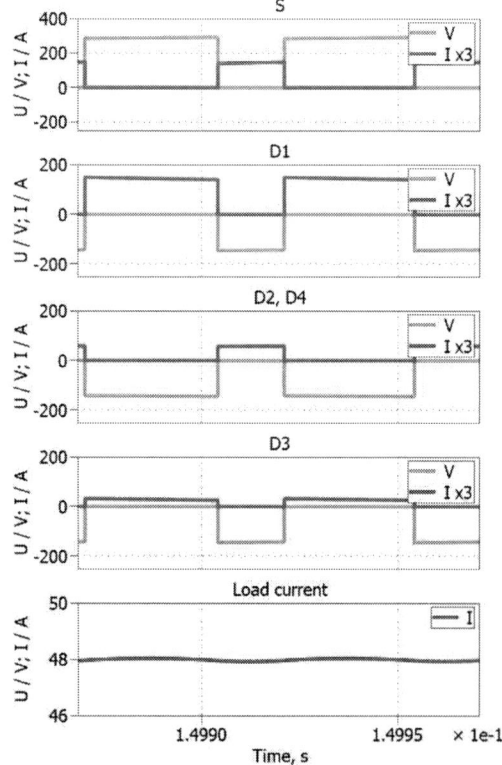

Fig. 14. Current and voltage waveforms of the switches and a waveform of the load current of the switched capacitor step-down converter

The efficiency of the switched capacitor step-down converter is reduced due to the influence of the equivalent series resistances of the capacitors. In addition, the voltage stress of the main switch is greater than that of the switched inductor step-down converter, Fig. 14. The control of the converter is the same as that of a conventional buck converter. Additional duty cycle expansion can be achieved by upgrading the converter topology with additional capacitors and diodes as described in [17].

E. Tapped inductor buck converter

Tapped inductor buck converter, [18], is made as a conventional buck converter with the additional inductor, i.e., the additional winding. Both inductors are wound on the same core, Fig. 15. The duty cycle of the converter can be increased by adjusting the ratio of the turns of these two windings.

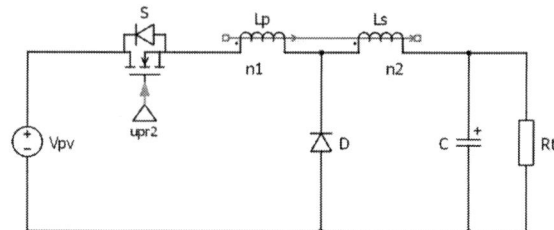

Fig. 15. Tapped inductor buck converter

In the interval from 0 to DT, the switch S is on. Both windings are connected in series between the source and the load and are accumulating the energy. In the interval from DT to T, switch S is off. The circuit is closed by diode D. The energy accumulated in both windings is delivered to the load through the secondary winding Ls. The current flows through only one winding and is multiplied by the turns ratio, Fig. 16.

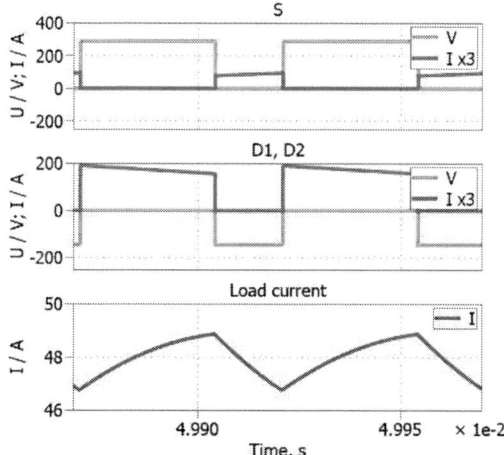

Fig. 16. Current and voltage waveforms of the switches and a waveform of the load current of the tapped inductor buck converter

Simplicity and a small number of elements are the main features of the tapped inductor buck converter. However, the leakage inductance of the inductor causes voltage spikes at the main switch, which further stresses the switch and reduces the efficiency (ideal elements were used in the simulation, so the voltage spike caused by leakage inductance is not shown in the waveforms in Fig. 16). Also, the emitter of the main switch is at the floating potential, which may introduce additional difficulties in generating the control signal for the switch. The output current ripple is increased due to the sudden changes in the current value. The current stress of the diode is also increased.

One of the possible modifications of the tapped inductor buck converter to reduce the voltage spikes at the main switch

978-1-6654-3236-8/21 $31.00 © 2021 IEEE

is the topology shown in Fig. 17. Two additional diodes Ds1 and Ds2 and capacitor Cs are used to accumulate the energy from leakage inductance during one switching interval and reuse it in another switching interval. The emitter of the switch S is not at the floating potential, so that the generation of the control signals is simplified [19].

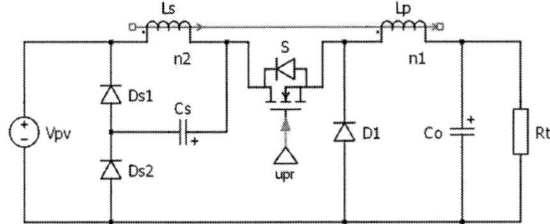

Fig. 17.   Tapped inductor step-down converter with lossless clamp

When switch S is on, both windings accumulate energy from capacitor Cs through diode Ds2 and from the source. When the switch S is off, the winding Lp delivers the energy to the load through the diode D1 while the energy from the leakage inductance is stored in the capacitor Cs through the diode Ds1.

The reduction of the losses of the tapped inductor buck converter can be achieved with the tapped inductor converter with series capacitor [20], [21]. However, the converter consists of three active switches which complicate the control, so it is not considered in this paper.

F. Combined step-down converters

Some of the mentioned topologies can be combined to further extend the duty cycle of the converter. The converter described in [22] was created by combining the quadratic buck converter and the switched inductor step-down converter, Fig. 18. The topology described in [23] realizes the duty cycle expansion by switching inductors and capacitors simultaneously, Fig. 19.

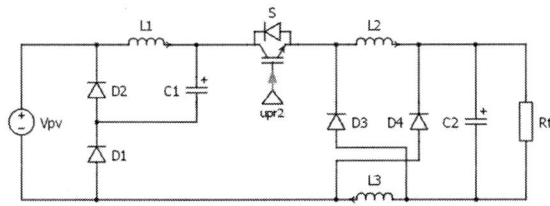

Fig. 18.   The combination of the quadratic buck converter and the switched inductor step-down converter

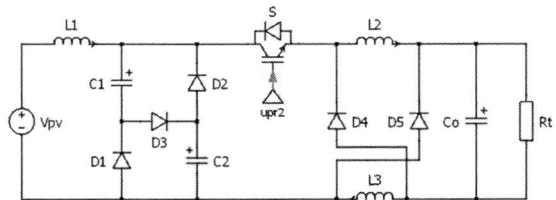

Fig. 19.   The combination of the switched capacitor step-down converter and the switched inductor step-down converter

IV.   SIMULATION EXPERIMENTS

Simulation models with ideal elements and electro-thermal models were implemented in the PLECS software package from Plexim for the previously mentioned step-down converter topologies. Using the ideal models, the current-voltage characteristics of the switches were determined. Based on these results, commercial components were selected to simulate the electro-thermal losses of the switches. The

efficiencies of the converters were calculated based on the simulations with the electro-thermal models. The operating points of the converters were set to the operating point found in the electric bike charging station. The rated voltage of the photovoltaic module is 240 V and the rated voltage of the battery storage is 48 V. The duty cycles of the mentioned topologies are given in Table I.

TABLE I.   RATIO CONVERSIONS OF THE ABOVE-MENTIONED CONVERTERS

Topologies	Ratio conversion (V_{out}/V_{in})
Conventional buck converter	D
Quadratic buck converter**	D^2
Double quadratic buck converter	D^2
Series capacitor buck converter **	$\dfrac{D}{2}$
Modified series capacitor step-down converter described in [15]	$\dfrac{D}{4}$
Switched inductor step-down converter **	$\dfrac{D}{2-D}$
Switched capacitor step-down converter**	$\dfrac{D}{2-D}$
Modified switched capacitor step-down converter described in [17]	$\dfrac{D}{a-(a-1)\cdot D}$
Tapped inductor buck converter**	$\dfrac{D}{D+n\cdot(1-D)}$
Tapped inductor step-down converter with lossless clamp	$\dfrac{D}{D+n\cdot(1-D)}$
Tapped inductor converter with series capacitor described in [20] and [21]	$\dfrac{D}{n}$
Step-down converter described in [22]	$\dfrac{D}{(2-D)^2}$
Step-down converter described in [23]	$\dfrac{D}{(2-D)^2}$

** – converters simulated in PLECS

D – duty cycle

a – number of switching capacitor cells

n – the winding ratio of tapped inductors

A. Ideal models of semiconductor elements

Fig. 20 shows a model of the switched inductor step-down converter, in which the main switch S and diodes D1 and D2 are presented with an ideal model. Table II shows the parameters of the simulation model of the switched inductor step-down converter.

TABLE II.   SIMULATION MODEL PARAMETERS OF THE SWITCHED INDUCTOR STEP-DOWN CONVERTER

Parameter	Symbol	Value
Input voltage	Vpv	240 V
Load voltage	Vo	48 V
Switching frequency	fsw	20 kHz
Inductor inductance	$L1, L2$	560 µH
Capacitor capacitance	Co	150 µF
Load resistance	Rt	1 Ω
The equivalent series resistance of a capacitor	Rl	1 mΩ
Duty cycle	D	1/3

978-1-6654-3236-8/21 $31.00 © 2021 IEEE

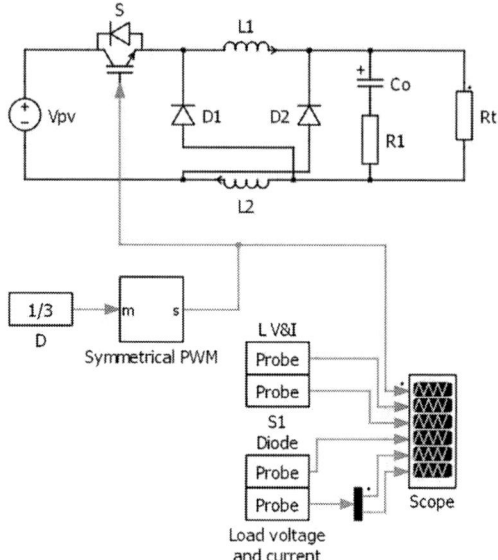

Fig. 20. PLECS simulation model of the switched inductor step-down converter

B. Simulation results with ideal models of semiconductor switches

Simulation results of a switched inductor step-down converter with ideal switches are shown in Fig. 21. Current-voltage relations of switches used for component selection are given in Table III. Based on these relations, the IGBT HGTG30N60A4 and the diode IDW75E60 are selected. The parameters of these elements are given in Table IV. Simulation of other topologies has confirmed that the selected semiconductor devices can be used in other converters mentioned above.

TABLE III. MAXIMUM CURRENT AND VOLTAGE VALUES OF SWITCHES MEASURED FROM THE SIMULATION

Switch	Maximum voltage	Maximum current
IGBT	288 V	31.4 A
Diodes D1 and D2	144 V	31.4 A

TABLE IV. PARAMETERS OF SEMICONDUCTOR SWITCHES

IGBT HGTG30N60A4	
U_{CE}	600 V
$I_{C,25°C}$	75 A
$I_{C,110°C}$	60 A
I_{cmax}	240 A
Dioda IDW75E60	
U_{RRM}	600 V
$I_{FAV,25°C}$	120 A
$I_{FAV,100°C}$	75 A
I_{fmax}	220 A
$U_{F,75A}$	1.65 V

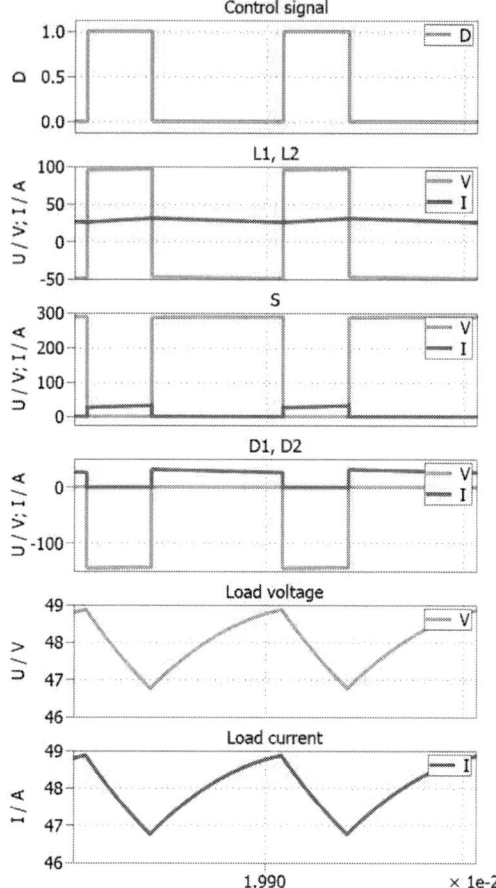

Fig. 21. Current and voltage waveforms on elements of the switched inductor step-down converter

C. Electro-thermal models of switches

In order to calculate the semiconductor losses, i.e. the efficiency of the analyzed DC converters, the idealized switch models in the previous simulation models were replaced by electro-thermal models. Electro-thermal models consist of a temperature-dependent electrical model of the switch and a model of the thermal system of the switch represented by an equivalent RC network, Fig. 22. In the electrical model, the losses of the switch are calculated and brought to the input of the thermal model in the form of a current source. The voltage response of the equivalent RC network, due to the excitation of this current source, represents the temperature of the switch. The temperature thus calculated is used as a parameter in the next step of the simulation to obtain temperature-dependent conduction losses and switching losses from the look-up table.

A look-up table for calculating conduction losses provides a temperature dependence of the static V-I characteristic of the switch. A look-up table for calculating switching losses provides a dependence of the switching energy on temperature, current and blocking voltage. The tables are created by importing graphs from a switch data sheet into a PLECS subroutine. The axes of the subroutine coordinate system are aligned with the axes of the graph. A sampling of the losses is performed by placing the mouse pointer on the characteristic. The coordinates of the point (current and voltage for conduction losses and current and energy for switching losses) are automatically calculated based on the position of the point between the axes.

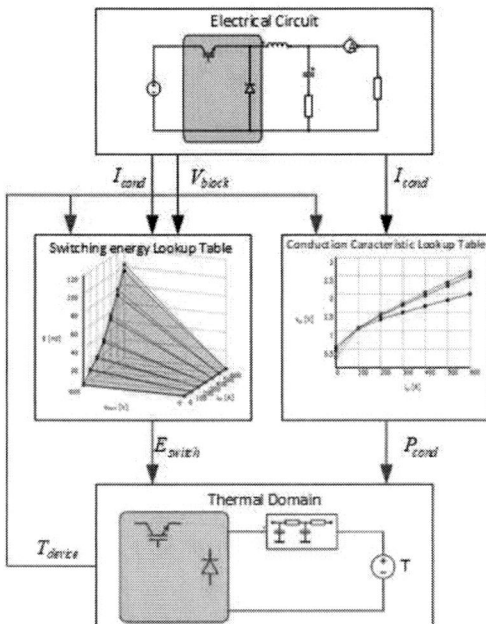

Fig. 22. Block diagram of electro-thermal switch model

Examples of conduction losses and switching losses are shown in Figs. 23 and 24. Values not included in the look-up tables are interpolated between the two closest points. When calculating the conduction losses, based on the current from the look-up table, the voltage drop across the switch is taken in each iteration. The product of this voltage and current gives the power of conduction losses. To calculate switching losses, the current values of voltage and current are used to read the switching energy from the look-up table. This energy is then multiplied by the switching frequency to obtain the switching losses averaged over the switching period.

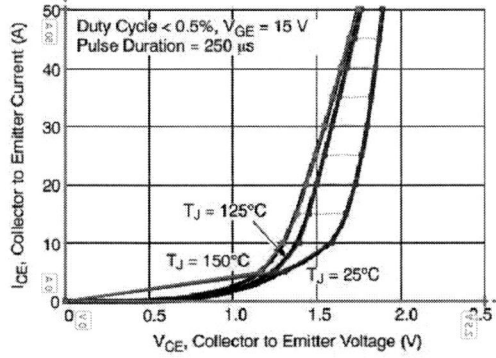

Fig. 23. Sampling of temperature-dependent V-I characteristics of the switch

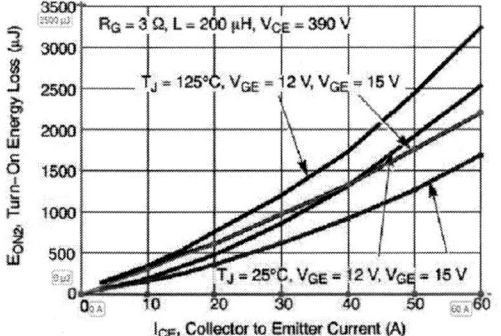

Fig. 24. Sampling of a temperature and current dependent switching energy characteristics

D. Simulation results with electro-thermal models of semiconductor switches

The simulation model of a switched inductor step-down converter with electro-thermal models of the switches is shown in Fig. 25. The values of the model parameters are given in Table V.

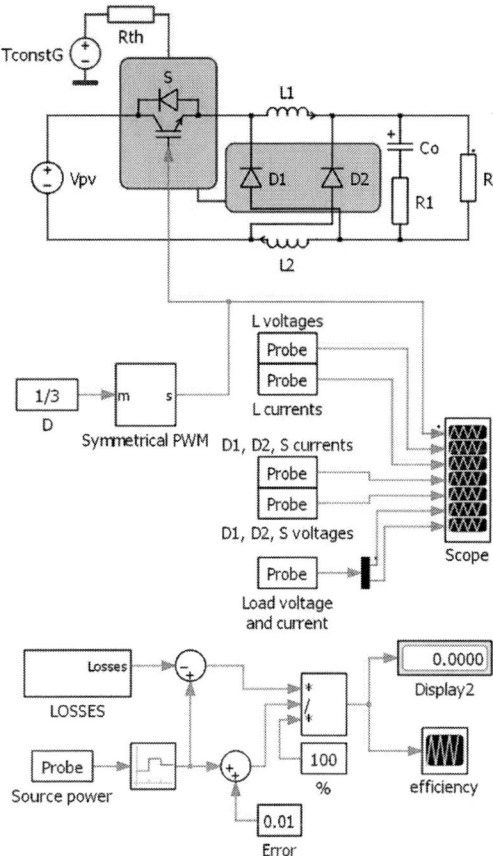

Fig. 25. Switched inductor step-down converter model with the electro-thermal models of switches

TABLE V. PARAMETERS OF ELECTRO-THERMAL MODELS

IGBT	
U_F	1.8 V
Initial temperature	25 °C
Diode	
U_F	1.4 V
Initial temperature	25 °C
Others	
The initial temperature of the heatsink	25 °C
The thermal capacity of the heatsink	0.05 J/K
Thermal resistance Rth	2 K/W
The temperature of an ambient TconstG	25 °C

The efficiency of all the above converters is calculated based on these electro-thermal models using this formula:

$$\eta = \frac{P_{PV} - P_{lossess}}{P_{PV}} \cdot 100\% \tag{1}$$

978-1-6654-3236-8/21 $31.00 © 2021 IEEE

where:

- P_{PV} is the power of the source (PV module),
- P_{losses} is the power loss in semiconductors.

The losses of the switches are measured and calculated in the subsystem "LOSSES". The simulation was carried out for five different topologies of converters, listed in Table I and marked with **.

V. COMPARATIVE ANALYSIS OF CONVERTERS

The result of the previously performed simulations are conduction losses and switching losses of semiconductor switches, Fig. 26.

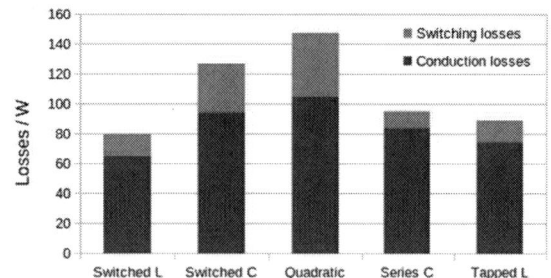

Fig. 26. Overview of losses of different converters

In Fig. 26, it can be seen that for the given operating point and switching frequency, the switched inductor step-down converter and the tapped inductor buck converter have the lowest losses (79.5 W and 89 W respectively). Moreover, these converters are the series capacitor buck converter, have minimum switching losses. The switching losses of the tapped inductor buck converter would be even larger if the influence of the leakage inductance of the coupled windings was included in the simulation.

Quadratic buck converter has the largest switching losses (42.22 W) due to the high voltage stress of the switches and the largest conduction losses (104.92 W) due to the longest duty cycle.

Despite a large number of switches (1 transistor and 4 diodes), the switched capacitor step-down converter has high efficiency due to the relatively low current and voltage stress on diodes D2, D3 and D4.

The configuration of the semiconductor switches of the switched inductor step-down converter, the switched capacitor step-down converter and the quadratic buck converter is the same as that of the conventional buck converter, allowing the use of the same control circuits. Additional flexibility in the control of quadratic buck converter is that the current of either inductor can be measured and used in the current control loop.

The series capacitor buck converter must be designed properly and used at high switching frequencies to avoid oscillation while its duty cycle is limited to 50 %. The emitter of the main switch of the tapped inductor buck converter is at a floating potential, which adds complexity to the control circuit. In addition, the design must account for voltage spikes caused by leakage inductance at the active switch.

Switched inductor step-down converters and tapped inductor buck converters suffer from high output current ripple (current ripple over 2 A), which affects filter element selection and increases the price of the converter.

Table VI contains the calculated efficiencies of the analyzed converters, the number of switches of the different converters, the complexity of the control and the ripple of the output current.

TABLE VI. BASIC CHARACTERISTICS OF SIMULATED CONVERTERS

Topology	No. of switches	η	Control	ΔI
Switched inductor step-down converter	Transistors: 1 Diodes: 2	96.54 %	+	2.14 A
Switched capacitor step-down converter	Transistors: 1 Diodes: 4	94.49 %	+	0.13 A
Quadratic buck converter	Transistors: 1 Diodes: 3	93.58 %	++	0.1 A
Series capacitor buck converter	Transistors: 2 Diodes: 2	95.85 %	+++	0.2 A
Tapped inductor buck converter	Transistors: 1 Diodes: 1	96.13 %	+++	2.13 A

VI. CONCLUSION

In this paper, five high step-down converter topologies were analyzed (quadratic buck converter, series capacitor buck converter, switched inductor step-down converter, switched capacitor step-down converter, tapped inductor buck converter) that can be used to connect a photovoltaic module to battery storage when the voltage level of a module is significantly larger than the voltage level of batteries. The main disadvantages of a conventional buck converter operating at a large conversion ratio are reduced efficiency due to increased switching losses and increased peak current through the switch and the inductor. Isolated step-down converters suffer from additional copper and core losses while increasing the volume and mass of the converter.

Analyzed topologies effectively expand the duty cycle and reduce the voltage stress of the switches, thereby increasing the efficiency of the converter. From the described simulations, it is concluded that the tapped inductor buck converter is the most efficient with the efficiency over 96 % and requires the least number of semiconductor elements (only two), but increased output current ripple and voltage spikes at the main switch should be considered in the converter design. Alternatives for this application include a series capacitor buck converter with the efficiency of 95.85 % as well as a switched inductor step-down converter with the efficiency of 96.54 %.

The disadvantages of these high step-down converters are a larger number of elements with more complex control, which ultimately increases the price of the converter.

Continuing the research of high step-down converter topologies, a physical model of a converter selected for installation in the electric bike charging station will be created. Measurements of this physical model will determine its characteristics and confirm the results from the simulation experiments.

REFERENCES

[1] Franklin, E., Types of Solar Photovoltaic Systems, The University of Arizona, Colledge of Agriculture & Life Sciences, Cooperative Extension, No. AZ1745-2017, 2017.

[2] Worku, Muhammed Y., Mohamed A. Hassan, and Mohamed A. Abido 2019. "Real Time Energy Management and Control of Renewable Energy based Microgrid in Grid Connected and Island Modes" *Energies* 12, no. 2: 276. https://doi.org/10.3390/en12020276

[3] Khatib, Tamer & Ibrahim, Ibrahim & Mohamed, Azah. (2016). A review on sizing methodologies of photovoltaic array and storage

battery in a standalone photovoltaic system. Energy Conversion and Management. 120. 430–448. 10.1016/j.enconman.2016.05.011.

[4] Raghavendra, K. V. G., Zeb, K., Muthusamy, A., Krishna, T. N. V., Kumar, P., Kim, D. H., Kim, M. S., Cho, H. G., Kim, H. J., A Comprehensive Review of DC–DC Converter Topologies and Modulation Strategies with Recent Advances in Solar Photovoltaic Systems, Electronics, 9(1), 31, DOI: 10.3390/electronics9010031, 2020

[5] G, Dileep & Singh, S.N.. (2017). Selection of non-isolated DC-DC converters for solar photovoltaic system. Renewable and Sustainable Energy Reviews. 76. 1230-1247. 10.1016/j.rser.2017.03.130.

[6] K. George and S. Ang, "Topology survey for GaN-based high voltage step-down single-input multi-output DC-DC converter systems," 2016 IEEE 4th Workshop on Wide Bandgap Power Devices and Applications (WiPDA), 2016, pp. 340-343, doi: 10.1109/WiPDA.2016.7799964.

[7] B. Soleymani and E. Adib, "A High Step-Down Buck Converter With Self-Driven Synchronous Rectifier," in IEEE Transactions on Industrial Electronics, vol. 67, no. 12, pp. 10266-10273, Dec. 2020, doi: 10.1109/TIE.2019.2959454.

[8] Esteki, M., Poorali, B., Adib, E. and Farzanehfard, H. (2015), High step-down interleaved buck converter with low voltage stress. IET Power Electronics, 8: 2352-2360. https://doi.org/10.1049/iet-pel.2014.0976

[9] R. Aguilar-Najar, F. Perez-Pinal, G. Lara-Salazar, C. Herrera-Ramirez and A. Barranco-Gutierrez, "Cascaded buck converter: A reexamination," 2016 IEEE Transportation Electrification Conference and Expo (ITEC), 2016, pp. 1-5, doi: 10.1109/ITEC.2016.7520211.

[10] Wang, Tao & Li, JiKun & He, Xiang. (2017). One-cycle controlled quadratic buck converter. International Journal of Circuit Theory and Applications. 46. 10.1002/cta.2488.

[11] F. L. de Sá, D. Ruiz-Caballero and S. A. Mussa, "Analysis of the double quadratic buck converter," 2016 18th European Conference on Power Electronics and Applications (EPE'16 ECCE Europe), 2016, pp. 1-10, doi: 10.1109/EPE.2016.7695460.

[12] P. S. Shenoy, M. Amaro, J. Morroni and D. Freeman, "Comparison of a Buck Converter and a Series Capacitor Buck Converter for High-Frequency, High-Conversion-Ratio Voltage Regulators," in IEEE Transactions on Power Electronics, vol. 31, no. 10, pp. 7006-7015, Oct. 2016, doi: 10.1109/TPEL.2015.2508018.

[13] I. Lee, S. Cho and G. Moon, "Interleaved Buck Converter Having Low Switching Losses and Improved Step-Down Conversion Ratio," in *IEEE Transactions on Power Electronics*, vol. 27, no. 8, pp. 3664-3675, Aug. 2012, doi: 10.1109/TPEL.2012.2185515.

[14] P. S. Shenoy et al., "Automatic current sharing mechanism in the series capacitor buck converter," 2015 IEEE Energy Conversion Congress and Exposition (ECCE), 2015, pp. 2003-2009, doi: 10.1109/ECCE.2015.7309943.

[15] *Ebrahim, F., & Mathew, T. (2016). An Interleaved High Step-Down Conversion Ratio Buck Converter with Closed Loop Control for Low Switch Voltage Stress, International Journ al of Engineering Science and Computing*

[16] Yao, Jia & Zheng, Kaisheng & Abramovitz, Alexander. (2019). Dynamic Analysis of the Switched Inductor Buck Converter. 1721-1725. 10.1109/IECON.2019.8927207.

[17] S. Xiong, S. Tan and S. Wong, "Analysis and Design of a High-Voltage-Gain Hybrid Switched-Capacitor Buck Converter," in IEEE Transactions on Circuits and Systems I: Regular Papers, vol. 59, no. 5, pp. 1132-1141, May 2012, doi: 10.1109/TCSI.2012.2191313.

[18] Chadha, Ankit, "Tapped-Inductor Buck DC-DC Converter" (2019). Browse all Theses and Dissertations. 2276. https://corescholar.libraries.wright.edu/etd_all/2276

[19] K. Nishijima et al., "A novel tapped-inductor buck converter for divided power distribution system," 2006 37th IEEE Power Electronics Specialists Conference, 2006, pp. 1-6, doi: 10.1109/pesc.2006.1711779.

[20] K. I. Hwu, W. Z. Jiang and Y. T. Yau, "Ultrahigh Step-Down Converter," in IEEE Transactions on Power Electronics, vol. 30, no. 6, pp. 3262-3274, June 2015, doi: 10.1109/TPEL.2014.2338080.

[21] K. I. Hwu & W. Z. Jiang (2017) Performance comparison between tapped-inductor buck converter and ultrahigh step-down converter, International Journal of Electronics Letters, 5:4, 475-490, DOI: *10.1080/21681724.2017.1279225*

[22] Khambuya, Ronnakorn & Khwan-on, Sudarat. (2016). A New High Step-down DC-DC Converter for Renewable Energy System Applications. Procedia Computer Science. 86. 349-352. 10.1016/j.procs.2016.05.094.

[23] O. Pelan, N. Muntean, O. Cornea and F. Blaabjerg, "High voltage conversion ratio, switched C & L cells, step-down DC-DC converter," 2013 IEEE Energy Conversion Congress and Exposition, 2013, pp. 5580-5585, doi: 10.1109/ECCE.2013.6647459.

[24] Selvan, Saravana & Nair, Pratap & Umayal,. (2016). A review on photo voltaic MPPT algorithms. International Journal of Electrical and Computer Engineering (IJECE). 6. 567-582. 10.11591/ijece.v6i1.9204.

[25] P. T. Szemes and M. Melhem, "Analyzing and modeling PV with "P&O" MPPT Algorithm by MATLAB/SIMULINK," 2020 3rd International Symposium on Small-scale Intelligent Manufacturing Systems (SIMS), 2020, pp. 1-6, doi: 10.1109/SIMS49386.2020.9121579.

[26] Abdulrazzaq, Ali & Hussein Ali, Adnan. (2018). Efficiency Performances of Two MPPT Algorithms for PV System With Different Solar Panels Irradiancess. International Journal of Power Electronics and Drive Systems (IJPEDS). 9. 10.11591/ijpeds.v9.i4.pp1755-1764.

[27] O. Kirshenboim, T. Vekslender and M. M. Peretz, "Closed-Loop Design and Transient-Mode Control for a Series-Capacitor Buck Converter," in IEEE Transactions on Power Electronics, vol. 34, no. 2, pp. 1823-1837, Feb. 2019, doi: 10.1109/TPEL.2018.2829932.

[28] P. S. Shenoy et al., "Automatic current sharing mechanism in the series capacitor buck converter," 2015 IEEE Energy Conversion Congress and Exposition (ECCE), 2015, pp. 2003-2009, doi: 10.1109/ECCE.2015.7309943.

[29] P. S. Shenoy, M. Amaro, D. Freeman and J. Morroni, "Comparison of a 12V, 10A, 3MHz buck converter and a series capacitor buck converter," 2015 IEEE Applied Power Electronics Conference and Exposition (APEC), 2015, pp. 461-468, doi: 10.1109/APEC.2015.7104391.

978-1-6654-3236-8/21 $31.00 © 2021 IEEE

UPS with PFC input stage for railway applications with improved immunity on input overvoltage and energy strikes

Ivan Šolc *
ivan.solc@fer.hr

Ante Hećimović **
ante.hecimovic@fer.hr

Viktor Šunde *
viktor.sunde@fer.hr

Željko Ban **
zeljko.ban@fer.hr

* Department of Electric Machines, Drives and Automation, University of Zagreb, Faculty of Electrical Engineering and Computing, Zagreb, Croatia
** Department of Control and Computer Engineering, University of Zagreb, Faculty of Electrical Engineering and Computing, Zagreb, Croatia

Abstract— The main topic of this article is uninterruptible power supply (UPS) in railway applications, focusing on the power factor correction (PFC) stage of UPS and its interaction with overvoltages and energy strikes. The PFC topology used in this article is a two-phase interleaved DC/DC boost converter with diode bridge rectifier connected to the 25 kV, 50 Hz contact line via 25 kV/0.23 kV transformer. Since overvoltages and energy strikes can occur in the contact line due to thyristor locomotive commutation or self-induction, these interferences can disturb the normal operation of the PFC module and in some cases even destroy the module. A simulation model of the PFC is created, tested with simulated overvoltages, and three modifications/improvements to the existing topology are proposed.

Keywords—railway, uninterruptible power supply (UPS), power factor correction (PFC), battery charger, overvoltage, energy strike, interferences

I. Introduction

Recently, railway transport has become increasingly important, mainly for economic, environmental and energy reasons. Further development of railway transport requires appropriate power supply and development of electric rail vehicles and the railway infrastructure in general.

The railway power system (catenary) consists of a supply line and an overhead line. The supply line connects the traction power substation (TPSS) to the overhead contact line. The overhead contact line consists of a catenary/wire suspended from a supporting cable [1]. The current collector ("pantograph") slides along the contact wire/catenary and supplies the locomotive's electric motor drive. With the help of the return line via rails and ground leading back to the TPSS, the electric circuit is closed. In Croatia, the overhead contact line operates at a voltage of 25 kV, 50 Hz. The energy required to supply the overhead contact line system comes from 110 kV transmission lines to the TPSS, where the 100 kV/25 kV transformer is located.

Railway transport infrastructure includes communication and control modules, electrically operated rail switches with remote control, light and sound signaling, as well as other equipment and devices.Most of this infrastructure equipment within the railway system require reliable and uninterrupted power supply for safety reasons. For this purpose, various types of UPS-es are used. Since the infrastructural equipment operates at low AC and DC voltage (230 V, 50 Hz; 3 x 400/230 V, 50 Hz; 3 x 380 V, 83,33 Hz; 12 V_{dc}; 24 V_{dc}; 110 V_{dc}), UPS-es are powered by a 25 kV/0.23 kV transformer, which is also located in the TPSS.

Difficulties encountered in the rail vehicle power supply system, as well as in the power supply of infrastructure equipment from the UPS system, are the result of various disturbances caused primarily by passage of the train through the section powered by TPSS. Some examples of such disturbances are overvoltages and distortions of basic waveform of the catenary voltage. All of the aforementioned have a negative (sometimes fatal) impact on the UPS systemand connected equipment. To ensure the functionality of these devices, a UPS that is resistant to the previously mentioned disturbances is required. One solution is to use more robust active and passive components within the UPS.

An important requirement for UPS systems used to power infrastructure equipment, in addition to interference immunity, is a built-in PFC input stage, i.e., compliance with power quality regulations [2]. This primarily refers to the reduction of Total Harmonic Distortion (THD) factor and electromagnetic interference (EMI). In recent years, researchers and engineers have been striving to improve PFC and ensure that the quality of the grid current complies with the stipulations of standard regulations such as IEC 61000-3-2 and IEEE 519 [2][3]. The IEC 61000-3-2 standard refers to the maximum value for harmonic currents from the second harmonic up to and including the 40th harmonic, while the IEEE 519 standard refers to the control of the higher harmonics. Depending on the response speed (i.e. dynamic limits), PFC can be designed with a boost DC/DC converter or as a combination of boost DC/DC converter in the first stage and any DC/DC converter in the second stage [3]. For higher power (> 1 kW) and singlephase applications, the interleaved technique is recommended [2], [10]. This topology is characterized by the parallel connection of two or more converters with the same switching frequency but with control signals that are phase-shifted. The advantage of this topology is the reduction of component stress and dimensions, as well as the reduction of ripple currents of the inductors due to anti-phase operation. Another advantage is the reduction of electromagnetic interference (EMI) [2]. In recent research, various ways of PFC control have been considered [11]-[13]. Some of the PFC control techniques are Average Current Mode Control (ACMC) and Peak Current Mode Control (PCMC). The PFC can operate in Continuous Conduction Mode (CCM), Discontinuous Conduction Mode (DCM) or Critical Conduction Mode (CrCM) [4], [5].

This article analyzes the characteristics and immunity to input overvoltages of a UPS, fed by a 25 kV transformer. UPS mentioned uses two boost converters connected in parallel in interleaved mode to correct the power factor. It operates in CCM and in ACMC. The article consists of six chapters and a conclusion. The second chapter describes

---

All the costs of publishing of this paper are co-financed by the "MARETONII" project co-funded under the Competitiveness and Cohesion Operational Program from the European Regional Development Fund.

978-1-6654-3236-8/21 $31.00 © 2021 IEEE

voltage and energy interferences in the 25 kV line caused mainly by train passing through the section powered by the TPSS. It also considers accompanying regulations/standards defining the amounts of voltage and current harmonics allowed in the railway network. In the third chapter, the used topology of the PFC circuit and its operating principle are briefly described. In the fourth chapter, a simulation model of the existing UPS topology is described. Moreover, the behavior of this topology in presence of irregular shapes of the input voltage is investigated through simulation experiments. At the end of this chapter, the reasons for the misbehavior of the UPS are identified. Considering the above points, a modified UPS topology resistant to the observed problems is proposed in the fifth chapter. At the end of this chapter, simulation experiments show improvements in the behavior of UPS compared to the original version when irregular shapes of the input voltage occur. At the end of the conclusion, guidelines for further research are given.

II. IRREGULAR WAVEFORMS OF THE INPUT VOLTAGE AND ASSOCIATED NORMS

On the low-voltage (secondary) side of the 25 kV/0.23 kV transformer that supplies the UPS system, a sinusoidal voltage waveform without higher harmonics and other disturbances is expected. In reality, this is not always the case due to disturbances in the 25 kV catenary induced when a train passes through a particular TPSS. Some of these disturbances significantly change the waveform and magnitude of the peak voltage on the low-voltage side of the transformer. This poses a hazard to the equipment that is powered by that transformer.

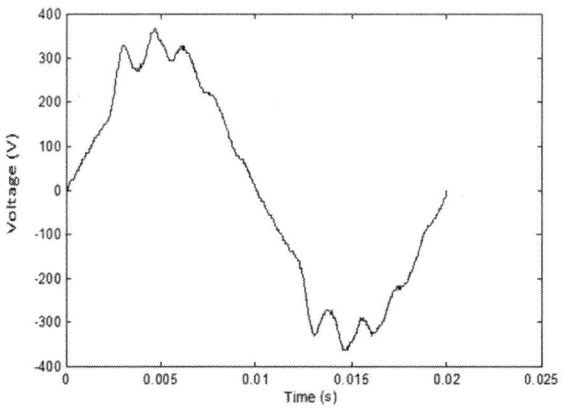

Figure 1: Distorted UPS input voltage

Figure 1 shows the distorted waveform of the transformer secondary voltage that occurs when the diode locomotive enters the TPSS [9]. This distortion is caused by the emphasized amplitudes of the 3rd, 5th, 7th, 9th, and especially the 11th and 13th harmonics. The measured values of the amplitudes of these harmonics are 14.89 V, 10.38 V, 7.56 V, 8.61 V, 18.58 V and 16.35 V, respectively. The waveform changes depending on how the locomotive accelerates, how it brakes, etc.

The second type of distorted input voltage manifests itself as a sinusoidal voltage with a peak value of 450 V that lasts between 5 and 15 seconds, as shown in Figure 2. The cause of this form of input voltage is self-induction in the overhead line when the locomotive leaves the TPSS. The longer the section of the supply line, the more pronounced/noticeable this phenomenon is. This is due to the higher value of parasitic inductance [9].

The third and most dangerous type of distorted input voltage manifests itself in the form of voltage spikes of about 600 V, Figure 3. This type of disturbance is a consequence of the passage of thyristor locomotives through the TPSS [9]. The following sections explain why this type of disturbance prevents proper operation of UPS.

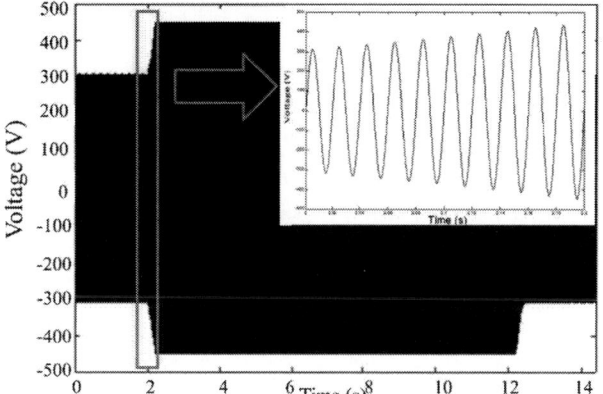

Figure 2: Input overvoltage with peak value of 450 V, duration approximately 12 seconds

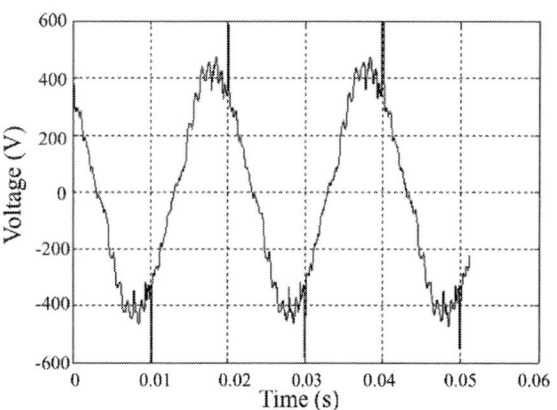

Figure 3: Input voltage with overvoltage spikes

There are a number of rules/norms that must be followed for the electrical subsystem of the railway infrastructure. For the purpose of this article, only the standards that concern on the low-voltage side of the transformer, where the input of UPS is connected, are considered.

The norm HR EN 50163:2007 contains information on permissible amounts (limits) of voltage and frequency [6]. According to point 4 of this standard, the permissible values of voltage and frequency of the 25 kV, 50 Hz network are given in Table 1. Among other standards that set requirements for power factor, it is worth mentioning the standard HR EN 50388:2013. This standard defines the condition that the power factor of trains with power >2 MW must be greater than 0.95. For trains of power ≤2 MW the power factor must be greater than 0.85, calculated during driving cycle between two stations.

978-1-6654-3236-8/21 $31.00 © 2021 IEEE

TABLE 1: VOLTAGE AND FREQUENCY LIMITS OF ELECTRIC TRACTION SYSTEM

Lowest intermittent voltage, duration up to 10 minutes	U_{min2}	17.5 kV
Lowest continuous voltage	U_{min1}	19.0 kV
Nominal/Rated voltage	U_n	25.0 kV
Maximum continuous voltage	U_{max1}	27.5 kV
Lowest intermittent voltage, duration up to 5 minutes	U_{max2}	29.0 kV
Maximum allowable peak value of voltage, duration up to 1 second	U_{max3}	38.75 kV
Frequency range	f	49 – 51 Hz

III. UNINTERRUPTIBLE POWER SUPPLY – TOPOLOGY AND PRINCIPLE OF OPERATION

Uninterruptible power supply systems can be implemented as passive, active or line-interactive [14]. In this paper, an active UPS is examined. Generally, UPS-es consist of an input stage, a battery used for energy storage and an output DC/DC or DC/AC converter, depending on the requirements. The batteries in the analyzed UPS system are Valve Regulated Lead-Acid(VRLA) type. The available UPS output voltages are as follows: 110 V_{dc}, 48 V_{dc}, 24 V_{dc} and 230 $V_{ac, rms}$.

A. Topology of the power factor correction module

The input stage of the UPS serves as a battery charger. In this case, the input stage has a power factor correction module and a DC/DC converter used for galvanic isolation and control of the battery charging process. The PFC module consists of a diode bridge rectifier and a two-phase interleaved boost converter, Figure 4. The DC/DC battery charging converter is an LLC type.

Parameters of the PFC input stage are given in the table below.

TABLE 2: PFC INPUT STAGE PARAMETERS

Description	Parameter	Value
Input voltage, rated	U_{in}	230 V_{rms}
Output votlage, rated	U_{out}	400 V_{dc}
Ratedpower	P	3000 W
Input current, rated	I_{in}	13 A_{rms}
Output current, rated	I_{out}	7.5 A
Switchingfrequency	f	68.1 kHz

The input voltage of the LLC is 400 V_{dc}, same as the PFC output voltage, and the LLC output voltage is 110V_{dc}.

The analyzed PFC module contains an EMI filter and an L - N varistor, as well as an L - N - GND star connection varistor protection between the transformer and the diode bridge rectifier. The varistor protection becomes significant at 700 V overvoltages. Since the analyzed overvoltages are below 700 V, the effect of varistor protection on the input voltage can be neglected.The chokes in EMI filter seen on Figure 4 are common mode chokes; one core (labeled L3:x, L4:x, L5:x) contains two separate windings (labeled 1 or 2 on each core).

Components D7, R1, and S1 are used to limit the inrush current of output capacitor C1. While the capacitor is being precharged, switch S1 and MOSFET-s Q1 and Q2 are turned off. The MOSFET-s are disabled by the PFC controller because it is in "soft start" mode during the precharge. When the capacitor voltage reaches approx. 300 V, switch S1 is closed and bypasses resistor R1, the controller exits the "soft-start" mode and starts normal operation.

B. PFC control stage

Commercially available integrated circuit PFC controllers are used to control the duty cycle of MOSFETs Q1 and Q2. In this case, the Texas Instruments UCC28070 controller is used [7]. There are many other commercially available PF controllers such as Infineon ICE2PCS01/02, On semiconductor FAN4800AU, STMicroelectronics L4981 or TI UCC3818A [16], [15], [17]. The UCC28070 is used to control two-phase interleaved boost converters in average current control in continuous conduction mode using an analog circuit based on transconductance operational amplifiers, Figure 5. It uses cascade control; the inner control loop is used to control the average inductor current, while the outer loop is the output voltage control loop which generates the reference current value for the current controller. The waveform of the reference current is obtained through a voltage divider (R11, R12), assuming that the input voltage has a sinusoidal waveform. PI regulators are used to control the output voltage and inductor currents and are tuned by the RC impedance connected to the outputs of the corresponding OP amplifiers. RC impedances are not part of the integrated circuit and are connected to the pin labeled VAO for the voltage PI regulator and CAOa or CAOb for each of the inductor current PI regulators of the boost converter, Figure 5. For proper operation of the PFC, some protection functions are implemented in the control stage, especially input overvoltage protection. The typical input overvoltage protection is implemented by defining the maximum input voltage (typically 264 - 275 V_{rms}, or 373 - 389 V_{max}[18]-[21]). If the input voltage exceeds the limit, the PFC is disconnected from the supply voltage and triggers a fault. In applications where relatively frequent overvoltages are expected, this type of input overvoltage protection is not acceptable, so a new PFC topology must be designed to withstand the overvoltages described in section II.

978-1-6654-3236-8/21 $31.00 © 2021 IEEE

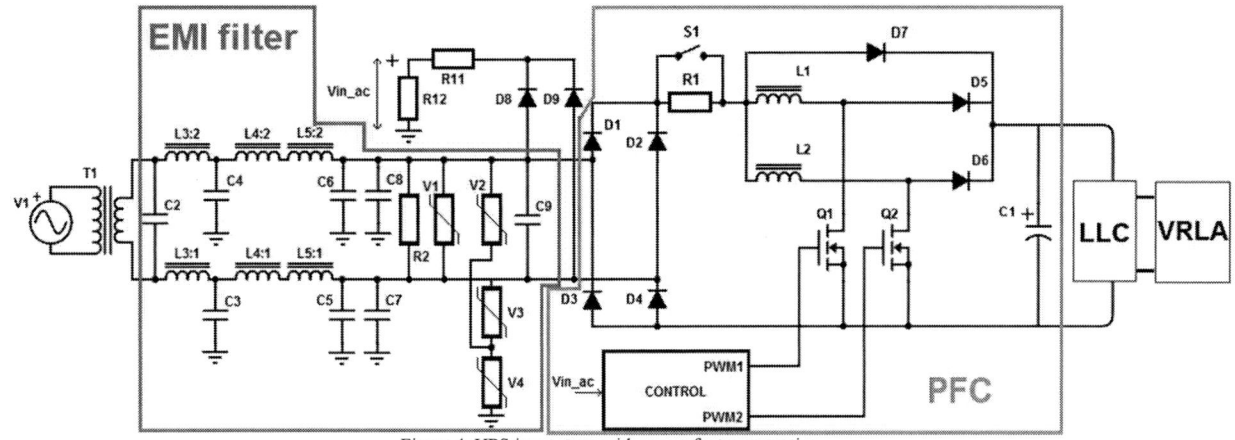

Figure 4: UPS input stage with power factor correction

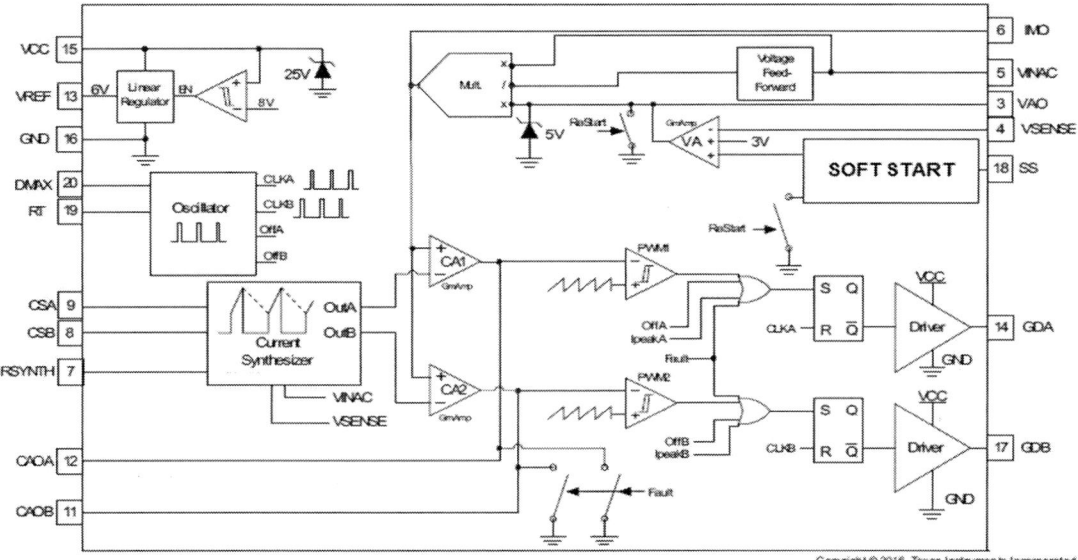

Figure 5: Block diagram of a UCC28070 PFC controller

IV. SIMULATION OF THE PFC MODULE OPERATION DURINGDISTURBANCES

In this section, a simulation model of the PFC module based on the topology shown in Figure 4 is presented. The overvoltages and waveforms described in Section II are reproduced and fed to the model. Some of the disadvantages of the current topology over the mentioned surges are identified. Solutions to these drawbacks are described and tested in Section V.

A. Simulation model

The simulation model of the analyzed PFC module from Figure 4 was created in PLECS. The model parameters are given in Table 3, and the model scheme is shown in Figure 6. The control block "TI UCC28070" is shown in Figure 7. EMI filter model is shown in Figure 8 based on a topology described in section III.A.

TABLE 3: SIMULATION MODEL PARAMETERS

Element	Parameter	Value
D1-D4	*Forward voltage* U_0	1.0 V
L1-L2	L	390 μH
D5-D6	U_0	1.55 V
FETD-FETD1	*On-resistance* R_{DS_on}	125 mΩ
C1-C5	C	330 μF
	R_{ESR}	0.01 Ω
C6, C7	C	150 nF
D7	*Forward voltage* U_0	1.15 V
R1	R	44 Ω
Input voltage divider gain		0.0076584
Output voltage divider gain		0.0076584
D8-D9	*Forward voltage* U_0	0 V
C8	C	330 nF

978-1-6654-3236-8/21 $31.00 © 2021 IEEE

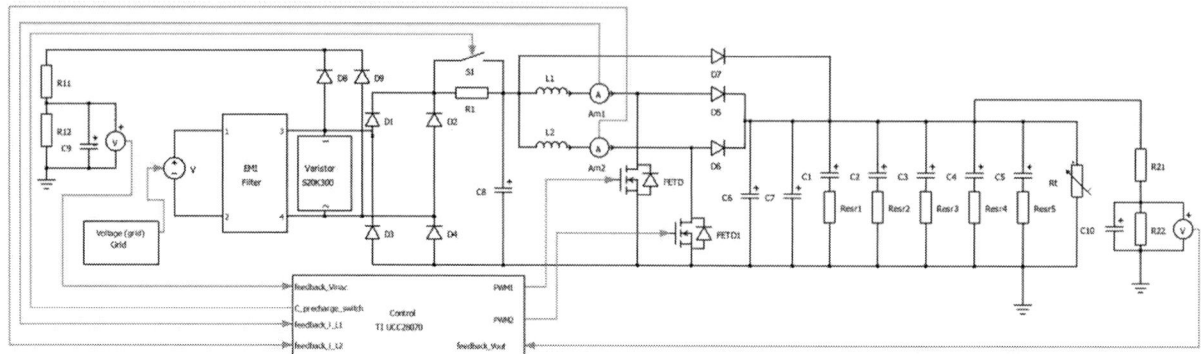

Figure 6: PLECS simulation model of PFC module

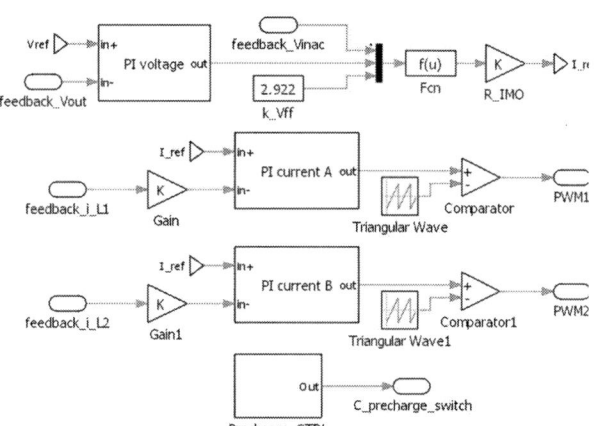

Figure 7: PLECS simulation model of the "TI UCC28070" control block

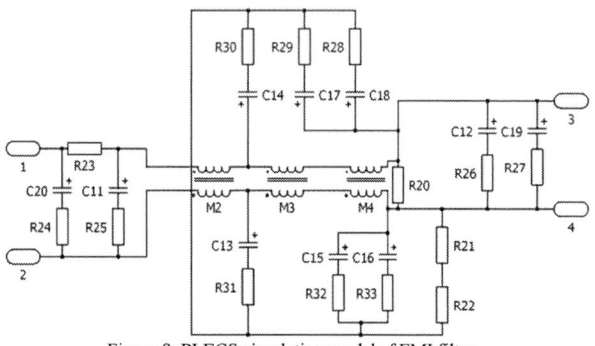

Figure 8: PLECS simulation model of EMI filter

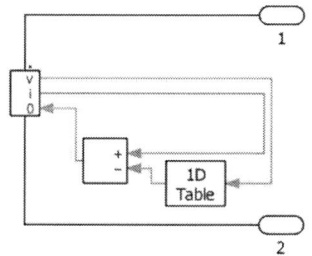

Figure 9: PLECS simulation model of a S20K300 varistor

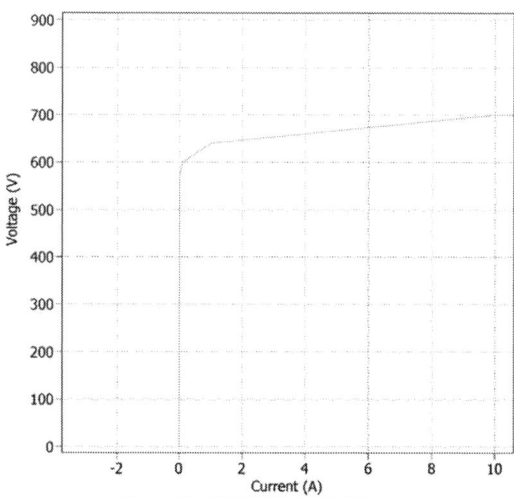

Figure 10: S20K300 model U-I curve

Varistor protection is modelled as a single L-N S20K300 varistor. The L-N-GND star connection is omitted since the L-N section is already modelled via S20K300 varistor, and L-GND and N-GND sections are not relevant for this topic. Since PLECS does not yet have a fully developed varistor model, one had to be created using a 1-D lookup table with input values being voltages across the varistor, and the output values being currents through the varistor, Figure 9. Detailed description of the model can be found in[23]. Simulated U-I curve is shown on Figure 10 and is based on datasheet listed in references [23].

B. Simulation results

When supplied with the input voltage containingwaveforms corresponding to the Section II, the PFC module cannot function properly. Three main problems are identified. First, the current waveform reference is no longer sinusoidal when overvoltages and higher order harmonics occur. This is due to the fact that UCC 28070 obtains the current reference waveform by taking the input voltage waveform directly from a voltage divider. The input voltage is rectified by diodes D8 and D9 and scaled down by a voltage divider consisting of resistors R11 and R12. This downscaled voltage signal is used directly as the waveform of the reference current. Since the current PI controllers ensure that

978-1-6654-3236-8/21 $31.00 © 2021 IEEE   92

the actual inductor currents follow the reference, the actual waveform of the input (grid) current will match that of the input (grid) voltage, Figure 11[7],[8]. The second problem in the operation of the PFC module occurs when the input voltage exceeds the output voltage (400 V in this case), due to the voltage spike in the input voltage waveform. During this event, diode D7 used for precharging the output capacitors is forward biased, shorting out the boost converters. The output capacitors are then exposed to the voltage spikes. Therefore, the output voltage can become too high and can damage the connected LLC converter or the capacitors themselves. This effect is somewhat reduced with the presence of the EMI filter, which reduces the amplitude of the spike and its chokes prevent fast and steep rise of the input current. Varistor's contribution is limited, since its threshold is shown to be around 600 V, Figure 10. Figure 12 shows that the output voltage is still acceptable, but without an EMI filter the output voltage would be close to the peak value of the voltage spike, which would damage the output capacitors and/or the LLC converter. Moreover, if the

voltage spikes peak value exceed 700 V, the EMI filter cannot dampen the overvoltage, and the output overvoltage would occur. The third problem is related to the boost converter topology itself. Boost converters cannot produce an output voltage that is lower than the input voltage. When an input overvoltage occurs, the PI current controllers generate the minimal duty cycle, which is zero. The MOSFET-s are turned off throughout the entire switching cycle, creating an LC filter consisting of the converter inductors and the output capacitors that can partially pass the input voltage spikes. Also, during overvoltages the precharge diode D7 is forward biased, which shorts out the boost converter. Therefore, the current spike, as a result of a voltage spike, will flow through the diode, instead of through the inductors. Inductors' currents will therefore slowly decrease until the overvoltage passes. When the output voltage becomes higher than the input voltage again, the PFC resumes its normal operation, Figure 14.

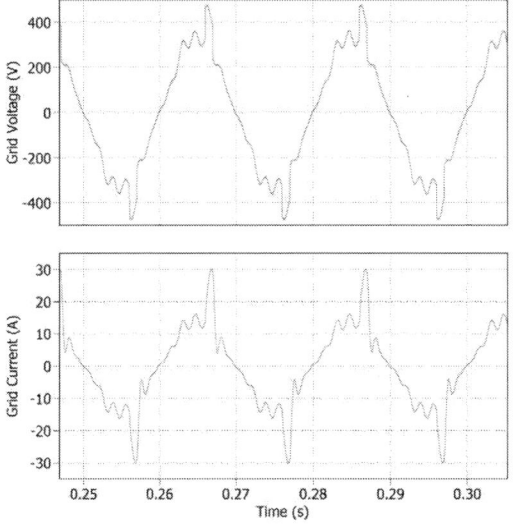

Figure 11:Grid voltage and current

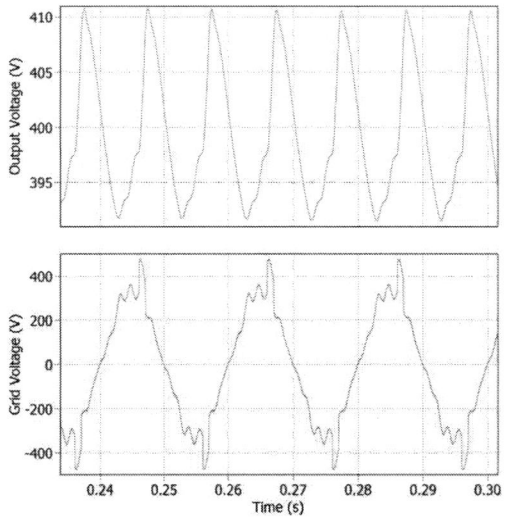

Figure 12: Outputvoltage vs. gridvoltage

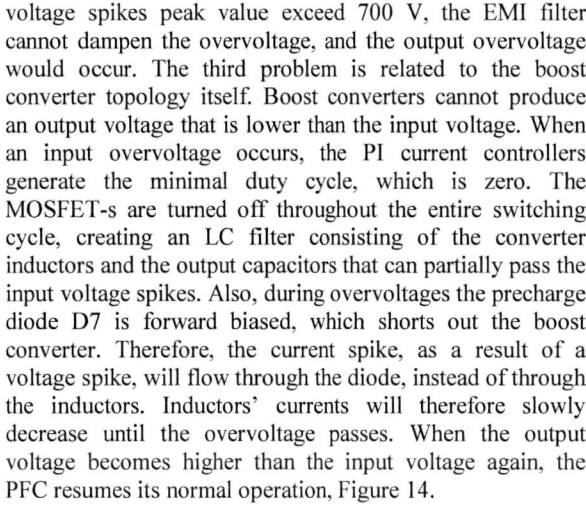

Figure 13: Inductor currents vs. grid voltage

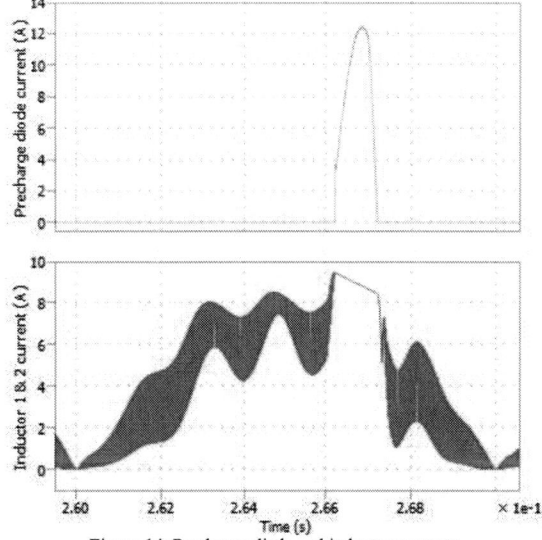

Figure 14: Precharge diode and inductor currents

V. MODIFIED UPS INPUT STAGE

The reference current waveform problem can be solved by using an internally generated sine wave reference instead of the input votlage waveform. The precharge diode problem can be solved by using a semiconductor switch such as a MOSFET or IGBT that can be turned on or off by the controller. During the precharge, the switch is turned on, allowing the inrush current to flow to the capacitors, and when precharging is complete, the switch is turned off so that an overvoltage cannot propagate through the converter. Under normal operating conditions, the switch would be turned off. A buck-boost converter topology will eliminate the problem of a zero duty cycle and current spikes during overvoltages.

A. ModifiedUPS input stage simulation model

Proposed solutions are integrated into the PLECS simulation model, resulting in a modified PFC module used as the input stage of the UPS, Figure 17. It uses the buck-boost converters in an interleaved mode of operation, a MOSFET switch instead of a diode for the output capacitor precharge, and the internally generated current reference waveform. Model parameters are given in Table 4.

TABLE 4: MODIFIED MODEL PARAMETERS

Element	Parameter	Value
D1-D4	*Forward voltage U_0*	1.0 V
L1-L2	*L*	6mH
D5-D6	*Forward voltage U_0*	1.55 V
FET1, FET2	*On-resistance $R_{DS_{on}}$*	125 mΩ
C1	*C*	1650 μF
	R_{ESR}	0.01 Ω
FET3	*On-resistance $R_{DS_{on}}$*	100 mΩ
R1	*R*	44 Ω

Simulation results of the modified PFC converter are shown below. It can be seen from Figure 15 that the input current of the PFC converter is indeed sinusoidal, despite the input voltage being distorted by higher-order harmonics. Due to interleaved mode of operation, grid current is not fully discontinuous. With duty cycles greater than D=0.5, the switch in the first phase of the converter is still turned on when the switch from the second phase turns on.

The ACMC control block is shown in detail in Figure 16. PI controllers are described directly as transfer functions, which can be easily implemented on a microcontroller and are much easier to modify, as opposed to the fixed RC constants used with UCC28070. Measured variables (output voltage and inductors' currents) are filtered with a low-pass filter to eliminate noise or other high-frequency interferences.

B. Modified PFC module simulation results

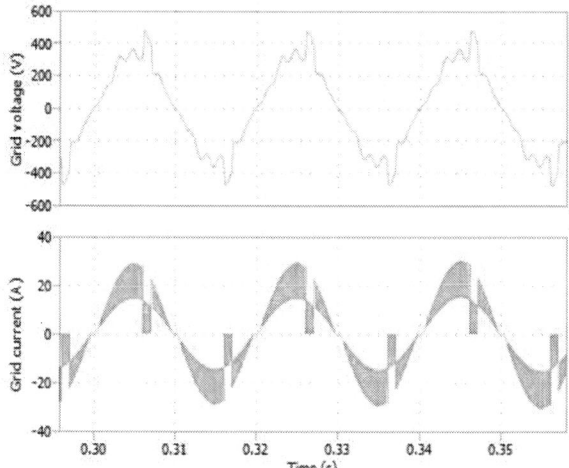

Figure 15: Input voltage and input current with internally generated sine wave reference

This results in a continuous grid current as long as the duty cycle is greater than 0.5. When an overvoltage spike is present, the duty cycle drops below 0.5, causing the input current to become discontinuous, Figure 18. When using the internally generated sine wave reference, it is critical to properly synchronize the sine wave with the actual input voltage waveform. The synchronization algorithm is not described in this article. From Figure 19, it can be seen that the output voltage remains at the desired level despite the input overvoltages.

The inductor currents are also unaffected by the input overvoltages when a buck-boost converter is used, as shown in Figure 20. The buck-boost converter topology relates to a completely new PFC converter design process, but also eliminates the disadvantages of the boost converter design.

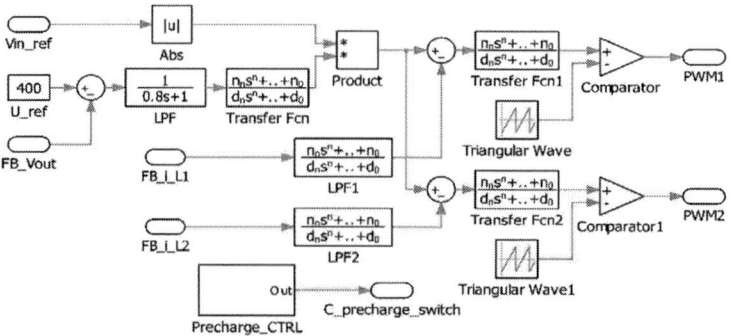

Figure 16: ACMC control block

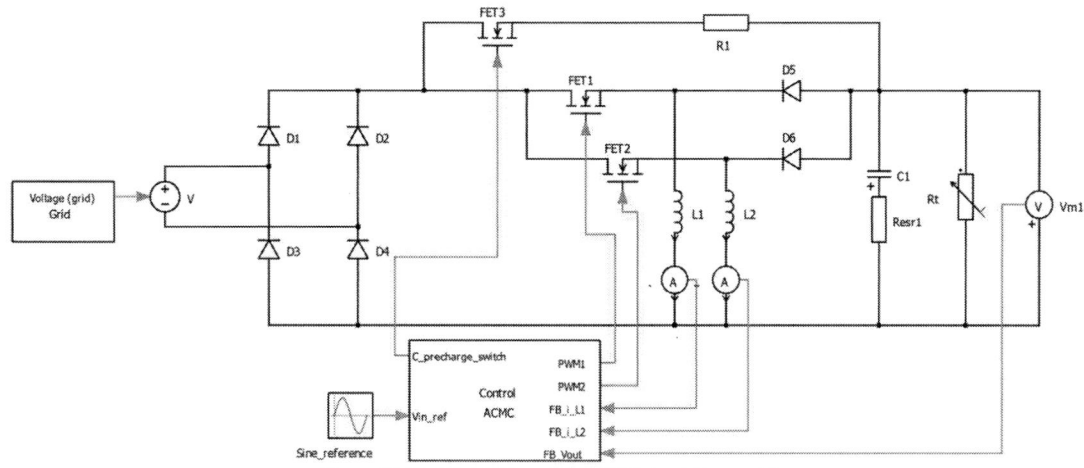

Figure 17: Modified PFC module PLECS simulation model

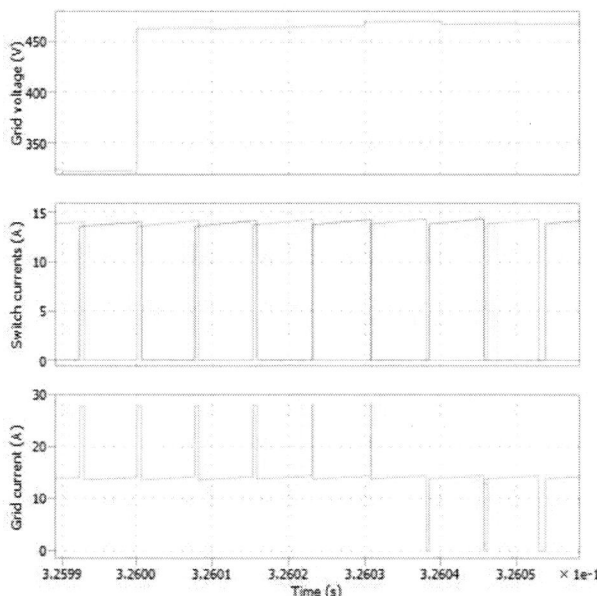

Figure 18: Grid voltage, grid current and switch currents during overvoltage spike

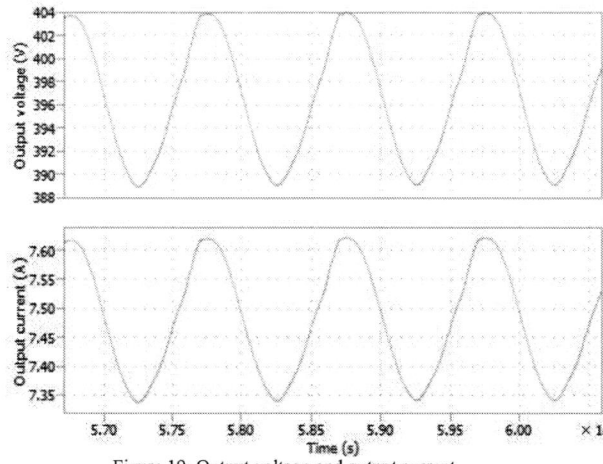

Figure 19: Output voltage and output current

Figure 20: Inductor current vs. input voltage of buck-boost PFC module

VI. CONCLUSION

Railway signalization and other infrastructure systems require an uninterruptible power supply (UPS) to function properly. The UPS itself is powered by a 25 kV, 50 Hz overhead contact line through a 25 kV/0.23 kV transformer. Since the contact line is susceptible to overvoltages, energy spikes, and other forms of disturbances, these disturbances are propagated from the transformer to the input stage of the UPS. The most serious disturbances are described. The UPS input stage has a power factor correction (PFC) module which is described in detail. The three most serious problems in the operation of the UPS module under the mentioned disturbances are presented: non-sinusoidal current draw due to distorted reference as a result of distorted input voltage, output overvoltage as a result of input overvoltage propagating through the precharge diode, and current spikes during input overvoltages. A modified version of the PFC module is proposed. The module consists of an internally generated sinusoidal reference, a MOSFET as a switch instead of the precharge diode, and a buck-boost converter topology to eliminate current spikes. Using PLECS model of this module and simulation experiments it is shown that these modifications solve the aforementioned problems regarding PFC module operation.

978-1-6654-3236-8/21 $31.00 © 2021 IEEE

REFERENCES

[1] I. Uglešić, M. Mandić, "Electric traction power supplies", pp. 54, Graphis d.o.o., Zagreb, 2015.

[2] A. Marcos-Pastor, E. Vidal-Idiarte, A. Cid-Pastor, L. Martinez-Salamero "Interleaved Digital Power Factor Correction Based on the Sliding-Mode Approach", IEEE Transaction on Power Electronics, Vol. 31, br. 6,pp. 4641-4653,June 2016.

[3] A. Fernández, J. Sebastián, P.Villegas, M. M. Hernando, D. G. Lamar, "Dynamic Limits of a Power-Factor Preregulator", IEEE Transactions on Industrial Electronics, Vol. 52, No. 51, pp. 77-87, February 2005.

[4] D. Stepins, J. Huang, "Effects of Switching Frequency Modulation on Input Power Quality of Boost Power Factor Correction Converter", International Journal of Power Electronics and Drive System (IJPEDS), Vol. 8, No. 2, pp. 882-899, June 2017.

[5] L. Rosetto, G. Spiazzi, P. Tenti, "Control techniques for power factor correction converters", University of Padova

[6] „Policy regarding technical specifications for electric railway infrastructure ", Narodne Novine broj 129/10 i 23/11

[7] "UCC28070 Interleaving Continuous Conduction Mode PFC Controller", Texas Instruments, April 2016.

[8] J.Schönberger, "Modeling a PFC controller using PLECS", Plexim GmbH

[9] M.Štetić, „Disturbances in 230 V power supply when using 25 kV catenary line as a power source", izvještaj s terena (field & measurements report - Croatia), December 2003.

[10] D. L. Woldegiorgis, „Design and control of three – channel interleaved boost power factor correction (PFC) converter", Researchgate publication 326719283, July 2018.

[11] D. K. Saini, "True-Average Current-Mode Control of DC-DC Power Converters: Analysis, Design, and Characterization", Wright State University, 2018.

[12] K. T. Wan, "Advanced current-mode control techniques for DC-DC power electronic converters", Missouri University of Science and Technology, doctoral dissertation, 2009.

[13] P. Frgal, „Average Current Mode Interleaved PFC Control", Freescale Semiconductor Inc, AN5257, February 2016.

[14] M. Aamir, K. Ahmed, K. S. Mekhilef, "Review: Uninterruptible Power Supply (UPS) system", Renewable and Sustainable Energy Reviews, Volume 58, Pages 1395-1410, May 2016.

[15] „Standalone Power Factor Correction (PFC) Controller in Continuous Conduction Mode (CCM)", Infineon technologies, datasheet, April 2017.

[16] „High-Voltage, Multimode Power Factor Controller", On semiconductor, datasheet, February 2021.

[17] „L4981 Power factor corrector", STMicroelectronics, datasheet, November 2001.

[18] www.synqor.com/product-overviews/industrial-power-supplies/isolated-power-factor-correction-modules, accessed 20.5.2021.

[19] absopulse.com/ac-dc-power-supply-with-active-pfc-and-redundant-operation-for-industrial-and-railway-applications/, accessed 20.5.2021.

[20] https://recom-power.com/en/applications/mobility/railway/railway.html?1#138, accessed 20.5.2021.

[21] https://www.directindustry.com/prod/siemens-power-supplies/product-17494-2042089.html, accessed 20.5.2021.

[22] https://forum.plexim.com/1142/how-to-simulate-a-device-with-known-v-i-equation, accessed 20.7.2021.

[23] "EPCOS SIOV metal oxide varistors" B722 series, datasheet, December 2011.

Observability Conditions for Speed Sensorless Induction Motor Models with Neglected or Included Iron Loss Representation

Krisztián Horváth
Department of Automation
Széchenyi István University
Győr, Hungary
krisztian.horvath@sze.hu

Abstract—Iron loss is usually neglected in induction motor models used for speed sensorless control and observer design. Thus, the complexity of control and estimator algorithms are reduced. However, the application of a more accurate model can improve the performance, hence the inclusion of iron loss in the machine model is becoming more widespread. But the effect of iron loss modeling on observability has not been analyzed yet. In this paper, observability conditions are presented for non-linear state-space models with included iron loss. Furthermore, the results are compared with the observability properties of the traditional models where iron loss is ignored.

Index Terms—iron loss, non-linear observability, speed sensorless model, squirrel-cage induction machine

I. Introduction

Speed sensorless induction machine (IM) drives have reduced cost, increased reliability and noise immunity as it is well-known. The control methods of these drives can be grouped in two main categories, which are discussed in detail in [1]. By exploiting anisotropies of IMs, improved speed estimation performance may be achieved when the excitation frequency is low. Nevertheless, the utilization of these methods are strongly depends on the machine design as written in [2]. Thus, the fundamental model-based estimators and observers are used more generally for sensorless IM applications. However, the low speed operation of these methods is usually poor. Besides IMs, the model-based approaches are also popular in case of other AC drives, as shown in [3] and [4] for instance.

In general, iron loss is neglected in IM models applied for control and observer design, as mentioned in [5]. But study [6] points out that the iron loss causes unwanted cross-coupling and performance deterioration in vector controlled IM drives. In addition, the uncompensated iron loss leads to inaccurate sensorless speed estimation, as presented in [7]. To reduce the error caused by iron loss, works [8]–[11] use IM models with incorporated iron loss for sensorless control and estimator design. But to the best of the author's knowledge, observability study for these non-linear models has not been presented

The research presented in this paper was funded by the "Thematic Excellence Program – National Challenges Subprogram – Establishment of the Center of Excellence for Autonomous Transport Systems at Széchenyi István University (TKP2020-NKA-14)" project.

yet. However, the observability properties of traditional state-space representations with neglected iron loss are discussed for example in studies [12]–[14].

This paper presents observability conditions for IM state-space models with incorporated iron loss. Moreover, the results are compared with the traditional models where the iron loss is ignored.

II. Mathematical Model of IM

The mathematical descriptions of IM dynamics are presented in this section. First, the traditional space vector model with neglected iron loss is discussed in the stator reference frame, which describes the electromagnetic behavior. Then, the space vector model is augmented by an iron loss resistance, which is connected in parallel with the magnetizing inductance. And finally, the mechanical behavior is modeled.

A. Space Vector Model with Neglected Iron Loss

The equivalent circuit of squirrel-cage IM can be seen in Fig. 1, which uses space vectors in complex form. The reference frame of the model is oriented to the stator, where the space vectors are given by α and β components as follows. $\boldsymbol{v}_s = v_{s\alpha} + jv_{s\beta}$ is the stator voltage vector, $\boldsymbol{i}_s = i_{s\alpha} + ji_{s\beta}$ and $\boldsymbol{i}_r = i_{r\alpha} + ji_{r\beta}$ are the stator and rotor current vectors, $\boldsymbol{\psi}_s = \psi_{s\alpha} + j\psi_{s\beta}$ and $\boldsymbol{\psi}_r = \psi_{r\alpha} + j\psi_{r\beta}$ are the stator and rotor flux vectors, respectively. In Fig. 1., ω_r denotes the rotor electrical velocity, and the parameters are the R_s stator and R_r rotor resistances, the $L_{s\sigma}$ stator and $L_{r\sigma}$ rotor leakage inductances, and the L_m mutual inductance.

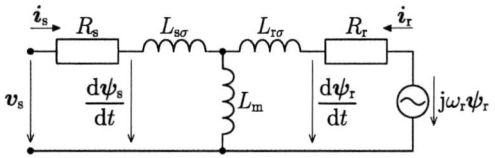

Fig. 1. Equivalent circuit of IM with neglected iron loss.

978-1-6654-3236-8/21 $31.00 © 2021 IEEE

Based on the equivalent circuit in Fig. 1., the voltage and flux equations may be written as

$$v_s = R_s i_s + \frac{d\psi_s}{dt}, \tag{1}$$

$$0 = R_r i_r + \frac{d\psi_r}{dt} - j\omega_r \psi_r, \tag{2}$$

$$\psi_s = L_s i_s + L_m i_r \tag{3}$$

and

$$\psi_r = L_r i_r + L_m i_s \tag{4}$$

where $L_s = L_{s\sigma} + L_m$ and $L_r = L_{r\sigma} + L_m$ are the stator and rotor inductances, respectively.

By using the rotor flux and stator current vectors, the electromagnetic torque can be determined as

$$T_e = \frac{3}{2} p \frac{L_m}{L_r} \Im(\psi_r^* i_s) \tag{5}$$

where p is the number of pole pairs, and the asterisk denotes complex conjugate value.

B. Space Vector Model with Included Iron Loss

In order to consider iron losses, the above model may be augmented by R_i iron loss resistance which is connected in parallel with L_m, as can be seen in Fig. 2. In this model, $i_i = i_{i\alpha} + j i_{i\beta}$ and $i_m = i_{m\alpha} + j i_{m\beta}$ are the iron loss current and the magnetizing current vectors, respectively.

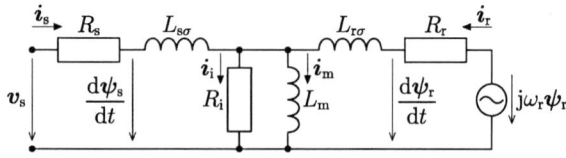

Fig. 2. Equivalent circuit of IM with included iron loss representation.

As in case of the previous model, the stator and rotor voltage equations are given by (1) and (2) in the augmented model as well. But the flux equations should be rewritten as follows:

$$\psi_s = L_{s\sigma} i_s + L_m i_m, \tag{6}$$

$$\psi_r = L_{r\sigma} i_r + L_m i_m. \tag{7}$$

In addition, the following equations can be written based on Fig. 2.:

$$R_i i_i = L_m \frac{d i_m}{dt}, \tag{8}$$

$$i_s + i_r = i_i + i_m. \tag{9}$$

Due to the iron loss modeling, the electromagnetic torque is

$$T_e = \frac{3}{2} p \frac{L_m}{L_r} \Im(\psi_r^* (i_s - i_i)) \tag{10}$$

in this augmented model.

C. Mechanical Model

Since, the electromagnetic time constants are usually smaller than the mechanical one, the angular velocity of the rotor can be given as a slowly varying or constant quantity:

$$\frac{d\omega_m}{dt} = 0 \tag{11}$$

where $\omega_m = \omega_r / p$ is the rotor mechanical speed.

A more accurate description may be obtained by using the equation of motion:

$$\frac{d\omega_m}{dt} = \frac{T_e - T_l}{J} \tag{12}$$

where J and T_l denote the total inertia and the load torque, respectively. Load is unknown disturbance, but it is usually modeled as slowly varying quantity:

$$\frac{dT_l}{dt} = 0. \tag{13}$$

By using equations (12)-(13) in the machine model applied for the design of a speed sensorless estimator or observer, the dynamic performance can be improved and it allows load estimation as well.

III. Observability Analysis

In this section, observability analysis is presented for IM state-space models based on the local weak observability theory introduced in [15]. First, the observability theory is briefly discussed. Then, observability conditions are given for two widely used state-space representations, where iron loss is neglected. In the first case, the model has 5 state variables and assumes slowly varying rotor velocity. The state vector of the second investigated model is extended by the load torque, and this 6-dimensional model contains the (12) equation of motion as well. Afterwards, the observability of two state-space models are analyzed, which include the iron loss. The first of these models is 7th order and assumes constant speed, while the second one is augmented by the load torque as state variable and includes the equation of motion.

A. Theory of Local Weak Observability

In continuous time, the state-space representation of a non-linear system may be written as

$$\frac{dx}{dt} = f(x, u), \tag{14}$$

$$y = h(x) \tag{15}$$

where x, u, y are the state, the input and the output vectors, respectively, and $f(\cdot)$, $h(\cdot)$ are non-linear functions.

In order to analyze the observability of system (14)-(15), the observability matrix should be defined as

$$O = \begin{bmatrix} \frac{\partial \mathcal{L}_f^0 h}{\partial x} & \frac{\partial \mathcal{L}_f h}{\partial x} & \cdots & \frac{\partial \mathcal{L}_f^{n-1} h}{\partial x} \end{bmatrix}^T \tag{16}$$

where $\mathcal{L}_f^k h$ denotes k^{th} order Lie derivative, and n is the number of state variables.

The model given by equations (14)-(15) is locally weakly observable at x_0, if the observability matrix has full-rank at that point.

B. Observability Conditions for 5-dimensional IM Model with Neglected Iron Loss

In the first analyzed IM model, the iron loss is neglected and constant rotor velocity is assumed. In this case, the 5-dimensional state-space representation is derived based on equations (1)-(4) and (11). The state, input and output vectors of this model are

$$\boldsymbol{x}_1 = \begin{bmatrix} i_{s\alpha} & i_{s\beta} & \psi_{r\alpha} & \psi_{r\beta} & \omega_r \end{bmatrix}^T, \quad (17)$$

$$\boldsymbol{u}_1 = \begin{bmatrix} v_{s\alpha} & v_{s\beta} \end{bmatrix}^T \quad (18)$$

and

$$\boldsymbol{y}_1 = \begin{bmatrix} i_{s\alpha} & i_{s\beta} \end{bmatrix}^T. \quad (19)$$

Due to the mechanical sensorless operation, the (19) output vector consists of the stator current components. In this model, the $\boldsymbol{f}(\cdot)$ and $\boldsymbol{h}(\cdot)$ functions are

$$\boldsymbol{f}_1(\boldsymbol{x}_1, \boldsymbol{u}_1) = \begin{bmatrix} \frac{R_r L_m \psi_{r\alpha}}{\sigma L_s L_r^2} - \frac{R_s L_r^2 + R_r L_m^2}{\sigma L_s L_r^2} i_{s\alpha} + \frac{L_m \omega_r \psi_{r\beta}}{\sigma L_s L_r} + \frac{v_{s\alpha}}{\sigma L_s} \\ \frac{R_r L_m \psi_{r\beta}}{\sigma L_s L_r^2} - \frac{R_s L_r^2 + R_r L_m^2}{\sigma L_s L_r^2} i_{s\beta} - \frac{L_m \omega_r \psi_{r\alpha}}{\sigma L_s L_r} + \frac{v_{s\beta}}{\sigma L_s} \\ \frac{R_r L_m}{L_r} i_{s\alpha} - \frac{R_r}{L_r} \psi_{r\alpha} - \omega_r \psi_{r\beta} \\ \frac{R_r L_m}{L_r} i_{s\beta} - \frac{R_r}{L_r} \psi_{r\beta} + \omega_r \psi_{r\alpha} \\ 0 \end{bmatrix} \quad (20)$$

and

$$\boldsymbol{h}_1(\boldsymbol{x}_1) = \begin{bmatrix} i_{s\alpha} & i_{s\beta} \end{bmatrix}^T. \quad (21)$$

In order to simplify the expression (20), $\sigma = 1 - \frac{L_m^2}{L_s L_r}$ has been introduced, which denotes the leakage coefficient.

Since, the state vector has 5 and the output vector has 2 elements, the dimensions of the observability matrix are 10×5 for the model (17)-(21). To guarantee the full-rank property of the observability matrix, there must be at least one regular matrix consisting of 5 different rows of the observability matrix. Based on the first 5 rows, $\boldsymbol{O}_{1a}$ matrix is created, which has the following determinant:

$$\det\{\boldsymbol{O}_{1a}\} = -\frac{L_m^3}{L_r^2(L_m^2 - L_s L_r)^3}(R_r^2 + L_r^2 \omega_r^2)\frac{d\psi_{r\beta}}{dt}. \quad (22)$$

If $\det\{\boldsymbol{O}_{1a}\} \neq 0$, the locally weakly observable property is guaranteed for the system described by expressions (17)-(21). And the determinant of $\boldsymbol{O}_{1a}$ is non-zero, if $\psi_{r\beta}$ is not constant. Similarly, another condition can be determined based on the determinant of $\boldsymbol{O}_{1b}$, which consists of the first 4 and the 6th rows of the observability matrix. The determinant of $\boldsymbol{O}_{1b}$ is

$$\det\{\boldsymbol{O}_{1b}\} = \frac{L_m^3}{L_r^2(L_m^2 - L_s L_r)^3}(R_r^2 + L_r^2 \omega_r^2)\frac{d\psi_{r\alpha}}{dt}. \quad (23)$$

Thus, the regularity of the observability matrix is provided in that case as well, when $\psi_{r\alpha}$ is not constant. Consequently, the 5th order model is locally weakly observable, if the rotor flux vector is not constant $\left(\frac{d\psi_r}{dt} \neq 0\right)$. In other words, the (17) state vector of the model can be observed based on the stator current and voltage vectors if the rotor flux vector varies. This result is the same as in works [12] and [13].

C. Observability Conditions for 6-dimensional IM Model with Neglected Iron Loss

By using equations (1)-(5), (12) and (13), a 6th order state-space model may be defined as

$$\boldsymbol{x}_2 = \begin{bmatrix} i_{s\alpha} & i_{s\beta} & \psi_{r\alpha} & \psi_{r\beta} & \omega_r & T_l \end{bmatrix}^T, \quad (24)$$

$$\boldsymbol{u}_2 = \begin{bmatrix} v_{s\alpha} & v_{s\beta} \end{bmatrix}^T, \quad (25)$$

$$\boldsymbol{f}_2(\boldsymbol{x}_2, \boldsymbol{u}_2) = \begin{bmatrix} \frac{R_r L_m \psi_{r\alpha}}{\sigma L_s L_r^2} - \frac{R_s L_r^2 + R_r L_m^2}{\sigma L_s L_r^2} i_{s\alpha} + \frac{L_m \omega_r \psi_{r\beta}}{\sigma L_s L_r} + \frac{v_{s\alpha}}{\sigma L_s} \\ \frac{R_r L_m \psi_{r\beta}}{\sigma L_s L_r^2} - \frac{R_s L_r^2 + R_r L_m^2}{\sigma L_s L_r^2} i_{s\beta} - \frac{L_m \omega_r \psi_{r\alpha}}{\sigma L_s L_r} + \frac{v_{s\beta}}{\sigma L_s} \\ \frac{R_r L_m}{L_r} i_{s\alpha} - \frac{R_r}{L_r} \psi_{r\alpha} - \omega_r \psi_{r\beta} \\ \frac{R_r L_m}{L_r} i_{s\beta} - \frac{R_r}{L_r} \psi_{r\beta} + \omega_r \psi_{r\alpha} \\ \frac{3}{2}\frac{p^2 L_m}{L_r J}(i_{s\beta}\psi_{r\alpha} - i_{s\alpha}\psi_{r\beta}) - \frac{p}{J}T_l \\ 0 \end{bmatrix} \quad (26)$$

and

$$\boldsymbol{h}_2(\boldsymbol{x}_2) = \begin{bmatrix} i_{s\alpha} & i_{s\beta} \end{bmatrix}^T. \quad (27)$$

The observability matrix of the 6-dimensional IM model has 12 rows and 6 columns. From the first 6 rows, the matrix $\boldsymbol{O}_2$ may be constructed, which has the following determinant:

$$\det\{\boldsymbol{O}_2\} = \frac{pR_r L_m^4(\psi_{r\alpha}^2 + \psi_{r\beta}^2)}{L_r^2(L_m^2 - L_s L_r)^3 J}\left(\frac{R_r^2 + L_r^2 \omega_r^2}{R_r L_r}\frac{d\varphi_{\psi_r}}{dt} + \frac{d\omega_r}{dt}\right) \quad (28)$$

where φ_{ψ_r} denotes the rotor flux space vector position, whose derivative is

$$\frac{d\varphi_{\psi_r}}{dt} = \omega_r + \frac{R_r L_m}{L_r}\frac{i_{s\beta}\psi_{r\alpha} - i_{s\alpha}\psi_{r\beta}}{\psi_{r\alpha}^2 + \psi_{r\beta}^2} = \omega_r + \frac{2}{3}\frac{R_r}{p}\frac{T_e}{\psi_{r\alpha}^2 + \psi_{r\beta}^2} \quad (29)$$

where T_e is given by (5). The 6-dimensional model is locally weakly observable, if $\det\{\boldsymbol{O}_2\}$ is not equal to zero. As a result, the observability condition is

$$\psi_{r\alpha}^2 + \psi_{r\beta}^2 \neq 0 \wedge \frac{R_r^2 + L_r^2 \omega_r^2}{R_r L_r}\frac{d\varphi_{\psi_r}}{dt} + \frac{d\omega_r}{dt} \neq 0 \Rightarrow \text{rank}\{\boldsymbol{O}_2\} = 6. \quad (30)$$

As can be seen in (30), a necessary condition is that the magnitude of the rotor flux vector cannot be zero. In addition, the following expression must be fulfilled:

$$\frac{R_r^2 + L_r^2 \omega_r^2}{R_r L_r}\frac{d\varphi_{\psi_r}}{dt} + \frac{d\omega_r}{dt} \neq 0. \quad (31)$$

Besides the condition of rotor flux magnitude, some special cases can be formulated to provide the locally weakly observable property based on expression (31):

- When the rotor angular velocity is constant, then the frequency of the rotor flux vector cannot be zero. Based on (29), the unobservable curve can be written as

$$T_e = -\frac{3}{2}\frac{p}{R_r}\left(\psi_{r\alpha}^2 + \psi_{r\beta}^2\right)\omega_r. \quad (32)$$

- When the rotor angular velocity equals to zero, then the slip frequency should not be zero. In other words, the electromagnetic torque should not be zero if the rotor speed is zero.

- If the rotor flux frequency is zero, then the rotor speed cannot be constant.

These results are also known in the literature, see for example [14].

D. Observability Conditions for 7-dimensional IM Model with Included Iron Loss

Due to the iron loss modeling, the dimension of the state-space model is increased by two compared to the former models where R_i is neglected. Assuming constant or slowly varying speed, the 7-dimensional state-space model with included iron loss can be given as follows:

$$\boldsymbol{x}_3 = \begin{bmatrix} i_{s\alpha} & i_{s\beta} & \psi_{r\alpha} & \psi_{r\beta} & i_{i\alpha} & i_{i\beta} & \omega_r \end{bmatrix}^{\mathrm{T}}, \quad (33)$$

$$\boldsymbol{u}_3 = \begin{bmatrix} v_{s\alpha} & v_{s\beta} \end{bmatrix}^{\mathrm{T}}, \quad (34)$$

$$\boldsymbol{f}_3(\boldsymbol{x}_3, \boldsymbol{u}_3) =$$
$$\begin{bmatrix} -\frac{R_s}{L_{s\sigma}} i_{s\alpha} - \frac{R_i}{L_{s\sigma}} i_{i\alpha} + \frac{1}{L_{s\sigma}} v_{s\alpha} \\ -\frac{R_s}{L_{s\sigma}} i_{s\beta} - \frac{R_i}{L_{s\sigma}} i_{i\beta} + \frac{1}{L_{s\sigma}} v_{s\beta} \\ \frac{R_r L_m}{L_r}(i_{s\alpha} - i_{i\alpha}) - \frac{R_r}{L_r}\psi_{r\alpha} - \omega_r\psi_{r\beta} \\ \frac{R_r L_m}{L_r}(i_{s\beta} - i_{i\beta}) - \frac{R_r}{L_r}\psi_{r\beta} + \omega_r\psi_{r\alpha} \\ \frac{R_r(L_m(i_{s\alpha}-i_{i\alpha})-\psi_{r\alpha})}{L_{r\sigma}L_r} - \frac{\omega_r\psi_{r\beta}}{L_{r\sigma}} - \frac{R_s i_{s\alpha}+R_i i_{i\alpha}}{L_{s\sigma}} - \frac{R_i L_r i_{i\alpha}}{L_m L_{r\sigma}} + \frac{v_{s\alpha}}{L_{s\sigma}} \\ \frac{R_r(L_m(i_{s\beta}-i_{i\beta})-\psi_{r\beta})}{L_{r\sigma}L_r} + \frac{\omega_r\psi_{r\alpha}}{L_{r\sigma}} - \frac{R_s i_{s\beta}+R_i i_{i\beta}}{L_{s\sigma}} - \frac{R_i L_r i_{i\beta}}{L_m L_{r\sigma}} + \frac{v_{s\beta}}{L_{s\sigma}} \\ 0 \end{bmatrix}$$
$$(35)$$

and

$$\boldsymbol{h}_3(\boldsymbol{x}_3) = \begin{bmatrix} i_{s\alpha} & i_{s\beta} \end{bmatrix}^{\mathrm{T}}. \quad (36)$$

Similarly to the 5th order model, $\boldsymbol{O}_{3a}$ and $\boldsymbol{O}_{3b}$ matrices are constructed from the observability matrix of the 7-dimensional model. $\boldsymbol{O}_{3a}$ consists of the first 7 rows of the observability matrix, and $\boldsymbol{O}_{3b}$ includes the first 6 and the 8th rows. The determinants of these matrices are

$$\det\{\boldsymbol{O}_{3a}\} = \frac{R_i^5}{L_{s\sigma}^5 L_{r\sigma}^3 L_r^2}(R_r^2 + L_r^2\omega_r^2)\frac{\mathrm{d}\psi_{r\beta}}{\mathrm{d}t} \quad (37)$$

and

$$\det\{\boldsymbol{O}_{3b}\} = -\frac{R_i^5}{L_{s\sigma}^5 L_{r\sigma}^3 L_r^2}(R_r^2 + L_r^2\omega_r^2)\frac{\mathrm{d}\psi_{r\alpha}}{\mathrm{d}t}. \quad (38)$$

According to these results, the 7-dimensional model is locally weakly observable, if the rotor flux vector is not constant. This condition is the same as in case of the 5th order model. Thus, the condition $\frac{\mathrm{d}\psi_r}{\mathrm{d}t} \neq 0$ provides locally weakly observable property to the 5- and the 7-dimensional models as well. Iron loss modeling in this case does not affect the observability condition.

E. Observability Conditions for 8-dimensional IM Model with Included Iron Loss

By using equations (1), (2), (6)-(10), (12) and (13), the 8-dimensional IM state-space model may be given as follows:

$$\boldsymbol{x}_4 = \begin{bmatrix} i_{s\alpha} & i_{s\beta} & \psi_{r\alpha} & \psi_{r\beta} & i_{i\alpha} & i_{i\beta} & \omega_r & T_l \end{bmatrix}^{\mathrm{T}}, \quad (39)$$

$$\boldsymbol{u}_4 = \begin{bmatrix} v_{s\alpha} & v_{s\beta} \end{bmatrix}^{\mathrm{T}}, \quad (40)$$

$$\boldsymbol{f}_4(\boldsymbol{x}_4, \boldsymbol{u}_4) =$$
$$\begin{bmatrix} -\frac{R_s}{L_{s\sigma}} i_{s\alpha} - \frac{R_i}{L_{s\sigma}} i_{i\alpha} + \frac{1}{L_{s\sigma}} v_{s\alpha} \\ -\frac{R_s}{L_{s\sigma}} i_{s\beta} - \frac{R_i}{L_{s\sigma}} i_{i\beta} + \frac{1}{L_{s\sigma}} v_{s\beta} \\ \frac{R_r L_m}{L_r}(i_{s\alpha} - i_{i\alpha}) - \frac{R_r}{L_r}\psi_{r\alpha} - \omega_r\psi_{r\beta} \\ \frac{R_r L_m}{L_r}(i_{s\beta} - i_{i\beta}) - \frac{R_r}{L_r}\psi_{r\beta} + \omega_r\psi_{r\alpha} \\ \frac{R_r(L_m(i_{s\alpha}-i_{i\alpha})-\psi_{r\alpha})}{L_{r\sigma}L_r} - \frac{\omega_r\psi_{r\beta}}{L_{r\sigma}} - \frac{R_s i_{s\alpha}+R_i i_{i\alpha}}{L_{s\sigma}} - \frac{R_i L_r i_{i\alpha}}{L_m L_{r\sigma}} + \frac{v_{s\alpha}}{L_{s\sigma}} \\ \frac{R_r(L_m(i_{s\beta}-i_{i\beta})-\psi_{r\beta})}{L_{r\sigma}L_r} + \frac{\omega_r\psi_{r\alpha}}{L_{r\sigma}} - \frac{R_s i_{s\beta}+R_i i_{i\beta}}{L_{s\sigma}} - \frac{R_i L_r i_{i\beta}}{L_m L_{r\sigma}} + \frac{v_{s\beta}}{L_{s\sigma}} \\ \frac{3}{2}\frac{p^2 L_m}{L_r J}\big((i_{s\beta} - i_{i\beta})\psi_{r\alpha} - (i_{s\alpha} - i_{i\alpha})\psi_{r\beta}\big) - \frac{p}{J}T_l \\ 0 \end{bmatrix}$$
$$(41)$$

and

$$\boldsymbol{h}_4(\boldsymbol{x}_4) = \begin{bmatrix} i_{s\alpha} & i_{s\beta} \end{bmatrix}^{\mathrm{T}}. \quad (42)$$

The observability matrix of the 8-dimensional IM model has 16 rows and 8 columns. From the first 8 rows, matrix $\boldsymbol{O}_4$ can be given. The determinant of $\boldsymbol{O}_4$ is

$$\det\{\boldsymbol{O}_4\} = \frac{pR_r R_i^6(\psi_{r\alpha}^2 + \psi_{r\beta}^2)}{L_{s\sigma}^4 L_{r\sigma}^6 L_r J}\left(\frac{R_r^2 + L_r^2\omega_r^2}{R_r L_r}\frac{\mathrm{d}\varphi_{\psi_r}}{\mathrm{d}t} + \frac{\mathrm{d}\omega_r}{\mathrm{d}t}\right) \quad (43)$$

where φ_{ψ_r} denotes the rotor flux vector position, whose derivative is

$$\frac{\mathrm{d}\varphi_{\psi_r}}{\mathrm{d}t} = \omega_r + \frac{2}{3}\frac{R_r}{p}\frac{T_e}{\psi_{r\alpha}^2 + \psi_{r\beta}^2}. \quad (44)$$

In (44), T_e is given by (10).

The observability condition for the 8-dimensional model is

$$\psi_{r\alpha}^2 + \psi_{r\beta}^2 \neq 0 \wedge \frac{R_r^2 + L_r^2\omega_r^2}{R_r L_r}\frac{\mathrm{d}\varphi_{\psi_r}}{\mathrm{d}t} + \frac{\mathrm{d}\omega_r}{\mathrm{d}t} \neq 0 \Rightarrow \mathrm{rank}\{\boldsymbol{O}_4\} = 8. \quad (45)$$

Conditions (30) and (45) are the same. Therefore, the same condition provides locally weakly observable property to the 6- and the 8-dimensional models as well. Iron loss modeling in this case does not affect the observability condition either.

IV. CONCLUSION

In this paper, observability analysis has been presented for IM state-space models. Due to the non-linear description of IMs, local weak observability theory has been used for the analysis and observability conditions have been given for each models.

The first investigated model with neglected iron loss uses stator current components, rotor flux components and rotor angular velocity as state variables. In case of this state-space representation, constant or slowly varying rotor speed is assumed. This model is locally weakly observable, if the rotor flux vector is not constant. The augmented counterpart of this model with included iron loss has the same observability condition. Then, observability condition has been given for the model with neglected iron loss, which contains the equation of motion. In this model, the load torque is an additional element of the state vector. The counterpart of this model with included iron loss has the same condition. Consequently, the iron loss modeling does not affect the observability condition in the analyzed cases.

978-1-6654-3236-8/21 $31.00 © 2021 IEEE

REFERENCES

[1] J. Holtz, "Sensorless control of induction machines—With or without signal injection?," *IEEE Transactions on Industrial Electronics*, vol. 53, no. 1, pp. 7–30, 2006.

[2] F. Briz and M. W. Degner, "Rotor position estimation," *IEEE Industrial Electronics Magazine*, vol. 5, no. 2, pp. 24–36, 2011.

[3] K. Kyslan, V. Šlapák, V. Petro, A. Marcinek, and F. Ďurovský, "Speed sensorless control of PMSM with unscented Kalman filter and initial rotor alignment," in *Proceedings of the International Conference on Electrical Drives & Power Electronics (EDPE)*, pp. 373–378, Sept. 2019.

[4] K. Kyslan and V. Petro, "Design and simulation of direct and indirect back EMF sliding mode observer for sensorless control of PMSM," *Power Electronics and Drives*, vol. 5, pp. 215–228, 2020.

[5] J. Holtz, "Sensorless control of induction motor drives," *Proceedings of the IEEE*, vol. 90, no. 8, pp. 1359–1394, 2002.

[6] E. Levi, "Impact of iron loss on behavior of vector controlled induction machines," *IEEE Transactions on Industry Applications*, vol. 31, no. 6, pp. 1287–1296, 1995.

[7] E. Levi and M. Wang, "Impact of iron loss on speed estimation in sensorless vector controlled induction machines," in *Proceedings of the 23rd International Conference on Industrial Electronics, Control, and Instrumentation (IECON)*, pp. 977–982, Nov. 1997.

[8] S.-D. Wee, M.-H. Shin, and D.-S. Hyun, "Stator-flux-oriented control of induction motor considering iron loss," *IEEE Transactions on Industrial Electronics*, vol. 48, no. 3, pp. 602–608, 2001.

[9] M. Tsuji, F. Xu, Y. Tsuruda, S.-i. Hamasaki, and S. Chen, "Characteristics of MRAS based induction motor sensorless vector control system taking into account iron loss," in *Proceedings of the International Conference on Electrical Machines and Systems (ICEMS)*, pp. 673–678, Nov. 2009.

[10] S. Yamamoto, H. Hirahara, A. Tanaka, T. Ara, and K. Matsuse, "Universal sensorless vector control of induction and permanent-magnet synchronous motors considering equivalent iron loss resistance," *IEEE Transactions on Industry Applications*, vol. 51, no. 2, pp. 1259–1267, 2015.

[11] M. Comanescu, "Sliding mode speed and load torque observer for core and copper loss minimization control of the induction motor drive," in *Proceedings of the IEEE International Conference on Industrial Technology (ICIT)*, pp. 534–539, Feb. 2018.

[12] P. Vaclavek, P. Blaha, and I. Herman, "AC drive observability analysis," *IEEE Transactions on Industrial Electronics*, vol. 60, no. 8, pp. 3047–3059, 2013.

[13] C. Canudas De Wit, A. Youssef, J. Barbot, P. Martin, and F. Malrait, "Observability conditions of induction motors at low frequencies," in *Proceedings of the 39th IEEE Conference on Decision and Control (CDC)*, pp. 2044–2049, Dec. 2000.

[14] M. Ghanes, J. De Leon, and A. Glumineau, "Observability study and observer-based interconnected form for sensorless induction motor," in *Proceedings of the 45th IEEE Conference on Decision and Control (CDC)*, pp. 1240–1245, Dec. 2006.

[15] R. Hermann and A. Krener, "Nonlinear controllability and observability," *IEEE Transactions on Automatic Control*, vol. 22, no. 5, pp. 728–740, 1977.

A Comparative Study of Different SMO Switching Functions for Sensorless PMSM Control

Viktor Petro, Karol Kyslan

Dept. of Electrical Engineering and Mechatronics
Technical University of Košice, Faculty of Electrical Engineering and Informatics
Košice, Slovak Republic
viktor.petro@tuke.sk, karol.kyslan@tuke.sk

Abstract—**This paper presents an experimental comparison of different switching functions for reduced-order sliding mode observers (ROSMO) for sensorless control of surface-mounted permanent magnet synchronous motor (SMPMSM). In ROSMO, a signum function is used to calculate feedback signals for the observer. Due to the chattering phenomenon and switching frequency limitations, the signum function can be substituted by the saturation function or by the sigmoid function. A comparative study of these functions is shown in the paper and their performance is evaluated by experimental verification.**

Index Terms—**permanent magnet synchronous machine, sensorless control, sliding mode control, switching function, estimation**

I. Introduction

Permanent magnet synchronous motors (PMSMs) have been gaining popularity in the last decades due to their high torque to inertia ratio, high power density, and efficiency. Because of these advantages, PMSMs are wildly used in the industry, especially for high-performance servo drives. According to the rotor construction, PMSMs can be divided into two groups: surface-mounted permanent magnet synchronous motors (SMPMSM) and interior permanent magnet synchronous motors (IPMSM). In the case of the SMPMSM, the quadrature and direct axis inductances are equal ($L_d = L_q$) whereas in IPMSM the direct axes inductance is always higher than the quadrature axis inductance ($L_d > L_q$) [1].

To achieve a high dynamic performance of PMSM, field-oriented control (FOC) strategy is adopted. It requires reliable and precise rotor position information. Incremental sensors, encoders, and resolvers mechanically mounted to the motor shaft are used for this purpose. However, the position sensor is sensitive to electromagnetic noise, decreases the overall mechanical reliability of the drive, requires mounting space, and increases the drive cost. The elimination of the sensor is preferable for some applications. In general, PMSM sensorless control approaches can be classified into two groups: magnetic saliency tracking and back electromotive force (EMF) detection [2]. In the first case, the variation of motor inductance due to saturation effects and/or geometrical saliency is observed. This variation is proportional to the rotor position. The "INFORM" [3], [4], and the high-frequency signal injection [5] are the most known methods in this group. These approaches can handle stand-still and low-speed regions. The other group of methods is based on the back-EMF

estimation. These methods can be used from above $5\% - 10\%$ of nominal speed. This group covers the Kalman filter [6], [7], reduced-order Luenberger observer [8], and the reduced or full order sliding-mode observer (SMO) [9].

One of the first applications of reduced-order sliding mode observer (ROSMO) for sensorless SMPMSM control was published in [10]. Since a discontinuous signum function was used, a low-pass filter (LPF) had to be included in the estimation algorithm. Because of the unwanted effects of LPF, the substitution of the signum function with other functions was further researched. In general, the saturation and sigmoid functions are often selected instead of signum function. In [11] the effect of switching functions in SMO was studied, however, only by MATLAB/Simulink. The goal of this paper is to compare these switching functions experimentally.

II. Mathematical Model of SMPMSM and SMO

The stator reference frame $\alpha\beta$ is used in most cases for the design of ROSMO. The state equation of SMPMSM using the current as a state variable is defined as:

$$\begin{bmatrix} \dot{i}_\alpha \\ \dot{i}_\beta \end{bmatrix} = \frac{1}{L_s} \begin{bmatrix} -R_s & 0 \\ 0 & -R_s \end{bmatrix} \begin{bmatrix} i_\alpha \\ i_\beta \end{bmatrix} + \frac{1}{L_s} \begin{bmatrix} u_\alpha - e_\alpha \\ u_\beta - e_\beta \end{bmatrix}, \quad (1)$$

where u_α and u_β are the applied stator voltage components, i_α and i_β are the stator current components, $e_\alpha = -\omega_e \lambda_{PM} \sin(\theta_e)$ and $e_\beta = \omega_e \lambda_{PM} \cos(\theta_e)$ are the stator back-EMF voltage components, where ω_e is the electrical angular velocity and $\lambda_{PM} = \frac{2}{3}\frac{k_t}{p} = \frac{k_e}{p}$ is the permanent magnet flux, where k_t is the motor torque constant, k_e is the EMF constant and p is the number of motor pole pairs, $R_s = R_{2ph}/2$ where R_{2ph} is the phase-to-phase resistance and $L_s = L_{2ph}/2 = L_d = L_q$ where L_{2ph} is the phase-to-phase inductance and L_d, L_q are the direct and quadrature axis inductances.

It is clear from previous equations that the back-EMF components include values of rotor position and velocity. Since the amplitude of back-EMF components cannot be directly measured during the control process, its estimation is required. ROSMO is used for the estimation in this paper. The mathematical model of ROSMO can be defined as [1]:

$$\begin{bmatrix} \dot{\hat{i}}_\alpha \\ \dot{\hat{i}}_\beta \end{bmatrix} = \frac{1}{L_s} \begin{bmatrix} -R_s & 0 \\ 0 & -R_s \end{bmatrix} \begin{bmatrix} \hat{i}_\alpha \\ \hat{i}_\beta \end{bmatrix} + \frac{1}{L_s} \begin{bmatrix} u_\alpha - \hat{e}_\alpha - z_\alpha \\ u_\beta - \hat{e}_\beta - z_\beta \end{bmatrix}, \quad (2)$$

978-1-6654-3236-8/21 $31.00 © 2021 IEEE

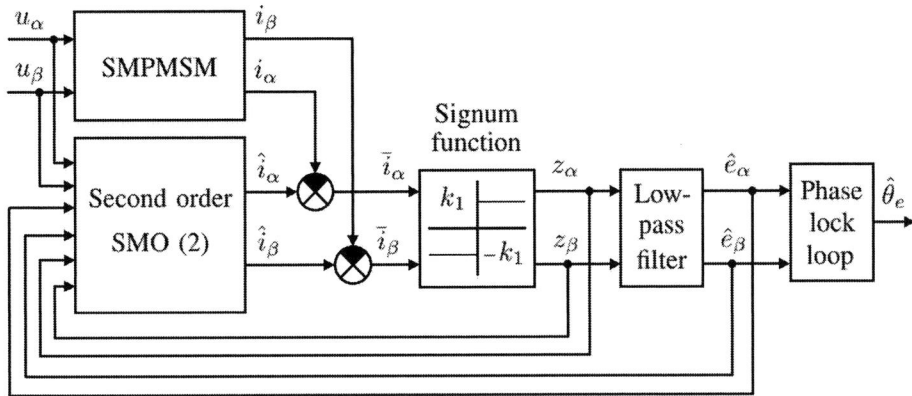

Fig. 1. Block diagram of reduced-order SMO with the signum function.

where superscript (ˆ) indicates an observed value and z_α, z_β are the observer feedback signals.

Assuming that all the real parameters of the motor i.e. the resistance and the inductance of windings are the same as in the observer equations, the applied voltage components are known and the current components are measured, the only unknown values are back-EMF components. Thus matching the observed currents to the measured currents will yield the actual back-EMF values. Since the observer's task is to force the observed current to match the measured current, the switching function is selected as follows:

$$\mathbf{s}(\mathbf{x}) = \begin{bmatrix} \bar{i}_\alpha \\ \bar{i}_\beta \end{bmatrix} = \begin{bmatrix} \hat{i}_\alpha - i_\alpha \\ \hat{i}_\beta - i_\beta \end{bmatrix}, \tag{3}$$

where $\bar{i}_\alpha$, $\bar{i}_\beta$ represents the error between the observed and measured current components. The switching action will occur when:

$$\mathbf{s}(\mathbf{x}) = 0. \tag{4}$$

Equation (4) defines the sliding surface (or the switching surface) of the SMO. The phase plane is divided into two areas in which the switching function has different signs.

A. Signum function

The trajectories of current differences, i.e. $\bar{i}_\alpha$ and $\bar{i}_\beta$ are forced towards the sliding surface. It is done using the following discontinuous feedback signal:

$$\begin{bmatrix} z_\alpha \\ z_\beta \end{bmatrix} = k_1 \begin{bmatrix} \text{sgn}(\hat{i}_\alpha - i_\alpha) \\ \text{sgn}(\hat{i}_\beta - i_\beta) \end{bmatrix}. \tag{5}$$

The sign of the current difference is determined by the signum function and it is multiplied by a suitable feedback gain k_1. The value of k_1 will be discussed later in the text.

The average value of feedback signals from (5) in a short time interval represents the estimated back-EMF values. Therefore its components can be obtained using a low-pass filter (LPF):

$$\begin{bmatrix} \hat{e}_\alpha \\ \hat{e}_\beta \end{bmatrix} = \frac{\omega_c}{s + \omega_c} \begin{bmatrix} z_\alpha \\ z_\beta \end{bmatrix}, \tag{6}$$

where ω_c is the cutoff frequency of the LPF. The cutoff frequency should be selected small enough to filter out the high-frequency component introduced by the discontinuous function. On the other way, ω_c should not affect the fundamental component of the feedback signal during the filtration process as observed back-EMF components will be further used for the position and speed calculation. LPF introduces a phase lag and amplitude damping and compensation of the observed values is necessary [12].

For ideal ROSMO, the signum function is executed with infinite frequency and the trajectories of the current differences reach and stay on the sliding hyperplane, i.e. $\mathbf{s}(\mathbf{x}) = 0$ and $\dot{\mathbf{s}}(\mathbf{x}) = 0$. However, the switching frequency calculation will be limited in a real application and the well-known chattering phenomenon will occur. To suppress the chattering of the observer, one of the available solutions is to use other suitable functions as a substitute for the signum function. In this article, the saturation and the sigmoid functions are substituted and experimentally verified.

B. Saturation function

The signum function can be substituted by the saturation function. If the absolute value of the current differences $\bar{i}_\alpha$ and $\bar{i}_\beta$ is lower than a predefined value E_{max} the feedback signal changes into a linear transient region [13]:

$$\begin{bmatrix} z_\alpha \\ z_\beta \end{bmatrix} = \frac{k_1}{E_{max}} \begin{bmatrix} (\hat{i}_\alpha - i_\alpha) \\ (\hat{i}_\beta - i_\beta) \end{bmatrix}. \tag{7}$$

The used saturation function can be seen in Fig. 2. The value of the feedback gain k_1 will be discussed later in the text.

C. Sigmoid function

In the case of the sigmoid function substitution, the discontinuous signum function is fully replaced by a continuous function F. Feedback signals can be written as follows:

$$\begin{bmatrix} z_\alpha \\ z_\beta \end{bmatrix} = k_1 \begin{bmatrix} F(\hat{i}_\alpha - i_\alpha) \\ F(\hat{i}_\beta - i_\beta) \end{bmatrix}, \tag{8}$$

978-1-6654-3236-8/21 $31.00 © 2021 IEEE

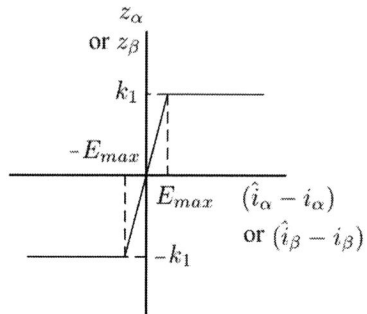

Fig. 2. The used saturation function.

where the function $F(x)$ is defined as [14]:

$$F(x) = \frac{2}{1 + e^{-\alpha x}} - 1, \qquad (9)$$

where α is the adjustable parameter. The sigmoid function is shown in Fig. 3 with different values of α parameter. The value of the feedback gain k_1 will be discussed later in the text.

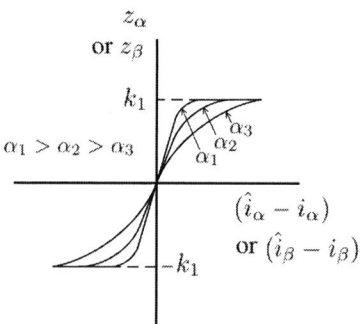

Fig. 3. The used sigmoid function.

For each type of switching function, the final observed back-EMF components are obtained by using the low-pass filter in (6). If the value E_{max} in saturation function is selected high enough, or if the value of α in sigmoid function is selected small enough, the continuous output (i.e. observed back-EMF values) is achieved even without LPF. This might simplify the system, however, the dynamic performance of the observer will be reduced.

The observed electrical position can be extracted from estimated back-EMF components using an arctangent function or a phase-lock loop (PLL). The estimated mechanical speed is calculated using the following equation:

$$\hat{\omega}_m = \frac{1}{p} \frac{d\hat{\theta}_e}{dt}. \qquad (10)$$

D. Feedback switching gain k_1

The value of the feedback gain k_1 plays an important role in the observer's performance. To prove the convergence of the observer a proper Lyapunov function candidate is selected. The derivation for the signum function can be found in [10], for saturation function in [2] and for sigmoid function in [14]. The following must stand to ensure stability for all three functions:

$$k_1 > \max(|e_\alpha|, |e_\beta|). \qquad (11)$$

III. EXPERIMENTAL RESULTS

The proposed switching functions (signum, saturation, and sigmoid) were tested experimentally with the same ROSMO topology and the only difference was the switching function. The goal was to outline their performance and to provide their mutual comparison.

The experimental setup consists of SMPMSM with parameters listed in Tab. 1, TMS320F28375D digital signal processor, and a prototype of voltage source inverter (VSI). Two different sampling rates were used. The ROSMO and the current loop were sampled at 20 kHz and the speed control loop was sampled at 1 kHz sampling frequency. The cut-off frequency for LPF given by (6) was set to 10 kHz. Decreasing the value of the cut-off frequency eliminates the noise in observed values but introduces a higher position observation error. To compare the estimation errors of the signum, saturation and sigmoid functions, the squared errors (SE) were calculated for each experiment at time step k as follows:

$$\text{SE}(k) = \left(\omega_{act}(k) - \omega_{obs}(k)\right)^2, \qquad (12)$$

where ω_{act} and ω_{obs} are the values of the actual and observed mechanical speed. Then the root-mean-square error (RMSE) values of estimated speeds and positions were calculated in the interval $0s \leq t \leq$ DL, where DL is the length of a dataset used for calculation of RMSE:

$$\text{RMSE} = \sqrt{\frac{1}{DL + 1} \sum_{N=1}^{DL} \text{SE}_N}. \qquad (13)$$

The rotor position and speed estimations using the back-EMF are reliable approximately above $5 - 10\%$ of the nominal speed. Therefore, some type of start-up procedure is required [15]. In this paper, the position sensor was used to reach a certain speed level. Then, the switch-over from sensored to sensorless control was executed. The switchover speed 300 rpm was selected and it represents 10% of the nominal speed. The observation fails below the selected switchover speed as the back-EMF amplitude has a critically low value to be observable. Therefore, the dataset for calculation of RMSE does not include values during the initial start-up period.

First, the basic signum function was verified. The value of the feedback gain was set to $k_1 = 200$ V and remained the same for all three switching functions. Experimental results of the speed and position observation performance and observation error for signum function are shown in Fig. 4 and Fig. 5, respectively. The RMSE speed observation error at steady state was about 77 rpm and the peak error was 116 rpm.

978-1-6654-3236-8/21 $31.00 © 2021 IEEE

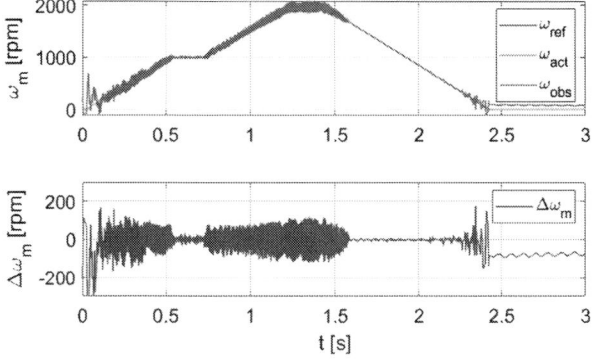

Fig. 4. Speed estimation using the signum function.

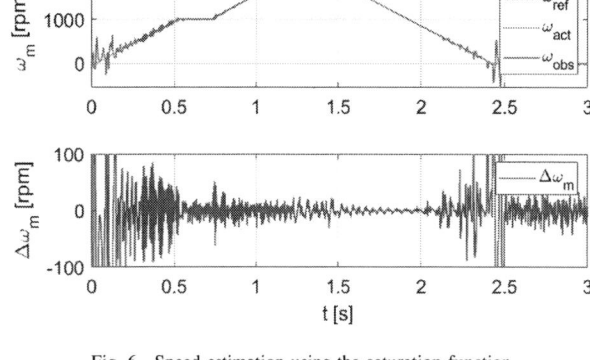

Fig. 6. Speed estimation using the saturation function.

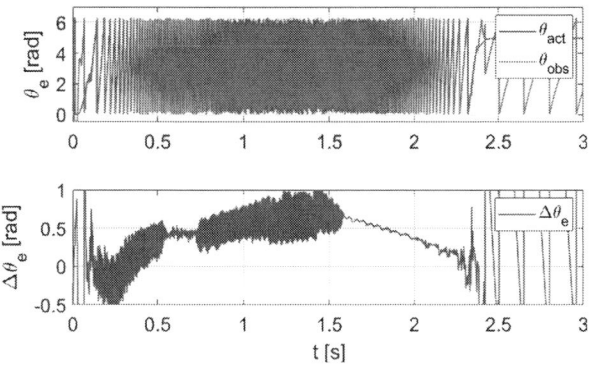

Fig. 5. Position estimation using the signum function.

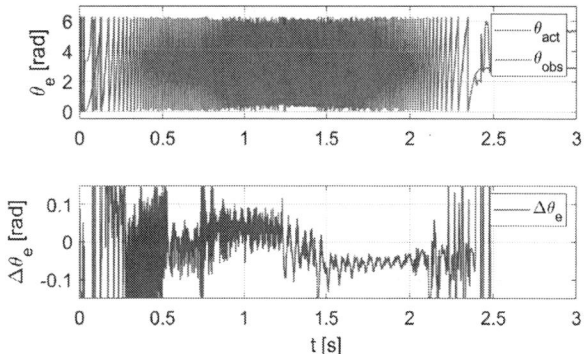

Fig. 7. Position estimation using the saturation function.

The RMSE position observation error at steady state was 0.6 electrical radians with a peak error of 0.98 rad.

Next, the saturation function was used with the same feedback gain k_1. In this case, the parameter E_{max} is user-defined and can affect the estimation performance. Selecting higher values of E_{max} suppresses the chattering phenomenon since the transient region becomes wider. On the other hand, higher values of E_{max} decrease the dynamic performance of the observer.

Experimental results for the speed and position observation performance and observation error for saturation function with $E_{max} = 20$ A are in Fig. 6 and Fig. 7, respectively. The RMSE speed observation error at steady state was around 9.4 rpm and the peak error was 52 rpm. The RMSE position observation error at steady state was 0.04 electrical radians with a peak error of 0.09 rad.

Finally, the sigmoid function was verified using the same feedback gain k_1. Here, the parameter α is user-defined. The wider is the transient region between k_1 and $-k_1$, the smoother are the output values of the observer. Selecting a high value for α leads to a wide transient region what suppresses the chartering but heavily decreases the dynamic performance of the observer. Experimental results of the speed and position observation performance and error for $\alpha = 0.15$ are in Fig. 8 and in Fig. 9. The RMSE speed observation error at steady state was around 8.9 rpm and the peak error was 20 rpm. The RMSE position observation error at steady state was 0.04 electrical radians with a peak error of 0.09 rad.

During experiments, the switchover from sensored to sensorless control was performed at the speed value 300 rpm. The observer is active above that speed. To express the effect of the α parameter and E_{max} parameter change on the observer's performance, the dataset for RMSE calculation was selected from $t = 0.5$ s to $t = 1.5$ s. This time window includes two steady states: 1000 rpm and 2000 rpm, and the transient state between them. The results are shown in Tab. 2.

The saturation function and the sigmoid function have both a transient region and achieve similar performance when the steady-state is reached. This can be seen when comparing Fig. 6 with Fig. 8. Using only the basic signum function introduces noise and chattering. As a result, the control performance is heavily decreased. To cope with the chattering phenomenon, the transient region is approached if the current error is less than a certain value. This is achieved by using the saturation and/or the sigmoid function.

Introducing the transient region near the zero current error

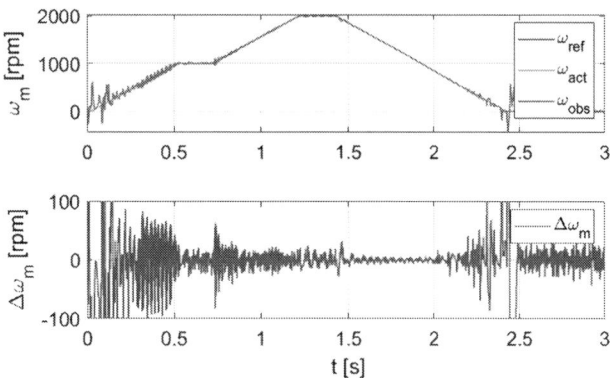

Fig. 8. Speed estimation using the sigmoid function.

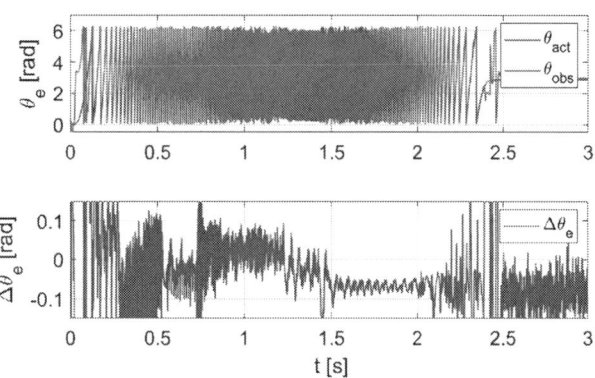

Fig. 9. Position estimation using the sigmoid function.

value improves the observation performance significantly.

Experimental results validate that utilization of the transient region is beneficial. Consider the example of sigmoid function and compare it with Fig. 3. Selecting $\alpha = 0.05$ yields to the wider transient region against selecting a higher value (e.g. $\alpha = 0.3$). It can be seen from Tab. II that for the higher value of α, thus the narrower liner region, the RMSE gets lower values. However, when selecting the transient region to be too wide, the overall dynamic performance of the sensorless drive is reduced.

IV. CONCLUSIONS

The experimental comparison of different switching functions for reduced-order sliding mode observer for sensorless control of surface-mounted permanent magnet synchronous motor has been presented in this paper. The results illustrate that the basic signum function for sliding mode observer should be rather replaced with saturation or sigmoid function. It is caused by utilisation of the transient region of these switching functions that grants the observer some additional "degree of freedom" during the observation process. The best results have been achieved for the sigmoid function with the highest value of parameter α.

TABLE I
SMPMSM PARAMETERS

Motor type	TGN3-0115-30-48/T1
DC link voltage	$U_{DC} = 48$ V
rated torque	$M_N = 1.13$ Nm
rated current	$I_N = 12.9$ A
torque constant	$k_t = 0.101$ Nm/A
EMF constant	$k_e = 6.12$ V/1000 rpm
number of pole pairs	$p = 5$
rated speed	$n_N = 3000$ rpm
stator resistance	$R_{2ph} = 0.258$ Ω
stator inductance	$L_{2ph} = 0.6$ mH

TABLE II
COMPARISON OF RMSE VALUES FOR DIFFERENT SWITCHING FUNCTIONS

Switching Function	RMSE of ω_m	RMSE of θ_e
signum	77 rpm	0.98 rad
saturation for $E_{max} = 10$ A	12.6 rpm	0.06 rad
saturation for $E_{max} = 40$ A	18.5 rpm	0.08 rad
sigmoid for $\alpha_1 = 0.30$	16.3 rpm	0.05 rad
sigmoid for $\alpha_2 = 0.15$	17.3 rpm	0.06 rad
sigmoid for $\alpha_3 = 0.05$	43.6 rpm	0.16 rad

It should be mentioned that temperature changes and inductance parameter mismatches will also affect the performance of the observer. The adaptive parameter estimation [16] can be used in that case. If the difference between d and q axis inductance is significant, a solution proposed in [1] should be used. The performance of switching functions depends also on the sampling time of the observer's control loop. Experimental results further indicate that some optimum value of the α parameter for the sigmoid switching function could exist. This should be a direction for future work. It would be readable to research how other classes of sigmoid function (e.g. those given in [17]) affect the performance of the observer.

ACKNOWLEDGMENT

This work was supported by the Scientific Grant Agency of the Ministry of Education of the Slovak Republic under the project VEGA 1/0493/19 and by the Slovak Research and Development Agency under the contract APVV-18-0436.

REFERENCES

[1] Wang, G., Zhang, G. and Xu, D. (2020). Position Sensorless Control Techniques for Permanent Magnet Synchronous Machine Drives. Springer: Singapore. ISBN 978-981-15-0049-7.

[2] S. Chi, Z. Zhang and L. Xu, "Sliding-Mode Sensorless Control of Direct-Drive PM Synchronous Motors for Washing Machine Applications," in IEEE Transactions on Industry Applications, vol. 45, no. 2, pp. 582-590, March-April 2009, doi: 10.1109/TIA.2009.2013545.

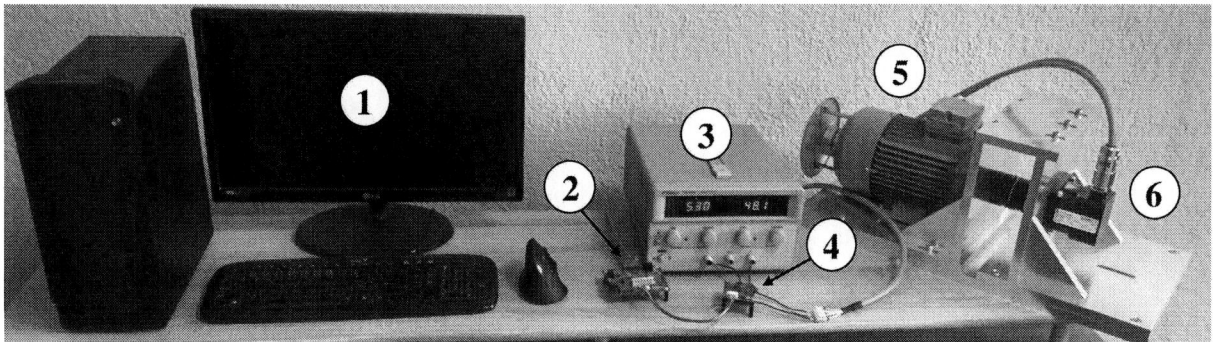

Fig. 10. The experimental setup: (1) PC with CCSTUDIO IDE, (2) resolver to IRC converter, (3) DC power supply, (4) VSI prototype and TMS320F28375D digital signal processor, (5) 3kW AC motor for loading, (6) SMPMSM for experimental verification.

[3] M. Schroedl, "Sensorless control of AC machines at low speed and standstill based on the "INFORM" method," IAS '96. Conference Record of the 1996 IEEE Industry Applications Conference Thirty-First IAS Annual Meeting, 1996, pp. 270-277 vol.1, doi: 10.1109/IAS.1996.557028.

[4] G. Xie, K. Lu, S. K. Dwivedi, J. R. Rosholm and F. Blaabjerg, "Minimum-Voltage Vector Injection Method for Sensorless Control of PMSM for Low-Speed Operations," in IEEE Transactions on Power Electronics, vol. 31, no. 2, pp. 1785-1794, Feb. 2016, doi: 10.1109/TPEL.2015.2426200.

[5] G. Wang, J. Kuang, N. Zhao, G. Zhang and D. Xu, (2018). Rotor Position Estimation of PMSM in Low-Speed Region and Standstill Using Zero-Voltage Vector Injection, in IEEE Transactions on Power Electronics, vol. 33, no. 9, pp. 7948-7958, Sept. 2018, doi: 10.1109/TPEL.2017.2767294.

[6] M. Nicola and C. I. Nicola, "Sensorless Control of PMSM using Fractional Order SMC and Extended Kalman Observer," 2021 18th International Multi-Conference on Systems, Signals & Devices (SSD), 2021, pp. 526-532, doi: 10.1109/SSD52085.2021.942937.

[7] K. Kyslan, V. Šlapák, V. Petro, A. Marcinek and F. Ďurovský, "Speed Sensorless Control of PMSM with Unscented Kalman Filter and Initial Rotor Alignment," 2019 International Conference on Electrical Drives & Power Electronics (EDPE), The High Tatras, 2019, pp. 373-378, doi: 10.1109/EDPE.2019.8883918.

[8] Microchip Technology Inc. (2017). Sensorless FOC for PMSM using Reduced Order Luenberger Observer [online]. Available at: https://www.microchip.com/content/dam/mchp/documents/OTH/ApplicationNotes/ApplicationNotes/00002590B.pdf

[9] Petro V, Kyslan K. Design and Simulation of Direct and Indirect Back EMF Sliding Mode Observer for Sensorless Control of PMSM. Power Electronics and Drives. 2020; 5(1): 215-228. https://doi.org/10.2478/pead-2020-0016.

[10] Utkin, V., Guldner, J., and Shi, J. (2009). Sliding Mode Control in Electro-Mechanical Systems (2nd ed.). CRC Press. ISBN 978-1-4200-6560-2.

[11] V. Srikanth and A. A. Dutt, "A comparative study on the effect of switching functions in SMO for PMSM drives," 2012 IEEE International Conference on Power Electronics, Drives and Energy Systems (PEDES), 2012, pp. 1-6, doi: 10.1109/PEDES.2012.6484351.

[12] Shuaichen Ye. (2019). Design and performance analysis of an iterative flux sliding-mode observer for the sensorless control of PMSM drives. ISA Transactions, Volume 94, 2019, Pages 255-264. https://doi.org/10.1016/j.isatra.2019.04.009.

[13] Microchip Technology Inc. (2010). Sensorless Field Oriented Control of a PMSM. AN1078. [online]. Available at: http://ww1.microchip.com/downloads/en/appnotes/01078b.pdf

[14] Kazraji S, Soflayi R, Sharifian M. Sliding-Mode Observer for Speed and Position Sensorless Control of Linear-PMSM. Electrical, Control and Communication Engineering. 2014;5(1): 20-26. https://doi.org/10.2478/ecce-2014-0003.

[15] S. Zossak, M. Stulraiter, P. Makys and M. Sumega, "Initial Position Detection of PMSM," 2018 IEEE 9th International Symposium on Sensorless Control for Electrical Drives (SLED), 2018, pp. 12-17, doi: 10.1109/SLED.2018.8486043.

[16] Y. Liu, J. Fang, K. Tan, B. Huang, and W. He, "Sliding Mode Observer with Adaptive Parameter Estimation for Sensorless Control of IPMSM," Energies, vol. 13, no. 22, p. 5991, Nov. 2020.

[17] T. M. Mitchell, (1997). Machine Learning (1st ed.). McGraw-Hill Science. ISBN 0070428077.

Contactless Energy Transmission using a Transformer with Movable Secondary

Mariusz Stepien

Department of Power Electronics, Electrical Drives and Robotics
The Silesian University of Technology
Gliwice, Poland
mariusz.stepien@polsl.pl

Abstract—Contactless energy transmission is a very convenient method of energy delivery to loads with variable location in respect of energy source. The most simple and most efficient transfer system is a transformer with an air gap. The paper is focused on analysis of properties of the transformer with relatively large air gap and the secondary winding which is a movable part. The transformer is used to deliver energy to an electronic equipment mounted on the head of drill machine. In order to increase energy efficiency and power density the capacitance compensation system for leakage inductances is applied. Since the secondary is movable, an analysis of generation of the energy resulting from spatial changes of magnetic field is carried out. The analysis presented in the paper is based on ANSYS software and is carried out mostly as harmonic one for linear models.

Index Terms—contactless energy transfer, coreless transformer, numerical modelling, leakage inductance.

I. Introduction

Recently, contactless energy transfer (CET) is a technology intensively developed for wide range of applications. Modern development of CETs results also with improved parameters of energy delivery, mostly increased efficiency and reduced flux leakage. Range of possible applications increase if the length of acceptable air gap increases. Also the range of power of CET systems varies in very wide range. The highest values are required for induction heating systems [1] and for charging systems in automotive industry [2]. Smaller output power range is required for different industrial electronics applications like supplying system for artificial heart [3] or multicoil systems with distributed load [4]. The smallest power levels are observed in electronic applications like control of semiconductor switches [5] or PCB galvanic separation [6].

The system analyzed in the paper is based on a coreless transformer with an air gap dedicated for supplying an electronics system mounted on the head of drill machine [7]. The transformer is able to deliver the power in a range between a few and several dozens of watts, so it can be classified as a device of low level of power. In order to increase level of power transferred to the output analysis of capacitive compensation of leakage flux is carried out. The transformer can operate at different locations of secondary winding in respect to the primary one (can rotate). In the quantitative analysis presented here only the arrangement of secondary

aligned to primaries is carried out. Intermediate locations are neglected.

Taking into account that the secondary winding operates as a rotating part it is possible to induce current in the secondary not only because of magnetic field varying in time. The other possibility is operation under constant magnetic field varying in a space – the primary can be supplied by DC current or replaced by permanent magnets in the another one. In the paper only time-varying magnetic field is considered.

II. Structure of the transformer

Energy transmission in the system is carried out using the coreless transformer with relatively large air gap (between 2 and 10 mm). The general structure of the transformer (without rectifier and load) is presented in Fig. 1. General dimensions of the transformer are as follows:

- diameter of the transformer: 40 mm,
- height of the transformer: 15 mm,
- length of single coil (approx.): 20 mm,
- coil depth (radius direction): 7.25 mm,
- air gap between primary and secondary: 5 mm,
- number of turns in each coil: 80,
- material of windings: copper,
- dedicated frequency of operation: 500 kHz.

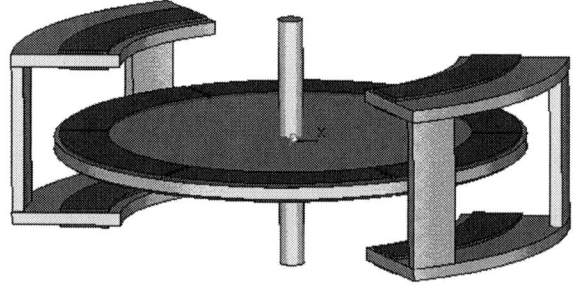

Fig. 1. Idea of the structure of coreless transformer with movable secondary.

The transformer has modular structure. The primary winding divided in two sections, each of two coils, top and bottom one. Each coil has planar structure and is wounded on the non-metallic support. Internal side of each coil is directly coupled

978-1-6654-3236-8/21 $31.00 © 2021 IEEE

to the secondary, while external layers are uncoupled. The secondary winding of the transformer contains six sections (six coils), distributed along the circumference of circle. Each coil is connected to the common load by own diode rectifier.

Presented in Fig. 1 transformer is dedicated for transmission of energy under variable in time magnetic field. It means that the transformer can operate correctly at any rotational speed of the secondary part (including operation at non-rotating secondary). In case of operation with magnetic field variable in a space (generator mode), when the secondary is rotating with given speed, the primary winding can be supplied by DC current or replaced by permanent magnets. Such a situation is not considered in the paper.

III. NUMERICAL MODEL AND ASSUMPTIONS FOR ANALYSIS

The structure of the transformer has an evident axial symmetry. Nevertheless for 2D modelling an axial symmetry is not useful because the current in windings flows in X direction. So, the best simplification for 2D modelling is an extension to the linear model. Geometry of the model with dimensions of base (initial) structure is presented in Fig. 2.

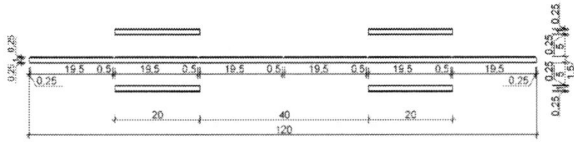

Fig. 2. Structure and dimensions of 2D transformer model.

The transformer is a coreless structure, so the analysis is possible using harmonic analysis with linear parameters. Windings of the transformer are made of copper and for assumed frequency of operation 500 kHz it produces quite strong skin effect. So, one of the most important aspects of modelling was mesh density. In order to calculated correctly the current density in windings, the finite elements were generated with the minimum characteristic size equal to half the skin depth. Mesh in windings region is presented in Fig. 3.

Calculations of transformer properties are carried out in three steps. The first one was calculations of the clamp parameters (input and output currents and voltages) at the idle state and the short-circuit state. Considering transformer as two-port network, the equivalent parameters were the result. In the second step, knowing the leakage inductance of both transformer coils, compensation capacitances have been determined. And in the third step properties of the transformer under nominal load have been calculated [8].

In order to perform above described calculation the FEM area was connected to the external circuit of excitation (voltage source) and load (resistance). The software ANSYS Mechanical APDL used for calculation is equipped by dedicated tool called as Electrical Circuit CIRCU124. It allows to

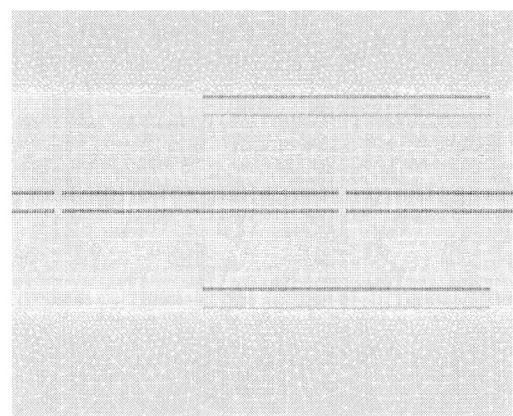

Fig. 3. FEM mesh in area of windings.

incorporate any passive elements like, resistors, coils, capacitor and also variety of excitations (voltage source, current source, controlled source, etc.). Such a possibility reduces strongly computational capacity required for calculations.

IV. RESULTS OF CALCULATIONS

A. Distribution of the magnetic field with and without capacitive compensation

Efficient contactless energy transfer system requires good coupling between windings, especially when the distance between winding is relatively large. Calculation of the transformer properties by a two port network at the idle and short-circuit state resulted with equivalent parameters, like wire resistances and leakage inductances. Knowing these parameters compensation capacitance has been calculated. The difference between uncompensated and compensated system can be observed in Fig. 4, where magnetic flux lines related to Z component of magnetic vector potential are presented. One can observe flux lines concentrated directly between windings.

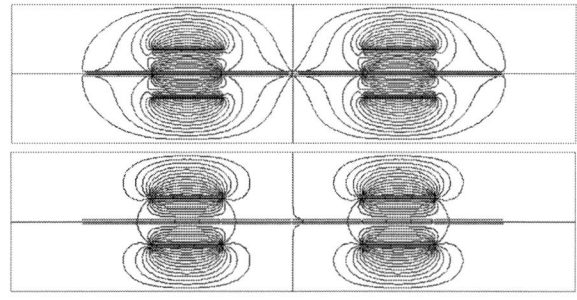

Fig. 4. Distribution of 2D flux lines in transformer with uncompensated (top) and compensated (bottom) leakage inductance.

The second quantity reflecting operation of the transformer is distribution of the magnetic flux, important mostly because of saturation phenomenon. In case of contactless transformer the saturation does not occur. Nevertheless, concentration of magnetic flux lines in area of windings results with very high local current density. It increases local power losses and influences on global power losses in transformer and its efficiency. In Fig. 5 one can observe magnetic flux density in the transformer.

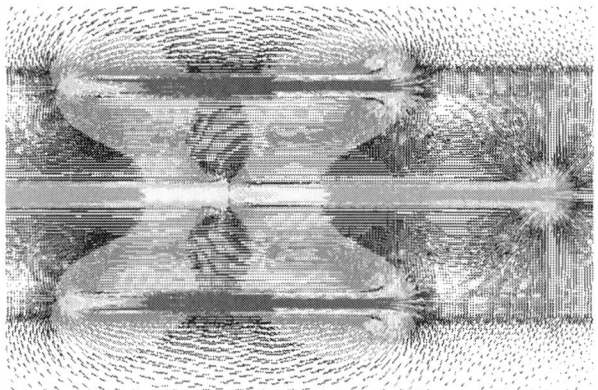

Fig. 5. Magnetic flux density in the transformer at nominal load and compensated leakage inductances.

B. Output power

Since capacitance compensate the leakage flux, the energy is more efficient delivered from the source to the load. The influence of compensation can be observed as an example on Fig. 6, where the transformer with air gap of 5 mm operates at 500 kHz with constant load of 0.5 Ω.

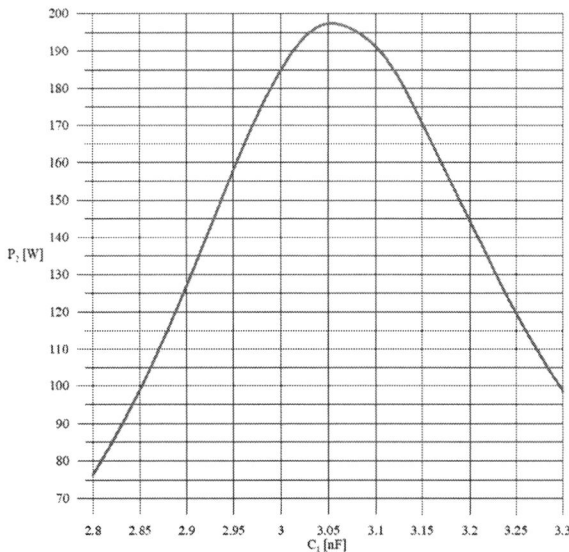

Fig. 6. Maximum output power of the transformer vs. serial capacitance.

The capacitors with the same capacitance are connected in series to the primary and secondary. One can observe that the output power increases in small range of capacitances

(between 2.8 nF and 3.3 nF) more than two times reaching maximum value about 200 W.

Presented in Fig 6 parameters were obtained at constant load resistance. Such an assumption is difficult to practical realization because the equivalent output resistance depends on output power and output current and varies during the operation. As was observed during analysis the transformer has quite narrow range of resistances resulting with high output power. The dependence of output power versus load resistance is presented in Fig.7. The results were obtained assuming capacitive capacitance with optimal capacitance.

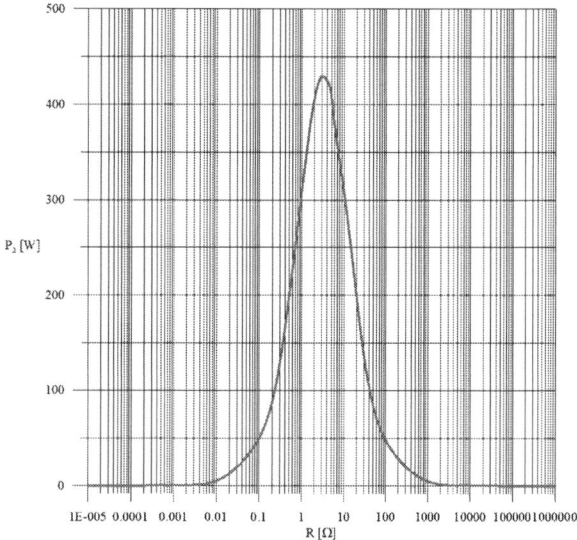

Fig. 7. Maximum output power of the transformer vs. load resistance.

C. Efficiency of the transformer

The transformer is dedicated to transfer a small amount of energy to the load located on the rotating part. Nevertheless, one of the most important parameters is the efficiency. Because of contactless structure and relative large air gap the efficiency of the transformer is relatively low. In order to increase the efficiency of the transformer a comparative analysis has been carried out in respect of the air gap length and the frequency of operation.

The dependence of efficiency on air gap length with the frequency as a parameter is presented in Fig 8. One can observe that the efficiency decreases with the air gap length. Because of coreless operation the best results are obtained for the highest frequency (operation with ferrite magnetic core produces at high frequencies additional power losses). For air gaps shorter than 7 mm and frequencies above 500 kHz the efficiency is above 10%.

In order to increase the frequency of the source more sophisticated device like resonant inverter is required. Switching of inverters produces power losses in transistors proportional to the frequency. So, the maximum frequency for the transformer should be limited. The dependence of efficiency on

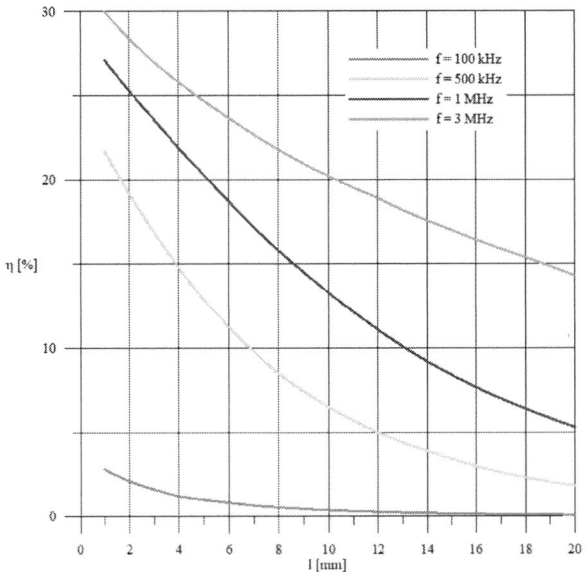

Fig. 8. Efficiency of the transformer as a function of air gap length with frequency as a parameter.

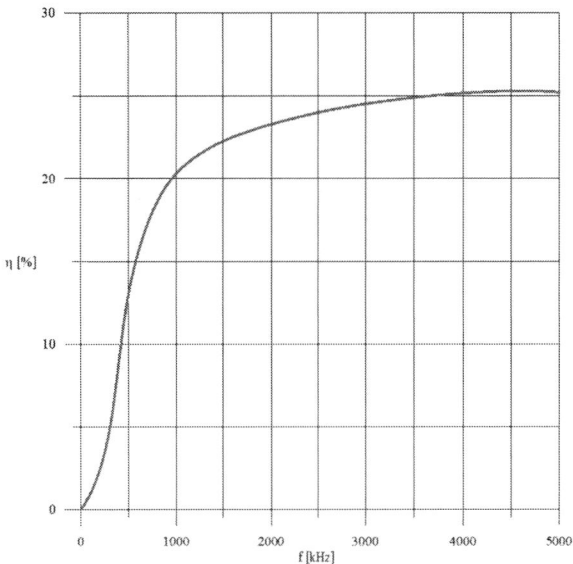

Fig. 9. Efficiency of the transformer vs. frequency at constant air gap.

the frequency for constant air gap is presented in Fig. 9. The frequency above 1 MHz increases efficiency slightly. So, the maximum frequency can be set as 1 MHz.

V. CONCLUSIONS

The transformer presented in the paper is dedicated to transfer an energy for rotational parts of machines. The required level of assumed energy to be delivered is relatively low. Proposed here transformer has modular design (it can be expanded to larger number of modules, e.g. for devices with larger diameter). Classical planar windings are used for energy transmission.

In order to reduce leakage flux, a compensation by capacitors is applied. The power delivered by the transformer increases more than two times in a range of capacitances varying in a range of +/-10%.

Important drawback of the transformer is relatively narrow range of load resistances related to high output power. In order to extend this range passive matching system can be applied.

Analysis of influence of air gap shows that the power is transmitted with the efficiency exceeding 10% for air gaps smaller than 7 mm and frequency higher than 500 kHz.

Analysis of efficiency vs. frequency shows that the frequency above 1 MHz increases efficiency in small range. The operation with large frequency is not effective because it causes increasing of power losses in a supplying inverter. Overall analysis shows that the transmission system operates correctly but requires additional optimization.

REFERENCES

[1] X. Duran-Reus, F. J. Quirós, D. Montesinos-Miracle, "Design of a high-frequency transformer for an induction heating system", 2014 IEEE 11th International Multi-Conference on Systems, Signals & Devices (SSD14), 2014, DOI: 10.1109/SSD.2014.6808858.

[2] K. W. Klontz, D.M. Divan, D.W. Novotny, "An actively cooled 120 kW coaxial winding transformer for fast charging electric vehicles", IEEE Transactions on Industry Applications, Vol.: 31 , Iss.: 6 , Nov/Dec 1995, DOI: 10.1109/28.475695.

[3] B. Grzesik, M. Stepien, "Analysis of coils equivalent parameters for biomedical TET applications", 15th International Power Electronics and Motion Control Conference and Exposition, EPE-PEMC 2012 ECCE Europe, Novi Sad, Serbia, 3-6.09.2012.

[4] Z. Kaczmarczyk, K. Bodzek, A. Ruszczyk, M. Kasprzak, A. Domoracki, "A multi-coil wireless power transfer (MC-WPT) system - analysis method and properties", Measurements, Automation, Monitoring, 2015 vol. 61 nr 10, s. 480-483.

[5] B. Sun, R. Burgos, D. Boroyevich, R. Perrin, C. Buttay, B. Allard, N. Quentin, M. Ali, "Two comparison-alternative high temperature PCB-embedded transformer designs for a 2 W gate driver power supply", 2016 IEEE Energy Conversion Congress and Exposition (ECCE), DOI: 10.1109/ECCE.2016.7855537.

[6] M. T. Carpenter, M. A. Broadmeadow, "Design of coreless PCB transformers for power and signal isolation in a modular ADC system for power quality data acquisition", 2014 Australasian Universities Power Engineering Conference (AUPEC), DOI: 10.1109/AUPEC.2014.6966600.

[7] P. Poroszewski, "Analysis of electrical energy transmission system with variable air gap" (in Polish: "Analiza systemu transferu energii przy zmiennej szczelinie powietrznej"), Master Thesis, Silesian University of Technology, Gliwice, Poland, 2008.

[8] B. Grzesik, K. Bodzek, M. Stepien, "Modular transformer with any turn to turn ratio", 13th International Conference on Power Electronics, Novi Sad, Serbia, 2005.

General approach of radial active magnetic bearings design and optimization

Cristina Adascalitei
Department of Electrical Machines
and Drives
Technical University of
Cluj-Napoca
Cluj-Napoca, Romania
cristina.adascalitei@emd.utcluj.ro

Radu Martis
Department of Electrical Machines
and Drives
Technical University of
Cluj-Napoca
Cluj-Napoca, Romania
radu.martis@emd.utcluj.ro

Claudia Martis
Department of Electrical Machines
and Drives
Technical University of
Cluj-Napoca
Cluj-Napoca, Romania
claudia.martis@emd.utcluj.ro

Abstract - The industries involved with high-speed rotational systems tend to replace the traditional mechanical bearing systems with magnetic suspension. This trend is due to various advantages of magnetic bearings, for example low mechanical losses, high reliability lubrication free operation and the ability to function in vacuum environments. This paper presents a general approach of the pre-sizing and optimal electromagnetic analysis of active magnetic bearings for high-speed machines.

Keywords— Magnetic bearings, design, electromagnetic analysis

I. INTRODUCTION

Magnetic bearings have many advantages with respect to their counterparts, roller and ball bearings, which are exploited in high-speed electrical machines for applications requiring for smaller volume, higher power density, direct connection and increased system efficiency.

According to the mechanism of producing magnetic force and their design, magnetic bearings can be classified into three types [1]:

- Active magnetic bearings (AMBs) developing the electromagnetic force(s) for levitating the rotor by using controlled electromagnets;

- Passive magnet bearings (PMBs), include permanent magnets to develop the levitation force(s);

- Hybrid magnetic bearings (HMBs) combining AMBs and PMBs, thus including both electromagnets and permanent magnets for developing the levitation force(s)[2-3].

PMBs have simple structure, but they suffer of low load carrying capacity and low stiffness. AMBs have better control ability and high stiffness, but they come with the disadvantage of higher losses due to the biased current and higher costs associated to the control system implementation (sensors, power amplifiers, etc.). HMB structure combines the advantages of AMBs and PMBs and it is widely used in several applications due to its compactness, affordability and short axial length [2-3].

The impact that equipping a high-speed electrical machine with a magnetic bearing has on the overall performance of the machine is important. Beside the increasing efficiency and reliability, by removing the need of lubrication and suppressing rotor vibration through an AMB will result in reducing the life cycle costs of the system. However, volume constraints, thermal behavior, stability control of the rotor, overall efficiency are important issues to be further analyzed and solved through a proper design and analysis. Regardless the machine type and topology, high-speed magnetic levitated electrical machines (HSMLEMs) have some specific design challenges as they ask for a design-integrated approach. Therefore, integrated optimal design and proper multi-physics analysis can provide the tools for high performance low cost HSMLEM.

This paper describes the steps of the radial AMB design and analysis procedure, considering both performances and dynamics, validated through the optimal design and analysis of two AMB topologies.

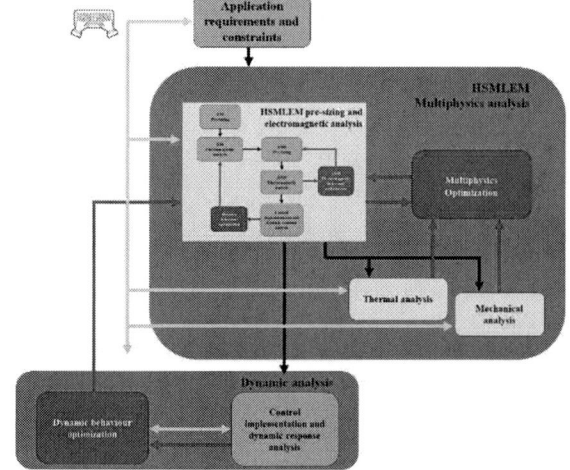

Fig. 1 HSMLEM design and analysis procedure

II. GENERAL DESIGN AND ANALYSIS APPROACH

The design and analysis of a HSMLEM follows the steps presented in Fig. 1. The procedure starts with the application requirements (rated power, rated/maximum speed, voltage source, etc.) and constraints (materials, geometrical dimensions, topology, etc.). It integrates a multiphysics and multilevel approach. Electromagnetic, thermal and mechanical design of the HSMLEM can be performed either independently or coupled, according to the application requirements. The dynamic model of the system is developed based on the parameters derived from multiphysics analysis and the dynamic response is evaluated for different operation scenarios. It should however be noticed that HSMLEM is in fact a system composed of minimum three electromagnetic components: the electrical machine and at least two magnetic bearings for levitating the rotating shaft and control the forces on x- and y-axis, respectively. Thus, the HSMLEM design has to be "decomposed" at the Pre-sizing and electromagnetic

978-1-6654-3236-8/21 $31.00 © 2021 IEEE

analysis step in two processes (one for the electrical machine and one for the AMB) which are interconnected.

AMB has to satisfy specific application constraints and at the same time, the electrical machine design provides part of the requirements and constraints for the AMB. The Pre-sizing and electromagnetic analysis step of the HSMLEM is presented in Fig. 2. The EM Pre-sizing follows the well-known procedure, and the electromagnetic analysis goal is to refine and optimize the machine geometry in order to obtain the defined electromagnetic performances. At the end, the process will provide inputs for AMB pre-sizing. The inputs are grouped into two categories. One category is represented by the application requirements (load capacity, stiffness and operation speed) and constraints (e.g. outer diameter, stack length, materials, maximum current, maximum magnetic field density in different regions, rotor etc.). The second category results from EM electromagnetic design as key parameters to be considered, like rotor part structure, dimensions, and material as well as the total mass that has to be driven by the levitation forces. The AMB pre-sizing step delivers all the necessary data to be used for AMB electromagnetic analysis and optimization. The AMB topology will be thus refined and optimized from electromagnetic point of view, and the parameters for dynamic model development are provided to the dynamic response module.

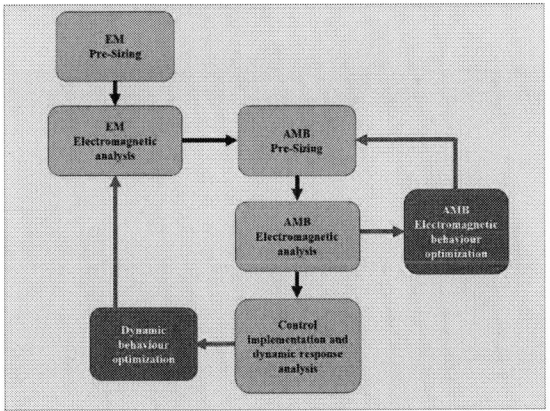

Fig. 2 HSMLEM pre-sizing and optimal electromagnetic analysis

Some preparatory work has to be carried out before starting the pre-sizing. This preparatory work consists of choosing the type and topology of the AMB and deriving the performance requirements (peak load capacity, bearing stiffness and damping, airgap length, etc.) and constraints (e.g. outer diameter, stack length, materials, maximum current, maximum magnetic field density in different regions, etc.).

Several heteropolar topologies with different poles number and different couplings between the two axes are considered. The 4-pole structure is simpler, but with coupled magnetic loops for x and y-axes, while the 8-pole one results in nearly uncoupled magnetic forces on x and y-axes. The 3-pole topology has cross-coupling problems due to the structure asymmetry, while 6-pole topology eliminates it.

The levitation force of the magnetic bearing corresponds to the magnetic field energy stored in the airgaps and can be written as:

$$F = \frac{B_g^2}{2 \cdot \mu_0} \cdot b_p \cdot L_a \qquad (1)$$

where B_g represents the air gap magnetic field density, μ_0 – air permeability, b_p – pole width and L_a – axial active length of the bearing. The airgap magnetic field density results from the superposition of bias and control flux density. The maximum value of the magnetic flux density occurs at the maximum current according to:

$$B_g = \frac{\mu_0 \cdot N \cdot I}{2 \cdot g} \qquad (2)$$

with N - the number of turns per coil, I - the current and g - the airgap length.

The main dimensions of the stator (inner diameter, height and width of the poles, width of the yoke) and the main data of the excitation windings (number of turns, wire diameter) are thus computed based on the maximum necessary bearing force and on the magnetic circuit topology.

Electromagnetic analysis is usually performed using a Finite Element based software for developing the AMB 2D/3D model and carrying out simulation in order to evaluate the device performances and characterize it. If needed, optimization goals are established and optimization procedure is implemented in order to obtain the optimal topology. The electromagnetic forces developed by the device, the variation of these forces with the rotor displacement on x- and y-axis respectively, the saturation and the cross-saturation effect, AMB efficiency, etc. are analyzed and evaluated against the application requirements.

The purpose of the control design is to provide a restoring force to attenuate the oscillations cause by the disturbance forces, by controlling the current of the coils to reach the reference input in a stable manner. The inputs to the control system are signals from the displacements and current sensors and the outputs are feeding the power amplifiers generating the voltage or current for the AMB coils. The dynamic analysis can be integrated into an optimization loop, with direct impact on the electromagnetic analysis.

The process will end by providing the final topology and data of the AMB, which will be further integrated in a system level analysis for HSMLEM multiphysics optimization.

As the aim of the present work is the optimal analysis of a radial AMB, the study will be limited to the pre-sizing and optimal electromagnetic analysis of two AMB topologies.

III. AMB TOPOLOGIES AND PRE-SIZING PROCEDURE

The application requirements and constraints are given in Table I. Two AMB topologies were considered, 4-pole and 8-pole, presented in Fig. 3 (a) and Fig. 3 (b). The airgap dimension for the two AMB is 0.5 mm.

TABLE I. APPLICATION REQUIREMENTS AND CONSTRAINTS

Parameter	Value
Requirements	
Maximum force (on x and y axis)	100 N
Maximum speed	30000 rot/min
Rotor maximum eccentric admissible displacement	0.05mm
Constraints	
Inner diameter of the rotor	37 mm
Outer diameter of the rotor	49 mm
Inner diameter of the stator	50 mm
Outer diameter of the stator	<130 mm
Axial length, L_a	15 mm
Maximum current, I_{max}	4 A
Maximum rotor pole magnetic field density, B_{max}	1.2 T

978-1-6654-3236-8/21 $31.00 © 2021 IEEE

Taking into account the maximum value of the force to be developed on x and y-axis, respectively, the width of the rotor pole is computed. The pre-sizing process continues with the computation of the slot area that has to accommodate the windings. Thus, the number of turns per coil is derived from (2) considering the maximum value of the magnetic field density and of the current. Then the area of the conductor, s_c results as the ratio between the maximum current and the current density, corresponding to a natural cooling of the device.

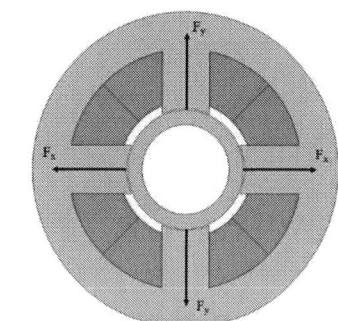

Fig. 3 (a) 4-pole AMB topology

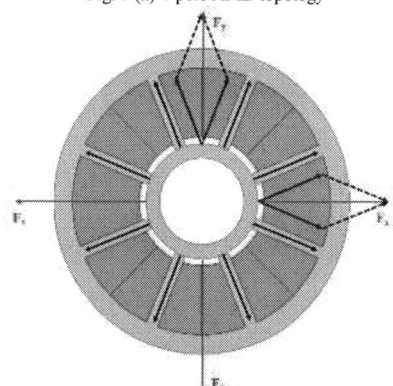

Fig. 3 (b) 8-pole AMB topology

The area of a slot will thus result as:

$$A_{slot} = 2 \cdot N \cdot s_c \cdot k_{fill} \qquad (3)$$

with k_{fill} the slot filling factor, depending on the coil construction and installation technology. Further, the width of the yoke and height of the pole are computed. The pre-sizing process results are given in Table II.

TABLE II. MAIN DIMENSIONS OF THE AMB TOPOLOGIES

Parameter	Value
Configuration: 4-pole AMB	
Pole height, h_p	25 mm
Pole width, b_p	17 mm
Coil height, h_b	20 mm
Coil width, g_b	12 mm
Number of turn coil, N	90
Configuration: 8-pole AMB	
Pole height, h_p	32 mm
Pole width, b_p	8 mm
Coil height, h_b	29 mm
Coil width, g_b	7 mm
Number of turn coil, N	75

IV. ELECTROMAGNETIC ANALYSIS AND OPTIMIZATION

A. Electromagnetic analysis

The electromagnetic analysis was used for a quick evaluation of the two topologies performances. Figures 4 to 7 present the magnetic field density distribution in the cross-section of the AMBs and the flux lines for each topology, as well as the radial forces distribution along the rotor circumference.

As it can be noticed from Fig. 4, for the 4-pole topology the flux lines of the field contributing to the development of x-axis and y-axis forces are not independent. Therefore, the cross effect has to be analyzed and considered at the control development. The radial force distribution (Fig. 5) shows that the average value of the force under each of the side poles (corresponding to x-axis force), as well as the one under each of the vertical poles (corresponding to the y-axis force) equals 102.67 N and 102.65 N, respectively.

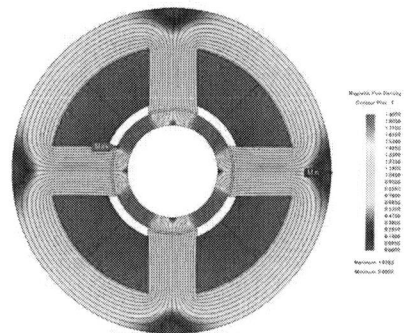

Fig. 4 Magnetic field density map and flux lines for 4-pole topology

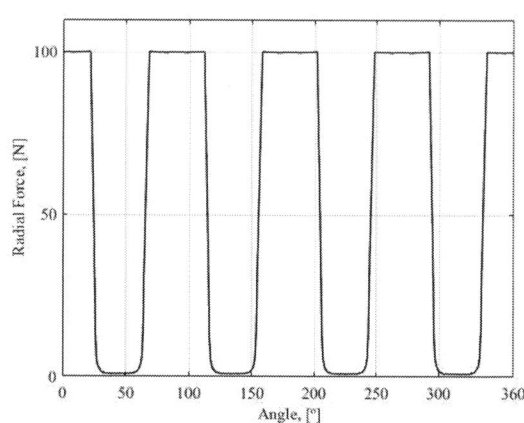

Fig. 5 Radial force distribution along the rotor circumference for 4-pole topology

For the 8-pole topology, the flux lines of the field contributing to the development of x-axis and y-axis forces are independent (Fig. 6).

This provides the means to control independently the displacement on x and y-axis during the operation. The radial force (Fig. 7) under each pole is around 80 N, thus the x- and y-axis forces are computed according to the representation in Fig. 3 (b):

$$F_x = \sqrt{2} \cdot F_{pole} \qquad (4)$$

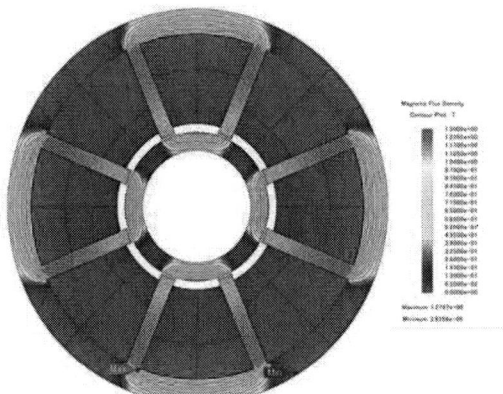

Fig. 6 Magnetic field density map and flux lines for 8-pole topology

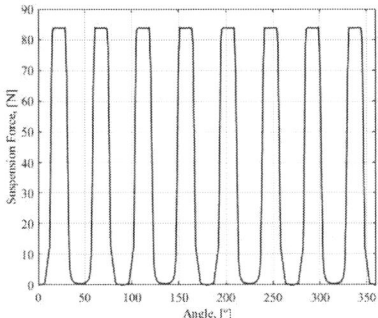

Fig. 7 Radial force distribution along the rotor circumference for 8-pole topology

B. Optimization

A multi-objective genetic algorithm optimization was applied on the structure of the stator part of the bearing. The objective is to reduce the volume of the active materials, therefore a parametrized model of the bearing was created for modifying the 3 main dimensions: back iron, pole height and pole width. These dimensions were modified in the optimization process in order to satisfy several criteria, such as the current density in the coils, the maximum magnetic flux density and the required force on the x and y axis. After running the optimization, the geometrical dimensions of the model are shown in Table III.

TABLE III. TOOTH WIDTH, BACK IRON WIDTH, STATOR OUTER RADIUS BEFORE AND AFTER THE OPTIMIZATION

		Tooth width [mm]	Back iron width [mm]	Outer stator radius [mm]
8-pole	Before	8	10	65
	After	7.83	9	55.8
4- pole	Before	17	17	65
	After	21.12	15.37	54.2

In order to compare the parameters obtained after the optimization, in Table IV are presented the values of the parameters for 8-pole AMB before the optimization and after that, and the Table V illustrates the parameters for the 4-pole AMB.

TABLE IV. PERFORMANCES OF 8-POLE AMB TOPOLOGY BEFORE AND AFTER OPTIMIZATION

		Parameter	Value
8-pole	Before	Suspension force in x-axis direction	107.1 N

		Suspension force in y-axis direction	107.1 N
		Tooth flux density	0.9 T
		Air gap flux density	0.64 T
8-pole	After	Suspension force in x-axis direction	108 N
		Suspension force in y-axis direction	106.8 N
		Tooth flux density	0.86 T
		Air gap flux density	0.64 T

TABLE V. PERFORMANCES OF 4-POLE AMB TOPOLOGY BEFORE AND AFTER OPTIMIZATION

		Parameter	Value
4-pole	Before	Suspension force in x-axis direction	115.1 N
		Suspension force in y-axis direction	112.7 N
		Tooth flux density	1.09 T
		Air gap flux density	0.58 T
4-pole	After	Suspension force in x-axis direction	105.4 N
		Suspension force in y-axis direction	102.5 N
		Tooth flux density	1.03 T
		Air gap flux density	0.58 T

It is clearly from the results that the optimization process fulfills the reduction of external radius of the magnetic bearings and reduce the volume of the active materials, while keeping the application specifications.

C. Performances evaluation

The efficiency of the AMB is given by:

$$\eta = \frac{P_1 - p_{losses}}{P_1} \tag{5}$$

where the electric power, P_1, is given by:

$$P_1 = \sum_1^m U_y \cdot I_y \tag{6}$$

with U_y and I_y the voltage and current of the yth coil of the AMB and m the total number of AMBs coils. p_{losses} represents the losses of the AMB as:

$$p_{losses} = p_{Joule} + p_{iron} \tag{7}$$

The Joule losses, p_{Joule}, results from:

$$p_{Joule} = \sum_1^m R_y \cdot I_y^2 \tag{8}$$

with R_y the yth stator coil resistance.

The iron losses, p_{iron}, of the AMB are of two different types, namely hysteresis losses and eddy current losses.

The hysteresis losses generated by the n harmonic of the magnetic flux density can be approximated by:

$$p_H = k \cdot V \cdot f_n \cdot B_{n\_max}^x \tag{9}$$

where the V represents the volume of the material, f_n and B_{n_max} represent the frequency and the magnitude of the n-harmonic of the magnetic flux density. The exponent x varies typically over [1.5, 2.5], k being an empirical constant.

The eddy currents losses can be calculated with:

$$p_{eddy} = \frac{\pi^2 \cdot V \cdot f_n^2 \cdot \Delta^2 \cdot B_{n\_max}^x}{6 \cdot \rho} \tag{10}$$

with Δ being the sheet thickness and ρ the sheet material resistivity.

The frequency of the magnetic flux density in the stator pole and yoke is zero, and the flux varies only during transients. However, the rotor core experiences a frequency that depends on the rotation speed and the number of pole pairs.

For the 4-pole AMB, for example, the magnetic flux density waveshapes in the rotor core for a 360° rotation at 30000 rpm and maximum current is given in Fig. 8. The magnetic field density in the rotor core has a continuous component equal to 1.28 T. Table VI presents the magnitude of its main time harmonics and the frequency of the 1st harmonic for several values of the rotation speed.

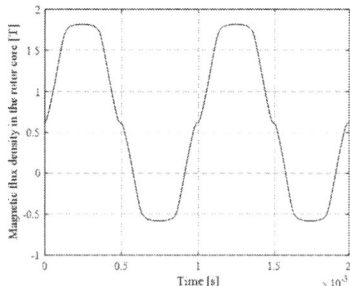

Fig. 8 Rotor magnetic flux density for a 360° rotation at 30000 rpm for 4-pole AMB topology

TABLE VI. MAGNITUDE OF THE MAGNETIC FIELD DENSITY MAIN TIME HARMONICS FOR SEVERAL VALUES OF THE ROTATION SPEED FOR 4- POLE AMB

Speed [rpm]	f_1 [Hz]	B_{1_max} [T]	B_{3_max} [T]	B_{5_max} [T]	B_{7_max} [T]
5000	166.67				
10000	333.33				
15000	500	1.28	0.016	0.1	0.03
20000	666.67				
25000	833.33				
30000	1000				

Table VII gives the losses of the 4-pole AMB for several values of the rotation speed. It can be noticed that the iron losses are very small and increase with the speed. The main losses are represented by the Joule losses.

TABLE VII. LOSSES AND EFFICIENCY FOR DIFFERENT ROTATION SPEEDS

Speed [rpm]	p_{Joule}[W]	p_{iron} [W]	p_{total} [W]	η[%]
5000	9.344	0.08	9.42	85,40
10000	9.344	0.24	9.58	85,40
15000	9.344	0.48	9.83	85,40
20000	9.344	0.80	10.15	85,40
25000	9.344	1.21	10.56	85,40
30000	9.344	1.70	11.05	85,40

For the 8-pole AMB, the magnetic flux density waveshapes in the rotor core for a 360° rotation at 30000 rpm and maximum current is given in Fig. 9. The magnetic field density in the rotor core has a continuous component equal to 1.04 T. Table VIII presents the magnitude of its main time harmonics and the frequency of the 1st harmonic for several values of the rotation speed.

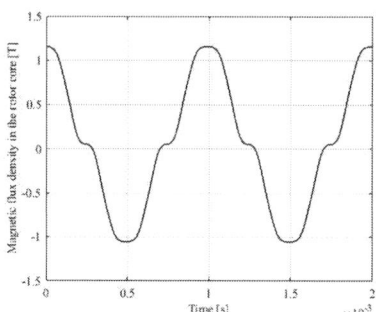

Fig. 9 Rotor magnetic flux density for a 360° rotation at 30000 rpm

TABLE VIII. MAGNITUDE OF THE MAGNETIC FIELD DENSITY MAIN TIME HARMONICS FOR SEVERAL VALUES OF THE ROTATION SPEED FOR 8-POLE AMB

Speed [rpm]	f_1 [Hz]	B_{1_max} [T]	B_{3_max} [T]	B_{5_max} [T]	B_{7_max} [T]
5000	166.67				
10000	333.33				
15000	500	1.04	0.163	0.09	0.0072
20000	666.67				
25000	833.33				
30000	1000				

Table IX gives the losses of the 8-pole AMB for several values of the rotation speed. It can be noticed that the iron losses are very small and increase with the speed. The main losses are represented by the Joule losses.

TABLE IX. LOSSES AND EFFICIENCY FOR DIFFERENT ROTATION SPEEDS

Speed [rpm]	p_{Joule}[W]	p_{iron} [W]	p_{total} [W]	η[%]
5000	6,4	0,005	6,405	90
10000	6,4	0,01	6,41	90
15000	6,4	0,016	6,416	90
20000	6,4	0,021	6,42	90
25000	6,4	0,028	6,428	90
30000	6,4	0,034	6,434	90

Comparing Table VII and Table IX it can be seen that the 4-pole topology has a lower efficiency, and the Joule losses are higher. The 8-pole active magnetic bearing can be used as it has a higher efficiency and, due to the topology, provides independent control for each axis.

D. Forces vs displacement

The operation of an AMB involves superposition of two fluxes: a steady-state flux induced by a bias current in the stator coils, named bias flux and a time-varying flux developed by a control current, named control flux. The control current, and thus a control flux is applied in order to attenuate the oscillations caused by the disturbance force. This results in the development of a force proportional to the displacement and the rotation speed of the levitated mass.

Fig. 10 presents the variation of the developed forces on x-axis and y-axis versus the displacement of the rotor on the x-axis direction for the 4-pole topology. It can be noticed that for a displacement in the positive x-axis, the developed force is higher in the same direction and smaller in the opposite one.

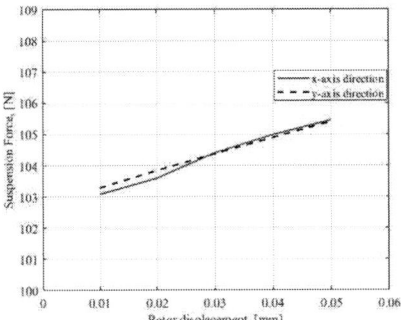

Fig. 10 The average values of the suspension force for different values of rotor displacements on x- axis, 4-pole AMB

Fig. 11 presents the variation of the developed forces on x-axis and y-axis versus the displacement of the rotor on the y-axis direction for the 4-pole topology. It is obvious that the displacement of the rotor in the y-axis direction causes greater fluctuations of the suspension force developed in x-axis direction.

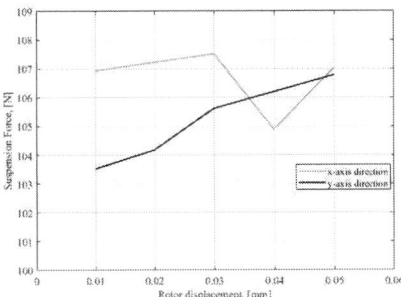

Fig. 11 The average values of the suspension force for different values of rotor displacements on y- axis, 4-pole AMB

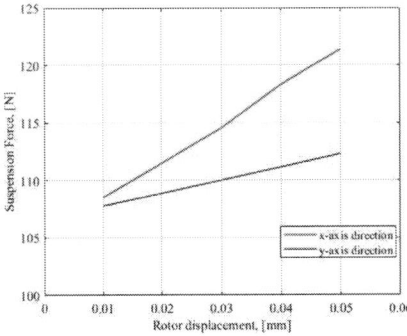

Fig. 12 The average values of the suspension forces for different values of rotor displacements on x- axis, 8-pole AMB

For the 8-pole AMB topology, the variation of the developed forces on the two directions versus the displacement of the rotor on the x-axis and y-axis are presented in Fig. 12, respectively Fig. 13.

In Fig. 13 it can be noticed that the developed forces on y-axis are greater than the developed forces in the x-axis because of the displacement of the rotor. In this case it is necessary to implement a solution for controlling the magnetic bearing for this kind of situations.

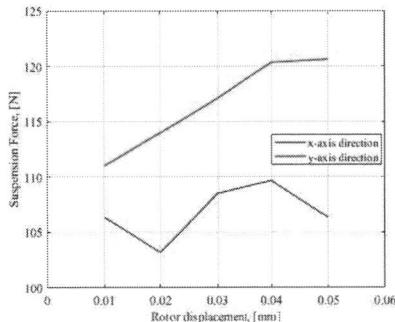

Fig. 13 The average values of the suspension forces for different values of rotor displacements on y- axis, 8-pole AMB

V. CONCLUSIONS AND PERSPECTIVES

The design and electromagnetic analysis of AMBs is an important step in the process of developing HSMLEM. The optimization process should be further carried out at the machine level, taking into account the multiphysics nature of the system, thus integrating thermal and mechanical aspects.

The results from the optimization process presented in this paper shows that a magnetic bearing more compact and with less volume of active materials can be achieved.

If the rotor moves in the direction of the x-axis or the direction of the y-axis, the values of the suspension forces will no longer be uniform but will increase on the direction where the airgap is smaller because of the rotor displacement. In this situation is necessary to implement a control method for magnetic bearings.

In future work, the authors aim to study the control of active magnetic bearings, and with the knowledge regarding the sizing of active magnetic bearings and their control, it is desired to achieve an entire system for operating a high-speed magnetic levitated electrical machine.

ACKNOWLEDGMENT

This work was supported by a grant of the Ministry of Research, Innovation and Digitization, CNCS/CCCDI – UEFISCDI, project number 6/2021, within PNCDI III".

REFERENCES

[1] G. Schweitzer, E. H. Maslen, Magnetic bearings: theory, design and application to rotating machinery, Springer Science & Business Media, 2009.

[2] F. Jiancheng, S. Jinji, X. Yanliang, and W. Xi, "A New Structure for Permanent-Magnet-Biased Axial Hybrid Magnetic Bearings,"IEEE Trans. Magn. Vol. 45, no. 12, pp. 5319-5325, Dec. 2009.

[3] X. Yanliang, D. Yueqin, W. Xiuhe, and K. Yu, "Analysis of hybridmagnetic bearing with a permanent magnet in the rotor by FEM," IEEE Trans. Magn., vol. 42, no. 4, pp. 1363–1366, Apr. 2006.

[4] W. Zhang, H.u Zhu, "Radial magnetic bearings: An overview", Results in Physics, 7 (2017).

[5] H. Eryong and L. Kun, "A novel structure for low-loss radial hybrid magnetic bearing," IEEE Trans. Magn., vol. 47, no. 12, pp. 4725–4733, Dec. 2011.

[6] J. Pyrhonen, T. Jokinen, V. Hrabocova, „Design of rotating electrical machines", 2008 John Wiley & Sons, Ltd. ISBN: 978-0-470-69516-6

The Application of Neural Network Metamodels for Interior Permanent Magnet Machine Performance Prediction

Zlatko Hanic
University of Zagreb
Faculty of Electrical Engineering
and Computing
Department of Electric Machines,
Drives and Automation
Zagreb, Croatia
zlatko.hanic@fer.hr

Ana Hanic
University of Zagreb
Faculty of Electrical Engineering
and Computing
Department of Electric Machines,
Drives and Automation
Zagreb, Croatia
ana.hanic@fer.hr

Marinko Kovacic
University of Zagreb
Faculty of Electrical Engineering
and Computing
Department of Electric Machines,
Drives and Automation
Zagreb, Croatia
marinko.kovacic@fer.hr

Abstract—To increase the computational efficiency of electrical machine optimization and to utilize transfer learning from one metamodel to another, metamodels based on neural networks seem to be a promising solution. This paper presents a methodology of applying neural networks for developing metamodel for the prediction of interior permanent magnet machine performance. Furthermore, it provides procedures and guidelines on design space sampling and developing neural-network-based metamodels to achieve good predicting performance. The proposed approach has been tested on a case of a six-phase 200 kW IPM motor.

Keywords—metamodels, neural network, electrical machines, design of experiments, permanent magnet motors

I. INTRODUCTION

Many applications like electric vehicles, traction propulsion, and aerospace push the electrical machine design and technology boundaries to the limits to optimize their performance [1], [2]. The optimization is often conducted using parameterized models, which optimize the electrical machine design in numerous, often thousands of iterations. That process is time-consuming since it is sequential, which means that the outcome of the current iteration determines the conditions of the next iteration.

A more computationally efficient approach is to use metamodels. Metamodels are simplified mathematical models that describe the input-output relation of the design space [3]-[6]. To build a metamodel for a specific design topology, it is required to investigate the design space, which is done by evaluating a couple of hundred random or pseudo-random designs. That data is then used to determine the metamodels for the design performance prediction. Metamodels can be analytical functions like polynomials, fitting functions like splines, Kriging model, etc.

Once metamodels are built with satisfying accuracy, they can be used for further calculations and optimizations. Evaluating values using metamodels is extremely fast (measured in milliseconds), so the optimization of the design using them is very computationally efficient. In addition, there are several benefits of using metamodels compared to classical parametric optimization. 1) The investigation of the design space, which represents the most challenging and time-consuming task, can be fully parallelized, and therefore, significantly faster. 2) If the requirements are changed in the middle of the project, the same metamodels can be used to run

the new optimization, and the new design can be obtained in minutes.

Neural networks are used for various computing tasks in the electrical machine domain. From fault diagnosis where convolutional neural networks (CNN) are used to diagnose induction machine rotor misalignment, rotor unbalance, broken rotor bars, bowed rotor and faulted bearings [7] or other faults related to the whole shaft assembly, including gearbox and bearings. [8] to tasks like online temperature monitoring in permanent magnet synchronous motor using deep residual neural networks [9].

In [10], CNN is used for interior permanent magnet motor topology optimization showing significant computing cost reduction compared to conventional topology optimization.

Neural networks are also great candidates for metamodels development. Using neural networks for metamodeling is applied to many different areas. For example, in [11], authors use neural networks for surrogate modelling of interior permanent magnet motor. The model is used for obtaining the torque waveform using the amplitude and frequency of stator current as input parameters. The authors also give an extensive review of surrogate electrical machine modelling techniques.

In [12], the authors compare metamodels for permanent magnet synchronous motor optimization using Kriging and neural network multilayer perceptron (MLP), showing MLP achieving better prediction performance. MLP metamodel is made of 4 separate neural networks, one for each output variable, not utilizing transfer learning from one neural network to the other.

In [13], optimization of a flux-intensifying permanent magnet machine design was conducted. The authors use a neural network with five input geometrical variables and three outputs to solve a multi-objective design problem.

This paper focuses on the metamodel of the electrical machine with one neural network fed by ten geometrical input quantities and fourteen outputs like back-EMF at maximum speed, flux densities in various parts of the machine, efficiency, corner speed, etc.

Comparing to the other neural network metamodels for electrical machines reported in the literature, the neural

978-1-6654-3236-8/21 $31.00 © 2021 IEEE

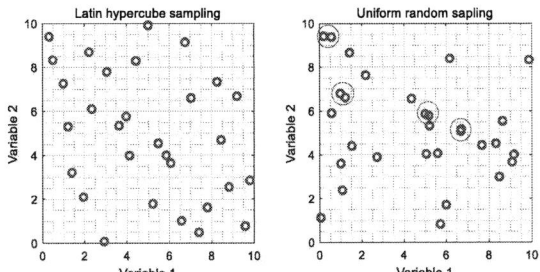

Fig. 1 The difference between latin hypercube sampling and uniform random sampling methods shown at the exaple of picking 30 samples in 2D space

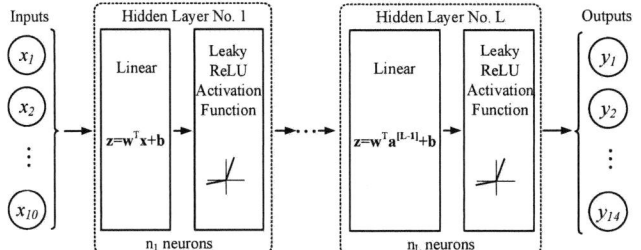

Fig. 2 Simplified structure of neural network with with fully connected layers

network described in this paper has a larger number of input and output variables. This enables usage of the proposed metamodel for widescale design space exploration rather than for fine-tuning of already pre-optimized design. Moreover, the proposed methodology is used for performance prediction for the six-phase interior permanent magnet motor. The procedures for investigating design space and training the neural network as well as prediction quality of the metamodel are presented.

II. METAMODELS BASED ON NEURAL NETWORKS

Metamodels based on neural networks are data-driven, so after defining input and output variables, the first step in developing the model is sampling the design space. The high quality of the data set that captures all the relevant features is essential for achieving the high accuracy of the metamodel.

After sampling the design space, finite element calculations are conducted for each sampled design. Input geometrical variables combined with results obtained by finite element method calculations represent the complete dataset. Dataset is then divided into three separate sets: training (70%), validation (15%), and test (15%). Training and validation sets are used for the development of the neural network metamodel, while the test set is used for non-bias testing and quality estimation of the metamodel.

A. Investigation of the Design Space

The investigation of the design space must catch as many as possible of the design features and characteristics. From one point, it is better if the number of points selected for the investigation of the design space is as big as possible. However, on the other hand, that investigation can become very time-consuming and often virtually impossible to conduct. For example, suppose the design space is defined by 10 independent variables discretized with only 10 points per dimension. In that case, there are 10^{10} design samples for the design space investigation, which practically cannot be evaluated. Therefore, the idea is to use fewer design samples and still have a proper exploration of the design space and get the required information about the design features and characteristics for creating accurate metamodels.

In this paper, the design space was investigated using Latin hypercube sampling (LHS) method. LHS uses random sampling and generates a specified number of data points in the multi-dimensional box-constrained hyperspace [14]. Compared to the simple uniform random sampling method, the LHS method does not produce point samples that are very close to each other, as shown with red circles in Fig. 1. Very close points in the design space do not bring any additional

information that can be used in the metamodel development process, and it should be avoided to save computational time.

B. Neural Networks

For this metamodel, the neural network with fully connected layers is selected, shown in Fig. 2. Given an input feature vector X, which represents geometrical design parameters, the neural network algorithm computes prediction vector Y, corresponding to output design quantities. The network is trained using an optimization algorithm and the training dataset until satisfactory accuracy is achieved.

Training a neural network involves setting up a number of hyperparameters, ranging from the learning rate, number of layers, number of hidden units in each layer, choice of activation function and optimization algorithm, etc. For the hyperparameter tuning process, the validation dataset is used.

Depending on a dataset, values of input and output variables in the model range around 1 mm for the air gap length, around a couple of hundred Nm for starting torque, and around a couple of hundred thousand watts for total power. When these features are on very different scales, weights and biases in hidden layers of the neural network also take values on different scales, resulting in poor performance. One way to enhance the performance of parameter learning and speed up neural network training is data normalization. Normalized inputs are calculated using

$$x_{i,normalized} = \frac{x_i - \mu_i}{\sigma_i} \qquad (1)$$

where μ_i and σ_i are the mean value and square root of the variance of the data set for each input variable x_i.

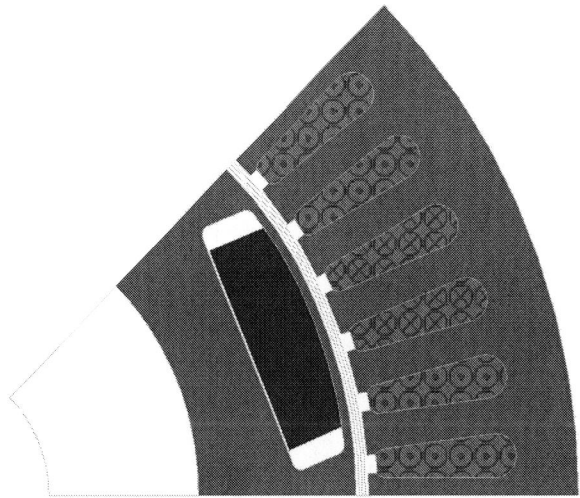

Fig. 3 One sample of the motor design

978-1-6654-3236-8/21 $31.00 © 2021 IEEE

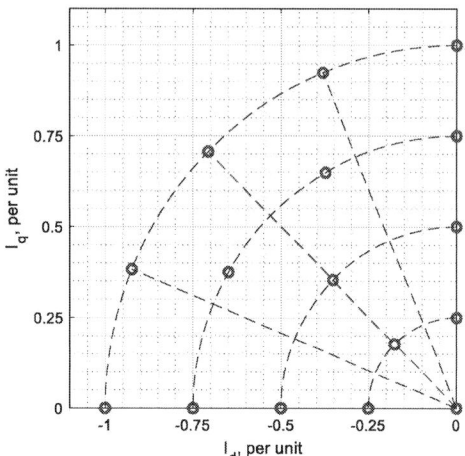

Fig. 4   Sample mapping in the I_d-I_q plane

In order to go through the same transformation before training and testing the neural network, training and test sets need to be normalized with the same values of μ and σ. It should be noted that the same values of μ and σ are also used for all other computations when the metamodel is used.

III. DEVELOPING METAMODEL FOR THE SIX-PHASE 200 kW IPM MOTOR

The motor used for the development of the described metamodel is a six-phase, 200 kW IPM motor with 48 slots and 8 poles. The motor has rotor with embedded bread loaf permanent magnets and single-layered armature winding. One design sample is shown in Fig. 3. The main fixed motor parameters are given in the table below. Before data sampling and running finite element simulations, the geometry of the model was parameterized.

For every motor design sample mapping in the I_d-I_q plane has been conducted using 16 operating points, as shown in Fig. 4. Obtained data is then used for calculating torque vs. speed characteristics and efficiency maps. Later, torque vs. speed characteristics and efficiency maps were used to obtain output variables for the metamodel, such as maximum torque, peak power, or maximum motor efficiency.

TABLE I.   FIXED MOTOR PARAMETERS

Parameter	Value	Unit
Number of slots	48	
Number of poles	8	
Number of phases	6	
Stator outer diameter	149.9	mm
Axial length	145.0	mm
Magnet material	N38UH	
Stator and rotor material	M235-35A	
Maximum rotor speed	16000	rpm

There are a total of ten input variables, including air gap length, stator inner diameter, slot depth, tooth width, tooth tip angle, number of turns per coil, inner rotor diameter, barrier span, barrier thickness, and percentage of coverage of the barrier with the permanent magnet.

There are a total of 14 output quantities used for the metamodeling. Output variables used for the metamodel are: no-load line-to-line induced voltage calculated at the maximum rotor speed of 16000 rpm ($BEMF$), maximum power at power vs speed characteristics (P_{max}), rotor speed where maximum power is obtained (n_{Pmax}), starting torque (T_{start}), torque at maximum speed ($T_{@nmax}$), corner speed (n_c), the maximum value of total copper losses calculated as the sum of AC and DC losses (P_{Cu}), the maximum value of the PM losses (P_{PM}), maximum efficiency (η_{max}), maximum value of the rotor iron losses (P_r), average value of the rotor iron losses at speed above 8000 rpm ($P_{r@hs}$). In addition, maximum flux densities in the stator yoke (B_{sy}), stator teeth (B_t), and rotor yoke (B_{ry}) are taken from the same operating points used for the calculation of the torque vs speed characteristics.

Varied hyperparameters were the number of hidden layers, the number of units in each hidden layer, and the neural network parameters optimization solver. The number of hidden layers was varied between 1 and 4. The number of units in each hidden layer was varied between 2^4 to 2^{13}, depending on the layer. While tested solvers were adaptive moment estimation Adam, stochastic gradient descent with momentum, and root mean square propagation. For the hyperparameter tuning process, Matlab Deep learning toolbox experiment manager was used. The toolbox enables parallel network training computations and easy comparison of network quality.

IV. RESULTS AND DISCUSSION

There were 400 different geometry design samples selected using the LHS method which were used for the investigation of the design space. For all 400 design samples, calculations were conducted for 9 combinations of the number of turns per coil, which gives a total of 3600 different designs.

As a result of the hyperparameter tuning process, 39 different neural networks were trained. Training six networks in parallel, each network took from 49 seconds to 1 minute 17 seconds to train. As a quality indicator, coefficients of determination for the six best trained neural networks are presented in TABLE II. , with varied hyperparameters for all six networks presented in TABLE III.

TABLE II.   COEFFICIENTS OF DETERMINATION FOR OUTPUT VARIABLES OF THE SIX BEST TRAINED NEURAL NETWORKS

Network	I.	II.	III.	IV.	V.	VI.
$BEMF$	0.997	0.997	0.997	0.986	0.992	0.988
B_{sy}	0.988	0.985	0.984	0.984	0.983	0.983
B_t	0.959	0.969	0.948	0.959	0.950	0.960
n_{Pmax}	0.930	0.923	0.943	0.921	0.925	0.929
n_c	0.993	0.994	0.994	0.983	0.986	0.984
$T_{@nmax}$	0.991	0.987	0.992	0.985	0.981	0.981
T_{start}	0.995	0.996	0.994	0.987	0.991	0.984
P_{Cu}	0.998	0.996	0.996	0.992	0.988	0.995
P_{PM}	0.954	0.960	0.948	0.949	0.949	0.937
P_{max}	0.992	0.989	0.991	0.986	0.980	0.978
η_{max}	0.907	0.906	0.901	0.913	0.900	0.901
P_r	0.989	0.988	0.984	0.985	0.989	0.989
$P_{r@hs}$	0.988	0.987	0.985	0.980	0.987	0.988
B_{ry}	0.971	0.967	0.964	0.963	0.966	0.964

978-1-6654-3236-8/21 $31.00 © 2021 IEEE

The coefficient of determination is calculated using:

$$R^2 = 1 - \frac{\sum_{i=1}^{N}(\widehat{y_i}-y_i)^2}{\sum_{i=1}^{N}(\widehat{y_i}-\bar{y})^2} \qquad (2)$$

Where $\widehat{y_i}$ is predicted value using neural network metamodel, y_i is calculated value using FEM and $\bar{y}$ is the average value of the y_i.

TABLE III. HYPERPARAMETERS OF THE SIX BEST TRAINED NEURAL NETWORKS

Network	I.	II.	III.	IV.	V.	VI.
Hidden layers	3	2	2	3	3	2
Hidden units in layer 1	512	1024	512	512	2048	2048
Hidden units in layer 2	1024	1024	512	512	1024	1024
Hidden units in layer 3	512	-	-	512	512	-

The requirement for the coefficient of determination was set to at least 0.9. As a result, the best match was achieved for predicting the BEMF and copper losses (up to 0.998), while the worst match was achieved for the calculation of the motor efficiency (up to 0.913).

Figures Fig. 5 - Fig. 8 show comparisons of different quantities obtained by finite element calculations and predicted by the neural network surrogate model number IV for 138 different motor designs, representing the testing dataset. Fig. 5 shows the maximum power, Fig. 6 shows the BEMF value on the maximum speed, Fig. 7 shows the maximum efficiencies, and Fig. 8 shows the rotor iron loss comparisons obtained using FEM calculations and neural network metamodel.

The proposed method has satisfactory accuracy and is computationally efficient, making it suitable for motor design optimization since evaluating the output values using neural network metamodel takes less than 5 ms. Furthermore, this enables more comprehensive design optimizations with more design evaluations than the parametric optimization with the finite element calculation in the optimization loop.

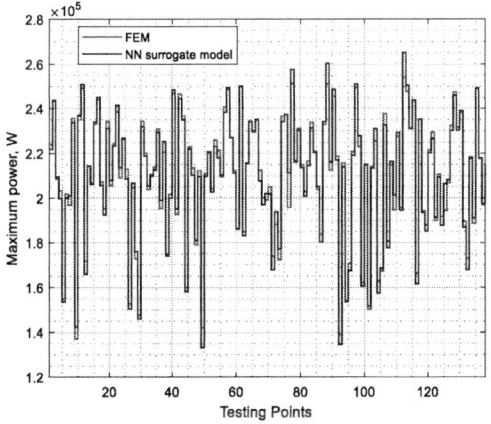

Fig. 5 Comparison of maximum power obtained using FEM and neural network suroggate model

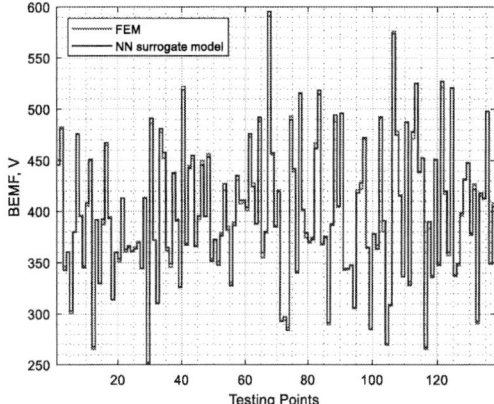

Fig. 6 Comparison of BEMF obtained using FEM and neural network suroggate model

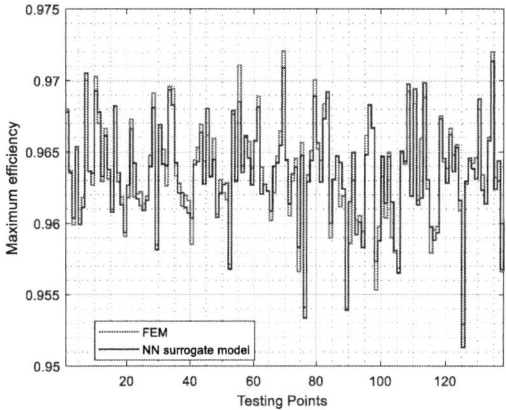

Fig. 7 Comparison of maximum efficiencie obtained using FEM and neural network suroggate model

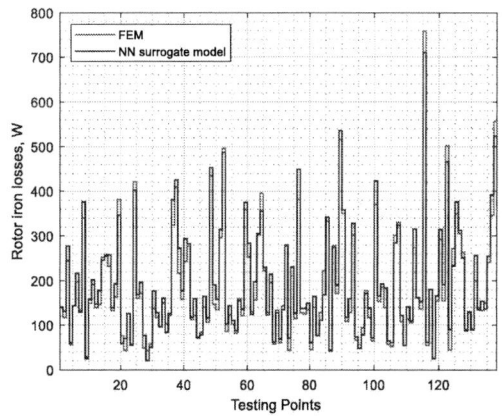

Fig. 8 Comparison of rotor iron losses obtained using FEM and neural network suroggate model

V. CONCLUSION

In this paper, we have presented an approach for developing neural network metamodels for predicting the performance of a six-phase, 200 kW interior permanent magnet motor. We have trained several neural networks, changing their hyperparameters and analyzed their performance. As a result, developed metamodels have good predicting quality, and they are evaluated in a matter of milliseconds. Furthermore, their computational efficiency makes them suitable for conducting optimizations. The same methodology can be utilized for more complex metamodels covering multiphysics phenomena and using output variables such as magnets temperatures, winding temperatures, or maximum mechanical stress in the rotor.

ACKNOWLEDGEMENT

This study was supported by the European Union through the European Regional Development Fund, under grant agreement No. KK.01.2.1.02.0041, project "Development of innovative polyphase electrical drive - PEP".

REFERENCES

[1] G. Bramerdorfer, "Mutiobjective electric machine optimization for highest reliability demands," in CES Transactions on Electrical Machines and Systems, vol. 4, no. 2, pp. 71-78, June 2020

[2] G. Bramerdorfer, J. A. Tapia, J. J. Pyrhönen and A. Cavagnino, "Modern Electrical Machine Design Optimization: Techniques, Trends, and Best Practices," in IEEE Transactions on Industrial Electronics, vol. 65, no. 10, pp. 7672-7684, Oct. 2018, doi: 10.1109/TIE.2018.2801805.

[3] S. D. Sudhoff and R. Sahu, "Metamodeling of Rotating Electric Machinery," in IEEE Transactions on Energy Conversion, vol. 33, no. 3, pp. 1058-1071, Sept. 2018, doi: 10.1109/TEC.2018.2793160.

[4] J. Dirba, L. Lavrinovicha, G. Jekabsons and S. Vitolina, "Metamodel for permanent magnet synchronous motor with outer rotor," 2012 Electric Power Quality and Supply Reliability, 2012, pp. 1-4

[5] Z. Ren, Y. Sun, B. Peng, B. Xia and X. Li, "Optimal design of electrical machines assisted by hybrid surrogate model based algorithm," in CES Transactions on Electrical Machines and Systems, vol. 4, no. 1, pp. 13-19, March 2020

[6] S. Orlova, V. Pugachov, J. Auzins and A. Rassõlkin, "Metamodel-based Optimization of Synchronous Reluctance Motor Rotor," 2021 28th International Workshop on Electric Drives: Improving Reliability of Electric Drives (IWED), 2021, pp. 1-6

[7] S. Shao, S. McAleer, R. Yan and P. Baldi, "Highly Accurate Machine Fault Diagnosis Using Deep Transfer Learning," in IEEE Transactions on Industrial Informatics, vol. 15, no. 4, pp. 2446-2455, April 2019, doi: 10.1109/TII.2018.2864759

[8] D. Xiao, Y. Huang, L. Zhao, C. Qin, H. Shi and C. Liu, "Domain Adaptive Motor Fault Diagnosis Using Deep Transfer Learning," in IEEE Access, vol. 7, pp. 80937-80949, 2019, doi: 10.1109/ACCESS.2019.2921480.

[9] W. Kirchgässner, O. Wallscheid and J. Böcker, "Estimating Electric Motor Temperatures With Deep Residual Machine Learning," in IEEE Transactions on Power Electronics, vol. 36, no. 7, pp. 7480-7488, July 2021, doi: 10.1109/TPEL.2020.3045596.

[10] H. Sasaki and H. Igarashi, "Topology Optimization Accelerated by Deep Learning," in IEEE Transactions on Magnetics, vol. 55, no. 6, pp. 1-5, June 2019, Art no. 7401305, doi: 10.1109/TMAG.2019.2901906

[11] M. Tahkola, J. Keränen, D. Sedov, M. F. Far and J. Kortelainen, "Surrogate Modeling of Electrical Machine Torque Using Artificial Neural Networks," in IEEE Access, vol. 8, pp. 220027-220045, 2020, doi: 10.1109/ACCESS.2020.3042834.

[12] You, Yong-min. "Multi-objective optimal design of permanent magnet synchronous motor for electric vehicle based on deep learning." Applied Sciences 10.2 (2020): 482.

[13] Q. AI, H. Wei and Y. Zhang, "Optimal design for Flux-intensifying Permanent Magnet Machine Based on Neural Network and Multi-objective optimization," 2020 4th CAA International Conference on Vehicular Control and Intelligence (CVCI), 2020, pp. 596-601, doi: 10.1109/CVCI51460.2020.9338647.

[14] J. Yin, W. Lu, X. Xin and L. Zhang, "Application of Monte Carlo sampling and Latin Hypercube sampling methods in pumping schedule design during establishing surrogate model," 2011 International Symposium on Water Resource and Environmental Protection, 2011, pp. 212-215

Investigation of the High Voltage GaN transistor module

Richard Zelnik, Michal Frivaldsky

Department of mechatronics and electronics, Faculty of electrical engineering and information technologies
University of Zilina
Zilina, Slovakia
richard.zelnik@feit.uniza.sk

Abstract— **There are very few manufacturers of GaN transistors on the market with a terminal voltage greater than 650V. For example, if we wanted to use these types of transistors in an inverter system in which the DC link reaches a voltage of 800V, we would need a transistor with a minimum blocking capacity of 1000V (including a margin of 20%). Therefore, it is possible to use SiC transistors or GaN transistors connected in so called stacked configuration, thus doubling their blocking ability. If 650V transistors are being used, their blocking capability in stacked configuration would be 1300V. There are several advantages to use this configuration, but it also exhibits negatives related to issues described within proposed paper. The paper deals with the design of a high voltage stacked configuration GaN module, while it describes simulation analyses, practical PCB design and initial experimental tests.**

Keywords—GaN transistor, switching performance, simulation, parametric, bidirectional converter, efficiency

I. INTRODUCTION

SiC (Silicone Carbide) and GaN (Galium Nitrid) are modern semiconductor materials which enable to increase the efficiency and power density of electronic devices. Using SiC technology, it is possible to achieve high values of blocking voltage of semiconductor devices up to several kilovolts, while GaN lags significantly behind this value, while the maximum optimized value of blocking voltage is 650V. Power supplies used in telecommunications systems, servers, and industrial power supplies (PSUs) convert AC power to isolated DC voltage. Considering 3-phase systems, the blocking capability of available GaN transistors is limited to 650V, which is insufficient because if PFC (Power Factor Corrector) stage would be equipped with GaN technology, at least 800V of blocking voltage is required [1] – [3]. Considering the need of a safety margin at the level of 20% of blocking voltage, a 1kV blocking capability of device is required. Due to this fact, there is requirement on investigations of the possibilities how to improve blocking capability of the GaN power transistors. One of the possible way is the use of series stacked connection of GaN eHEMT transistors, what enables to double the blocking voltage to 1.3 kV, if transistors with 650 V are being used.

GaN and SiC technologies are already in great demand on the market and from their development they have been spread within wide range of various applications. GaN transistors are being used in applications with voltages reaching 400 V, while SiC are more suitable for voltages above 600 V up to 1.2 kV. For mid- and low-voltage applications (below 1200V), GaN's switching losses are at least three times lower than SiC at 650V. SiC has some product availability at 650V but is generally designed for 1200V and higher [4-5].

The main advantage of using GaN transistors is the reduction in system size, weight, and the cost of other system components (such as capacitors, inductors and heatsinks). For example, changing from Si to GaN in a power supply can slash the size of magnetic components such as transformers up to half of the original solution. All of this can be accomplished while achieving better efficiency, or better power density, and possibly both phenomena.

From previous description is possible to conclude that main disadvantage of GaN technology is the ability to block voltages below 650 V. This is caused due to complicated technology of the manufacture of substrates for higher blocking voltages. Therefore, this paper describes possibility how to overcome low blocking voltage ability using series stacked configuration [6], while focus is given on details related to practical design approach of high voltage GaN transistor module, whereby two alternatives which differ by driving circuit are being evaluated.

II. SiC VS GaN SEMICONDUCTOR MATERIAL

These two materials have great potential for use in power applications for high switching speeds. Compared to the most common Si transistors, these materials show much better static and dynamic characteristics. For example, bandwidth refers to the energy required for an electron to cross the valence band of a semiconductor material. Materials that require energy typically greater than one or two electro-volts (eV) are referred to as wide-bandgap materials. SiC and GaN semiconductors are referred to as composite semiconductors because they are composed of multiple elements from the periodic table of elements. Principal comparison of the properties of semiconductor materials are listed in Table I.

TABLE I. COMPARISON OF THE SEMICONDUCTOR MATERIAL PROPERTIES

Material property	Si	SiC-4H	GaN
Energy Band-gap (eV)	1,1	3,2	3,4
Electric field strength 10^6 V/cm	0,3	3	3,5
Carrier Mobility (cm²/Vsec)	1450	900	2000
Saturation Velocity (10^6 cm/sek)	10	22	25
Thermal Conductivity (watt/cm² K)	1,5	5	1,3
Maximum working temperature (°C)	150	1000	900

978-1-6654-3236-8/21 $31.00 © 2021 IEEE

III. HIGH VOLTAGE GaN TRANSISTOR MODULE IN SERIES STACKED CONFIGURATION

As already mentioned, the market available GaN transistors does not reach the required blocking voltage for the use in power electronic systems with DC link voltage above 600 V. Therefore, investigation on a high voltage GaN eHEMT module was performed [7]. The main goal is to benefit from its excellent properties, which are: low losses, high efficiency, the ability to operate at high frequencies. The module consists of two GaN eHEMT transistors connected in series and an integrated excitation network (SSD). Within the mentioned study, the authors evaluated influence of the components of SSD network on the switching performance and voltage balance on the power transistors. How ever more detailed discussion are missing related to practical design and issue relevant for designing such stacked configuration. As will be seen, the proper driving circuit is crucial, when safe and reliable operation of such configuration is required.

The high-voltage GaN transistor module consists of two series-connected GaN eHEMTs and an integrated gate controller. The module has three terminals as a classic MOSFET (gate, drain and source) and works similarly to one switching device (Figure 1). An external gate signal is applied to the gate terminal of the module to drive both HEMTs. The lower device in the module is driven directly via an external gate controller, while the integrated auxiliary gate circuits, referred to here as SSDs (the main components of the SSD are Cd1 and Cd2), are responsible for controlling the upper device.

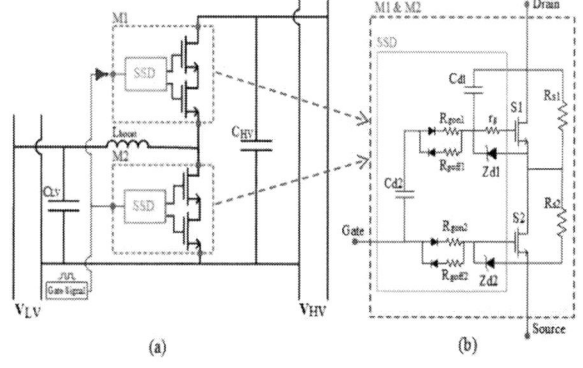

Fig. 1. Example of use HV GaN module based bidirectional boost converter. (a) Converter with the proposed GaN switching modules M_1 and M_2. (b) Internal circuity of the GaN module and the SSD [7]

IV. SELECTED GaN TECHNOLOGY AND TRANSISTOR TYPE

For testing and simulation of high voltage GaN module transistor GS66506T from Gan Systems company was selected. It is an enhancement mode GaN-on-silicon power transistor. It has better efficiency than a Si MOSFET because it has a smaller gate charge and a smaller recovery charge. GaN Systems implements patented Island Technology® cell layout for high-current die performance & yield. GaNPX® packaging enables low inductance & low thermal resistance in a small package. The GS66506T is a top-side cooled transistor that offers very low junction-to-case thermal resistance for demanding high-power applications [8].

Fig. 2. GaNPX technology and system of packaging [13]

The combination of these excellent features provides very high switching efficiency. Table II lists the basic operating parameters of the selected transistor [2].

TABLE II. TRANSISTOR´S PARAMETERS

Material composition	GaN	
Type	GS66506T	
V_{DS}	650	V
$R_{DS(ON)}$	90	mΩ
	105 at 150°C	mΩ
I_D	22.5	A
V_{GS}	-10 to +7	V
$V_{GS(th)}$	1.3	V
R_G	1.1	Ω
V_{PLAT}	3	V
C_{ISS}	195	pF
C_{OSS}	49	pF
$Q_{G(tot)}$	4.4	nC
Q_{RR}	0	nC

V. HV GaN MODULE SWITCHING ANALYSIS THROUGH PSPICE SIMULATION

GaN Systems® provides PSpice and LTspice models for simulation, so we decided to use OrCAD PSpice 16.6 simulation software. Because typical switching frequencies achievable with GaN Systems transistors are comparable with the response times associated with thermal phenomena, a complete electro-thermal model has been developed [4-6].

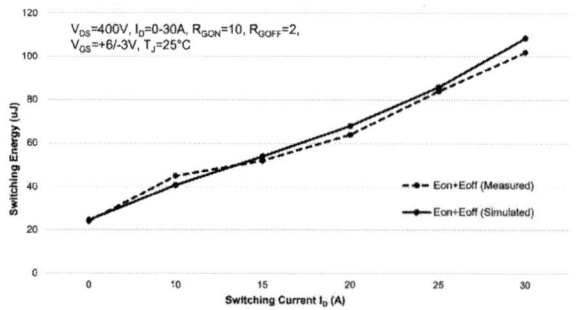

Fig. 3. Example of comparison between switching loss simulation and experimental measurement[13]

978-1-6654-3236-8/21 $31.00 © 2021 IEEE   124

This approach ensures reliable and accurate behavior compared to the physical sample (Fig. 3). The thermal modeling is implemented with a separate thermal network, which allows for self-heating simulations to take place. A package-level model is provided, and the thermal impedance of the package is included in the model. As a simulation software OrCAD PSpice 16.6 was used.

A. HV GaN module simulation circuit

The simulation analysis of switching performance of GaN GS66506T was provided under one supply voltages i.e. 800 Vdc. The current was set to $I_{DMAX} = 15$ A and switching frequency of 500 kHz. The selected value of switching frequency shall confirm the ability of the operation of GaN transistors at several times higher switching frequencies compared to Si transistors, while maintaining low switching losses. The circuit schematic of testing circuit is shown on Fig. 4, while it represents experimental test circuit for dynamic performance of power transistors [9] – [12]. As freewheeling diode, SDT12S60 was selected. Initial simulation study shall serve to identify values of the circuit components (for example SSD network) to verify proper operation of GaN module (similar distribution of drain-source voltages on stacked configuration), for the purposes of practical design.

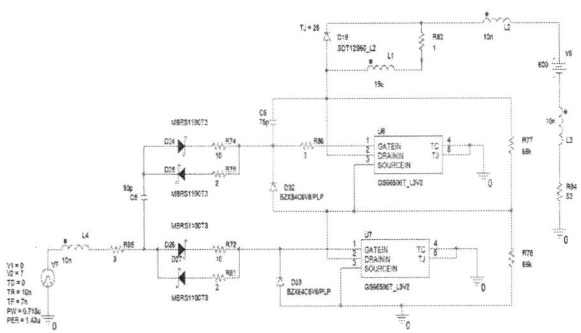

Fig. 4. Simulation circuit of switching performance analysis under hard-switching

Fig. 5. shows the switching waveforms of voltage and current of the module operated at described conditions. A very important issue during simulation circuit optimization is the optimal setting of transistor´s driving components [3]. They have clinical impact on the distribution of the supply voltage between series-stacked configuration of the transistors. If one of the transistors shall sustain higher voltage then other, it can result in higher degradation and reduction of reliability or lifetime.

Fig. 6 shows the voltage waveforms V_{DS} on the individual transistors. In this case, it is able to see that the voltage redistribution is almost identical but considering experimental sample where many parasitic components could affect operation, the goal would be to reach less than 10% of voltage imbalance.

Fig. 7 is showing time waveforms on the gates of both transistors. The upper transistors has driving signals with a threshold of 1V during off period. This is related to R-C networks connected to gate terminal of upper transistor. However, the dynamics and switching performance is not affected by this fact.

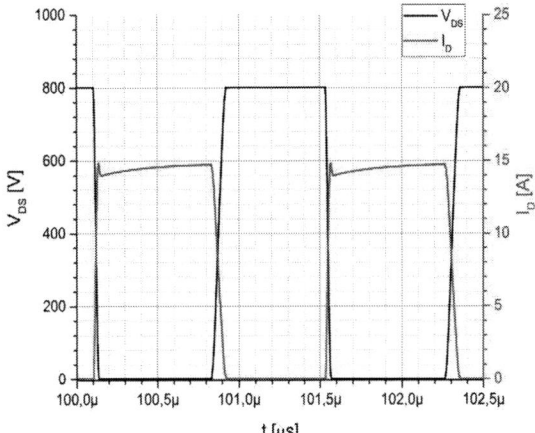

Fig. 5. Switching waveforms of HV GaN module during steady-state operation

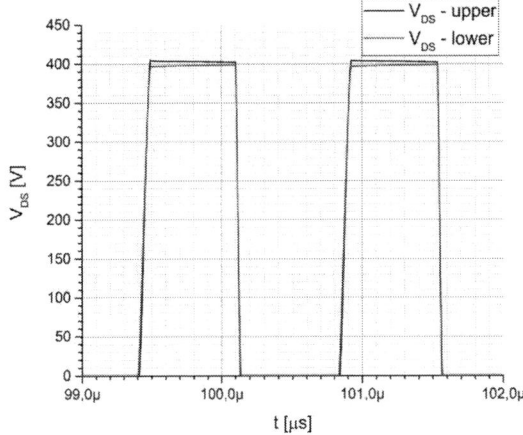

Fig. 6. Detail of sharing of supply voltage between series-stacked transistors

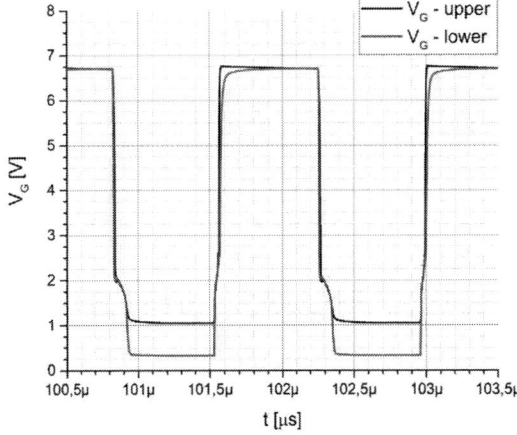

Fig. 7. Detail of V_{GS} voltage (black-upper transistor, red-lower transistor)

VI. DESIGN OF THE HIGH VOLTAGE GAN MODULE

In the next chapter we will focus on the practical design of the HV GaN module. The design consists of two parts. The first part is the power converter design for the verification purposes [13]-[15]. Therefor boost converter was designed (Figure 8) where, only passive elements on the board such as inductance, input/output capacitors, balancing resistors and diode are placed.

978-1-6654-3236-8/21 $31.00 © 2021 IEEE   125

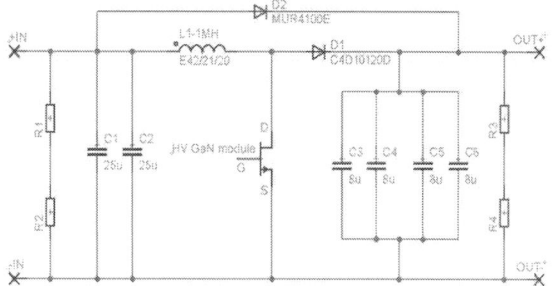

Fig. 8. Boost converter power board for module evaluation

The second part is the "switch" high voltage GaN module, which contains the driving circuit, passive excitation elements (SSD network) and two power transistors connected in series.

Two versions of the module have been developed, which differs with the type of the driving circuit. HV GaN module v2 contains a driver from Infineon 1EDI20N12AF, a single-channel isolated driver with the possibility of driving with negative voltage (Single-channel isolated gate driver ICs for high voltage GaN switches). In this case the driving signal is -5V/+ 5V.

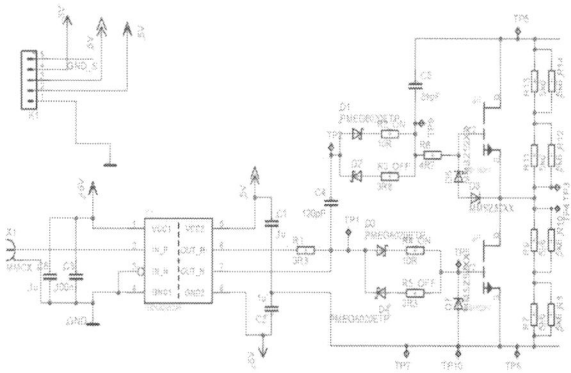

Fig. 9. HV GaN module v2 with isolated driver from Infineon

The second type HV GaN module v3 contains driver from Texas Instruments UCC27611, a single-channel, high-speed, gate driver optimized for 5V drive, specifically addressing enhancement mode GaN FETs.

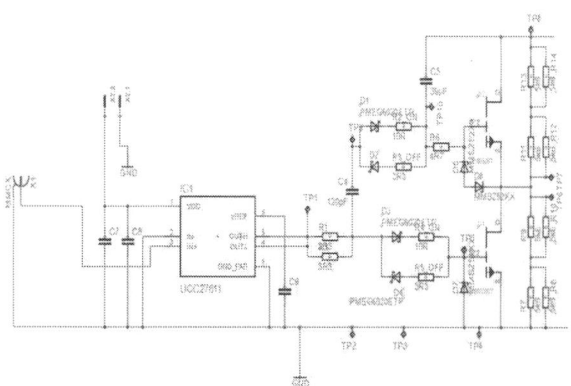

Fig. 10. HV GaN module v3 with low side driver from TI

For a better visualization of the HV GaN module, 3D visualization was designed using the Fusion program (Fig. 11

- Fig. 13). The PCB was exported directly from the Eagle to Fusion PCB design program, while other components were added separately, including the design of the heatsink. In Fig. 11 layout of individual module sections on the PCB are visible. The first section consists of input connectors for power and driving signals. The second section presents the elements for transistor driving components, including the power part (Cd1, Cd2, Rs). Due to the location of cooling and better availability of measuring points, GS66506T transistors are placed on the bottom part of the module board [16].

In Fig.12 a side view of the module is reported. The heatsink is installed from the bottom on the transistors, which have the possibility of cooling from the top. In Fig. 14 is showing manufactured experimental sample of the designed high voltage GaN, which had undergone initial testing.

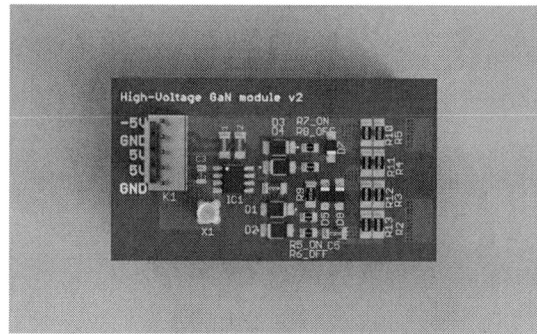

Fig. 11. 3D model of the HV GaN module (Top view)

Fig. 12. 3D model of the HV GaN module (Side view)

Fig. 13. Experimental prototype of the HV GaN module

VII. PRELIMINARY EXPERIMENTAL RESULTS

As mentioned above, the module has been tested in a boost converter topology connection. Two samples with different drivers were tested under the same input and output conditions. Subsequent measurement results were compared. The aim of the measurements was to evaluate the voltage distribution on the HV GaN module and compare this performance for tested versions of driving circuits. The conditions of experimental testing were as follows:

- Input voltage: 200 V
- Switching frequency: 100 kHz
- Duty cycle: 30 % (refers to blocking voltage on transistor to app. 300 V)
- Output power of converter: 50 W

Both designed modules have been preliminary tested to verify difference of voltage imbalance considering different types of drivers. Also, evaluation of the efficiency of testing boost converter was realized to see the effect of optimized driving on this performance. The measured input/output values are given in the table III, while "k" is the percentage evaluation of the voltage imbalance between transistors. From the measured values we can conclude that we have met the requirement on voltage imbalance below 10% for the module v2.

TABLE III MEASUREMENT RESULTS FOR PRELIMINARY MODULES TESTING

Ver.	Vin [V]	Iin [mA]	Vout [V]	Iout [mA]	Pin [W]	Pout [W]	Eff [%]	k [%]
v2	400	574	574	387	229,6	222,14	96,75	8
v3	400	-	-	-	-	-	-	-

Fig. 14 and Fig. 15 shows the voltage waveforms testing conditions. The blue waveform represents the voltage at the gate V_G. The green waveform refers to drain-source voltage of the lower transistor and light blue waveform refers to upper transistor´s V_{DS}. In the case of a light blue waveform (upper transistor), oscillations were generated during switching, which were caused by a differential voltage probe. It is clear from the waveforms that the voltage redistribution is within the specified range and no problem has occurred in terms of excitation that would affect the operation of the converter.

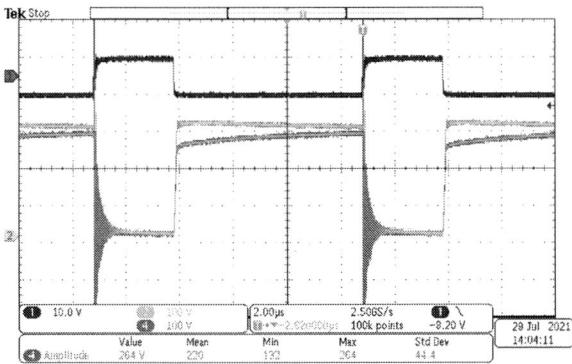

Fig. 14. Waveform at Vin=200/Vout=287V for module v2

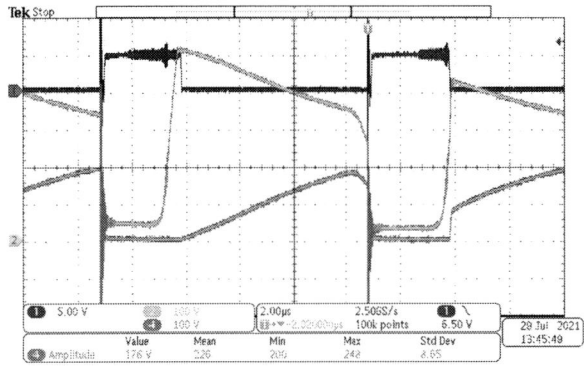

Fig. 15. Waveform at Vin=400 V/Vout=V for module v3

The critical difference of the voltage imbalance could be observed for module v3. According to analysis of the stability of the operation, module v3 results in faulty conditions when the value of the input voltage exceeded 350 V. Subsequently, as the input voltage increased, there was a problem with the driving, where the upper transistor began to take over most of the supply voltage (Fig. 15).

In future, the extensive experimental testing will be performed focusing on the optimization of the stability of operation at high voltages (above 800 V).

VIII. CONCLUSION

This paper described the procedure of the design, simulation and practical realization of perspective high voltage GaN transistor module, which should be used in application requiring high-performance switching operated at systems with DC link higher than 800 V. It contains a verified proposal for series connection of two GaN transistors (stacked configuration), transistor technology selection, and simulation verification of voltage balancing. This approach represents useful tool for optimizing circuit components for the purposes of practical realization of the experimental prototype. Next, two alternatives of proposed HV GaN module were realized, while the main difference between them is the type of the driving circuit. Experimental results have been performed using boost power converter, while preliminary tests were focused on the comparisons of the voltage imbalance between both versions. It was observed, that driving circuit has major impact on the stability of operation, while it strongly affects the voltage imbalance. Future experiments will be therefore focused on precise evaluation of the driving circuit configuration, targeting the lowest voltage distribution difference as possible.

ACKNOWLEDGMENT

This research was funded by national grant agency VEGA 1/0063/21. Authors would like to thank to project support funded by APVV under Nr. 0500-20.

REFERENCES

[1] Juncheng L.; Di Maso P., GaN enables efficient, cost-effective 800V EV traction, EDN – electronics community for engineers, Date: MAY 22, 2020, Avalaible at: https://www.edn.com/gan-enables-efficient-cost-effective-800v-ev-traction-inverters/

[2] R. Zelnik and M. Pipiska, "Simulation analysis of switching performance of GaN power transistors in a high-voltage configuration," 2020 21st International Scientific Conference on Electric Power

Engineering (EPE), 2020, pp. 1-6, doi: 10.1109/EPE51172.2020.9269164.

[3] Koscelnik, J.; Prazenica, M.; Frivaldsky, M.; et al. Design and Simulation of Multi-element Resonant LCTLC Converter with HF Transformer, Conference: ELEKTRO 10th International Conference Location: Rajecke Teplice, SLOVAKIA Date: MAY 19-20, 2014

[4] Dobrucky, B. ., Kascak, S. ., Prazenica, M. ., & Jarabicova, M. . (2019). Improving Efficiency of Hybrid Electric Vehicle Using Matrix Converters. Elektronika Ir Elektrotechnika, 25(4), 29-35. https://doi.org/10.5755/j01.eie.25.4.23967

[5] Slavomir Kascak, Michal Prazenica, Miriam Jarabicova, Marek Paskala, Interleaved DC/DC Boost Converter with Coupled Inductors, In: Advances in Electrical and Electronic Engineering, DOI: 10.15598/aeee.v16i2.2413

[6] M. Shojaie, N. Elsayad, H. Moradisizkoohi and O. A. Mohammed, "Design and Experimental Verification of a High-Voltage Series-Stacked GaN eHEMT Module for Electric Vehicle Applications," in *IEEE Transactions on Transportation Electrification*, vol. 5, no. 1, pp. 31-47, March 2019. doi: 10.1109/TTE.2018.2888476

[7] M.Shojaie, N. Elsayad, and O. A. Mohammed, "Design of an all-GaN bidirectional DC-DC converter for medium voltage DC ship power systems using series-stacked GaN modules", in Proc. IEEE Appl. Power Electron. Conf. Expo. (APEC), San Antonio, TX, USA, Mar. 2018, pp. 2155-2161

[8] GN008: GaN Switching Loss Simulation Using LTSpice 05/2018 available at https://gansystems.com/

[9] By: Spanik, P.; Dobrucky, B.; Frivaldsky, M.; et al. Measurement of swtitching losses in power transistor structure, ELEKTRONIKA IR ELEKTROTECHNIKA Issue: 2 Pages: 75-78 Published: 2008

[10] Frivaldsky, Ing Michal; Drgona, Peter; Spanik, Pavol, Experimental analysis and optimization of key parameters of ZVS mode and its application in the proposed LLC converter designed for distributed power system application, INTERNATIONAL JOURNAL OF ELECTRICAL POWER & ENERGY SYSTEMS Volume: 47 Pages: 448-456 Published: MAY 2013

[11] Frivaldsky, Michal; Dobrucky, Branislav; Prazenica, Michal; et al. Multi-tank resonant topologies as key design factors for reliability improvement of power converter for power energy applications, ELECTRICAL ENGINEERING Volume: 97 Issue: 4 Pages: 287-302 Published: DEC 2015

[12] Kandrac, J.; Frivaldsky, M.; Prazenica, M.; et al. Design and Verification of proposed Operation Modes of LLC Converter, ELEKTRONIKA IR ELEKTROTECHNIKA Volume: 18 Issue: 8 Pages: 27-30 Published: 2012

[13] V. Augutis, D. Gailius, D. Styra, A. Dumčius, Transistor Control with Additional Charge Pumping Circuit, Vol 78, No 6 (2007)

[14] S. Kascak, T. Laskody, M. Prazenica and R. Konarik, "Current control contribution to a single-phase induction motor fed by single-leg voltage source inverter," *2016 ELEKTRO*, 2016, pp. 172-175, doi: 10.1109/ELEKTRO.2016.7512059.

[15] Hruska, K., Kindl, V. & Pechanek, R. Design of a high-speed permanent magnet synchronous motor for electric kart. Electr Eng 99, 1141–1150 (2017). https://doi.org/10.1007/s00202-017-0623-2

[16] P. Varecha, P. Makys, M. Pacha and S. Zossak, "Effect of MOSFET lifetime on reliability of low-side MOSFET current sensing technique," *2020 ELEKTRO*, 2020, pp. 1-6, doi: 10.1109/ELEKTRO49696.2020.9130248.

Unbalanced Load Modeling and Control in Microgrid with Isolation Transformer

Hao, Jiang
Electrical Power Engineering
Singapore Institute of Technology, Singapore
Hao.Jiang@singaporetech.edu.sg

Shuyu, Cao
Electrical Power Engineering
Singapore Institute of Technology, Singapore
Shuyu.Cao@singaporetech.edu.sg

Chew Beng, Soh
Electrical Power Engineering
Singapore Institute of Technology, Singapore
ChewBeng.Soh@singaporetech.edu.sg

Feng, Wei
Electrical and Electronic Engineering
Singapore Polytechnic, Singapore
Fred_Wei@sp.edu.sg

Abstract—Renewables and energy storage rely heavily on power electronics. The control of power electronics in a microgrid with renewables and energy storages become more challenging under unbalanced load operation condition. The unbalance in current magnitude and phase will cause ripples in normal dq0 conversion, further causing voltage unbalance in the power converter current control loop. In this paper, a simulation test microgrid is studied as a good representation of a practical microgrid with local ancillary single loads. The double second-order generalized integrator-phase lock loop (DSOGI-PLL) technique for frequency and phase angle detection is modified to address the unbalanced current issue due to local single-phase load connected through an isolation transformer. The proposed method is verified through software simulation in both grid-connected and islanded modes. From the result, the modified DSOGI method proves to be a viable technique for improving power converter current control loop performance under single load or unbalanced load.

Keywords— *microgrid, unbalanced load, phase lock loop (PLL).*

I. INTRODUCTION

Microgrid (MG) has been widely used in power system study. It brings about flexibility in the operation of energy sources. Microgrid supports grid-connected mode and islanded mode and usually involves renewable energy sources such as wind, solar. The integration and control of such renewable sources or energy storage devices produce new challenges. Power electronics such as voltage source converter (VSC) are widely used in MG. The MG structure and ratings of individual components are usually limited by practical constraints and the power quality can be compromised as compared to the regional grid. Besides, the existence of single-phase load in a three-phase MG is often neglected in research studies. The usage of single line diagram also does not give a proper representation of the unbalanced load issue. Even though the assumption of balanced three-phase MG does not affect the general conclusion in many studies, it remains important to investigate the power quality and stability in MG under unbalanced load conditions [1], [2]. In fact, the unbalanced load condition is more likely to happen in MG. Within an MG, there are unavoidable needs for single-phase load such as power adaptors for controllers, illumination lights, displays, cooling fans, etc. They cause significant unbalance current in the system, especially when the three-phase, 3-wire distribution system is adopted.

The unbalanced load currents will result in unbalanced voltages in the VSC control. This phenomenon is more adverse when the MG is lightly loaded or in an idle state. In addition, over-current events might occur especially under the heavily loaded phase when it is running close to full capacity.

The problem of the unbalanced load has been investigated in the literature. In [3], a robust control method for islanded MG is given. The unbalanced currents are assumed to be measurable and the problem of unbalanced current and harmonics are converted to a H_∞ control optimization problem. It is then reformulated into a convex linear matrix inequality (LMI) problem. Another control strategy is proposed in [4], the droop control strategy, negative sequence impedance compensator and adaptive resonance frequency are combined to achieve the control objective in a multi-bus medium voltage MG. However, the above two methods have high complexity in implementation. A four-leg VSC connected in a PV system is controlled by designing the currents in AC reference to compensate load unbalance [5]. Another control strategy is proposed to adopt a new topology with only 4 switches as compared with the normal 6 switches topology of three-phase inverter in [6]. A Lyapunov function-based nonlinear control method is used to trace the inverter current directly to lower the harmonic distortion in an unbalanced system [6]. The unbalanced current problem of 3-leg inverter interface is more challenging as compared with 4-leg inverter [7], [8]. Model predictive control (MPC) has also been deployed to minimize voltage unbalance due to single-phase loads with fast dynamic response [9]. Other advanced methods such as compensatory neural fuzzy network with asymmetric membership function have also been investigated in [10] to eliminate the usage of active power filter.

The problem of unbalanced currents will cause inaccurate detection of the positive-sequence detection in VSC control. It will cause the current control loop to output distorted voltage reference. Positive-sequence detection is widely used in PLL. SRF-PLL and its derivatives have the advantage of simplicity but lack harmonic rejection ability [11], [12]. More advanced methods such as SOGI, DSOGI, MRF-PLL, MCCF-PLL are proposed with different levels of frequency-adaptiveness and harmonics attenuation [13]–[16].

The focus of this paper is to find a novel method to minimize the negative impact of unbalanced currents on voltage distortion in a three-phase, 3-wire MG system. The contribution of this study is to apply DSOGI-PLL technique to design a simple, fast, robust, and accurate control system

978-1-6654-3236-8/21 $31.00 © 2021 IEEE

for VSC-based power sources. The proposed method is applicable for 3-wire system with isolation transformer and single-phase loads. It is also capable of harmonics attenuation which will be shown in later sections.

This paper is organized in the following way. Section II presents the mathematics and theory in the proposed methods. In Section III, the studies on MG system are explained. The simulation results are explained in Section IV. Section V draws the conclusion of this paper.

II. POSITIVE SEQUENCE DETECTION AND PROPOSED DESIGN METHOD

Suppose the AC currents are unbalanced and have harmonics. It can be described in the general equations as shown in (1).

$$
\begin{aligned}
i_a(t) &= \sum_{h=1}^{+inf} [I_h^+(\cos(h\omega t + \theta_h^+) + I_h^-(\cos(h\omega t + \theta_h^-)] \\
i_b(t) &= \sum_{h=1}^{+inf} [I_h^+(\cos\left(h\omega t + \theta_h^+ - \frac{2}{3}\pi\right) \\
&\quad + I_h^-(\cos\left(h\omega t + \theta_h^- + \frac{2}{3}\pi\right)] \\
i_c(t) &= \sum_{h=1}^{+inf} [I_h^+(\cos\left(h\omega t + \theta_h^+ - \frac{4}{3}\pi\right) \\
&\quad + I_h^-(\cos\left(h\omega t + \theta_h^- + \frac{4}{3}\pi\right)]
\end{aligned}
\tag{1}
$$

where, ω is the actual frequency; θ_h^+, θ_h^- are the phase angle of the hth harmonics in positive/negative sequence; I_h^+, I_h^- are the phase angle of the hth harmonics in positive/negative sequence.

By applying abc-to-$\alpha\beta$ transformation as shown in (2) and (3), the unbalanced AC current can be represented in $\alpha\beta$ stationary reference frame as shown in (4) for positive sequence components and (5) for negative sequence components.

$$
i_\alpha(t) = T_{\alpha\beta} \begin{bmatrix} i_a(t) \\ i_b(t) \\ i_c(t) \end{bmatrix} = \begin{bmatrix} i_\alpha^+(t) \\ i_\beta^+(t) \end{bmatrix} + \begin{bmatrix} i_\alpha^-(t) \\ i_\beta^-(t) \end{bmatrix}
\tag{2}
$$

where

$$
T_{\alpha\beta} = \frac{2}{3} \begin{bmatrix} 1 & -\dfrac{1}{2} & -\dfrac{1}{2} \\ 0 & \dfrac{\sqrt{3}}{2} & -\dfrac{\sqrt{3}}{2} \end{bmatrix}
\tag{3}
$$

$$
\begin{bmatrix} i_\alpha^+(t) \\ i_\beta^+(t) \end{bmatrix} = \begin{bmatrix} \sum_{h=1}^{+inf} i_{\alpha,h}^+(t) \\ \sum_{h=1}^{+inf} i_{\beta,h}^+(t) \end{bmatrix} = \begin{bmatrix} \sum_{h=1}^{+inf} I_h^+\cos(h\omega t + \theta_h^+) \\ \sum_{h=1}^{+inf} I_h^+\sin(h\omega t + \theta_h^+) \end{bmatrix}
\tag{4}
$$

$$
\begin{bmatrix} i_\alpha^-(t) \\ i_\beta^-(t) \end{bmatrix} = \begin{bmatrix} \sum_{h=1}^{+inf} i_{\alpha,h}^-(t) \\ \sum_{h=1}^{+inf} i_{\beta,h}^-(t) \end{bmatrix} = \begin{bmatrix} \sum_{h=1}^{+inf} I_h^-\cos(h\omega t + \theta_h^-) \\ -\sum_{h=1}^{+inf} I_h^-\sin(h\omega t + \theta_h^-) \end{bmatrix}
\tag{5}
$$

After applying dq0 transformation shown in (6) and (7), in (6). the 3-phase unbalanced AC currents can be decomposed

into positive/negative sub-dq components shown in (8) and (9) [16].

$$
\begin{bmatrix} i_d(t) \\ i_q(t) \end{bmatrix} = T_{dq} \begin{bmatrix} i_\alpha(t) \\ i_\beta(t) \end{bmatrix} = \begin{bmatrix} i_d^+(t) \\ i_q^+(t) \end{bmatrix} + \begin{bmatrix} i_d^-(t) \\ i_q^-(t) \end{bmatrix}
\tag{6}
$$

where

$$
T_{dq} = \begin{bmatrix} \cos(\omega_v t + \theta_v) & \sin(\omega_v t + \theta_v) \\ -\sin(\omega_v t + \theta_v) & \cos(\omega_v t + \theta_v) \end{bmatrix}
\tag{7}
$$

-θ_v is the phase angle of the voltage from PLL.

$$
\begin{aligned}
\begin{bmatrix} i_d^+(t) \\ i_q^+(t) \end{bmatrix} &= \begin{bmatrix} \sum_{h=1}^{+inf} i_{d,h}^+(t) \\ \sum_{h=1}^{+inf} i_{d,h}^+(t) \end{bmatrix} \\
&= \begin{bmatrix} \sum_{h=1}^{+inf} I_h^+\cos((h\omega - \omega_v)t + \theta_h^+ - \theta_v) \\ \sum_{h=1}^{+inf} I_h^+\sin((h\omega - \omega_v)t + \theta_h^+ - \theta_v) \end{bmatrix}
\end{aligned}
\tag{8}
$$

$$
\begin{aligned}
\begin{bmatrix} i_d^-(t) \\ i_q^-(t) \end{bmatrix} &= \begin{bmatrix} \sum_{h=1}^{+inf} i_{d,h}^-(t) \\ \sum_{h=1}^{+inf} i_{d,h}^-(t) \end{bmatrix} \\
&= \begin{bmatrix} \sum_{h=1}^{+inf} I_h^-\cos((h\omega + \omega_v)t + \theta_h^- - \theta_v) \\ -\sum_{h=1}^{+inf} I_h^-\sin((h\omega + \omega_v)t + \theta_h^- - \theta_v) \end{bmatrix}
\end{aligned}
\tag{9}
$$

When the estimated frequency ω_v from PLL is within a close range to the actual frequency ω, it can be described as quasi-locked condition. Note that PLL is applied upon the voltage signal instead of the current signal. If a quasi-locked condition is assumed, the dq-frame current can be simplified as (10), which can be derived from (6) via changing of variables and separating into DC components and disturbance components.

$$
\begin{bmatrix} i_d(t) \\ i_q(t) \end{bmatrix} = \begin{bmatrix} \bar{\imath}_d \\ \bar{\imath}_q \end{bmatrix} + \begin{bmatrix} \tilde{\imath}_d(t) \\ \tilde{\imath}_q(t) \end{bmatrix}
\tag{10}
$$

As shown in (10), i_d, i_q has a DC component $\bar{\imath}_d, \bar{\imath}_q$ which are the fundamental frequency positive sequence current $i_{d,1}^+, i_{q,1}^+$. The DC components can be expressed in (11).

$$
\begin{bmatrix} \bar{\imath}_d \\ \bar{\imath}_q \end{bmatrix} = \begin{bmatrix} i_{d,1}^+ \\ i_{q,1}^+ \end{bmatrix} \approx \begin{bmatrix} I_1^+\cos(\theta_h^+ - \theta_v) \\ I_1^+\sin(\theta_h^+ - \theta_v) \end{bmatrix}
\tag{11}
$$

The higher-order positive sequence components and all negative sequence components are disturbances to the dq0 transformation. Evidently, the disturbance terms $\tilde{\imath}_d(t), \tilde{\imath}_q(t)$ have to be properly eliminated or filtered in the context of unbalanced load currents.

Figure 1 shows the schematic of the second-order generalized integrator (SOGI). i is the input alpha-beta component and the i' is the output signal. qi' is the quadrature phase signal. It has the following transfer function [13].

978-1-6654-3236-8/21 $31.00 © 2021 IEEE

$$D(s) = \frac{i'(s)}{i(s)} = \frac{K\hat{\omega}s}{s^2 + K\hat{\omega}s + \hat{\omega}^2} \tag{12}$$

$$Q(s) = \frac{qi'(s)}{i(s)} = \frac{K\hat{\omega}^2}{s^2 + K\hat{\omega}s + \hat{\omega}^2} \tag{13}$$

where, $\hat{\omega}$ set the resonance frequency and should be frequency adaptive by feeding the ω_v signal from PLL; K is the damping factor of SOGI.

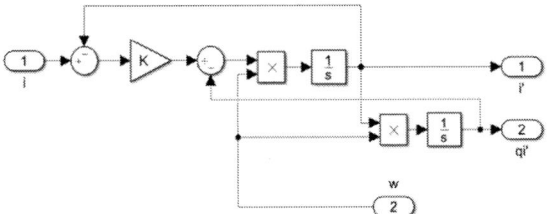

Figure 1 SOGI schematic

Figure 2 shows the D(s) bode response from SOGI input signal to output signal and the Q(s) bode response from SOGI input signal to quadrature output signal.

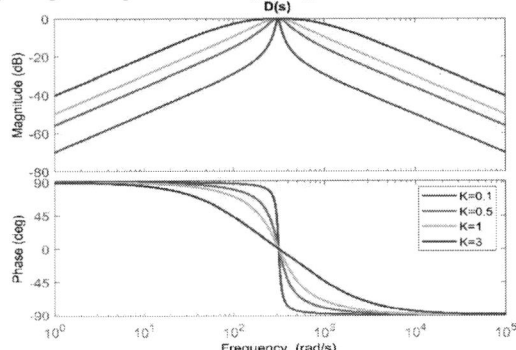

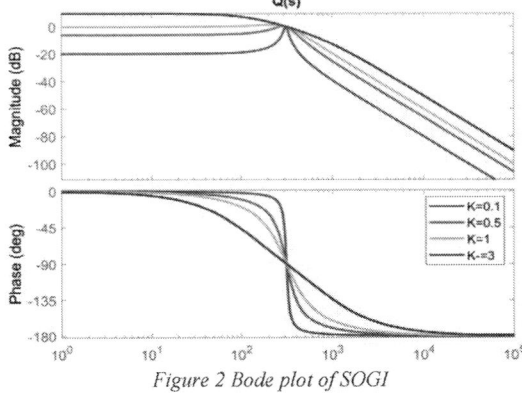

Figure 2 Bode plot of SOGI

From the bode plot of SOGI, the stability is always assured as the gain margin is infinite regardless of the damping factor value of K. The gain at the resonant frequency is 0 db. If it coincides with the fundamental frequency, then any higher order harmonics are damped. The larger value of K provides better damping of harmonics.

DSOGI quadrature signal generator (DSOGI-QSG) is shown in Figure 4.

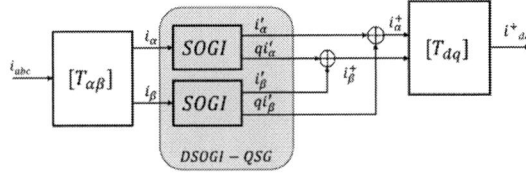

Figure 3 DSOGI – QSG schematic to generate i_{dq}^+

The resulting output form DSOGI-QSG $i_{\alpha\beta}^+$ is the fundamental positive sequence components. The Laplace transfer function of $i_{\alpha\beta}^+$ is shown in (14).

$$
\begin{aligned}
i_{\alpha\beta}^+(s) &= \frac{1}{2}\begin{bmatrix} D(s) & -Q(s) \\ Q(s) & D(s) \end{bmatrix} i_{\alpha\beta}(s) \\
&= \frac{1}{2}\frac{K\hat{\omega}}{s^2 + K\hat{\omega}s + \hat{\omega}^2}\begin{bmatrix} s & -\hat{\omega} \\ \hat{\omega} & s \end{bmatrix} i_{\alpha\beta}(s)
\end{aligned}
\tag{14}
$$

III. TEST MG AND SIMULATION SETUP

The paper study the specific problem of positive sequence detection of unbalanced current for the current control loop in a general VSC controller. Hence, it can be applied in any VSC enabled DG or MG.

Figure 4 shows the system block diagram of VSC battery inverter for MG application used as a test case for the proposed DSOGI-QSG method for unbalanced load control.

The lithium-ion battery system (rated 50kW/50kAh) is integrated with a VSC interface. The PWM switching harmonics of the power electronics are filtered by an LC filter. Then a step-up transformer is connected in series to transform the line-line voltage to the 400V bus. A delta-wye grid transformer is connected in between the point of common coupling (PCC) and VSC. A three-phase load is added to the system to model the energy loss and power consumption due to symmetrical loads locally. A single-phase load is connected to the MG via an isolation transformer.

The combination of symmetrical three-phase load and a single-phase load describes general MG structure under unbalanced conditions. It also provides a good representation of practical MG where both three-phase loads and single-phase load are important and unavoidable. The single-phase load models the power consumption from power adaptors for controllers, industry PC, illumination lights, displays, cooling fans that actually exist in the practical system. It can be found that these loads are time-variant and non-linear and post extra challenges to the power quality in a weak MG.

The isolation transformer primary side is connected to phase A & C (400V) and the secondary side is 230V. The isolation transformer provides single-phase voltage in a three-phase, 3-wire system and it also protects potential electric shock at the secondary side.

In the VSC control function shown in Figure 4, the PLL takes the v_{abc} signals to generate v_{dq} and phase angle ωt.

978-1-6654-3236-8/21 $31.00 © 2021 IEEE

ωt is used in DSOGI to generate i_{dq} free of unbalanced load disturbance.

The MG test system can be operated in both islanded mode where the V/f control is enabled. Battery VSC inverter is the grid former in islanded mode. When in grid connected mode, Battery VSC inverter is a grid follower with PQ control function. The transition from islanded mode to grid-connected mode is through active synchronization. Frequency, phase angle, & voltage magnitude is adjusted to match the grid in a close range before connection. The MG operating mode transition is not discussed here as it is not the focus of this paper.

The proposed DSOGI VSC control method works for MG system to operate at grid tied application and islanded mode operation in unbalanced load condition. DSOGI function block is applied to eliminate the high order harmonics in dq frame current. It requires VCS voltage phase angle as it input signal. In MG grid tied application, the VCS voltage phase angle is generated by the PLL block. In MG islanded operation condition, the VSC voltage phase angle can be either generated internally from frequency setting signal or using PLL phase measurement with voltage sensor installed on the grid side of the grid connection/disconnection circuit breaker.

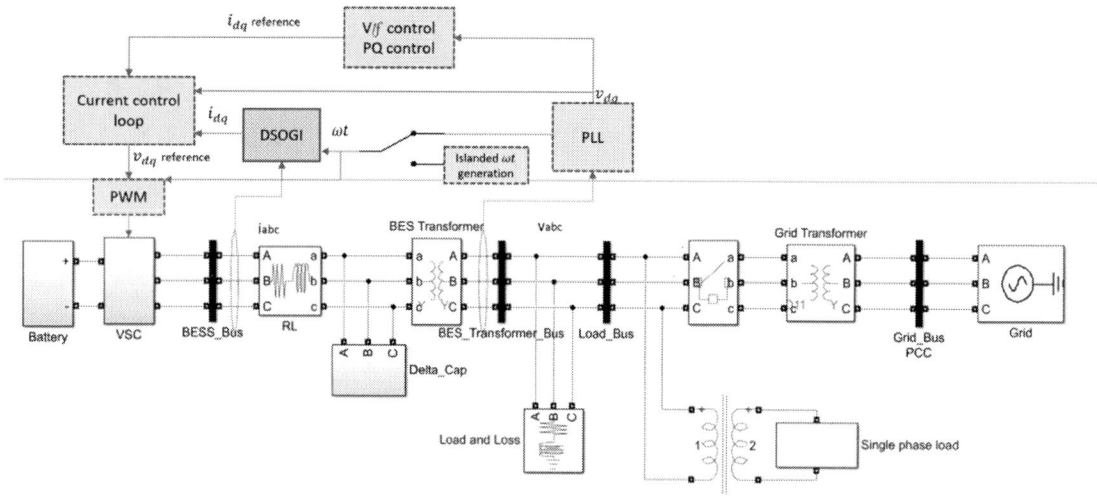

Figure 4 Battery VSC Control System with Unbalanced Load for MG Application.

The current control shown in Figure 5 is widely used in VSC inner loop control. The I_{dref} and I_{qref} input can be designated as the control output from different outer controllers such as PQ controller, droop control, DC-link voltage regulation, etc.

The proposed DSOGI based dq0 generator can replace the standard abc-dq0 transformation. The standard abc-dq0 transformation block is used as a base comparison in the simulation.

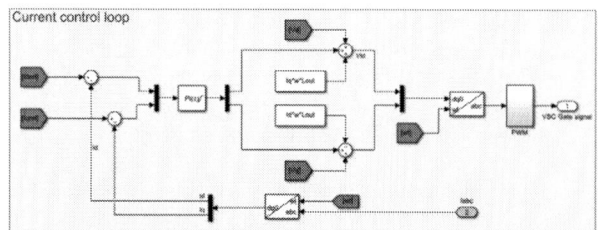

Figure 5 Current control loop of VSC

IV. SIMULATION RESULTS

A. Unbalanced current control in islanded mode

The simulation is carried out in MATLAB/Simulink environment. The single-phase load is set at 650W/400VAr.

The MG is in islanded mode. The MG is disconnected before the delta-wye grid transformer.

There are limited harmonics in this situation as the LC filter is designed properly in the test MG. The harmonics rejection capability will be shown in Section IV (C) where an intentional rejection of harmonics is applied to the system.

In the simulation, the MG is in a lightly loaded condition. The frequency of PLL is fed into DSOGI. The damping factor K is set to 0.5 and the system frequency is at 50Hz.

Figure 6 shows the simulation results comparison with DSOGI disabled and enabled under significant unbalanced load. Initially the standard dq0 transformation is applied. At time= 0.1s, the DSOGI is applied.

It can be seen from Figure 6 that the i_{dq} signal oscillation is large when standard dq0 transformation is applied. The proposed method can quickly damp the negative sequence current components and achieve i_{dq} in DC form. Without the proposed method, the V_d, V_q outputs from the current control loop have distorted voltage signals. The oscillation of V_d, V_q will cause unbalanced phase voltage magnitude. The simulation results show the effectiveness and fast dynamic response of the proposed method in Id, Iq and Vd , Vq.

978-1-6654-3236-8/21 $31.00 © 2021 IEEE

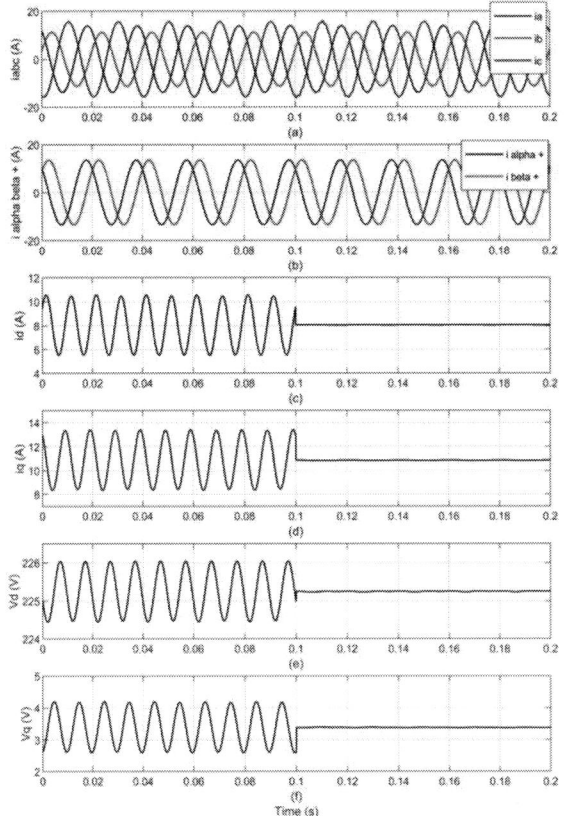

Figure 6 Simulation results of proposed current control in islanded mode for DSOGI disabled vs. enabled

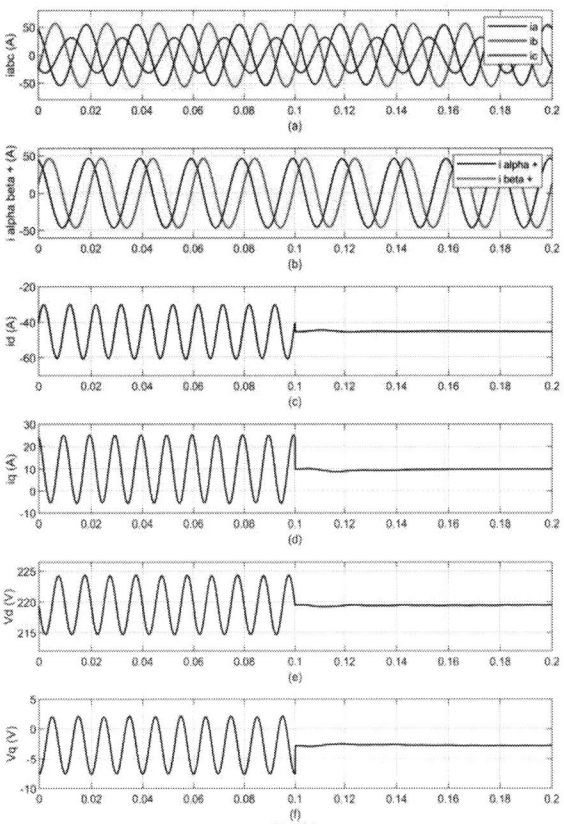

Figure 7 Simulation results of w/wo proposed current control method in grid-connected mode for DSOGI disabled vs. enabled

B. Unbalanced current control in grid-connected mode

Figure 7 shows the simulation results with DSOGI disabled and enabled for battery VSC inverter control with 650W/400Var single-phase load in MG grid-connected operation condition. At time=0.1s, the dq0 transformation is switched from standard control method to the proposed DSOGI control method. The DSOGI damping factor K is set to 0.5.

The current unbalance is significant as shown in Figure 7. The micro-grid current at phase C is significantly lower than the other two phases. The $i_{\alpha\beta}^+$ obtained at the fundamental frequency is also shown in Figure 7.

It can be seen from Figure 7 that the i_{dq} signal oscillation becomes larger without the proposed DSOGI current control method.

The proposed current control method can quickly damp the negative sequence current components within a few cycles. Without the proposed method, the V_d, V_q outputs generated from current control loop which are shown in Figure 7 have distorted voltage signals. The oscillation of V_d, V_q causes larger unbalanced phase voltage magnitude. The simulation results show that the proposed controller is also valid when the MG is in grid-connected mode or there is current unbalance from the grid.

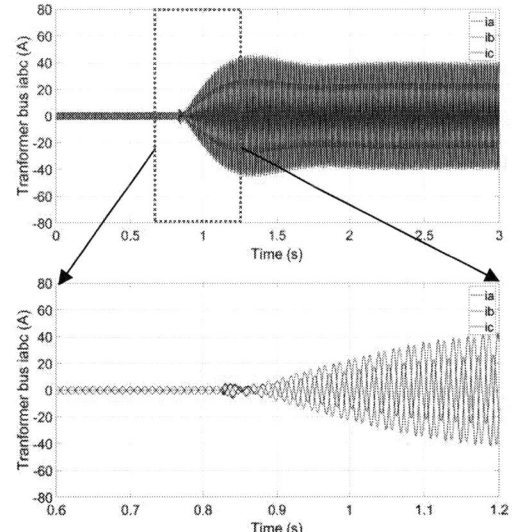

Figure 8 Grid transition from islanded mode to grid connected mode with DSOGI current control

Figure 8 demonstrates the transient response of the proposed current control method when the MG goes through grid integration. At time=0, a command to synchronize to grid is issued. After 0.82 s, the circuit breaker is closed. The MG changes from grid former to PQ control mode.

The steady state current charge in BES transformer bus is due to different network structure and power flow. In islanded mode, only i_a, i_c current exist with same magnitude and opposite direction. The current in phase b is 0. This is a result of single-phase load connected to the battery VCS inverter with an isolation transformer in a three-phase, three wire system. After the MG is grid-connected, the single-phase load causes severe current unbalance in the system.

C. Unbalanced current control in grid-connected mode with harmonics

To test and demonstrate the harmonic tolerance of the proposed method. Harmonics are added in the current measurement on the top of the previous test.

Figure 9 shows the hormonic current injection simulation results of -3-phase current waveforms, DOSIGI output $i_{\alpha\beta}^+$ at the fundamental frequency, i_d. I_q, V_d, V_q with proposed DSOGI control disabled and enabled for Battery VSC inverter control. Initially the standard dq0 transformation is applied. At time= 0.1s, the DSOGI is applied.

In this simulation, the damping factor K for SDOGI is set to 0.1 with more aggressive damping. The injected 3rd order harmonic current magnitude is 5 A. The injected 4th order harmonic current magnitude is 4A.

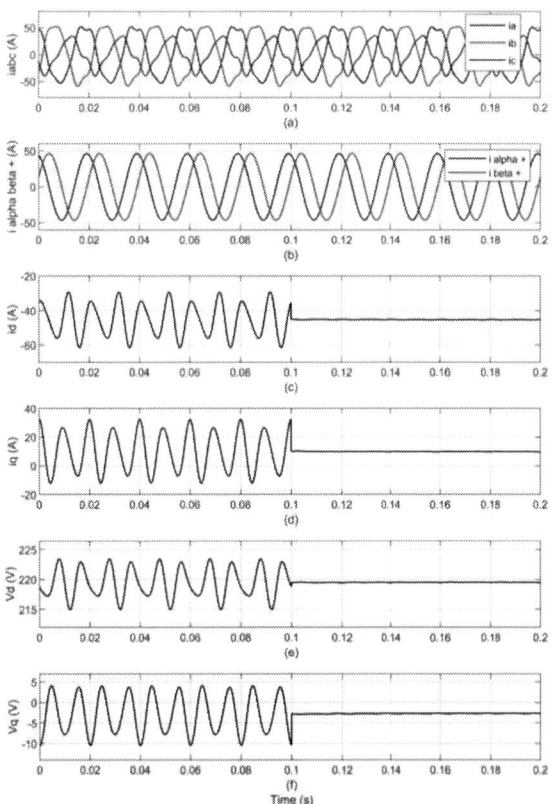

Figure 9 Simulation results of current control in grid-connected mode with current harmonics for DSOGI disabled vs. enabled

Without the proposed method, the i_{dq} waveforms in Figure 9 are no longer DC values. This further deteriorates the voltage output from the current control loop. Hence, worst voltage quality will happen.

With the proposed method, the i_{dq} waveforms can reject the negative impact from harmonics. It further proves that the proposed method is robust, accurate, fast, and harmonic resilient.

V. CONCLUSION

In this paper, the specific problem from unbalanced current in the context of MG with single-phase loads. An accurate, fast, robust, and harmonic resilient method is proposed with the DSOGI technique. The performance of the proposed method is validated through a simulation model of the MG with unbalanced load and battery VSC inverter control system. The simulation result shows that it can improve the voltage quality in VSC inner current control loop under unbalanced loads.

ACKNOWLEDGMENT

The research work described in this paper is part of " Micro-grid Digital Twin Development for Effective Energy Management and Deployment" project. The project reference number is EDGE-GC2018-001.

REFERENCES

[1] T. Ma and G. Cheng, "A novel control method for islanding mode microgrid with unbalanced load," in 2015 5th International Conference on Electric Utility Deregulation and Restructuring and Power Technologies (DRPT), 2015, pp. 2134–2138, doi: 10.1109/DRPT.2015.7432602.

[2] H. Cai, P. Zhang, H. Zhao, J. Shi, W. Yao, and X. He, "Controller design for three-phase inverter with power unbalanced loads applied in microgrids," in *2015 IEEE Energy Conversion Congress and Exposition (ECCE)*, 2015, pp. 4588–4593, doi: 10.1109/ECCE.2015.7310309.

[3] M. Hamzeh, S. Emamian, H. Karimi, and J. Mahseredjian, "Robust Control of an Islanded Microgrid Under Unbalanced and Nonlinear Load Conditions," *IEEE J. Emerg. Sel. Top. Power Electron.*, vol. 4, no. 2, pp. 512–520, 2016, doi: 10.1109/JESTPE.2015.2459074.

[4] M. Hamzeh, H. Karimi, and H. Mokhtari, "A New Control Strategy for a Multi-Bus MV Microgrid Under Unbalanced Conditions," *IEEE Trans. Power Syst.*, vol. 27, no. 4, pp. 2225–2232, 2012, doi: 10.1109/TPWRS.2012.2193906.

[5] V. F. Pires, O. Husev, D. Vinnikov, and J. F. Martins, "A control strategy for a grid-connected PV system with unbalanced loads compensation," in *2015 9th International Conference on Compatibility and Power Electronics (CPE)*, 2015, pp. 154–159, doi: 10.1109/CPE.2015.7231065.

[6] S. Dasgupta, S. N. Mohan, S. K. Sahoo, and S. K. Panda, "Application of Four-Switch-Based Three-Phase Grid-Connected Inverter to Connect Renewable Energy Source to a Generalized

Unbalanced Microgrid System," *IEEE Trans. Ind. Electron.*, vol. 60, no. 3, pp. 1204–1215, 2013, doi: 10.1109/TIE.2012.2202350.

[7] D. Vyawahare and M. Chandorkar, "Compensator-less stand-alone microgrid with 4-leg inverters for unbalanced and nonlinear loads," in *2016 IEEE International Conference on Power Electronics, Drives and Energy Systems (PEDES)*, 2016, pp. 1–6, doi: 10.1109/PEDES.2016.7914494.

[8] G. H. Kim, C. Hwang, J. H. Jeon, G. S. Byeon, J. B. Ahn, and C. H. Jo, "Characteristic analysis of three-phase four-leg inverter based load unbalance compensator for stand-alone microgrid," in *2015 9th International Conference on Power Electronics and ECCE Asia (ICPE-ECCE Asia)*, 2015, pp. 1491–1496, doi: 10.1109/ICPE.2015.7167976.

[9] M. S. Golsorkhi and D. D. Lu, "A Decentralized Control Method for Islanded Microgrids Under Unbalanced Conditions," *IEEE Trans. Power Deliv.*, vol. 31, no. 3, pp. 1112–1121, 2016, doi: 10.1109/TPWRD.2015.2453251.

[10] F. Lin, K. Tan, Y. Lai, and W. Luo, "Intelligent PV Power System With Unbalanced Current Compensation Using CFNN-AMF," *IEEE Trans. Power Electron.*, vol. 34, no. 9, pp. 8588–8598, 2019, doi: 10.1109/TPEL.2018.2888732.

[11] T. A. Youssef and O. Mohammed, "Adaptive SRF-PLL with reconfigurable controller for Microgrid in grid-connected and stand-alone modes," in *2013 IEEE Power & Energy Society General Meeting*, 2013, pp. 1–5, doi: 10.1109/PESMG.2013.6673028.

[12] A. Nouralinejad, A. Bagheri, M. Mardaneh, and M. Malekpour, "Improving the Decoupled Double SRF PLL for grid connected power converters," in *The 5th Annual International Power Electronics, Drive Systems and Technologies Conference (PEDSTC 2014)*, 2014, pp. 347–352, doi: 10.1109/PEDSTC.2014.6799398.

[13] P. Rodríguez, R. Teodorescu, I. Candela, A. V Timbus, M. Liserre, and F. Blaabjerg, "New positive-sequence voltage detector for grid synchronization of power converters under faulty grid conditions," in *2006 37th IEEE Power Electronics Specialists Conference*, 2006, pp. 1–7, doi: 10.1109/pesc.2006.1712059.

[14] F. Xiao, L. Dong, L. Li, and X. Liao, "A Frequency-Fixed SOGI-Based PLL for Single-Phase Grid-Connected Converters," *IEEE Trans. Power Electron.*, vol. 32, no. 3, pp. 1713–1719, 2017, doi: 10.1109/TPEL.2016.2606623.

[15] S. Prakash, J. K. Singh, R. K. Behera, and A. Mondal, "Comprehensive Analysis of SOGI-PLL Based Algorithms for Single-Phase System," in *2019 National Power Electronics Conference (NPEC)*, 2019, pp. 1–6, doi: 10.1109/NPEC47332.2019.9034724.

[16] S. Golestan, M. Monfared, and F. D. Freijedo, "Design-Oriented Study of Advanced Synchronous Reference Frame Phase-Locked Loops," *IEEE Trans. Power Electron.*, vol. 28, no. 2, pp. 765–778, 2013, doi: 10.1109/TPEL.2012.2204276.

Modeling and Predictive Control of LLC Resonant Converter for Solar Powered E-Bicycle Charging Station

Karla Draženović, Ante Perić, Željko Jakopović, Viktor Šunde
Department of Electric Machines, Drives and Automation
University of Zagreb Faculty of Electrical Engineering and Computing, Zagreb, Croatia
karla.drazenovic@fer.hr, ante.peric@ fer.hr, zeljko.jakopovic@fer.hr, viktor.sunde@fer.hr

Abstract — **This paper presents a predictive control approach for controlling the output voltage of an LLC resonant converter. The half-bridge LLC converter is used as a DC battery charger in a solar powered charging station for electric bicycles. The LLC resonant converter was selected for this application due to its inherent soft switching, high efficiency, ability to operate at high switching frequencies, and has a relatively wide voltage range. The predictive control algorithm provides regulation of the output voltage to the set point. The switches of the converter are controlled directly, i.e. without using a modulator which implies a variable switching frequency.**

Keywords — *LLC resonant converter, predictive control, first harmonic approximation, solar-powered charging station, electric bicycle*

I. Introduction

Electric bicycles are playing an increasingly important role in the personal mobility sector. Electric bicycles offer a solution to the problems of excessive CO_2 emissions, excessive air pollution, and traffic congestion in cities. To ensure a viable future for e-mobility, bicycle-friendly infrastructure such as charging stations is needed [1].

Normally, the e-bicycle battery is charged by plugging it into a power outlet at home or at work. This method of charging is slow and impractical as the user has to carry a charger. To solve the aforementioned problem, this paper proposes an e-bicycles charging station powered by photovoltaic panels and the AC grid. The charging station allows direct DC charging of the e-bicycle from the DC power of the solar panels, without the need to use additional adapters. This solution represents a significant simplification for the users of electric bicycles. Moreover, a good distribution of charging stations in the urban area would allow a significant increase in the range of electric bicycles. Another important advantage of the charging station presented in this paper is the charging of the bicycle battery directly with the energy of the photovoltaic panels, which is in line with the objective of sustainable development and reduction of emissions.

The e-bicycle charging station is designed as a DC nano-grid with two different power sources: a photovoltaic panel and the AC grid. The charging station is equipped with a battery storage system that enables off-grid operation. All components within the nano-grid are connected to the DC bus via appropriate power converters. In this paper the DC-DC converter used to charge the battery of an e-bicycle is analyzed. The requirements of the battery charger are high power density, galvanic isolation, buck-boost operation and unidirectional power flow. According to the requirements, a half-bridge LLC resonant converter was chosen.

The half-bridge LLC resonant converter is suitable for applications which require a wide range of output voltages, such as is charging of the battery of an electric bicycle, and also it meets the requirements for unidirectional power flow and galvanic isolation. The LLC resonant converter achieves higher efficiency, power density, and charging power compared to the flyback converter used to charge the battery of an electric bicycle in [2]. Phase Shifted Full-Bridge (PSFB) and Dual Active Bridge (DAB) [3], [4], [5], [6], [7] are widely used for charging electric vehicles. The PSFB is characterized by a constant switching frequency, a wide range of output voltages and high efficiency. On the other hand, a DAB converter features bidirectional power flow, high efficiency and power density. However the resonant topologies such as LLC can offer better performances in terms of efficiency and power density in this power range.

The LLC resonant converter is usually controlled by changing the switching frequency so that the duty cycle is constant (50%). The switching frequency of the converter tends to be controlled so that its value is close to or equal to the resonant frequency of the LC resonant tank. Advanced control algorithms for LLC resonant converters are usually developed to increase efficiency and improve the dynamic response. Control algorithms used to increase the efficiency of the converter can be divided into control algorithms for light load conditions, control algorithms for precise resonant frequency tracking, and synchronous rectification algorithms. Under light load conditions, the switching frequency is much higher than the resonant frequency and the efficiency of the converter is low. Therefore, in order not to degrade the efficiency of the converter, it is necessary to use a suitable control method [8], [9]. Control algorithms for precise resonant frequency tracking help to improve the efficiency as the value of circulating current is minimal in the vicinity of the resonant frequency and soft-switching is achieved on the primary and secondary sides of the converter [10], [11], [12]. Wide band gap semiconductors and proper control of the synchronous rectifier offers significant reduction of losses compared to the classical diode rectifier [13], [14], [15]. In this paper, a method for predictive output voltage control of a half-bridge LLC resonant converter is proposed. The output voltage reference is obtained from the external charging control algorithm. The battery charging algorithm estimates the state of the connected battery and based on its value, the voltage reference for the predictive control is determined.

978-1-6654-3236-8/21 $31.00 © 2021 IEEE

This paper is organized as follows. In Section II the design of the nano-grid of an e-bicycle charging station is described. The topology and operation of a half-bridge LLC resonant converter is presented in Section III. In Section IV, an overview of the existing control methods of the LLC resonant converter is given. In Section V the design procedure of the LLC converter is shown. In Section VI a predictive algorithm and the simulation model developed in MATLAB/Simulink is presented. The simulation results obtained by the proposed predictive method of output voltage control are also presented in Section VI.

II. NANO-GRID OF SOLAR E-BICYCLE CHARGING STATION

The electric bicycle charging station presented in this paper is designed for fast charging of light electric vehicles. The presented charging station has four charging points with different types of bicycle plugs, which allows AC and DC charging. The charging algorithm detects the battery type, the state of charge and adjusts the charging speed. The charger offers the possibility of charging for different battery types and capacities without the use of additional adapters, which simplifies the use of the charging station.

The charging station is powered from the photovoltaic system and the electrical grid. Power converters enable the control of the energy flow between the different components of the DC nano-grid. The charging station contains an integrated battery energy storage, which allows the autonomous operation of the charging station at low power generation from photovoltaic modules.

The DC nano-grid of the charging station is shown in the block diagram in Fig. 1. The photovoltaic module is connected to the DC bus of the nano-grid through a step-down converter. The step-down converter has a built-in algorithm to track the maximum power to achieve the maximum energy utilization of the photovoltaic module. In case the solar energy is not available and the energy stored in the battery is not sufficient to supply the load, energy from the AC grid is used. The inverter that provides energy from the AC grid is bidirectional, which means that the charging station can supply power to the grid. In the proposed charging station configuration, the e-bicycle batteries are charged using an isolated DC converter. The DC chargers for the e-bicycles are to provide galvanically isolated DC power that is adjustable in both output voltage level and maximum charging current depending on the e-bicycle battery. A half-bridge LLC resonant converter was selected as the battery charger converter for the proposed e-bicycle charging station. The half-bridge LLC resonant converter meets the requirements of high efficiency, high power density, galvanic isolation, and unidirectional power flow.

III. TOPOLOGY AND CIRCUIT ANALYSIS OF LLC CONVERTER

A. Topology of the LLC Half-Bridge Converter

Resonant converters have lower switching losses compared to PWM converters due to their soft-switching capability [16]. Soft-switching allows operation at higher switching frequencies and reduces the dimensions of the filter components as well as the transformer. LLC resonant converters are used in a variety of applications due to their high efficiency, high power density and low EMI. Compared to a series resonant converter, an LLC resonant converter has

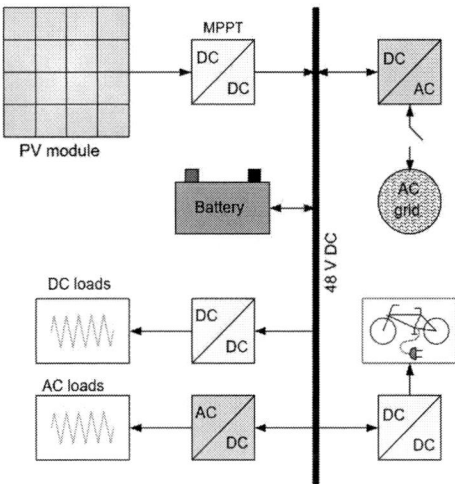

Figure 1. DC nano-grid of electric bicycles charging station

better light load regulation and is more suitable for applications that require a wide range of input and output voltages [17]. On the other hand, compared to a parallel resonant converter, the LLC converter has higher efficiency due to lower circulating energy [16].

The schematic of the half-bridge LLC resonant converter studied in this work is shown in Fig. 2. The converter consists of a controlled switching network, a resonant tank, a transformer and a single-phase center-tapped diode rectifier. The DC input voltage U_{in} is fed to an inverter consisting of fully controllable switches (S_1 and S_2). The transformer magnetizing inductance, resonant inductance and resonant capacitance form an LLC resonant circuit. When the frequency of the square-wave voltage is near the resonant frequency of the tank, the tank network oscillates with approximately sinusoidal waveforms of the switching frequency. The square wave voltage is rectified with a full-wave diode rectifier.

The LLC resonant converter has two resonant frequencies. The first resonant frequency f_0 corresponds to the frequency of the series resonant tank of L_r and C_r. The second resonant frequency f_p is lower than the series resonant frequency and is determined by the resonant inductance, resonant capacitance and the magnetizing inductance of the transformer. The following expressions apply to the resonant frequencies:

$$f_{r1} = f_0 = \frac{1}{2\pi \cdot \sqrt{L_r \cdot C_r}}$$
$$f_{r2} = f_p = \frac{1}{2\pi \cdot \sqrt{(L_r + L_m) \cdot C_r}} \tag{1}$$

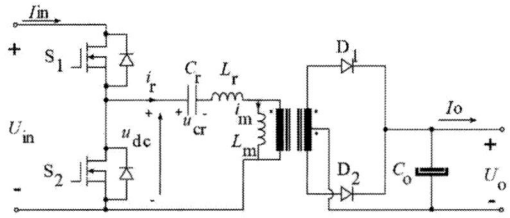

Figure 2. Topology of half-bridge LLC resonant converter

978-1-6654-3236-8/21 $31.00 © 2021 IEEE   137

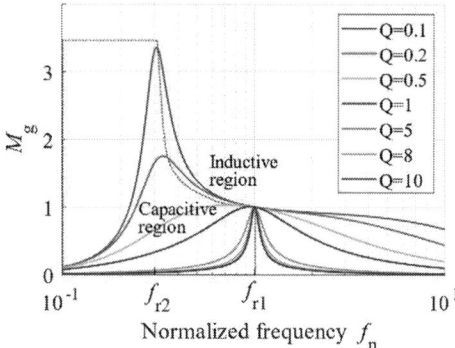

Figure 3. DC characteristic of LLC resonant half-bridge converter

The DC characteristic of the LLC resonant converter can be divided into an inductive region and a capacitive region, as shown in Fig. 3. The converter operates in the inductive region when the switching frequency is greater than the resonant peaks frequency. On the left side of the boundary line, which closely corresponds to the resonant peaks, the converter operates in the capacitive region. Only in the inductive region the converter achieves zero voltage switching. The most common and most favorable choice of switching frequency is close to the resonant frequency f_0 in the inductive operating range of the converter. This creates favorable switching conditions, resulting in low switching losses and high switching frequencies of the LLC resonant converter.

B. Operation Modes of LLC Half-Bridge Converter

Depending on the input voltage and load current conditions, the operation of the LLC resonant converter can be divided into three different modes, namely: operation at the resonant frequency, operation below the resonant frequency, and operation above the resonant frequency. The waveforms of the resonant current and the magnetizing current for all three modes of operation are shown in Fig. 4.

In the first mode of operation, when the switching frequency is equal to the resonant frequency f_0, the primary-side switches have zero voltage switching and soft-switching occurs on the secondary side. This is the optimum operating point of the converter, but depending on the variations of the input and output voltage, it is not always possible to ensure that the switching frequency is equal to the resonant frequency. If the switching frequency is lower than the resonant frequency f_0, the switches on the primary side of the converter still switch at zero voltage, but the switching losses of the primary switches and output diodes are higher due to

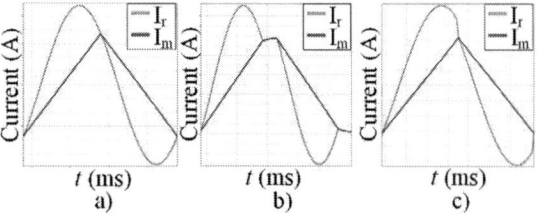

Figure 4. Resonant current and magnetizing current for the three modes of operation: a) $f_s = f_0$, b) $f_s < f_0$, c) $f_s > f_0$

the higher circulating current. When operating at switching frequencies lower than the resonant frequency, the converter should be prevented from entering the capacitive region. When the switching frequency is higher than the resonant frequency f_0, the primary switches achieve zero voltage switching, but the rectifier diodes have higher reverse recovery losses. From the presented analysis, it is concluded that the optimal selection of the switching frequency is near the resonant frequency f_0.

IV. REVIEW OF EXISTING CONTROL METHODS

The control characteristic of the LLC resonant converter is very complex as a result of fast dynamic of the resonant tank [18]. Several control methods are reported in the literature with the aim of achieving loss reduction, better dynamic response, and converter robustness. In general, LLC resonant converter control strategies can be classified as voltage-mode control and current-mode control [19]. Voltage-mode control offers only satisfactory performance within a limited range of switching frequencies [20]. Based on the small-signal model obtained using one of the possible methods, the parameters of the linear compensator are determined. The disadvantage of this type of control is a satisfactory dynamic response only close to the steady state operating point. On the other hand, current-mode control methods offer a better transient response but with a rather complex control structure [20].

Another classification of the control methods of the LLC converter is variable switching frequency control and fixed switching frequency control. Variable switching frequency control requires the use of a wide range of switching frequencies to achieve output voltage regulation, which leads to a complex design of the converter's magnetic components and degradation of electromagnetic interference performance. To solve this problem, advanced control algorithms such as the observer-based control method are developed [21]. The model of the LLC converter for which the observer-based control was developed was determined using the extended describing function method. The dynamic performance obtained by the observer-based control method is significantly better compared to the control using a PID controller. LLC converters controlled with a fixed switching frequency are more suitable for applications where a wide voltage range is required compared to converters controlled with a variable switching frequency. Fixed switching frequency control methods are divided into phase shift modulation, pulse width modulation (PWM) and resonant frequency modulation [22]. The phase-shifting control method is applicable to the full-bridge topology of an LLC converter, and the disadvantage of this method is that the soft-switching operation for the primary switch is difficult to achieve when the phase shift angle is large [23], [24]. The PWM method achieves output voltage regulation, but using an additional bidirectional semiconductor switch on the primary side of the converter [25]. In [26], a hybrid method combining PWM and pulse frequency modulation (PFM) control is proposed for precise output voltage regulation to provide a solution to the problem of limited DSP resolution for high switching frequency operation. The operating point is determined by the PFS, and the precise output voltage control is achieved by the PWM method. The resonant frequency control is achieved by changing the parameters of the resonant circuit by switching additional switches on and off [26], [27].

Recently, nonlinear control methods have been increasingly developed to address the requirements of fast dynamic processes such as LLC resonant converters. Nonlinear control methods improve the dynamic behavior of the converter, robustness and stability of the system compared to linear control methods. One of the nonlinear control methods that has shown the best results is the optimal trajectory control method. The optimal trajectory control increases the efficiency of the converter and better dynamic response is obtained [28].

In this paper, a predictive algorithm for output voltage control is proposed to achieve better dynamic performance of the converter which is used for charging electric bicycles. The parameters of the converter were determined based on the requirements of the specific application using the first harmonic approximation method. This method is described in more detail in the next section.

V. DESIGN OF AN LLC HALF-BRIDGE CONVERTER

A. FHA model of LLC resonant converter

In the design of the LLC resonant circuit, the first harmonic approximation (FHA) method is used. In FHA method, only the fundamental components of square wave voltage and resonant current are considered. The first harmonic approximation method is characterized by ease of use and satisfactory accuracy. The disadvantage of this method is the reduction of accuracy for switching frequencies much lower than the resonant frequency. In this paper, the FHA method is used to determine the electrical specifications of the LLC resonant converter, according to which the simulation model in PLECS is built.

Using the FHA, simple equivalent circuits of the switch network, the resonant network, and the rectifier of the series resonant converter were derived. The FHA method replaces the switch network with a sinusoidal voltage source, while the diode rectifier is replaced with an equivalent resistor, simplifying the nonlinear system to a linear system, Fig. 5.

The square-wave voltage of the input inverter can be described using the Fourier series. Assuming that the higher harmonics are negligible, the voltage of the input inverter is approximated only by the fundamental harmonic:

$$u_{s1}(t)=\frac{2U_{dc}}{\pi}\cdot \sin (2\pi f_s t) \qquad (2)$$

where U_{dc} is the input voltage and f_s is the switching frequency.

The current in the resonant tank is sinusoidal and can be described by the following expression:

$$i_r(t)=I_{r1}\cdot \sin (2\pi f_s t-\varphi_s) \qquad (3)$$

where js is the phase angle of the current with respect to the voltage.

The fundamental harmonic of the square-wave voltage at the input of the diode rectifier and the sinusoidal current of the diode rectifier are in phase, so that the diode rectifier can be replaced by an equivalent resistance R_{eq}:

$$R_{eq}=\frac{8n^2}{\pi^2}\cdot R \qquad (4)$$

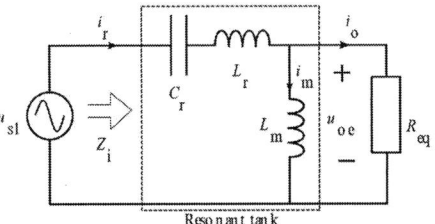

Figure 5. Linearized equivalent circuit of the LLC resonant converter

where R is the resistance of the load, and n is the transformer turns ratio.

The transfer function of the converter, i.e. the voltage gain of the converter determined by the FHA method can be written as the ratio of the fundamental harmonic of the output voltage to the fundamental harmonic of the input voltage. The transfer function of the converter obtained using FHA method is:

$$M_g=\frac{U_{oe}}{U_{s1}}=\left|\frac{jX_{Lm}\|R_{eq}}{(jX_{Lm}\|R_{eq})+j(X_{Lr}-X_{Cr})}\right|=$$
$$=\left|\frac{L_n f_n^2}{[(L_n+1)\cdot f_n^2-1]+j[(f_n^2-1)\cdot f_n\cdot Q_e\cdot L_n]}\right| \qquad (5)$$

where $L_n=L_m/L_r$ is the normalized inductance, $f_n=f_s/f_0$ the normalized switching frequency and $Q_e = \sqrt{(L_r/C_r)}/R_{eq}$ the quality factor. The voltage gain M_g of the LLC resonant converter as a function of f_n and Q_e is shown in Fig. 6.

B. LLC Resonant Converter Simulation Parameters

The initial parameters for the converter design are: input voltage value range, output voltage value range, rated power and rated switching frequency of the converter. Based on the known values of the initial parameters and the FHA analysis, the transformer turns ratio, resonant inductance, resonant capacitance, magnetizing inductance and quality factor are determined. The converter parameters determined by the FHA method are listed in Table 1. Based on the parameters determined in this way, a simulation model of an LLC resonant converter is developed in PLECS.

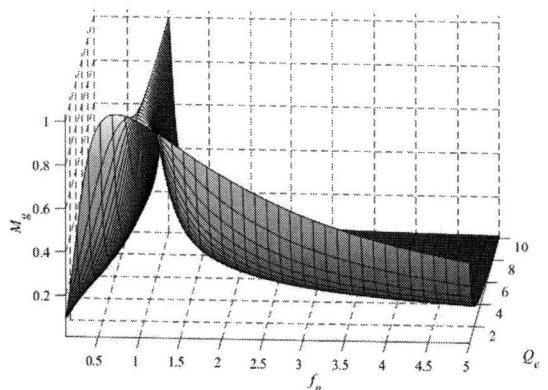

Figure 6. LLC resonant converter transfer function

978-1-6654-3236-8/21 $31.00 © 2021 IEEE

TABLE I. LLC RESONANT CONVERTER PARAMETERS

Simulation parameters	Value
Nominal input voltage U_{in}	48 V
Nominal output voltage U_O	24 V
Resonant capacitance C_r	1.42 μF
Resonant inductance L_r	0.8 μH
Resonant frequency	150 kHz
Magnetizing inductance	5 μH
Transformer turns ratio n	1

VI. PREDICTIVE CONTROLLER DESIGN

A. LLC converter modeling methods for control circuit synthesis

Classical methods such as state-space averaging have been successfully applied to PWM converters. The switching frequency of the PWM converter is usually much higher than the natural frequency of the LC filter and therefore the switching harmonics can be neglected and only the DC component is considered, resulting in an accurate linear model. On the other hand, for frequency controlled resonant converters, the switching frequency is close to the resonant tank's natural frequency and to obtain an accurate model, the switching harmonics cannot be neglected. A common modelling approach for resonant converters is the small-signal analysis based on the extended describing function [28], [29]. In this method, all switching harmonics are considered to achieve the highest possible accuracy.

In this paper, a discrete-time model of the converter is derived for the design of a predictive controller for an LLC resonant converter. The physical behavior of the LLC converter is modeled such that each mode of operation is described by the state-space equations. The continuous-time model in the state space is given by the following equation:

$$\frac{dx(t)}{dt} = A \cdot x(t) + B \cdot u(t)$$
$$y(t) = C \cdot x(t) \tag{6}$$

where the state vector is defined as $x(t) = [i_r(t)\ u_{cr}(t)\ i_m(t)\ u_o(t)]^T$, where $i_r(t)$ is the resonant current, $u_{cr}(t)$ is the resonant capacitor voltage, $i_m(t)$ is the magnetizing current, and $u_o(t)$ is the output voltage. The output vector is defined as $y(t) = [u_o(t)]$.

The state space equations of the converter are derived from the electrical equations using Kirchhoff's law for each of the state variables. The state space equations of the LLC resonant converter are as follows:

$$L_r \cdot \frac{di_r}{dt} = u_{dc} - i_r \cdot r_s - u_{cr} - \text{sign}(i_r - i_m) \cdot n \cdot U_0$$
$$C_r \cdot \frac{du_{cr}}{dt} = i_r$$
$$L_m \cdot \frac{di_m}{dt} = \text{sign}(i_r - i_m) \cdot n \cdot U_0 \tag{7}$$
$$C_0 \cdot \frac{du_o}{dt} = n \cdot |i_r - i_m| - \frac{U_0}{R}$$

where the input square-wave voltage u_{dc}, sign $(i_r - i_m)$ and $|i_r - i_m|$ are nonlinear terms. The proposed LLC resonant converter model includes the resistor (r_{esr}), which represents parasitic series resistance of the resonant capacitance and the internal resistance of the resonant inductance. The ESR of the output capacitor is assumed to be negligible, and the voltage of the output filter capacitor u_{C0} is assumed to be equal to the output voltage u_0.

Depending on which switch combination is currently on, the inverter can operate in 6 different switching states, which are listed in Table II. For each individual switching state listed in Table II and according to the equations (7), the corresponding system matrices A_1-A_6, B_1-B_6 and C_1-C_6 are assigned.

TABLE II. LLC RESONANT CONVERTER SWITCHING STATES

State	S1	S2	D1	D2
1	On	Off	Off	On
2	On	Off	On	Off
3	On	Off	Off	Off
4	Off	On	Off	On
5	Off	On	On	Off
6	Off	On	Off	Off

The discrete mathematical model of the LLC resonant converter was obtained using the ZOH discretization with sampling time T_s:

$$x(k+1) = E \cdot x(k) + F \cdot u(k)$$
$$y(k) = G \cdot x(k) \tag{8}$$

where the matrices E correspond to the discretized matrices A, the matrices F to the discretized matrices B, and the matrices G to the discretized matrices C. The obtained system matrices were used within the predictive algorithm to obtain the optimal switching state in the next step.

The goal of the predictive control presented in this paper is to accurately follow the given output voltage reference, i.e., to minimise the output voltage error by appropriate switching. Moreover, the output voltage must be adjusted to its new reference during the transient in the shortest possible time and with the least possible overshoot. The quadratic objective function is chosen:

$$J(k) = \sum_{i=k}^{k+N-1} (u_{o,err}(i+1|k))^2 \tag{9}$$

which penalizes the square of the output voltage error:

$$u_{o,err} = u_{o,ref} - u_o(k) \qquad (10)$$

In the next step, the optimal sequence of control variables is determined. The optimal sequence is the one that minimizes the objective function:

$$U(k) = \arg\min J(k) \qquad (11)$$

Constraints on the model are introduced using a set of permissible switching states. Considering all possible switching states and the current switching state, a set of possible switching states is defined for the next step depending on the current value of the switching frequency (equal to the resonant frequency, above the resonant frequency, below the resonant frequency). In each simulation step, only the first element is taken from the optimal sequence of control variables, and in each subsequent step, the optimization problem is solved again, but with updated values of the measured quantities.

B. Simulation Model

The simulation model developed in MATLAB/Simulink and PLECS Blockset is shown in Fig. 7 and Fig. 8. The simulation model consists of a MATLAB function block in which a predictive control algorithm is implemented and a PLECS circuit block with an LLC resonant converter. The predictive algorithm is tested on the hybrid battery model shown in Fig. 9 [30]. The model consists of a C_{cap} capacitor and a current-controlled current source that describe the state of charge of the battery. To model the self-discharge of the battery, an additional leakage resistor $R_{self\text{-}dis}$ can be used in the battery model, which is a function of the state-of-charge, temperature, and often the number of charge cycles. In practical applications, $R_{self\text{-}dis}$ can be neglected or a resistor with a very high resistance value can be used. Similar to the Thevenin-based battery models, the hybrid model has a RC network that is used to model transient battery behavior. The parameters of the hybrid model R_s, $R_{tr\,S}$, R_{tr_L}, C_{tr_S}, C_{tr_L}, U_{OC} are determined using experimental measurements on a particular battery tank. For the analysis of the proposed predictive algorithm, the parameters of the Li-ion battery model derived in [30] are used in this paper.

Within the simulation model in PLECS, electro-thermal models of semiconductor switches are used for the calculation of conduction losses and switching losses. The thermal part of

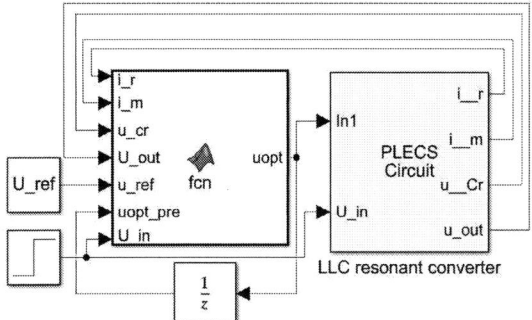

Figure 8. Simulation model of LLC resonant converter (MATLAB – control circuit, PLECS – power converter)

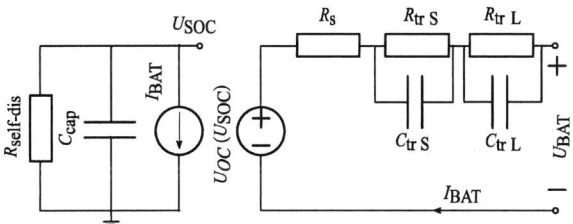

Figure 9. Hybrid electrical model of a battery

the model is defined by the RC network, while the electrical, temperature-dependent part of the model is given in terms of multidimensional look-up tables obtained from semiconductor data sheets. In the electrical model, the switching losses are calculated and set as the input of the thermal model in the form of a current source. The voltage response of the equivalent RC network, which is due to the current excitation of the current source, represents the temperature of the switch. The temperature calculated in this way is used as a parameter in the next step of the simulation in order to calculate temperature-dependent conduction losses and switching losses within the electrical model using look-up tables. The conduction losses are calculated as the product of on-state voltage drop set within PLECS in the form of a 2D look-up table as a function of current and temperature of the semiconductor switch $U_{on} = f(I, T_j)$. The switching losses are given in a 3D lookup table as a function of device current, temperature and blocking voltage $E_{on\backslash off} = f(U_{block}, I, T_j)$. The parameters of the LLC converter for the simulation model in PLECS are given in Table I in Section V.

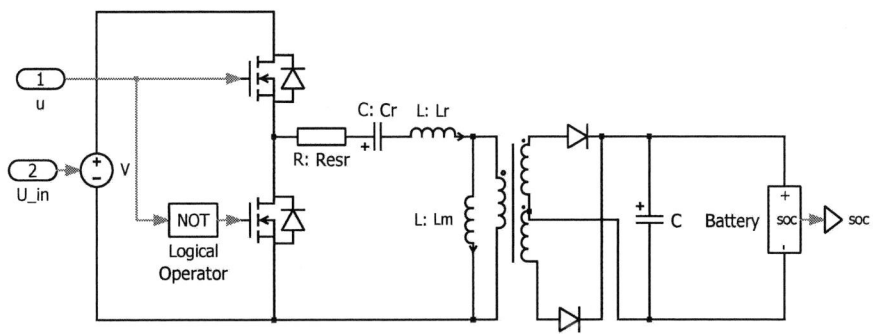

Figure 7. PLECS model of the half-bridge LLC resonant converter

978-1-6654-3236-8/21 $31.00 © 2021 IEEE

The MOSFET Wolfspeed C3M0030090K and the Infineon IDW75E60 diode, were selected for the semiconductor switches. The MOSFET and diode parameters are listed in Table III. The configuration parameters of the MATLAB / Simulink simulation model are given in Table IV.

TABLE III. SEMICONDUCTOR SWITCH PARAMETERS

MOSFET C3M0030090K	
Drain-Source Breakdown Voltage (U_{DS})	900 V
Continuous Drain Current (I_d)	63 A
Drain-Source Resistance (R_{ds-on})	30 mΩ
Diode IDW75E60	
Forward Voltage (U_f)	2 V
Forward Current ($I_{FAV,25°C}$)	120 A

TABLE IV. MATLAB/SIMULINK CONFIGURATION PARAMETERS

MATLBA/Simulink Configuration Parameters	
Solver	ode45 (Dormand-Prince)
Solver Max Step Size	1e-8
Solver Relative Tolerance	1e-6

C. Simulation Results

In this section, simulation results are presented to demonstrate the operation of the proposed predictive control. The main objective of the predictive control presented in this paper is to determine the appropriate switching frequency so that the output voltage follows the set point.

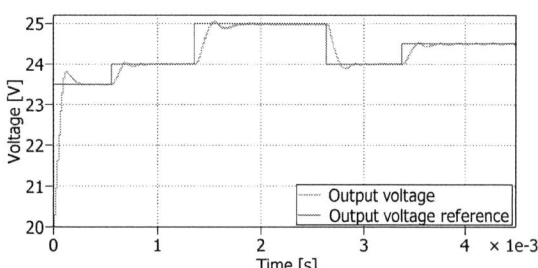

Figure 10. Output voltage reference tracking for the proposed predictive algorithm

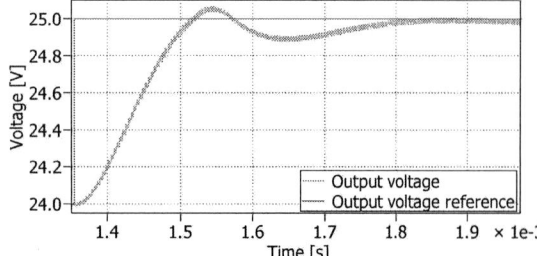

Figure 11. Step-up change in the output voltage reference

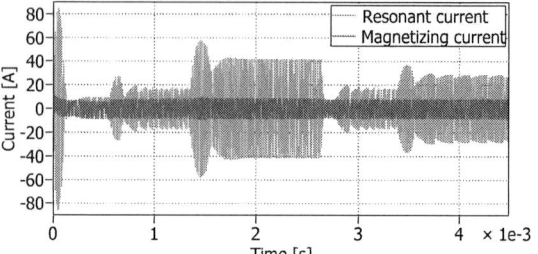

Figure 12. Resonant current and magnetization current for the proposed control method

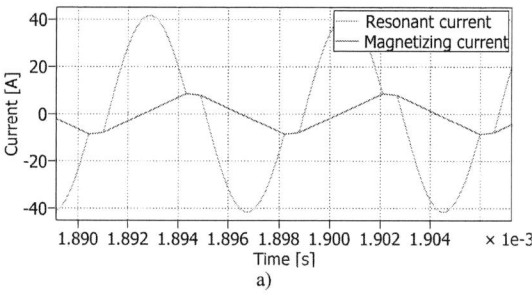

a)

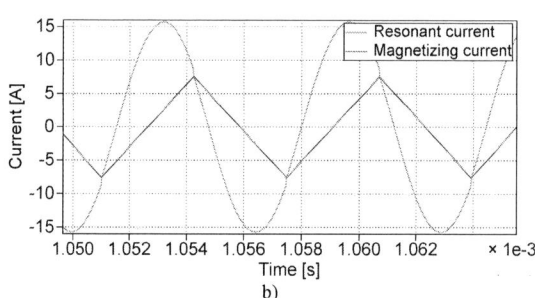

b)

Figure 13. Detail of steady state resonant current and magnetization current a) for 25 V output voltage reference, b) for 23.5 V output voltage reference

The simulation results for output voltage control are shown in Fig. 10. The output voltage reference is set as a step function with values of 23.5 V, 24.5 V, and 25 V, which corresponds to the expected voltage of the electric bicycle battery. The simulation results show that the output voltage has an overshoot of less than 0.4 V during the transient phase and the steady state is reached in less than 0.5 ms. The output voltage ripple is relatively small as shown in Fig. 11. The resonant circuit current and magnetizing current are shown in Fig. 12. To achieve a given output voltage reference, the predictive algorithm increases or decreases the resonant current depending on the magnitude of the output voltage reference, as shown in Fig. 12. During the transient process, the resonant current increases and when the output voltage reference is reached, the algorithm reduces the amount of current to a steady state value. The converter is controlled by switching frequencies near the resonant frequency, and depending on the output voltage reference the converter operates above or below the resonant frequency, Fig. 13.a) and 13.b). For the proposed predictive control algorithm, a loss analysis was performed which showed the highest converter efficiency of 91% in the operating mode below the resonant frequency and 94% in the operating mode above the resonant frequency. The efficiency of the converter achieved by this method is relatively low and the existing algorithm needs to be improved to increase the efficiency. The simulation results confirm that the predictive algorithm achieves very fast dynamic response and a small overshot.

VII. CONCLUSION

This paper presents an LLC resonant converter used in a solar powered e-bicycle charging station. The proposed converter was selected for its soft-switching capability, high efficiency, and high power density. Predictive control was implemented to improve the output voltage regulation and dynamic performance of the converter. Simulation results show that the predictive control provides fast and high quality dynamic response. Compared to other LLC converter control methods, the synthesis of the control circuit is relatively simple due to the simple objective function. The proposed predictive control algorithm achieves a relatively low converter efficiency of 91% for switching frequencies below the resonant frequency and 94% for switching frequencies above the resonant frequency, which needs to be increased by improving the existing algorithm. The main drawback of the proposed method is the need for relatively large computational power when implementing the method in an actual control system.

REFERENCES

[1] Tim Jones, Lucas Harms, Eva Heinen, 2016. "Motives, perceptions and experiences of electric bicycle owners and implications for health, wellbeing and mobility"

[2] G. R. Chandra Mouli, P. Van Duijsen, F. Grazian, A. Jamodkar, P. Bauer, and O. Isabella, "Sustainable E-Bike Charging Station That Enables AC, DC and Wireless Charging from Solar Energy," Energies, vol. 13, no. 14, p. 3549, Jul. 2020.

[3] S. Habib et al., "Contemporary trends in power electronics converters for charging solutions of electric vehicles," in CSEE Journal of Power and Energy Systems, vol. 6, no. 4, pp. 911-929, Dec. 2020, doi: 10.17775/CSEEJPES.2019.02700.

[4] S. Dusmez, A. Cook and A. Khaligh, "Comprehensive analysis of high quality power converters for level 3 off-board chargers," 2011 IEEE Vehicle Power and Propulsion Conference, 2011, pp. 1-10, doi: 10.1109/VPPC.2011.6043096.

[5] A. Khaligh and S. Dusmez, "Comprehensive Topological Analysis of Conductive and Inductive Charging Solutions for Plug-In Electric Vehicles," in IEEE Transactions on Vehicular Technology, vol. 61, no. 8, pp. 3475-3489, Oct. 2012, doi: 10.1109/TVT.2012.2213104.

[6] Y. Du, S. Lukic, B. Jacobson and A. Huang, "Review of high power isolated bi-directional DC-DC converters for PHEV/EV DC charging infrastructure," 2011 IEEE Energy Conversion Congress and Exposition, 2011, pp. 553-560, doi: 10.1109/ECCE.2011.6063818.

[7] D. S. Gautam, F. Musavi, W. Eberle and W. G. Dunford, "A Zero-Voltage Switching Full-Bridge DC--DC Converter With Capacitive Output Filter for Plug-In Hybrid Electric Vehicle Battery Charging," in IEEE Transactions on Power Electronics, vol. 28, no. 12, pp. 5728-5735, Dec. 2013, doi: 10.1109/TPEL.2013.2249671.

[8] Lee, J.; Kim, J.; Kim, J.; Baek, J.; Moon, G., "A high-efficiency PFM Half-bridge converter utilizing a half-bridge LLC converter under light load conditions," IEEE Trans. Power Electron. 2015, 30, 4931–4942.

[9] Kim, J.; Kim, C.; Kim, J.; Lee, J.; Moon, G., "Analysis on load-adaptive phase-shift control for high efficiency full-bridge LLC Resonant converter under light-load conditions," IEEE Trans. Power Electron. 2016, 31, 4942–4955.

[10] Kundu, U.; Sensarma, P., "Gain-relationship-based automatic resonant frequency tracking in parallel LLC converter," IEEE Trans. Ind. Electron. 2016, 63, 874–883

[11] Feng, W.; Mattavelli, P.; Lee, F.C., "Pulsewidth locked loop (PWLL) for Automatic resonant frequency tracking in LLC DC–DC transformer (LLC-DCX)," IEEE Trans. Power Electron. 2013, 28, 1862–1869.

[12] Li, H.; Jiang, Z. "On automatic resonant frequency tracking in LLC Series resonant converter based on zero-current duration time of secondary diode," IEEE Trans. Power Electron. 2016, 31, 4956–4962.

[13] Kim, K.; Youn, H.; Baek, J.; Jeong, Y.; Moon, G., "Analysis on synchronous rectifier control to improve regulation capability of highfrequency LLC resonant converter," IEEE Trans. Power Electron. 2018, 33, 7252–7259.

[14] Feng, W.; Lee, F.C.; Mattavelli, P.; Huang, D. A universal adaptive driving scheme for synchronous rectification in LLC resonant converters. IEEE Trans. Power Electron. 2012, 27, 3775–3781.

[15] Fei, C.; Li, Q.; Lee, F.C. Digital implementation of adaptive synchronous rectifier (SR) driving scheme for high-frequency LLC converters with microcontroller. IEEE Trans. Power Electron. 2018, 33, 5351–5361

[16] Bo Yang, F. C. Lee, A. J. Zhang and Guisong Huang, "LLC resonant converter for front end DC/DC conversion," APEC. Seventeenth Annual IEEE Applied Power Electronics Conference and Exposition (Cat. No.02CH37335), 2002, pp. 1108-1112 vol.2, doi: 10.1109/APEC.2002.989382.

[17] E. C. Chang, C. A. Cheng, H. L. Cheng and S. C. Lin, "Small signal modeling of LLC resonant converters based on extended describing function", Proc. Int. Symp. Comput. Consumer Control, pp. 365-368, Jun. 2012.

[18] E. Kim, J. Lee, Y. Heo, T. Marius, J. Ju and Y. Kook, "LLC resonant converter with wide output voltage control characteristics according to operating mode transition," 2018 IEEE Applied Power Electronics Conference and Exposition (APEC), 2018, pp. 2117-2123, doi: 10.1109/APEC.2018.8341309.

[19] J. Jang, M. Joung, S. Choi, Y. Choi and B. Choi, "Current mode control for LLC series resonant dc-to-dc converters," 2011 Twenty-Sixth Annual IEEE Applied Power Electronics Conference and Exposition (APEC), 2011, pp. 21-27, doi: 10.1109/APEC.2011.5744570.

[20] C. Buccella, C. Cecati, H. Latafat, P. Pepe and K. Razi, "Observer Based Control of LLC DC/DC Resonant Converter Using Extended Describing Functions," in IEEE Transactions on Power Electronics, vol. 30, no. 10, pp. 5881-5891, Oct. 2015, doi: 10.1109/TPEL.2014.2371137.

[21] Y. Wei, D. Woldegiorgis and A. Mantooth, "Control Strategies Overview for LLC Resonant Converter with Fixed Frequency Operation," 2020 IEEE 11th International Symposium on Power Electronics for Distributed Generation Systems (PEDG), 2020, pp. 63-68, doi: 10.1109/PEDG48541.2020.9244388.

[22] W. Liu, B. Wang, W. Yao, Z. Lu and X. Xu, "Steady-state Analysis of the Phase Shift Modulated LLC Resonant Converter", 2016 IEEE Energy Conversion Congress and Exposition (ECCE), pp. 1-5, 2016.

[23] H. Wu, T. Mu, X. Gao and Y. Xing, "A Secondary-Side Phase-Shift Controlled LLC Resonant Converter With Reduced Conduction Loss at Normal Operation for Hold-Up Time Compensation Application", IEEE Trans. Power Electron., vol. 30, no. 10, pp. 5352-5357, Oct. 2015.

[24] H. Wang and Z. Li, "A PWM LLC type resonant converter adapted to wide output range in PEV charging applications", IEEE Trans. on Power Electron., vol. 33, no. 5, pp. 3791-3801, May 2018.

[25] Park, H.; Jung, J., "PWM and PFM hybrid control method for LLC Resonant converters in high switching frequency operation," IEEE Trans. Ind. Electron. 2017, 64, 253–263.

[26] U. Kundu, S. Chakraborty and P. Sensarma, "Automatic Resonant Frequency Tracking in Parallel LLC Boost DC–DC Converter," in IEEE Transactions on Power Electronics, vol. 30, no. 7, pp. 3925-3933, July 2015, doi: 10.1109/TPEL.2014.2344021.

[27] S. K. Chung, B. G. Kang and M. S. Kim, "Constant frequency control of LLC resonant converter using switched capacitor", Electronics Letters, vol. 49, no. 24, pp. 1556-1558, Nov. 2013.

[28] W. Feng, F. C. Lee and P. Mattavelli, "Optimal Trajectory Control of LLC Resonant Converters for LED PWM Dimming," in IEEE Transactions on Power Electronics, vol. 29, no. 2, pp. 979-987, Feb. 2014, doi: 10.1109/TPEL.2013.2257864.

[29] C. Buccella, C. Cecati, H. Latafat, P. Pepe and K. Razi, "Linearization of LLC resonant converter model based on extended describing function concept", Proc. IEEE Int. Workshop Intell. Energy Syst., pp. 131-136, Nov. 2013.

[30] Min Chen and G. A. Rincon-Mora, "Accurate electrical battery model capable of predicting runtime and I-V performance," in IEEE Transactions on Energy Conversion, vol. 21, no. 2, pp. 504-511, June 2006.

978-1-6654-3236-8/21 $31.00 © 2021 IEEE

Quality Evaluation of Jointly Used Modular Multilevel Converters and Battery Energy Storages

Alexander Bubovich[1], Maxim Vorobyov[2], Ilya Galkin[3], Tenuun Dovudon[4]

Riga Technical University, Institute of Industrial Electronics and Electrical Engineering, Riga, Latvia

[1]aleksandrs.bubovics@rtu.lv, [2]maksims.vorobjovs@rtu.lv, [3]gia@eef.rtu.lv

Abstract—**Battery energy storages nowadays are utilized in various applications. The similarity of these applications and multilevel nature of the batteries logically encourages building of the interface converters based on a multilevel scheme. This paper is devoted to the choice of the configuration of this scheme and its operation mode. Tentative quality evaluation is made based on the root-mean declination of output voltage from its reference (THD). MATLAB-Simulink simulation is chosen as a reasonable data collection method. The results of the simulation show that a trade-off configuration may include 3-4 levels with minor inter-level pulse width modulation. In the same time further investigation must be done in order to study the influence of other technical (losses) and economical parameters.**

Keywords— Modular multilevel converters, DC-AC power converters, Energy storage, batteries, circuit simulation

I. INTRODUCTION.

Nowadays electrochemical Battery Energy Storages (BES or just batteries) are considered as the most available, reliable and practical storage of electrical energy for broad range of applications. There exist several BES types at different level of development of their electrochemistry/ technology [1], [2] Traditionally, one of the oldest and most "polished" technologies is Lead-Acid BES technology, which in spite of its drawbacks (lower specific energy and shorter lifetime) is still attractive due to its low cost and high efficiency. Another quite advanced BES technology – Nickel-Metal Hydride – provides good specific power, energy and lifetime, but has higher self-discharge and efficiency. The most recent and still developing group of BES technologies is based on Li-Ion chemistry. The corresponding batteries, depending on the particular chemical process show good specific energy (LiNiCoAlO2), specific power (LiNiMnCoO2), lifetime (LiFe2PO4), efficiency and their combinations providing these parameters at reasonable price.

The above-mentioned high-level parameters of the modern BES enables them for use in practically all fields of electrical engineering. In particular, the use of Li-Ion batteries revolutionized the technical and market parameters of portable electronics. Similarly, the use of these batteries accelerated development of plug-in electric vehicles (PEVs) or all-electric vehicles (AEVs). In addition, the increased specific energy of modern BES makes them applicable in energy supply systems. Recent engineering practices of the authors of this report, concentrated in the field of energy systems [3] and small electric vehicles [4], shows certain similarities in BES use that, first of all, regards their charging/discharging electronics, considered in this paper. This topic is essential for the evaluation of the power losses in the mentioned applications and understanding of their thermal behavior.

At the moment, numerous national and international regulations in the energy sector limit the use of fossil fuels for energy production and push on the use of renewable energy sources [5]. However, collecting of energy from renewable energy sources is uneven and production pattern of renewable energy sources does not always match the corresponding consumption demand. This mismatch can be compensated by Battery Energy Storage Systems (BESSs), appearing on the market [6]–[9]. The electronic interface of these BESS is a bidirectional charger/discharger tied to AC grid [3].

Another example of BES application, investigated by the authors, is electrical drive of small electric vehicle for needs of orthopedic rehabilitation. According to information, provided by World Health Organization's World report of disability [10], approximately 15% of world population is suffering from some sort of disability of whom 200 million have mobility difficulties. Wider integration of these disabled persons into social and economic life, assumes the use of modern technological equipment, for example, Power Assist Wheelchairs (PAWs). One of possible concepts of such wheelchairs is described in [4]. Its electrical drive is based on a low-speed permanent magnet electrical machine (AC) with segmented stator [11]–[14] and power electronic converter. The charging part of this converter is a rectifier, supplied from AC grid. In turn, its discharging part is an inverter, which supply specific AC load with potential demand of energy return that requires bidirectionality. The electric drive of the considered PAW operates in the closed compartment. Therefore, the accuracy of BES converter operation and its losses are interdependent factors that define the thermal mode of the drive.

As it can be seen, a bidirectional DC-AC inverter is essential part of the interface converter of the battery in both considered applications. This inverter may be built as a single-level pulse-mode circuit that forms pulse width or frequency modulated (PWM/FM) voltage. Alternatively, the inverter may generate level modulated (LM) voltage that requires the use of a multilevel topology. This topology is particularly suitable in both considered applications because they both contains a battery that consists of several cells providing the necessary voltage levels. The objective of this study is to find better configuration of the BES interface inverter considering the quality of the generated voltage. The criteria for the choice is the root-mean declination of the generated voltage from its reference curve. If the reference is sinusoidal, this declination is known as the Total Harmonic Distortion (THD) – in the case of BESS low THD is a real grid requirement. The tasks of the work, therefore, are to evaluate THD for different kinds of modulation (PWM vs. LM vs. mixed) and different number of levels, compare the results and provide conclusions on the better choice, in the same time keeping in mind that these

978-1-6654-3236-8/21 $31.00 © 2021 IEEE

matters affect also the losses and thermal mode of BES converters.

II. REVIEW OF MULTILEVEL CONVERTERS UTILISED JOINTLY WITH BATTERIES.

As it has already been mentioned BESs consist of several galvanic cells, forming a natural set of voltage levels. These levels can be processed by a multilevel converter (MLC). There are three main MLC topologies and each of them utilizes BES cells in a different way. Below a review of some MLC for BES is given.

The first MLC type, semiconductor clamped converters also known as neutral point clamped converters, provides commutation of internal BES levels to the output. There are two kinds of such MLC: diode clamped and active clamped [15]. Since MLC of this type deals with internal levels BES configuration is standard – series connection of several galvanic cells. This simplifies the use of BES at system level. On the other hand, the loading of separate cells is uneven that requires some extra Battery Management System (BMS). The use of neutral point clamped multilevel converter in BESSs is described in [16].

The next MLC type is known as converters with flying capacitors. Here voltage levels are taken from the voltage sources, which both terminals are commuted by converter switches. Typically, these sources are capacitors, but in the case of BES, they may be replaced by cell groups. Paper [17] describes and experimentally verifies the design of multilevel converter with flying capacitors for aircraft applications. Paper [18] provides an example of 13-level flying capacitor multilevel converter, based on GaN switches. If the voltage levels provided from the BES cell groups the entire battery is split into several uneven parts that complicates the use of this topology.

Finally, the third type of MLC utilizes independent voltage sources and is also known as cascaded H-bridge converter with separate DC sources. These MLC consist of several single-phase full bridge converters connected in series [19], [20]. Inputs of single-phase full bridge converters are connected to separate or independent DC sources, which, in case of implementation of multilevel converters in BESs, are battery cells or their groups. The output waveform is generated from such sources through full bridges designated in Figure 1(a) as "Power converter and energy storage building block". These blocks connect series of sources to the output. A unipolar version of this MLC includes also a common polarity inverter that allows simplification of cell converters – Figure 1(b).

Different configurations of the building blocks can be found in literature [21], [22]. Some alternatives are shown in Figure 2. The most commonly used and simple cell is a single-phase full transistor bridge (H-bridge), shown at Figure 2(a). It is possible to get three voltage levels on the output of the H-bridge: $+V_g$ (battery voltage), $-V_g$ and $0V$. Thus, the H-bridge is generating AC voltage and it is possible to control charging and discharging process of the connected battery. The second type of cell has a two-stage configuration – Figure 2(b). The part, forming output voltage – H-bridge – remains without changes. The second converter, added between H-bridge and the battery, is a pre-regulator that controls charging or discharging energy flow from the battery If the MLC forms unipolar voltage the cell converter is simpler, for example, it may be a DC/DC chopper or even a couple of switches.

There are also structures, where transformer is used in order to electrically isolate battery from the converter [21]. However, by using transformers it is possible to increase or decrease voltage, and, therefore, the advantages of cascaded H-bridge converter are being negotiated.

References [23]-[34] provide examples of usage of modular MLCs with independent sources in BESSs. In [23] self-adaptation control of second-life batteries (batteries that were previously used in transportation, the use of second-life batteries makes the BESS more affordable, as the batteries are the most expensive part of the BESS) is provided. This control strategy is based on online capacity estimation of the batteries, thus preventing them from overcharging and over-discharging. In [24] a unified control scheme of the multilevel converter with independent source based BESS is provided, which simplifies the control of state of charge (SOC) balancing and fault tolerant control of the system.

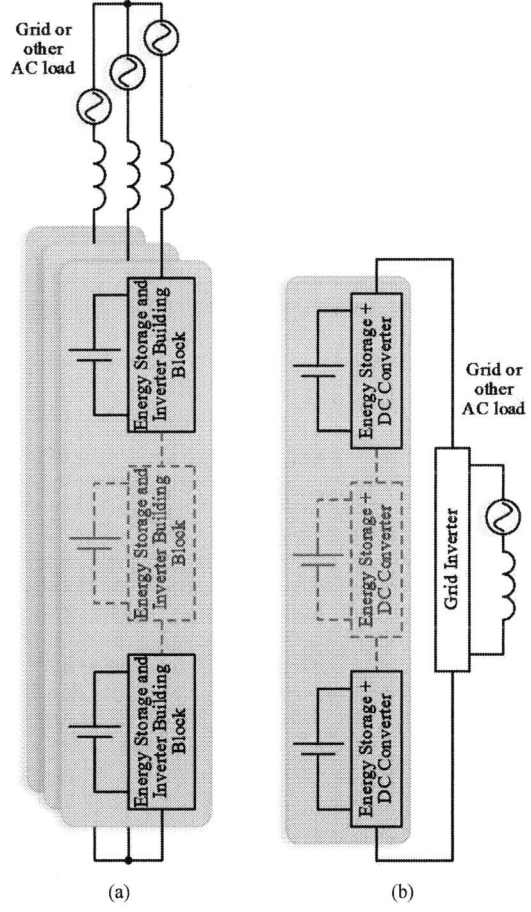

Fig. 1. Structure of multilevel converter with independent sources: (a) typical, (b) unipolar with a frontend (single phase).

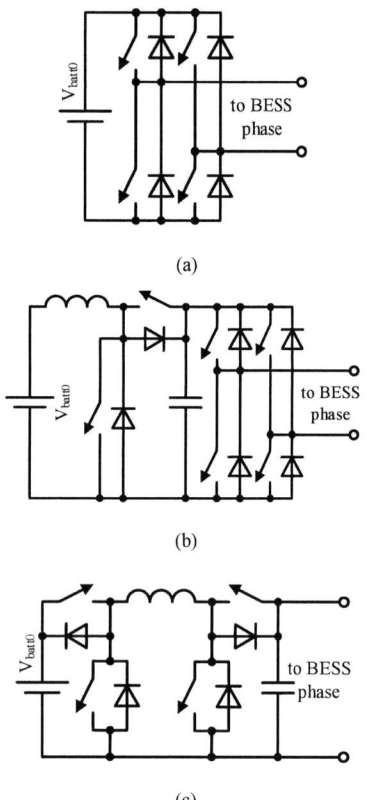

(a)

(b)

(c)

Fig. 2. Types of power electronic converters used in cascaded H-bridge multilevel converters: (a) H-bridge; (b) two-stage converter; (c) cell converter (buck-boost) for unipolar topologies.

Papers [25]–[29] also provide SOC balancing control of the multilevel converter with independent sources. In [30] the design of battery management system (BMS) for multilevel converter with independent source-based BESS is provided. In paper [31] the comparative analysis of SiC and Si switches is given, prototype of multilevel converter with independent sources is presented for the experimental verification. Reference [32] describes a 6.6 kV multilevel converter with independent sources with pulse width modulation (PWM) based BESS, experimental verification of downscaled prototype is also provided. Paper [33] provides analysis of usage modular multilevel converter with independent sources in static synchronous compensator combined with battery energy storage system – STATCOM/BESS for wind farm application. In [34] an example of modular multilevel converter with independent sources, where quasi-Z-source inverter is used as power electronic converter and energy storage building block, with photovoltaic (PV) panels connected to each converter, thus creating combined system of generation of energy from PV panels, and storage of generated energy.

Modular multilevel converters with independent sources provide the ability of active battery balancing, by "swapping" modules that are being turned on more frequently with the ones, that are being turned on less frequently, therefore, providing more even discharge of the cells in the battery pack, and, also more even degradation of the battery cells.

III. OPERATION OF MODULAR MULTILEVEL CONVERTERS WITH INDEPENDENT SOURCES.

There are two types of operation of multilevel converters – with or without the PWM technique. If the PWM technique is used, then commutation losses are higher, considering that the main source of losses of transistors are during the commutations. If the PWM voltage forming technique is not used, then output waveforms are stepped, however, with sufficient level count, it is possible to have output waveforms closer to the sinewave [35]. If the PWM technique is not used, it is possible to reduce commutation losses of switches to minimum, as the commutation of the transistors will be less frequent. There will be higher conduction losses, but, taking into account, that switching losses are much lower, in some cases these can be even neglected.

IV. EVALUATION.

In order to evaluate the performance of the multilevel converter with independent sources, MATLAB-Simulink models of different modular multilevel converters with independent sources, with H-bridges as power electronic converters connecting battery to the phase, were created. The main subject of the modulation is the estimation of necessity of using PWM in modular multilevel converters with independent sources. Models of H-bridges were created quite ideal (with ideal switches), however accurate models of batteries were created using Li-ion Samsung 30Q 18650 3000mAh 15A Battery INR18650-30Q batteries, at each H-bridge input batteries were connected in such way, that the amplitude of output voltage is equal to approximately 24V that is equal to operating voltage of the PAW, described in [4]. Output frequency is 50 Hz. Three cases were modulated: without PWM technique, With PWM technique, with different modulation frequencies: 2 kHz and 5 kHz. On Figure 4 example of single level voltage source inverter (VSI) is shown. Figure 5 shows the sub-model of H-bridge that is used to form phase cascades. Figure 6 shows the sub-model of control block.

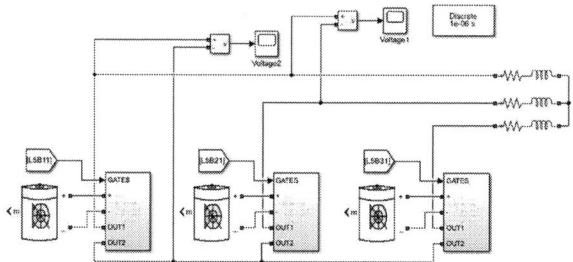

Fig. 3. Model of single level modular multilevel converter with independent sources

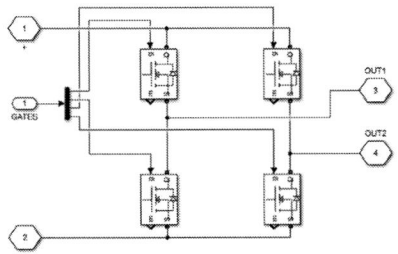

Fig. 4. Model of full bridge cell

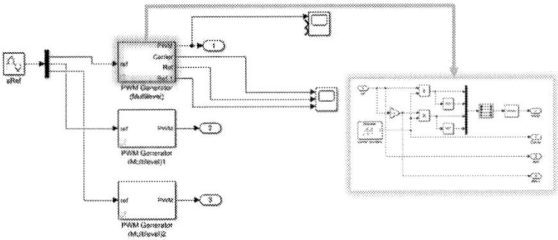

Fig. 5. Model of control block of single level modular multilevel converter with independent sources

Models of two, three and six-level modular multilevel converters with independent sources are made similarly, by increasing number of series connected H-bridges in phase cascades. Figure 6 shows the example of 6 level MLC with independent sources.

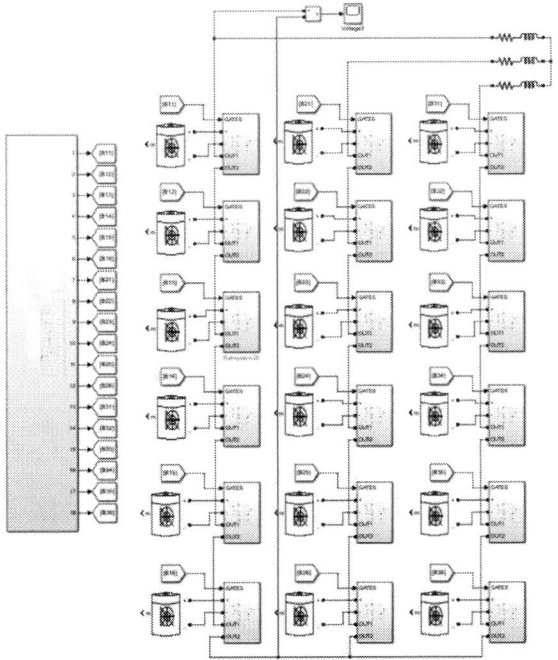

Fig. 6. Model of single level modular multilevel converter with independent sources

V. RESULTS AND DISCUSSION.

Output voltages of multilevel converters with independent sources were received in two ways – with and without PWM technique. The results for single level VSI (one H-bridge in each phase cascade), two, three and six level modular multilevel converters with independent sources are given.

Figure 7 shows output phase voltage of single level VSI with PWM modulation frequency equal to 5 kHz, Figure 8 shows output phase voltage single level VSI without PWM modulation.

Figure 9 shows the harmonics of phase voltage for single level modular multilevel converter with modulation frequency equal to 5 kHz, Figure 10 shows harmonics of output phase voltage with modulation frequency 2 kHz and Figure 11 shows harmonics for output phase voltage without PWM.

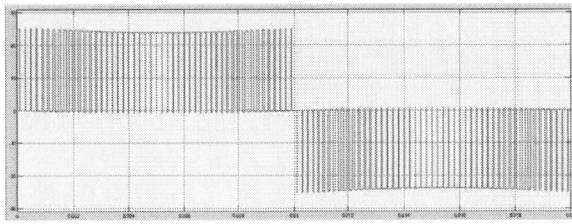

Fig. 7. Output phase voltage of single level VSI with PWM modulation frequency equal to 5 kHz

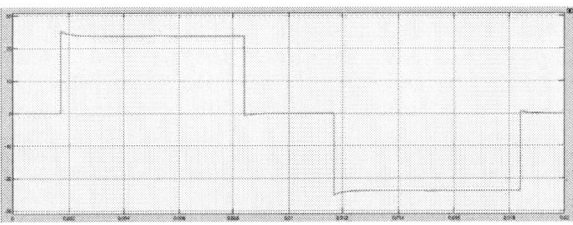

Fig. 8. Output phase voltage of single level VSI without PWM modulation.

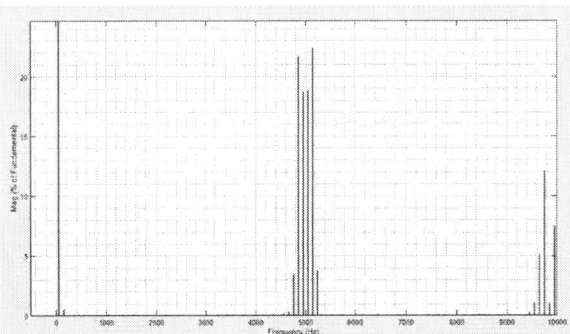

Fig. 9. Harmonics of phase voltage for single level VSI with modulation frequency equal to 5 kHz

978-1-6654-3236-8/21 $31.00 © 2021 IEEE

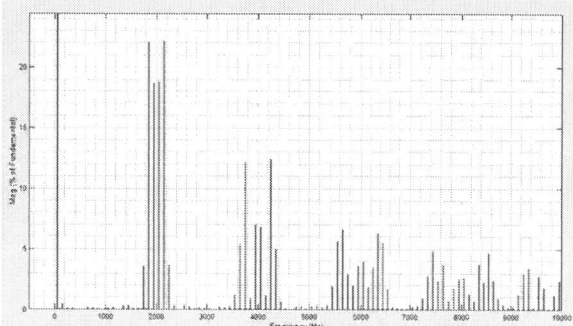

Fig. 10. Harmonics of phase voltage for single level VSI with modulation frequency equal to 2 kHz

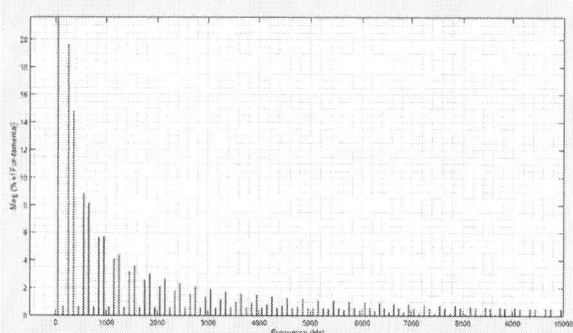

Fig. 11. Harmonics of phase voltage for single level VSI without PWM.

From Figures 9-11 it can be seen, that the modulation frequency is affecting the purity of the output waveform and the total harmonic distortion (THD). For 5 kHz modulation frequency the greatest impact gives exactly harmonics close to switching frequency and harmonics that are the multiplications of the modulation frequency. The same is in cases with 2 kHz modulation frequency. In case when PWM is not used, it is seen that harmonic decomposition is proportional.

The same analysis can be provided also for multilevel converters with more levels – for two, three and six-level MLCs with independent sources.

Figure 12 shows the output voltage of 3-level MLC with modulation frequency equal to 2kHz, Figure 13 shows the output voltage of 6-level MLC without PWM modulation.

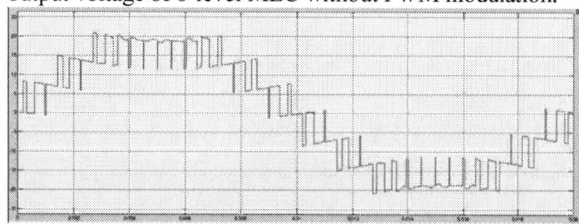

Fig. 12. Output phase voltage of 3-level MLC with independent sources with PWM modulation frequency equal to 2 kHz.

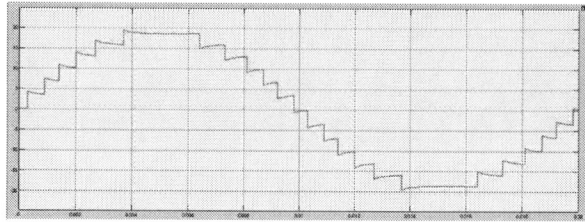

Fig. 13. Output phase voltage of 6-level MLC with independent sources without PWM modulation.

The results are shown at Figures 14-16 for two-level MLC, Figures 17-19 for three-level MLC, and Figure 20-22 – for six-level MLC.

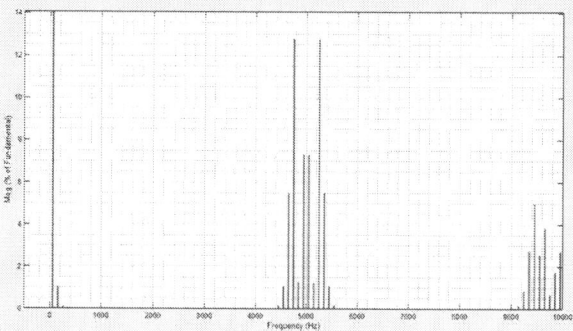

Fig. 14. Harmonics of phase voltage for 2-level MLC with modulation frequency equal to 5 kHz.

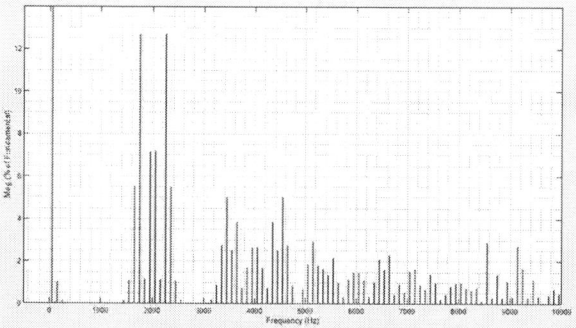

Fig. 15. Harmonics of phase voltage for 2-level MLC with modulation frequency equal to 2 kHz.

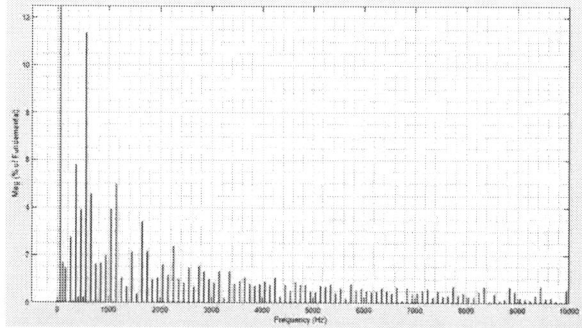

Fig. 16. Harmonics of phase voltage for 2-level MLC without PWM modulation.

978-1-6654-3236-8/21 $31.00 © 2021 IEEE   148

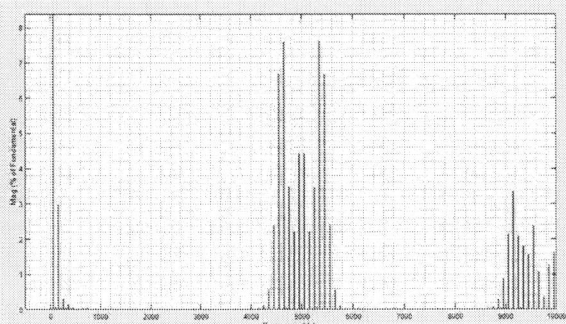

Fig. 17. Harmonics of phase voltage for 3-level MLC with modulation frequency equal to 5 kHz.

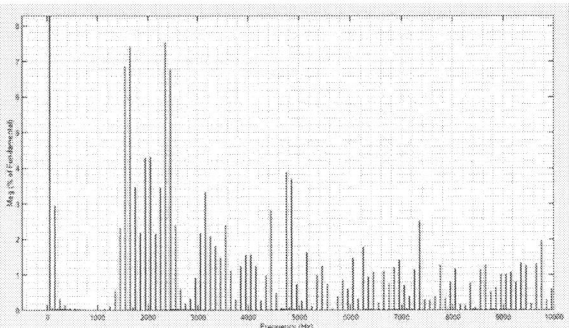

Fig. 18. Harmonics of phase voltage for 3-level MLC with modulation frequency equal to 2 kHz.

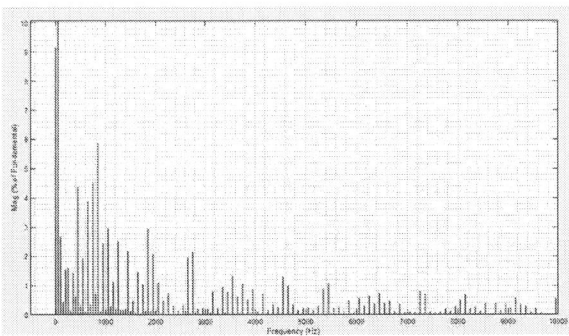

Fig. 19. Harmonics of phase voltage for 3-level MLC without PWM modulation.

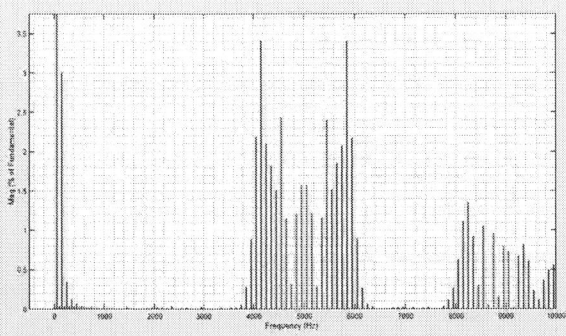

Fig. 20. Harmonics of phase voltage for 6-level MLC with modulation frequency equal to 5 kHz.

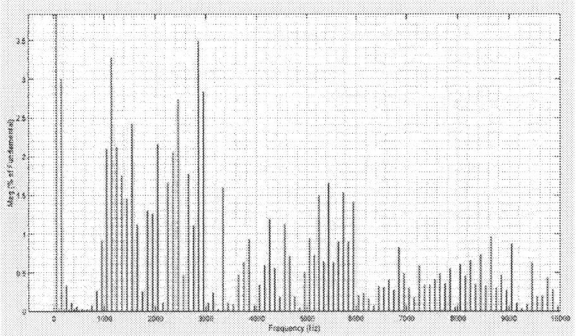

Fig. 21. Harmonics of phase voltage for 6-level MLC with modulation frequency equal to 2 kHz.

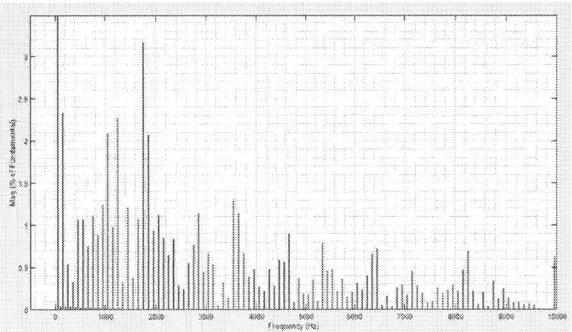

Fig. 22. Harmonics of phase voltage for 6-level MLC without PWM modulation.

In all cases with PWM modulation it can be seen, that modulation frequency is affecting the output voltage. Especially this effect is seen at Figures 10, 14, 17 and 20, when modulation frequency is chosen to be 5 kHz. For output voltage frequency 50 Hz (50 Hz is fundamental frequency), it can be seen, that the 5 kHz and harmonics that are multiples of 5 kHz have great effect. This effect is less noticeable for 2 kHz modulation frequency, however, this effect can also be seen. In case, when PWM is not used, it is seen (Figures 11, 16, 19 and 22) that the harmonic distribution is more proportional.

Table I shows the THD of output voltages for all modulated cases. It can be seen, that THD of output voltage waveforms when PWM is not used is lower, than in all cases, when PWM is used. However, it can be said, that exactly the modulation frequency of PWM is increasing distortion, making THD larger in cases when PWM is used to form output voltage. In the Table I for cases, when PWM is used THD is divided to Low Frequency Harmonic Distortion (LFHD) – Harmonic Distortion, that appears for frequencies $f = (0, \frac{1}{2} f_{\text{mod}}]$ and High Frequency Harmonic Distortion, that appears for frequencies $f > \frac{1}{2} f_{\text{mod}}$ (where f_{mod} - modulation frequency).

TABLE I.   THD OF OUTPUT PHASE VOLTAGES OF MULTILEVEL
CONVERTERS WITH INDEPENDENT SOURCES

	Modulation frequency 2 kHz			Modulation frequency 5 kHz			THD without PWM
	THD	LFHD	HFHD	THD	LFHD	HFHD	
Single-level VSI	54.46%	0.60%	53.86%	54.35%	0.66%	53.69%	31.44%
2-level MLC	28.67%	1.05%	27.62%	29.16%	1.06%	28.10%	18.50%
3-level MLC	22.17%	2.97%	19.20%	22.43%	3.00%	19.43%	13.78%
6-level MLC	11.59%	3.17%	8.42%	11.73%	3.04%	8.69%	7.97%

From Table I it can be seen, that in spite of the fact, that THD of output voltage waveforms is larger when PWM is used, it can also be said, that exactly higher harmonics (larger, than the half of modulation frequency) are giving that impact to THD. It means, that output voltages can be filtered in order to reduce this negative impact. In cases, when PWM is not used, harmonic distribution is proportional, which means, that even with output filter THD of output voltage will not decrease so significantly, then in case of when PWM is used.

By increasing level count of the MLC it is possible to decrease THD, however, it means, that converter will have higher conduction losses – larger switch count will be opened more time, than in case of lower level count. By implementing PWM + filtering output voltages it is also possible to reduce THD significantly, however, PWM implementation will result in increase of switching losses. One of the optimal solutions of implementation of MLC can be 3 - 4 levels with low frequency inter-level PWM.

VI. CONCLUSIONS AND FUTURE WORKS.

The above given study of the BES interface converters allows synthesis of their best configuration based on MLC topologies. The provided brief review of MLC shows that all three MLC types are compatible with BES. However, the neutral point clamped MLC may deal with simpler battery, but require BMS. On the other hand, modular MLCs with independent sources provides intrinsic management of the battery cell groups, but the battery itself is divided into as many parts as there are levels.

In the same time, the obtained results of the MATLAB-Simulink simulation show that good quality of the generated output voltage, which criteria is the root-mean declination of the voltage from its reference (in the case of sinusoidal reference known as THD), can be achieved in many ways. One way – increasing the number of MLC levels providing pure LM – leads to the higher conduction losses, but the other – the use single-level VSI and PWM – leads to the higher switching losses. Both solutions in pure form leads to thermal complications. However, a trade-off can be found. The optimal configuration may include 3 - 4 levels with minor inter-level pulse width modulation.

Finally, it become obvious, that the evaluation of the actual link between the quality of the output voltage of an MLC and its losses, as well as its thermal mode, requires further dedicated research.

ACKNOWLEDGMENT

This research work is funded by the Latvian Council of Science, project "Enhanced Thermal Management of Electric Drives in Orthopedic Rehabilitation Vehicles for their Better Reliability and Functionality", project No. lzp-2020/2-0390 and Riga Technical University's Doctoral Grant program.

REFERENCES

[1] S. Robert, *Introduction to Batteries—An IEEE Course*. NY, USA: IEEE: New York, 2013.

[2] X. Hu, C. Zou, C. Zhang, and Y. Li, "Technological Developments in Batteries: A Survey of Principal Roles, Types, and Management Needs," *IEEE Power Energy Mag.*, vol. 15, no. 5, pp. 20–31, Sep. 2017, doi: 10.1109/MPE.2017.2708812.

[3] I. A. Galkin, A. Blinov, M. Vorobyov, A. Bubovich, R. Saltanovs, and D. Peftitsis, "Interface Converters for Residential Battery Energy Storage Systems: Practices, Difficulties and Prospects," *Energies 2021, Vol. 14, Page 3365*, vol. 14, no. 12, p. 3365, Jun. 2021, doi: 10.3390/EN14123365.

[4] I. Galkin, A. Podgornovs, A. Blinov, K. Vitols, M. Vorobyov, and R. Kosenko, "Consideretions on Cost effective power assisted wheelchair subsystems," *Electr. Control Commun. Eng.*, vol. 14, no. 1, pp. 71–80, 2018, doi: 10.2478/ecce-2018-0008.

[5] "Energy strategy | Energy." https://ec.europa.eu/energy/topics/energy-strategy-and-energy-union_en (accessed May 10, 2021).

[6] "SolarEdge | A World Leader in Smart Energy Solutions." https://www.solaredge.com/ (accessed May 17, 2021).

[7] "Best Home Wind Turbines Updated: 2021." https://www.10powerup.com/best-home-wind-turbine/ (accessed May 17, 2021).

[8] "Powerwall | Tesla." https://www.tesla.com/powerwall?redirect=no (accessed May 17, 2021).

[9] "Energy supply of the future. Today. | to sunbathe." https://sonnen.de/ (accessed May 17, 2021).

[10] "W.H.O., World Report on Disability, 1st edition. World Health Organization, 2011, p. 350."

[11] R. Geidarovs, A. Podgornovs, and I. Galkin, "Simulation and Initial Evaluation of Modular Motor-Generator for Cost-Effective Power-Assist Wheelchair," in *2018 IEEE 59th International Scientific Conference on Power and Electrical Engineering of Riga Technical University (RTUCON)*, Nov. 2018, pp. 1–5, doi: 10.1109/RTUCON.2018.8659877.

[12] A. Podgornovs and I. Galkin, "Evaluation of Configurations of Modular Motor for Power-Assist Wheelchair," in *2019 26th International Workshop on Electric Drives: Improvement in Efficiency of Electric Drives (IWED)*, Jan. 2019, pp. 1–4, doi: 10.1109/IWED.2019.8664279.

[13] A. Podgornovs, R. Geidarovs, and I. Galkin, "Considerations on Selection of Modular Electric Drive for Power-Assist Wheelchair," Oct. 2019, doi: 10.1109/RTUCON48111.2019.8982291.

[14] I. Galkin, A. Podgornovs, and R. Geidarovs, "Data Evaluation of Novel Modular Wheelchair Motor," Jan. 2020, doi: 10.1109/IWED48848.2020.9069507.

[15] M. Bragard, N. Soltau, S. Thomas, and R. W. De Doncker, "The balance of renewable sources and user demands in grids: Power electronics for modular battery energy storage systems," *IEEE Trans. Power Electron.*, vol. 25, no. 12, pp. 3049–3056, 2010, doi: 10.1109/TPEL.2010.2085455.

[16] U. Abronzini *et al.*, "A Dual-Source DHB-NPC Power Converter for Grid Connected Split Battery Energy Storage System," in *2018 IEEE Energy Conversion Congress and Exposition, ECCE 2018*, Dec. 2018, pp. 2483–2488, doi: 10.1109/ECCE.2018.8557563.

[17] T. Modeer, N. Pallo, T. Foulkes, C. B. Barth, and R. C. N. Pilawa-Podgurski, "Design of a GaN-Based Interleaved Nine-Level Flying Capacitor Multilevel Inverter for Electric Aircraft Applications," *IEEE Trans. Power Electron.*, vol. 35, no. 11, pp. 12153–12165, Nov. 2020, doi: 10.1109/TPEL.2020.2989329.

[18] C. B. Barth *et al.*, "Design and control of a GaN-based, 13-level, flying capacitor multilevel inverter," Aug. 2016, doi:

978-1-6654-3236-8/21 $31.00 © 2021 IEEE

10.1109/COMPEL.2016.7556770.

[19] L. M. Tolbert, F. Z. Peng, and T. G. Habetler, "Multilevel converters for large electric drives," *IEEE Trans. Ind. Appl.*, vol. 35, no. 1, pp. 36–44, 1999, doi: 10.1109/28.740843.

[20] S. Khomfoi and L. M. Tolbert, "Multilevel Power Converters," in *Power Electronics Handbook*, 2nd ed., Elsevier Inc., 2006, pp. 451–482.

[21] V. Fernão Pires, E. Romero-Cadaval, D. Vinnikov, I. Roasto, and J. F. Martins, "Power converter interfaces for electrochemical energy storage systems - A review," *Energy Convers. Manag.*, vol. 86, pp. 453–475, Oct. 2014, doi: 10.1016/j.enconman.2014.05.003.

[22] M. Adest, L. Handelsman, Y. Galin, A. Fishelov, G. Sella, and Y. Binder, "SolarEdge patent for HD-Wave Inverters - Distributed power system using direct current power sources, US9368964B2," 2016.

[23] C. Liu, X. Cai, and Q. Chen, "Self-Adaptation Control of Second-Life Battery Energy Storage System Based on Cascaded H-Bridge Converter," *IEEE J. Emerg. Sel. Top. Power Electron.*, vol. 8, no. 2, pp. 1428–1441, Jun. 2020, doi: 10.1109/JESTPE.2018.2886965.

[24] Q. Chen, N. Gao, R. Li, X. Cai, and Z. Lu, "A unified control scheme of battery energy storage system based on cascaded H-bridge converter," in *2014 IEEE Energy Conversion Congress and Exposition, ECCE 2014*, Nov. 2014, pp. 2561–2566, doi: 10.1109/ECCE.2014.6953743.

[25] Z. Ling, Z. Zhang, Z. Li, and Y. Li, "State-of-charge balancing control of battery energy storage system based on cascaded H-bridge multilevel inverter," in *2016 IEEE 8th International Power Electronics and Motion Control Conference, IPEMC-ECCE Asia 2016*, Jul. 2016, pp. 2310–2314, doi: 10.1109/IPEMC.2016.7512657.

[26] A. S. Gadalla, X. Yan, and H. Hasabelrasul, "Active power analysis for the battery energy storage systems based on a modern cascaded multilevel converter," in *Proceedings - 2nd IEEE International Conference on Energy Internet, ICEI 2018*, Jul. 2018, pp. 111–116, doi: 10.1109/ICEI.2018.00028.

[27] A. S. Gadalla, X. Yan, and H. Hasabelrasul, "State-of-charge balancing control strategy for battery energy storage systems based on a modern cascaded multilevel PWM converter," in *Proceedings*

- 2018 IEEE 12th International Conference on Compatibility, Power Electronics and Power Engineering, CPE-POWERENG 2018, Jun. 2018, pp. 1–6, doi: 10.1109/CPE.2018.8372534.

[28] K. Kandasamy, D. M. Vilathgamuwa, and G. Foo, "Inter-module SoC balancing control for CHB based BESS using multi-dimensional modulation," in *Proceedings of the IEEE International Conference on Industrial Technology*, 2013, pp. 1630–1635, doi: 10.1109/ICIT.2013.6505917.

[29] J. Asakura and H. Akagi, "State-of-Charge (SOC)-Balancing Control of a Battery Energy Storage System Based on a Cascade PWM Converter," *IEEE Trans. Power Electron.*, vol. 24, no. 6, pp. 1628–1636, 2009, doi: 10.1109/TPEL.2009.2014868.

[30] M. Chen, B. Zhang, Y. Li, G. Qi, and J. Liu, "Design of a multi-level battery management system for a Cascade H-bridge energy storage system," in *Asia-Pacific Power and Energy Engineering Conference, APPEEC*, Mar. 2014, vol. 2015-March, no. March, doi: 10.1109/APPEEC.2014.7066076.

[31] K. Mordi, H. Mhiesan, J. Umuhoza, M. M. Hossain, C. Farnell, and H. A. Mantooth, "Comparative Analysis of Silicon Carbide and Silicon Switching Devices for Multilevel Cascaded H-Bridge Inverter Application with Battery Energy storage," Aug. 2018, doi: 10.1109/PEDG.2018.8447775.

[32] L. Maharjan, S. Inoue, and H. Akagi, "A transformerless energy storage system based on a cascade multilevel PWM converter with star configuration," *IEEE Trans. Ind. Appl.*, vol. 44, no. 5, pp. 1621–1630, 2008, doi: 10.1109/TIA.2008.2002180.

[33] L. Zhang, Y. Wang, H. Li, and P. Sun, "Hybrid power control of cascaded STATCOM/BESS for wind farm integration," in *IECON Proceedings (Industrial Electronics Conference)*, 2013, pp. 5288–5293, doi: 10.1109/IECON.2013.6699995.

[34] B. Ge, Y. Liu, H. Abu-Rub, and F. Z. Peng, "State-of-Charge Balancing Control for a Battery-Energy-Stored Quasi-Z-Source Cascaded-Multilevel-Inverter-Based Photovoltaic Power System," *IEEE Trans. Ind. Electron.*, vol. 65, no. 3, pp. 2268–2279, Mar. 2018, doi: 10.1109/TIE.2017.2745406.

[35] A. Bubovich and I. Galkin, "Evaluation of Optimal Switching of Modular Multilevel Inverter with Independent Voltage Sources," Nov. 2020, doi: 10.1109/RTUCON51174.2020.9316581.

Harmonic content of the input current of the boost converter in quasiperiodicity

Željko Stojanović
Department of electrical engineering
Zagreb University of Applied Sciences
Zagreb, Croatia
zeljko.stojanovic@tvz.hr

Denis Pelin
Faculty of Electrical Engineering, Computer Science and
Information Technology Osijek
University of Osijek
Osijek, Croatia
denis.pelin@ferit.hr

Abstract—**The paper deals with the harmonic content of the input current of the voltage mode controlled boost converter in quasiperiodicity. By decrease of the input voltage the converter exhibits quasiperiodicity caused by Hopf bifurcation and border collision bifurcation. In the area of these bifurcations mode locking, quasiperiodic steady-states and chaotic steady-states alternate. Characteristic waveforms, trajectories and harmonic contents of the input currrent in the area of Hopf bifurcation and border collision bifurcation are presented. The quasiperiodicity caused by both bifurcations has so far not been recorded in the boost converter. Additional harmonics in the input current of the converter are introduced by the quasiperiodicity. These harmonics are spread to the area of frequencies lower than the switching frequency as well as between the multiple integers of the switching frequency. Low frequency AC components in quasiperiodicity caused by border collision bifurcation are higher than low frequency AC components caused by Hopf bifurcation. These low frequency harmonics can be harmful to the source of the input voltage.**

Keywords—*border collision bifurcation, boost converter, harmonic content, Hopf bifurcation, quasiperiodicity*

I. Introduction

A common part of a DC-DC converter is an input filter. It has dual function. It filters unwanted harmonic components in both directions, from the DC voltage source to the converter and from the converter to the DC voltage source. A DC-DC converter operates at the switching frequency. That means the period of converter currents and voltages is equal to the period of a PWM driving signal. That mode of operation is so called period-one steady-state. Knowledge of harmonic content of the input current of the converter is necessary for the design of the input filter. Although the designer's goal is to ensure converters mode of operation in period-one steady-state it may not happened for some combination of converter parameters.

A DC-DC converter is nonlinear time varying circuit usually of the second order or a higher. Therefore, converters are prone to bifurcations. Beside period-one steady-state three different types of steady-states are possible in the converter. These are subharmonic, quasiperiodic and chaotic steady-states [1-6]. These steady-states can appear when the converter parameter changes out of the range of period-one steady-state. The change of the converters steady-state worsen various converters characteristics. Bifurcations can cause higher current and voltage ripple and peak values and consequently the effective values [6, 7]. Undesirable effects of bifurcations in chaotic steady-states also include unwanted

harmonic content [8] and audible sound [9]. Different steady-state cause different worsens of converters characteristics.

The quasiperiodic steady-state appears when two or more incommensurate frequencies governs the converters dynamics. Two different mechanisms - Hopf bifurcation and border collision bifurcation can lead to quasiperiodicity [4] [10]. The quasiperiodicity caused by Hopf bifurcation is treated in the boost converter [11] in the buck converter [4, 10, 13] and in the buck-boost converter [14]. The quasiperiodicity caused by border collision bifurcation is treated in the buck converter [4, 10, 12]. As a consequence of Hopf bifurcation the converter can operate in quasiperiodic steady-states, mode-locking and chaotic steady-states The same apply to border collision bifurcation. Researches on bifurcations in a DC-DC converters are mostly focused on physical mechanism of bifurcations. Therefore, design oriented approach to bifurcations and steady-states of a DC-DC converters is neglected. Wider power electronics community is not familliar with methods and tools for prediction and identification of bifurcations and nonregular steady-states like subharmonic, quasiperiodic and chaotic steady-states.

This paper deals with the harmonic content of input current of the voltage mode controlled boost converter in quasiperiodicity. The converter under consideration exhibits quasiperiodicity caused both by Hopf bifurcation and border collision bifurcation. The quasiperiodicity caused by both bifurcations has so far only been shown in the buck converter and not in the boost converter. In both cases considerable amount of harmonics lower than the fundamental harmonic of the input current appears. Experimental results of waveforms in quasiperiodicity caused by both mechanisms are presented.

II. Physical Realization of the Boost Converter

The schematic of the boost converter under consideration is shown in Fig. 1. The converter control is based on PWM modulator control circuit MC34060A [15]. The schematic recommended by the manufacturer is intentionally adjusted to make the converter prone to bifurcations. This was done by removing LC output filter.

Parameters of the converter were: The average value of the output voltage $V_d = 20.0$ V, PWM ramp voltage period $T = 500$ µs, the range of the input voltage $E = 7 - 18$ V, inductor $L = 3,60$ mH/0,81 Ω, load capacitance $C_d = 255$ µF and the load resistance $R_d = 81$ Ω. The load resistance is chosen to ensure that the converter operates only in period-one steady-state and quasiperiodicity.

978-1-6654-3236-8/21 $31.00 © 2021 IEEE

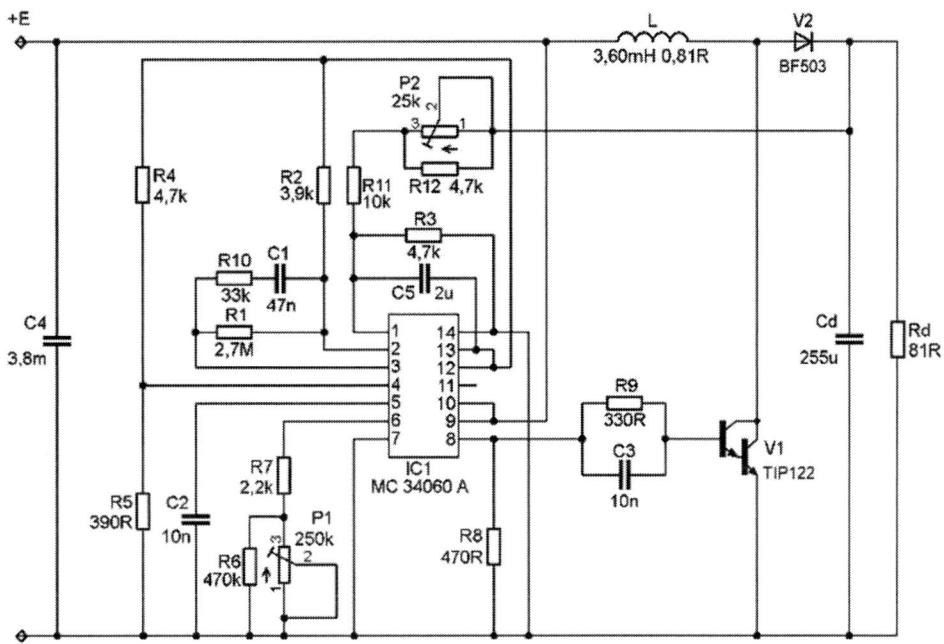

Fig. 1. Schematic of the boost converter under consideration

TABLE I.   STEADY-STATES OF THE CONVERTER DUE TO CHANGE OF THE INPUT VOLTAGE

Quasiperiodicity		Period-one steady-state
Input voltage 7.0 V – 11.3 V		
Border collision bifurcations	Hopf bifurcations	*Input voltage 11.3 V – 18.0 V*
Mode-locking	Mode-locking	
Quasiperiodic steady-states	Quasiperiodic steady-states	
Chaotic steady-states	Chaotic steady-states	

III. MEASUREMENT SETUP

The input voltage is changed in a full range $E = 7 - 18$ V. By decrease of the input voltage the converter undergoes bifurcation from period-one steady-state to various steady-states in the area of quasiperiodicity, Tab. 1. Firstly, the period-one steady-state ceases and Hopf bifurcations takes place. After the series of Hopf bifurcations border collision bifurcations takes place.

All shown steady-states are represented by characteristic waveforms, trajectory and harmonic content of the input current. Characteristic waveforms consist of an AC component of the output (capacitor) voltage $\tilde{v}_C$, ramp voltage (pin 5 of integrated circuit) v_{ramp}, control voltage (pin 3 of integrated circuit) v_{con}, and the input (inductor) current i_L. Respective channel sensitivites are $\tilde{v}_C$ - 2 V/div, v_{ramp} - 1 V/div, v_{con} - 1 V/div, i_L - 1 A/div, and time base - 2 ms/div.

IV. MEASUREMENT RESULTS

A. Period-one steady-state

The converter is designed to operate in period-one steady-state. In proper operation the converter operates in continuous conduction mode, discontinuous conduction mode or both. Regardless of the mode of operation the harmonic contents of the input current consists of switching frequency $f_s = 1/T = 2$ kHz and its integer multiples nf_s. Characteristic waveforms and harmonic content of the input current are shown in Fig. 2. According to the known harmonic content of the input current and to the required harmonic content of the input current the designer is able to design a suitable input filter.

B. Quasiperiodicity obtained by Hopf bifurcation

By decrease of the input voltage the converter undergoes bifurcations. Starting from the period-one steady-state the two most common bifurcations in DC-DC converters are period-doubling bifurcation and Hopf bifurcation. Period doubling did not occur, but converter undergo to the dynamics on torus. There are two ways in which the converter can undergo to the dynamics on torus. The one way is smooth bifurcation when the converter does not change the switching sequence. That bifurcation is Hopf bifurcation. Another way is nonsmooth bifurcation when the converter change the switching sequence. That bifurcation is border collision bifurcation. In

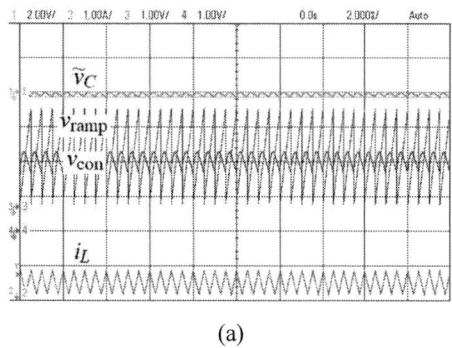

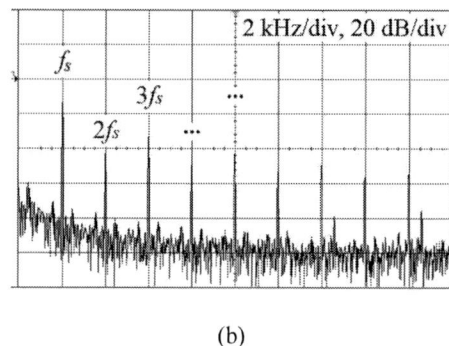

Fig. 2. Period-one steady-state at input voltage $E = 12{,}0$ V. (a) Characteristic waveforms. (b) Harmonic content of input current. Sensitivity 2 kHz/div, 20 dB/div.

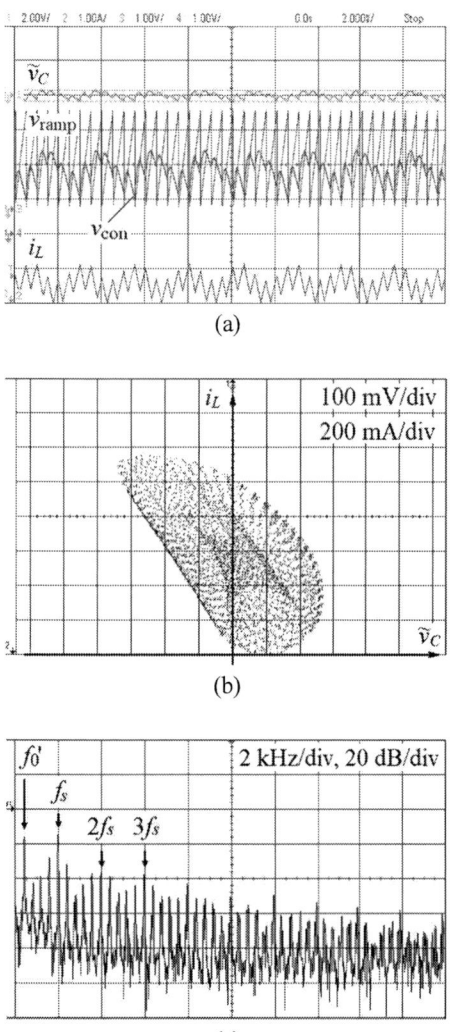

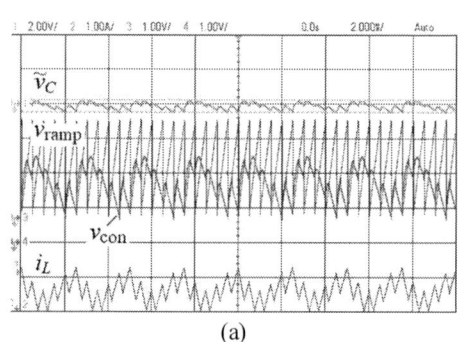

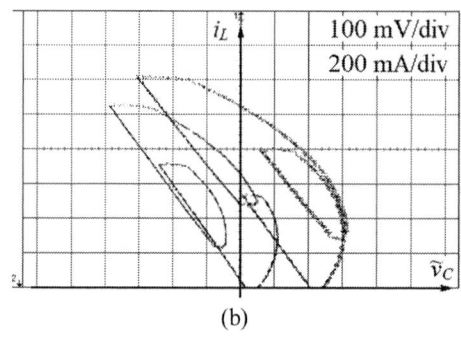

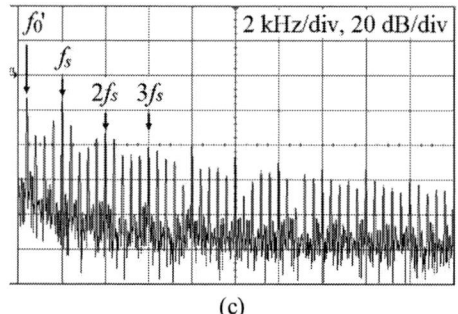

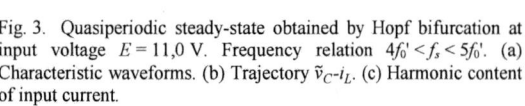

Fig. 3. Quasiperiodic steady-state obtained by Hopf bifurcation at input voltage $E = 11{,}0$ V. Frequency relation $4f_0' < f_s < 5f_0'$. (a) Characteristic waveforms. (b) Trajectory $\tilde{v}_C$-i_L. (c) Harmonic content of input current.

Fig. 4. Mode-locking obtained by Hopf bifurcation at input voltage $E = 10{,}5$ V. Frequency ratio $f_s = 5f_0'$. (a) Characteristic waveforms. (b) Trajectory $\tilde{v}_C$-i_L. c) Harmonic content of input current.

978-1-6654-3236-8/21 $31.00 © 2021 IEEE

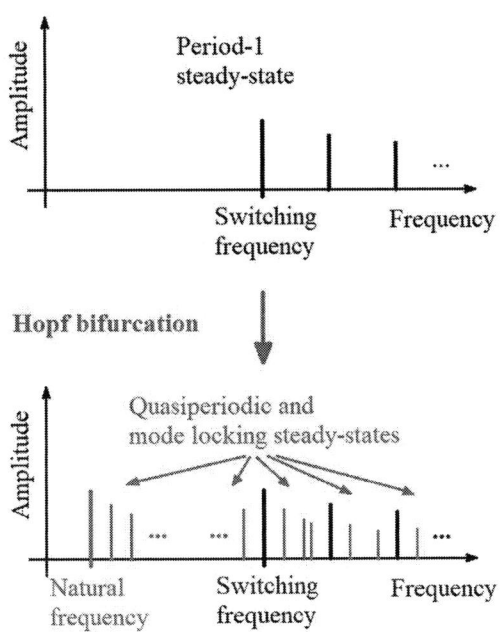

Fig. 5. Apearance of new frequencies in the first Hopf bifurcation.

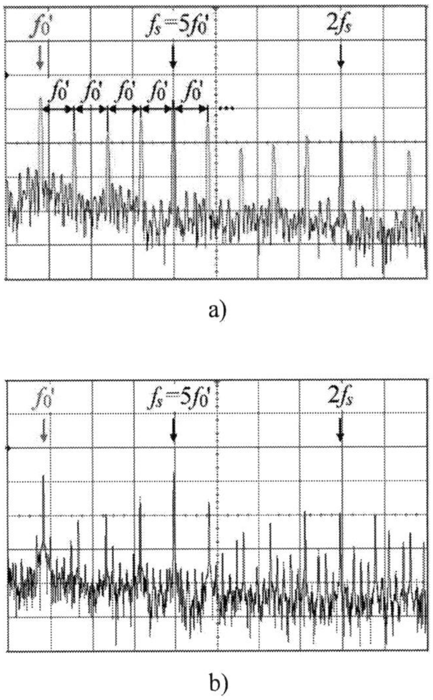

Fig. 6. Harmonic components of input current obtained by Hopf bifurcations. (a) Mode-locking with frequency ratio $f_s = 5f_0'$. (b) Quasiperiodic steady-state with frequency relation $4f_0' < f_s < 5f_0'$.

the observed case the switching sequence did not change. Thus, Hopf bifurcation occured. Characteristic waveforms, trajectory and harmonic content of the input currrent in the area of Hopf bifurcations are shown in Fig. 3 and Fig. 4.

Comparing the waveforms, Fig. 3 and Fig. 4 with the period-one steady-state, one can perceive appearance of new frequency components which are considerably lower than the switching frequency. These frequency components are caused by the natural frequency f_0 of the converter [14]. Total harmonic content consists of combination of frequencies $f_{mn} = mf_s + nf_0'$ where f_0' is close to f_0 [14]. When the frequencies f_s and f_0' are not commensurate the converter operates in quasiperiodic steady-state, Fig. 3. and when the frequencies f_s and f_0' are commensurate the converter operates in mode-locking Fig. 4. In both cases, the new frequency components are spread in frequency domain as shown in Fig. 5. By changing the input voltage, the ratio f_s/f_0' also changes, so mode locking and quasiperiodic steady-state alternates each other.

Identification of the steady-states by characteristic waveforms, trajectory and harmonic content was available. Of these methods for distinguishing quasiperiodic steady-state from mode-locking the most suitable is identification by trajectories. The trajectory of quasiperiodic steady-state is a surface, Fig. 3b, while the trajectory of mode-locking is closed curve Fig. 4b. Distinguishing quasiperiodic steady-state from mode-locking by comparing their harmonic content is not reliable, Fig. 3c and Fig. 4c because both harmonic contents looks similar. Harmonic content of mode-locking around switching frequency f_s is shown enlarged in Fig. 6a. The new harmonic components caused by dynamics on torus are clearly different from the harmonic components f_s and $2f_s$ caused by

switching frequency in period-one steady-state. The distinction is facilitated by an integer ratio $f_s = 5f_0'$. Harmonic content of quasiperiodic steady-state around switching frequency f_s is shown enlarged in Fig. 6b. The harmonic content is somewhat different than in Fig. 6a. In both cases, the new frequency components under the switching frequency appears.

C. Quasiperiodicity obtained by border collision bifurcation

By further decrease of the input voltage the converter undergoes border collision bifurcations. The bifurcations are triggered by the jump of the control voltage over the highest peak of the ramp voltage. Therefore, the switching sequence is different from the switching sequence in the area of Hopf bifurcations. The bifurcations are identified by comparing ramp voltage and control voltage.

Characteristic waveforms, trajectory and harmonic content of the input currrent in the area of border collision bifurcations are shown in Fig. 7 and Fig. 8. Mode locking $f_s = 16f_0'$ is shown in Fig. 7. The trajectory is a closed curve. The harmonic content consists of 15 harmonic components lower than the switching frequency. That is why this harmonic content is more dense then the harmonic content in Fig. 4c. Chaotic steady-state is shown in Fig. 8. The waveforms are similar to the previous waveforms od mode locking. In this case waveforms are not reliable for distinguishing chaotic steady-state from mode-locking. The trajectory is a surface and the harmonic content is continuous, and this is a sure sign of chaotic steady-state. In both steady-states, the new frequency components lower than the switching frequency appears. By changing the input voltage, the moment of jump

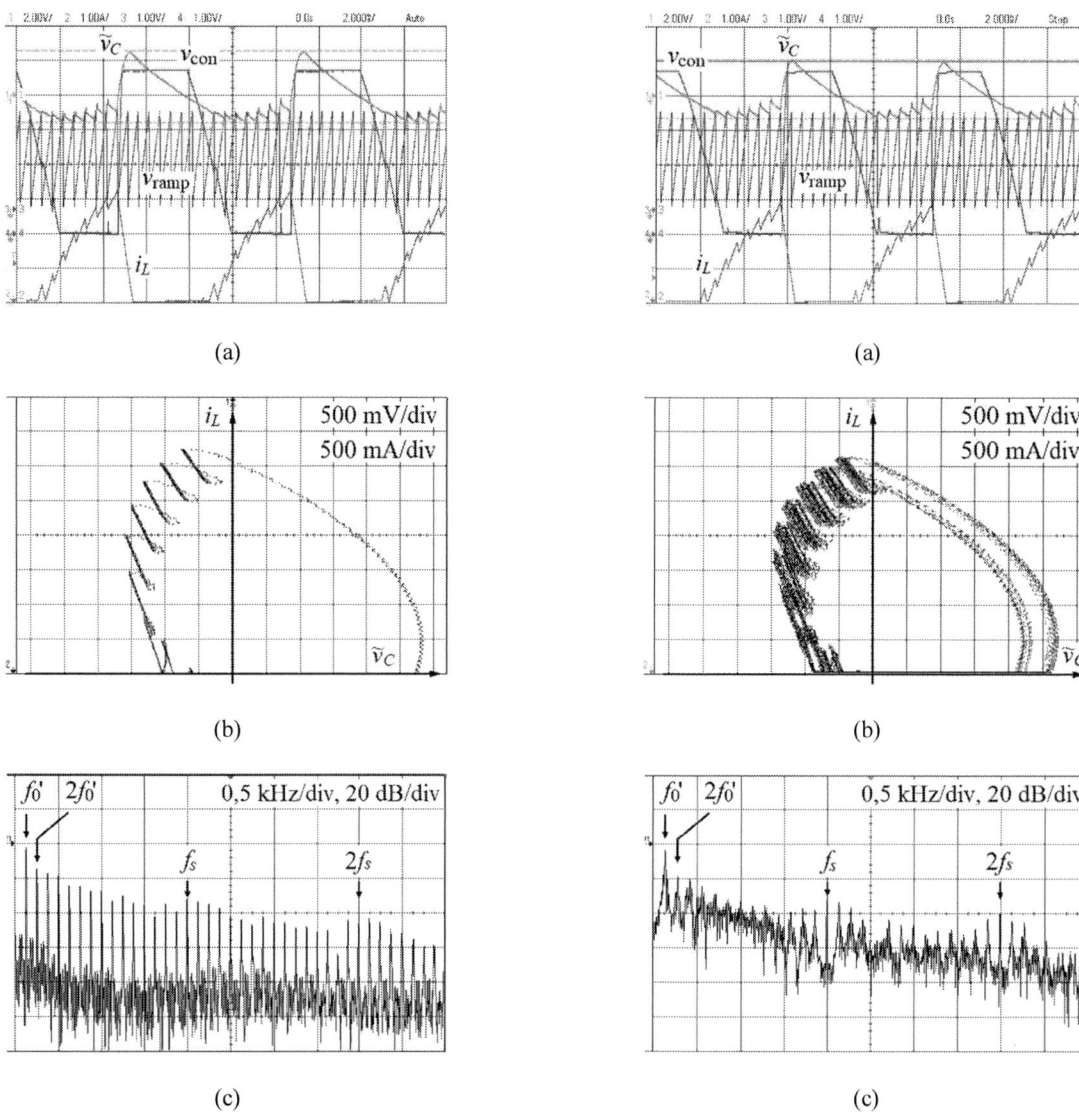

Fig. 7. Mode-locking obtained by border collision bifurcation at input voltage $E = 9,7$ V. Frequency ratio $f_s = 16f_0'$. (a) Characteristic waveforms. (b) Trajectory $\tilde{v}_C$-i_L. (c) Harmonic content of input current.

Fig. 8. Chaotic steady-state obtained by border collision bifurcation at input voltage $E = 9,0$ V. (a) Characteristic waveforms. (b) Trajectory $\tilde{v}_C$-i_L. (c) Harmonic content of input current.

of control voltage v_{con} over ramp voltage v_{ramp} also changes, so mode locking and chaotic steady-state alternates each other.

D. Comments

Regarding the harmonic content of the inductor current it can be stated:

- Appearance of quasiperiodicity introduce harmonics lower than the switching frequency f_s.

- The low frequency AC components in quasiperiodicity are higher than the low frequency AC components in period-one steady-state.

- Low frequency AC components in quasiperiodicity caused by border collision bifurcation are higher than low frequency AC components caused by Hopf bifurcation.

- The harmonic content in chaotic steady-state is continuous.

These low frequency harmonics can be harmful to the source of the input voltage.

The borders between the same type of bifurcations as well as between the different type of bifurcations are not fixed i.e. there are hystereses between steady-states. The causes of hysteresis have not been elucidated and require further research.

V. CONCLUSION

The harmonic content of the input current of the voltage mode controlled boost converter in quasiperiodicity is presented. The converter exhibits quasiperiodicity caused both by Hopf bifurcation and border collision bifurcation. In

978-1-6654-3236-8/21 $31.00 © 2021 IEEE

the area of these bifurcations mode locking, quasiperiodic steady-states and chaotic steady-states alternate. In both cases considerable amount of harmonics lower than the switching frequency of the input current appears. These harmonics can be harmful to the source of the input voltage. The authors hope that presented waveforms can help power electronics designers in identification of unwanted steady-states.

REFERENCES

[1] L. Benadero, A. El Aroudi, G. Olivar, E. Toribio, and E. Gomez, "Two-Dimensional bifurcation Diagrams: Background Pattern of Fundamental Dc-dc converters with PWM Control," International Journal of Bifurcation and Chaos, vol. 13, no. 2 pp. 427–451, 2003

[2] A. El Aroudi, M. Debbat, R. Giral, G. Olivar, L. Benadero, and E. Toribio, "Bifurcations in DC-DC Switching Converters: Review of Methods and Applications, " International Journal of Bifurcation and Chaos, vol. 15, no. 5, pp. 1549–1578, 2005

[3] Y. Chen, C. K. Tse, S.-S. Qui, L. Lindenmüller and W. Schwarz, "Coexisting fast-scale and slow-scale instability in current-mode controlled DC/DC converters: Analysis, simulation and experimental results," IEEE Transactions on circuits and systems—I, Fundamental theory and applications, vol. 55, no. 10, pp. 3335–3348, 2008.

[4] Z. T. Zhusubaliyev, E. Mosekilde, S. Maity, S. Mohanan, and S. Banerjee, "Border-collision route to quasi-periodicity: Numerical investigation and experimental confirmation," Chaos, vol. 16, no. 2, p. 023122, Jun. 2006.

[5] D. Pikulins, S. Tjukovs, and J. Eidaks "Effects of Control Non-idealities on the Nonlinear Dynamics of Switching DC-DC Converters," In: Stavrinides S., Ozer M. (eds) Chaos and Complex Systems. Springer Proceedings in Complexity. Springer, Cham. 2020

[6] I. Flegar and D. Pelin, "Steady-state responses of the boost converter," Proceedings of IEEE International Conference of Circuits and Systems, vol. 2, pp. 830-835, Dubrovnik, Croatia, July 2005.

[7] Ž. Stojanović, and D. Pelin, "Impact of quasiperiodic steady-state on boost converter current stress and inductor copper losses" // Book of abstract ; 20th International Symposium on Power Electronics / Katić, Vladimir (ur.). Novi Sad : GRID_FTN, Novi Sad, 2019.

[8] J. H. B. Deane, P. Ashwin, D. C. Hamill, and D. J. Jefferies, "Calculation of periodic spectral components in a chaotic dc-dc converter," IEEE Transactions on circuits and systems—I, Fundamental theory and applications, vol. 46, pp. 1313–1321, 1999.

[9] W. C. Y. Chan and C. K. Tse, "Study of Bifurcations in Current Programmed DC/DC Boost Converters From Quasi Periodicity to Period Doubling", IEEE Transactions on circuits and systems—I, Fundamental theory and applications, vol. 44, no. 12, 1997, pp. 1129–1142.

[10] Z. T. Zhusubaliyev, E. A. Soukhoterin, and E. Mosekilde, "Quasi-Periodicity and Border-Collision Bifurcations in a DC–DC Converter With Pulsewidth Modulation," IEEE Transactions on Circuits and Systems I: Fundamental Theory and Applications, vol. 50, no. 8, pp. 1047-1057, August 2003.

[11] A. El Aroudi and R. Leyva, "Quasi–Periodic Route to Chaos in a PWM Voltage-Controlled DC–DC Boost Converter," IEEE Transactions on Circuits and Systems I—Fundamental Theory and Applications, vol. 48, no. 8, pp. 967-978, August 2001.

[12] S. Maity, D. Tripathy, T. K. Bhattacharya, and S. Banerjee, "Bifurcation Analysis of PWM-1 Voltage-Mode-Controlled Buck Converter Using the Exact Discrete Model," IEEE Transactions on Circuits and Systems I: Regular Papers, vol. 54, no. 5, 2007.

[13] S. Banerjee, D. Giaouris, P. Missailidis, and O. Imrayed, "Local bifurcations of a quasiperiodic orbit", International Journal of Bifurcation and Chaos, vol. 22, no. 12, Dec. 2012.

[14] A. El Aroudi, L. Benadero, E. Toribio, and S. Machiche, "Quasiperiodicty and chaos in the DC–DC buck-boost converter," International Journal of Bifurcation and Chaos, vol. 10, pp. 359–371, 2000.

[15] MC34060A datasheet, ON Semiconductor

978-1-6654-3236-8/21 $31.00 © 2021 IEEE

Determination of material parameters using video extensometry during tensile testing

1st Jaroslav Bulava
Department of Mechatronics and Electronics, Faculty of Electrical Engineering and Information Technology
University of Žilina
Univerzitná 1, Žilina, Slovakia
jaroslav.bulava@feit.uniza.sk

2nd Libor Hargaš
Department of Mechatronics and Electronics, Faculty of Electrical Engineering and Information Technology
University of Žilina
Univerzitná 1, Žilina, Slovakia
libor.hargas@feit.uniza.sk

3rd Dušan Koniar
Department of Mechatronics and Electronics, Faculty of Electrical Engineering and Information Technology
University of Žilina
Univerzitná 1, Žilina, Slovakia
dusan.koniar@feit.uniza.sk

4th Silvia Štefůnová
Department of Mechatronics and Electronics, Faculty of Electrical Engineering and Information Technology
University of Žilina
Univerzitná 1, Žilina, Slovakia
silvia.janisova@feit.uniza.sk

Abstract— **This paper deals with an overview of non-contact measuring methods for measuring material stress. In the beginning, non-contact measuring methods such as video extensometry and the current state of use of these practice methods are described. The next part of the paper describes the methods of image processing, which are used in non-contact measuring methods of measuring the stress of the tested sample. Finally, the article describes a video extensometry algorithm to determine the basic parameters of the measured sample.**

Keywords—video extensometry, non-contact measuring methods, tensile testing, DIC correlation

I. INTRODUCTION

The determination of the mechanical properties of materials and structures is of great importance in the scientific and technical fields. Several methods have been developed to measure mechanical properties (single-axis tensile tests, bending tests, bulge tests, and high-speed dynamic tests). Of these methods, the most commonly used method is uniaxial tensile or compressive testing using a universal testing machine (UTM) to determine properties such as Young modulus, Poisson's ratio, elongation percentage, yield strength, and failure strength. To determine these mechanical properties, the surface deformations of the test specimen must be accurately measured. Both electrical resistance strain gauges and mechanical extensometers [1] may be used when testing materials to measure the surface tensions of a test specimen. Several types of methods and measuring instruments can be used to obtain information on the deformation of the surface of the measured material. However, in the detection of surface deformation of materials, these are usually microcracks. Also in these measurements, there is a tendency to use the smallest possible samples due to material savings and therefore it is often not possible to use direct measurement methods. For these and many other reasons, there is a tendency to use non-contact measuring methods. These measurement methods usually use either one or more cameras or a laser to detect and subsequently measure cracks on the surface of the measured materials.

II. STATIC TENSILE TEST

The tensile test is one of the basic mechanical tests which, due to its principle, simplicity, and efficiency, is the most widespread and most recognized test method for evaluating the mechanical properties of predominantly metallic materials. Its advantage is that it can break any material while maintaining the law of geometric similarity [9].

It consists of the fact that a smooth test body of a simple shape (mostly circular or rectangular cross-section) is attached to the jaws of the testing machine.

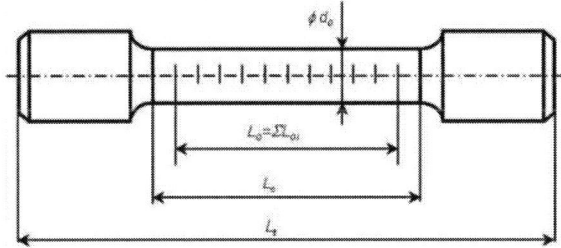

Fig. 1.   Testing sample[3]

The output is therefore a graph of this dependence and also the values of stress and strain characteristics of the tested material.

978-1-6654-3236-8/21 $31.00 © 2021 IEEE

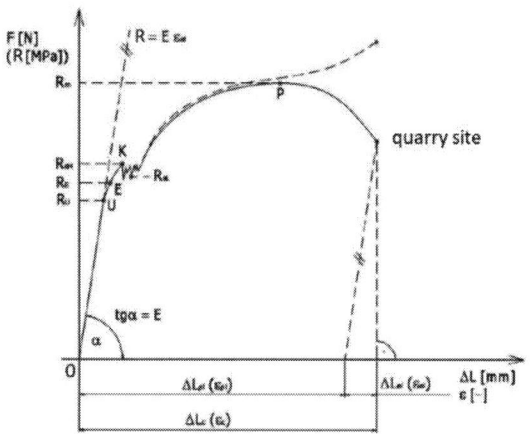

Fig. 2. Graph of static tensile test[1]

The first part of the diagram in Fig. 2 - from point 0 to point U - forms a line representing the elastic deformation. In the case of metallic materials, the reversible elastic deformation is determined by the amount of deformation that can be induced in the individual metal bonds before breaking. This value is usually below 1%. The mentioned section describes Hooke's law, expressed in the form:

$$R = E \cdot \varepsilon_{el} [MP_a] \qquad (1)$$

The elongation of the test rod is measured and recorded by a sensor located directly on it. If the extension of the test body is derived from the movement of the cross member of the tearing machine, the inclination of this part of the tensile diagram includes not only elastic deformation of the test bar but also elastic deformation of test machine parts - machine frame, dynamometer and jaws. Therefore, module E cannot be evaluated in this case [1].

In the section from point U to point P, the deformation remains uniform, but due to the beginning of the accumulation of plastic deformation, the linearity is lost and there is a deviation from the original linear trend. The stress keeps increasing with increasing deformation up to the highest point of the diagram, which the given dependence reaches. The process taking place in this part of the diagram is called strain hardening. At point P after the end of the uniform narrowing of the measuring part of the test bar, i. extensometer of the measured section on the test specimen, then a neck occurs and further deformation is associated with a decrease (relaxation) of stress. Upon completion of this uneven plastic deformation, i. after the exhaustion of possible dislocation sliding systems, the failure of the test bar occurs. The tensile test is considered to be successful only if the fracture occurs in a specific part of the sample and does not break, for example at the point of clamping.

III. VIDEO EXTENSOMETRY

Under the term video extensometry we can imagine image processing algorithms that detect various surface spots of materials, cracks, or other deformations of the measured material, but also artificially created marks such as circular points, lines or various shapes that can be painted on the test specimen, or otherwise attached. In addition to these property-based registration techniques, intensity-based image correlation algorithms using digital image correlation (DIC) can be used.

Fig. 3. Video extensometry

IV. DIC CORRELATION

DIC is a powerful optical technique widely used in the community of experimental mechanics to measure displacement and deformation over the entire field and is generally used as a post-processing technique with higher registration accuracy but high computational demands. The principle of the method itself provides the possibility to observe various phenomena during the deformation or movement of an object of any shape, while it applies to testing a wide range of materials. Correlation systems can monitor a wide range of surface points, which allows you to visualize the measured quantities over the entire monitored area [2,3].

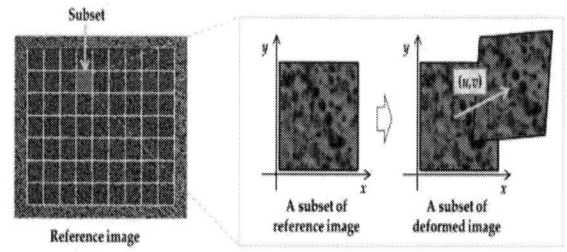

Fig. 4. DIC correlation[10]

DIC works on the principle of comparing digital photographs of the tested sample in different phases of deformation. The system divides the image into small pixels and subpixels. By tracking blocks of pixels, the system can measure surface displacement and create two-dimensional and shaped deformation vector fields and deformation maps. For the DIC to work, the pixel blocks must be helpful and different so that there are different levels of contrast and intensity. They do not require any special lighting and in many cases, the natural surface texture or component has sufficient image texture for the DIC to function without the need for further special surface preparation. However, in the case of a homogeneous uniform surface of the test sample, it is necessary to adjust this sample before the measurement itself and usually, by spraying two contrasting colors on the pattern, it is determined by white and black[4].

978-1-6654-3236-8/21 $31.00 © 2021 IEEE

V. IMAGE PROCESSING METHODES IN VIDEO EXTENSOMETRY

The goal of digital image processing is to obtain useful information from images without human assistance. The process of segmenting images with a complicated structure is one of the most difficult problems in image processing and has been for several decades. active area of research. The main goal of image segmentation is to divide the image into parts that have a strong correlation with the objects or areas of interest contained in the image. When processing an image, useful pixels are separated from others. The result of image segmentation is a group of segments that together cover the entire image or a group of outlines extracted from the image. One of the categories of object segmentation is the division of an image based on the intensity of changes, such as the edges in the image. The second category is based on dividing the image into areas that are similar according to predefined criteria, such as brightness [5].

Thresholding is one of the commonly used methods for image segmentation. This is useful for distinguishing objects from the background. By selecting an appropriate threshold value T, the gray level image can be converted to a binary image. The binary image should contain all the basic information about the position and shape of the objects of interest. The advantage of obtaining the first binary image is that it reduces the complexity of the data and simplifies the recognition and classification process. The most common way to convert a gray level image to a binary image is to select one threshold (T). Then all values of the gray level below this T will be classified as black (0) and the values above T will be white (1)[6].

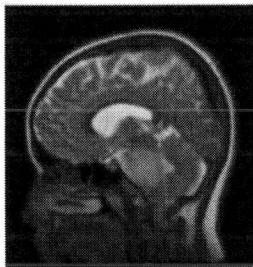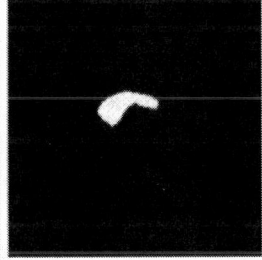

Fig. 5.   Tresholding

Edge detection plays a crucial role in image processing because the main role of image information is stated in the edges of objects. The main part of the information consists of photometric, geometric, and physical properties of objects. The successful completion of the subsequent processing phase is that the edge detection must be consistent. The boundary between overlapping objects and also the boundary between the object and the background is called the edge. Different surfaces of an object may receive different amounts of light, which in turn reflects changes in intensity levels, which are the most important information in an image when detecting edges. The basic step in edge detection processing is performed using one-dimensional and two-dimensional operators such as Roberts, Sobel, Laplacian, Laplacian of Gaussian (LoG), Prewitt, and many others. The most used and effective edge detectors include the Canny detector or the use of mathematical morphology[7,8].

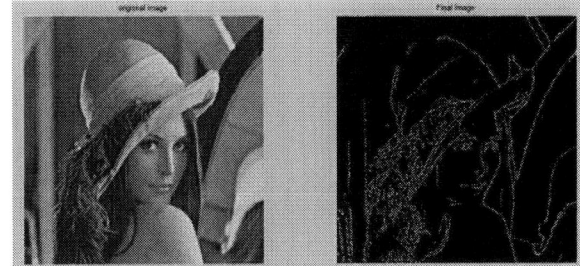

Fig. 6.   Canny edge detection

VI. HARDWARE

When creating a system for video extensometry, one of the key tasks is to choose the right hardware, without which this work could not exist. For video extensometry, it is very important to use the right type of camera, because the accuracy of the entire measurement is largely conditioned by the parameters of the camera.

To obtain as much data as possible during the tensile testing of the material using UTM, we decided to use two cameras. A high-resolution camera along with a high-speed camera. The high-resolution camera provides us with the details of the measured sample, while the high-speed camera captures very fast events during the tensile test that the first camera would not have to capture. Since one of the ideas of this work is to make the solution cost-effective, the cameras we use will be affordable industrial cameras, unlike most video extensometry systems, which use specialized cameras designed for this purpose. When choosing a camera with a high image resolution, we chose the Basler ace acA4600-10uc camera, which is shown in Fig. 7. This camera has a resolution of 4608 x 3288 pixels with a pixel size of 1.4 x 1.4 μm and a frame rate of 10 frames per second. Another advantage of this camera is its small dimensions of 41 x 29 x 29 mm, thanks to which it is easily portable, and also the fact that its exposure time is freely adjustable. This camera is designed for industrial, medical and transportation applications [11].

Fig. 7.   Camera Basler ace acA4600-10uc[11]

On the other hand, a camera with a lower resolution but with a high scanning speed, which is the Ximea MQ003MG CM camera, will also be used in experimental measurements. Although its resolution is only 648 x 488 pixels, its maximum frame rate is up to 500 frames per second, which allows us to observe very small changes in the measurement, which occur

978-1-6654-3236-8/21 $31.00 © 2021 IEEE

in a short time that we would not have noticed with the previous camera[11].

Fig. 8.   Camera Ximea MQ003MG CM[12]

Since we decided to use 2 cameras when designing the system, namely a high-resolution camera and a camera with a high image capture speed, it is very important to synchronize these two cameras correctly, as both cameras have different image capture speeds. To synchronize the cameras, we decided to use an external trigger source for both cameras, as both cameras have the option of an external trigger. This solution will ensure that during the measurement, the recordings from the cameras will be created at the same time and thus there will be no delay in the recording of individual camera images, which would be undesirable during further data processing[12].

VII. ALGORITHM

When obtaining basic information about the measured sample such as its height and width, the basic methods of image processing will be performed, namely, in the first step, thresholding of images, subsequent performance of boundary detection, and finally the calculation itself. Image thresholding is the first step that takes place at the beginning of the data processing itself, where the parameters can be set using the image histogram so that the background data is removed as best as possible and only the data of the measured sample remains. This setting can be difficult, but it is performed only once and constant parameters are set throughout the measurement. The basic step in obtaining basic information will be to use border detectors to produce the contours of the measured sample. The Canny Detector appears to be the most promising edge detector. Because after obtaining information about the sample boundaries in the image, we have information about each contour point of the measured sample, it is not necessary to add various artificial marks on the sample to calculate the dimensional change during the test itself. After obtaining the boundary information, the last step of the first part of the calculation of the sample dimensions will be a graphical record of the elongation and possible narrowing of the sample during the uniaxial tensile test using UTM. The calculation of the length and width of the measured sample during the tensile test depends on obtaining the length and width of the test sample before the measurement. After capturing the sample with a camera, we get the number of pixels that represent the length and width of the real sample,

from which we can determine the dimensions of one pixel in the image. If we know this value, we can determine the length and width of any part of the image at the same distance as the sample from the camera. The initial experimental design of the first part of the algorithm described above, together with a graphical record of the sample elongation, was given in LabVIEW and is shown in Fig. 9.

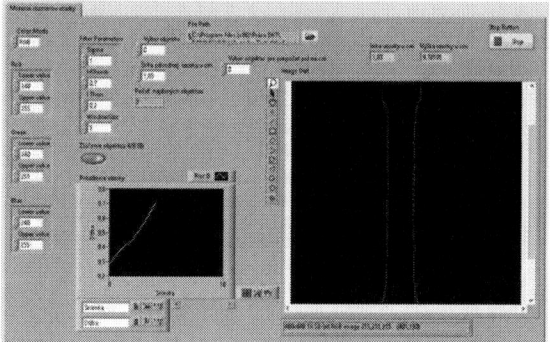

Fig. 9.   Experimental design of the first part of the algorithm

Fig. 9 shows the front panel of a program for measuring the length and width of a test sample. In this part of the program it is possible to set the image threshold as well as to select the edge detection parameters. After setting the threshold and edge detection, the algorithm calculates the dimensions of the test sample from the image.

The second part of the algorithm, which is used to find cracks and artificial patterns on the test sample, is shown in Fig. 10.

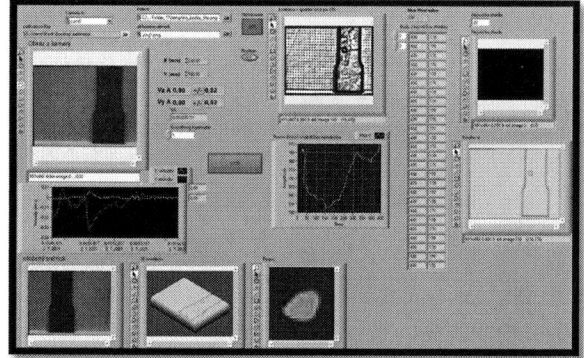

Fig. 10.   Experimental design of the second part of the algorithm

This part of the algorithm consists of the interconnection of DIC correlation and optical flow. After thresholding and edge detection, the DIC algorithm searches for the position of the added points on the tested sample, or the position of cracks from the images from the high-resolution camera. After obtaining the position of the selected points, the optical flux for the ROI is calculated from the center of which the position was calculated using DIC correlation. The calculation of the optical flux is performed on the images from the camera with a high scanning speed, which allows us to calculate the movement and propagation of cracks on the tested sample in real-time.

978-1-6654-3236-8/21 $31.00 © 2021 IEEE

VIII. EXPERIMENTAL MEASUREMENT

The tension of the experimental sample was recorded with the help of cameras, which was then stretched in the vertical direction. This sample contained various additional points to simulate motion. First, the position of these additional points was calculated using DIC correlation, as can be seen in Fig. 11.

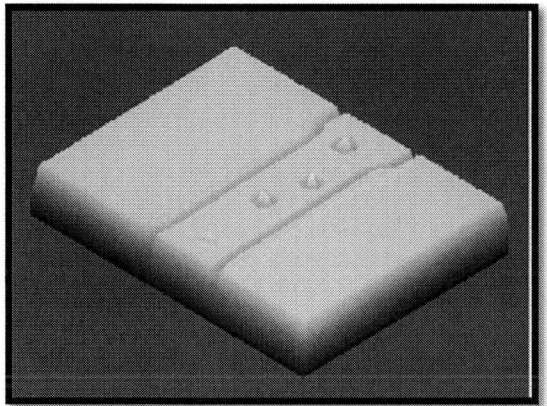

Fig. 11. Experimental design of the second part of the algorithm

Subsequently, the movement of these points was determined using optical flow, which can be seen in Fig. 12.

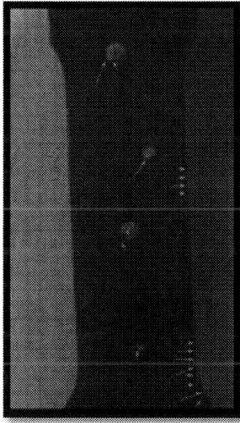

Fig. 12. Experimental design of the second part of the algorithm

Because the position of the selected points was determined using DIC correlation, it is not necessary to perform the calculation of the optical flow for the whole image, but only for the selected area, which significantly shortens the calculation time.

IX. CONCLUSION

When creating a new non-contact measuring method based on special visual systems for measuring the strength of the tested material, an overview of the current conditions of these measuring methods is required.

One or more image processing methods are required for the non-contact measurement method to work. In our work, we focused on the combination of DIC correlation and optical flow and the use of two cameras. The DIC correlation itself is accurate but difficult to calculate. On the other hand, the optical flow method is fast but less accurate than the DIC correlation method. By combining these two methods, we have achieved a relatively accurate detection of material parameters that can be performed in real-time using standard industrial cameras.

ACKNOWLEDGMENT

This study was funded by grants:

APVV 15-0462 - Research on sophisticated methods for analysing the dynamic properties of respiratory epithelium's microscopic elements.

APVV 17-0218 - Research of interaction mechanism of biological tissues with high-frequency electromagnetic field and its application in the development of new procedures in the design of electrosurgical devices.

APVV 20-0500 - Research of methodologies to increase the quality and lifetime of hybrid power semiconductor modules.

REFERENCES

[1] M.Kašuba, 'ANALYSIS OF THE MINIATURE TEST SPECIMEN RESULTS WITH VARIABLE GEOMETRIES'', available on the Internet: Íhttps://www.vutbr.cz/www_base/zav_prace_soubor_verejne.php?file _id=15103

[2] N. McCormick, J. Lord, '' Digital image corelation'', MATERIALSTODAY, 2010, Vol. 13 available on the Internet: https://www.sciencedirect.com/science/article/pii/S136970211070235 2

[3] Q. B. Tao, L. Benabou, N. H. Tran and D. B. Luu, "A Digital Image Correlation setup for the analysis of lead-free solder alloys," 2017 International Conference on System Science and Engineering (ICSSE), Ho Chi Minh City, 2017, pp. 404-407, doi:10.1109/ICSSE.2017.8030906.

[4] J. Pelegrin, W. O. S. Bailey and D. A. Crump, "Application of DIC to Extract Full Field Thermo-Mechanical Data From an HTS Coil," in IEEE Transactions on Applied Superconductivity, vol. 28, no. 4, pp. 1-5, June 2018, Art no. 4604305, doi: 10.1109/TASC.2018.2807376.

[5] J. Singaraju, K. Bhargavi,'' A Survey on Threshold Based Segmentation Technique in Image Processing'', ResearchGate, Vol.3, (2014), available on the Internet: https://www.researchgate.net/publication/309209325_A_Survey_on_ Thr eshold_Based_Segmentation_Technique_in_Image_Processing

[6] E. F. Aqeel, '' The Use of Threshold Technique in image segmentation'', Al-Mustansryah University, Collage of Education Computer Department, available on the Internet: https://www.iasj.net/iasj?func=fulltext&aId=101683

[7] D. Wang, H. Liu, "Edge detection of cord fabric defects image based on an improved morphological erosion detection methods," 2010 Sixth International Conference on Natural Computation, Yantai, 2010, pp. 3943-3947, doi: 10.1109/ICNC.2010.5584778.

[8] Lihui Jiang, "Image edge detection based on fuzzy weighted morphological filter," 2009 International Conference on Machine Learning and Cybernetics, Hebei, 2009, pp. 690-693, doi: 10.1109/ICMLC.2009.5212384.

[9] MatNet Slovakia, '' Static tensile test '', available on the Internet: http://www.matnet.sav.sk/index.php?ID=526

[10] Lee and col.,'' Stress Estimation Using Digital Image Correlation with Compensation of Camera Motion-Induced Error'',Sensors, 19 vol., 2019, available on the Internet: https://www.mdpi.com/1424-8220/19/24/5503/htm

[11] Edmund Optics Inc,, ''Basler ace acA4600-10uc Color USB 3.0 Camera'', dostupné na internete: https://www.edmundoptics.com/p/basler-ace-aca4600-10uc-color-usb30-camera/32436/

[12] Ximea, ''Ximea MQ003MG-CM'',available on the Internet: https://www.ximea.com/en/products/usb3-vision-cameras-xiq-line/mq003mg-c

ICE Vehicle Energy Consumption Measurement and Calculation Methodology for the Purpose of EV Battery Pack Design

Antonio Peršić
Department of Electric Machines, Drives and Automation
Faculty of Electrical Engineering and Computing
University of Zagreb
Zagreb, Croatia
antonio.persic@fer.hr

Vladimir Peršić
Chief Executive Officer
Novatec d.o.o.
Labin, Croatia
v.persic@novatec.hr

Hrvoje Kristek
Department of Electric Machines, Drives and Automation
Faculty of Electrical Engineering and Computing
University of Zagreb
Zagreb, Croatia
hrvoje.kristek@fer.hr

prof. dr. sc. Mario Vražić
Department of Electric Machines, Drives and Automation
Faculty of Electrical Engineering and Computing
University of Zagreb
Zagreb, Croatia
mario.vrazic@fer.hr

Abstract—**This paper presents a standard vehicle's energy consumption calculation for the purpose of an electric vehicle (EV) battery pack design procedure. The vehicle dynamics parameters were measured on a route located on a mountainous terrain with a standard driving profile. Vehicle used to conduct the measurements, was a standard internal combustion engine (ICE) vehicle. Afterwards the measured parameters have been used as inputs to a vehicle dynamics model to calculate the resistive forces "on the wheels" that the vehicle had to overcome to achieve the desired driving profile on the specified route. Those forces are then used to calculate the average energy consumption of the vehicle on the mentioned route, only to be scaled to the battery pack with the use of the powertrain model. Furthermore, the consumptions of the vehicle's onboard electrical devices are added to receive the complete energy consumption of the vehicle on the specified route with the desired driving profile. The before mentioned average energy consumption (in kWh/km), can be then used for further application of this vehicle model in different scenarios.**

Keywords— electric vehicle, battery pack, powertrain, gps module, vehicle dynamics, auxiliary on board devices, energy consumption.

I. INTRODUCTION

Today, transportation accounts for almost 30% of all greenhouse gas (GHG) emissions in the United States [1]. In 2017, 27% of total EU greenhouse gas emissions came from the transport sector (22% if international aviation and maritime emissions are excluded). CO2 emissions from transport increased by 2,2% compared with 2016 [2].

To prevent further damage to the environment, the transportation industry needs to switch from an ICE to other means of propulsion. Combined with the battery package, the electric motor can be a worthy successor to the ICE. Instead of gasoline, it uses electrical energy to achieve propulsion. The completion of the electrification process is only a small part of the process, along with an energy charging infrastructure, the crucial transformation needs to be done to the electrical power plant system. The only way to make an electric vehicle fully eco-friendly is to switch from fossil fuel power plants to renewable energy power plants as a charging medium. The vehicle is as eco-friendly as the energy that powers it.

Although the further increase in the number of electric vehicles with its current battery cell chemical composition proves to be harmful to the environment, the further development in the battery cell manufacturing technology and recycling process looks promising. The further increase in the number of electric vehicles in the world, with more renewable energy on the grid and new breakthroughs in the battery cell industry, will result in in a global reduction of CO2 and GHG [3].

With this in mind, this paper is based on the process of measuring and calculating the energy consumption of an ICE vehicle for the purpose of designing a battery package for is equivalent counterpart.

II. EXPERIMENTAL MEASUREMENT

This chapter covers the methods, assumptions and devices used to conduct the vehicle dynamics measurement and form the vehicle dynamics model.

A. Electric Vehicle Configuration

The two main parts of an electric vehicle is the battery package along with its charger and the powertrain which is responsible not only to generate mechanical energy to propel the vehicle but also receive the potential energy that can be used for regenerative braking and charging the battery pack. The initial specification for the electric vehicle is shown in Fig.1.

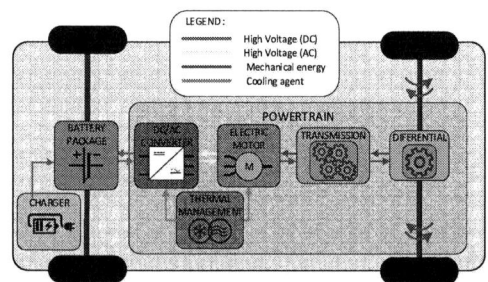

Fig. 1. Electric vehicle power-train

The powertrain consists of a DC/AC converter that converts DC power to controllable AC power, and a

978-1-6654-3236-8/21 $31.00 © 2021 IEEE

synchronous AC motor with permanent magnets connected to the wheel shaft via a differential and transmission block. [4,5]

The electric vehicle configuration will prove important in the following chapters, where the energy consumption calculated on the wheels is recalculated to the battery pack.

B. Driving Profile and Route

The conventional way of designing a battery pack begins with a simulation of the route and a driving profile, as a result the designers use a simulated global homologation drive cycle for urban and extra urban areas. But these can result as inaccurate, because most of the widely available documented drive cycles, like the New European Driving Cycle, fail to include elevation changes and overcoming steep inclines. So the designers usually simulate the incline with a linear function. [6,7]

The more accurate way would be to measure an actual drive cycle with the desired driving profile adapted to a specific route, terrain and at the end to the electric vehicle type.

The selected route covered in this paper is located on a mountainous terrain with sudden elevation change, with steep incline and decline levels. This results in a great increase in energy consumption even at low velocity, not to mention the high current peaks caused by the motor to overcome the before mentioned loads. The maximum current parameter needs to be carefully calculated so that the designed battery pack can discharge enough instantaneous current to enable the selected motor to overcome those load. The best experiment to carry out these calculation was done by accelerating a fully stopped vehicle on a steep incline level.

The selected terrain is located in Croatia on the peninsula of Istria, and the circular route is driven between two small cities, one situated at sea level and the other on top of a mountainous terrain. This results with an elevation difference of 220 m and the total distance of 26 km, driven in 1h 8min with the average velocity of 30 km/h that peaks at 60 km/h, with intermittent stops.

C. Measurement Device

The initial data was acquired using a high frequency GPS module named "PhidgetGPS" manufactured by Phidgets Inc. The GPS module's antenna was mounted to the vehicle's roof to ensure undisturbed communication to the satellite.

The GPS module itself is connected to a PC via USB, where the data was logged to a CSV (Comma Separated Value) file in the LABVIEW monitoring environment. With a sample rate of 10 Hz and a Circular Probable Error (Best Case) of 2,5 m which means that every sample has a 50% chance of being accurate in a circle of that radius. The maximum error in velocity measurement is 100 mm/s with a maximum timing error of 300 ns.

The following variables were extracted from the "PhidgetGPS" module: latitude, longitude, altitude, velocity, heading and time, as they proved sufficient for the energy consumption calculation. The data logs have been imported into the MATLAB programming environment in which the traveled distance, acceleration, road incline, resistive forces, developed power, energy consumption etc. have been calculated.

III. Energy Consumption Calculation Method

To calculate the total and average energy consumption of the vehicle, the measured data needs to be processed into total traction force that the vehicle is acted upon and/or also needs to overcome. After which the power and energy consumption can be calculated.

A. Force Calculation

The goal of this subchapter is to calculate the resultant traction force $F_{tot}[N]$ that acts on the vehicle and/or the loads that the vehicle needs to overcome. The following expression shows that the resultant traction force is the sum of all the resistive forces, gravitational and acceleration forces that the vehicle is subjected to,

$$F_{tot} = F_a + F_s + F_r + F_a \qquad (1)$$

where F_a [N] is the acceleration/deceleration force, F_s [N] the road slop force, F_r [N] is the road load resistive force, F_a [N] aerodynamic drag resistive force. [5]

The acceleration/deceleration force F_a [N], acording to Newton's second law is calculated with the following equation,

$$F_a = m_v \cdot a_v \qquad (2)$$

where m_v [kg] is the total vehicle mass and a_v [m/s2] the vehicle's acceleration.

The vehicle's acceleration/deceleration can be calculated with the following equation,

$$a_v = \frac{\Delta v}{\Delta t} \qquad (3)$$

where Δv [m/s] is the velocity difference and Δt [s] the time difference in a short time segment between samples.

The road slope force F_s [N] is calculated by transferring the gravitational force that acts on the vehicle to the axis of the vehicle's movement, as follows,

$$F_s = m_v \cdot g \cdot \sin(\alpha_s) \qquad (4)$$

where g [m/s2] is the gravitational acceleration and α_s [rad] road slope angle.

The road load (friction) force F_r [N] is given with the following expression,

$$F_r = m_v \cdot g \cdot c_{rr} \cdot \cos(\alpha_s) \qquad (5)$$

where c_{rr} [-] is the road rolling resistance coefficient.

The aerodynamic drag force is given by the expression:

$$F_a = \frac{1}{2} \cdot \rho \cdot c_d \cdot A \cdot v_v^2 \qquad (6)$$

Where ρ [kg/m3] is the air density at 20 °C, c_d [-] is the air drag coefficient, A [m2] the vehicle frontal area and v_v [m/s] the velocity of the vehicle.

The vehicle frontal area A [m2] is calculated with the electric vehicle height h_v [m] and width w_v [m].

$$A = h_v \cdot w_v \qquad (7)$$

978-1-6654-3236-8/21 $31.00 © 2021 IEEE

B. The Efficiency of the Electric Vehicle

Before this chapter introduces the energy consumption calculation, the electric vehicle efficiency needs to be mentioned. This efficiency calculation is based on the electric vehicle configuration shown in Fig. 1.

The measured vehicle dynamic forces and calculated power and energy, need to be recalculated from the position they were measured (the wheels) to the battery pack, this way the electrical energy consumption and regeneration from and into the battery pack can be calculated. This is done using the efficiency of each component that transfers the energy from the wheels to the battery pack, and taking each component's energy loss into account.

It should be taken into account that there are two components of the energy when talking about EV's: the energy that the vehicle consumed to drive the route and the potential energy that the vehicle received on the mentioned route.

During energy consumption the total efficiency is calculated as in the following equation,

$$\eta_{tot,p} = \eta_{dif} \cdot \eta_{red} \cdot \eta_{mot} \cdot \eta_{inv} \cdot \eta_{bat.dis} \qquad (8)$$

where η_{dif} [-] is the efficiency of the differential, η_{red} [-] is the efficiency of the gearbox, η_{mot} [-] is the electric motor efficiency in drive mode, η_{inv} [-] is the efficiency of the frequency converter in drive mode, $\eta_{bat,dis}$ [-] is the efficiency of the battery pack during discharge. [4]

In energy regeneration mode the total efficiency is given in the following equation,

$$\eta_{tot,r} = \eta_{dif} \cdot \eta_{red} \cdot \eta_{mot,gen} \cdot \eta_{inv,gen} \cdot \eta_{bat,chg} \qquad (10)$$

where $\eta_{mot,gen}$ [-] is the electric motor efficiency in generator mode, η_{inv} [-] is the efficiency of the frequency converter in generator mode, $\eta_{bat,chg}$ [-] is the battery pack charging efficiency.

C. Power and Energy

The total power P_{tot} [W] that the vehicle needs to deliver/consume on the wheels is calculated as the product between the resultant traction force and the vehicle velocity as shown in the following expression.

$$P_{tot} = F_{tot} \cdot v_v \qquad (11)$$

The total energy that the vehicle consumes and receives on the given route E_{tot} [Wh] is calculated as an integral of the total power in the driven period of time, as in the following expression.

$$E_{tot} = \int P_{tot} \cdot dt \qquad (12)$$

When in the discrete domain, the integration function is substituted with a summation function, as in the equation 10.

$$E_{tot} = \sum P_{tot} \cdot \Delta t \qquad (13)$$

The energy consumed by the vehicle on the driven route $E_{tot,p}$[N] is calculated by integrating (summing) the positive values of the total power in the driven period of time, as seen in equation (13). Positive values are taken into account because they represent the power that was consumed to achieve movement.

$$E_{tot,p} = \sum P_{tot} \cdot \Delta t \qquad | P_{tot} > 0 \qquad (14)$$

On the other hand, the potential energy that the vehicle receives from inertia and negative road slope $E_{tot,n}$[N] is calculated by integrating (summing) the negative values of the total power in the driven period of time, as seen in the following equation. Negative values are taken into account

$$E_{tot,n} = \sum P_{tot} \cdot \Delta t \qquad | P_{tot} < 0 \qquad (15)$$

because they represent the potential energy that acts on the vehicle as a force that "helps" the vehicle to move.

Furthermore, the before mentioned potential energy that is not wasted on mechanical braking, can be used in the function of regenerative braking, where the electric motor acts as a generator to produce power that can be used to charge the battery pack. Therefore, the regenerative energy that is not wasted on braking is calculated in the following expression,

$$E_{tot,r} = E_{tot,n} \cdot k \qquad (9)$$

where k [−] is the regenerative braking factor (0 = the total potential energy is consumed for mechanical braking, 1 = the total potential energy is used for regenerative braking). This factor is used to estimate the share of the potential energy that is not wasted on mechanical braking.

The energy consumption from equation (13) and its regenerative counterpart from equation (15) combined with the corresponding efficiencies, result in the following expression.

$$E_{tot,bat} = \frac{E_{tot,p}}{\eta_{tot,p}} + E_{tot,r} \cdot \eta_{tot,r} \qquad (16)$$

This is the total energy consumed from the battery package to complete the mentioned route $E_{tot,bat}$[W]. Keep in mind that the energy is divided into two components regenerative and consumed.

D. Energy Consumption of Auxiliary Devices

To calculate the total energy consumption of an electric vehicle, the energy used to run the onboard auxiliary devices E_{aux}[Wh] needs to be taken into consideration,

$$E_{aux} = P_{aux} \cdot t \qquad (17)$$

and it is calculated as a product of their power consumption P_{aux}[W] over the driving time. [9]

Furthermore, like in equation (16), the auxiliary devices energy consumption needs to be scaled to the battery pack.

$$E_{aux,bat} = \frac{E_{aux}}{\eta_{bbat,dis} \cdot \eta_{DCDC}} \qquad (18)$$

where η_{DCDC} [-] is the DC/DC converter efficiency and $\eta_{bat,dis}$ [-] is the efficiency of the battery pack during discharge.

With that in mind, the total energy consumed on the driven route $E_{EV,bat}$ [Wh] in the battery pack is calculated in the following expression.

$$E_{EV,bat} = E_{tot,bat} + E_{aux,bat} \tag{19}$$

By scaling the total energy consumed with the total driven distance, it results with the average energy consumption of the vehicle in the battery pack E_{avg} [Wh/km], as shown in the following equation.

$$E_{avg} = \frac{\max(E_{EV,bat}(t))}{\max(s(t))} \tag{20}$$

This value contains the averaged driver's profile, vehicle auxiliary and drive consumption, along with regenerative braking on the desired route. [8]

By multiplying the value from equation (20) with the desired vehicle range, an accurate energy consumption estimate would be calculated for that route, i.e. the battery pack's required energy $E_{bat,req}$ [Wh], as shown in the following equation.

$$E_{bat,req} = E_{avg} \cdot s_{req} \tag{21}$$

IV. EXPERIMENTAL RESULTS

This chapter contains the results of the experiment. The following figure contains the satellite view of the driven route.

Fig. 3. Route satelite view

It can be seen that Fig. 3 matches the plotted route in MATLAB, as shown in the figure below.

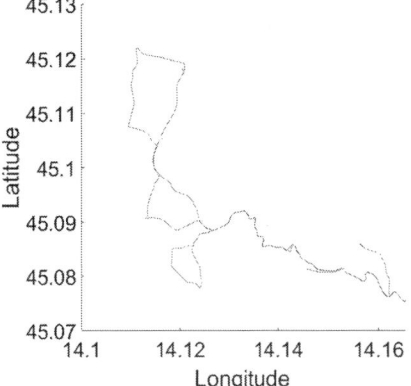

Fig. 4. Route plot in MATLAB

The calculated velocity was filtered with a moving average filter, and it can be seen in the following figure. With the average velocity recorded at 30 km/h, and the maximum velocity at 60 km/h.

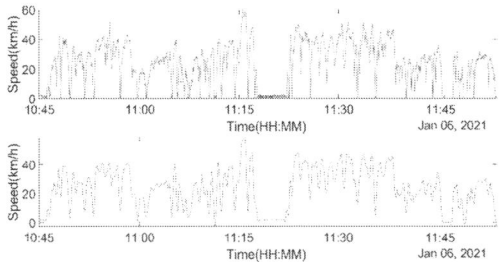

Fig. 2. Velocity on the route

The measured altitude was filtered with a moving average filter, and it can be seen in the following figure. With the minimum altitude recorded at 0 m at sea level, and the maximum altitude at 220 m above sea level.

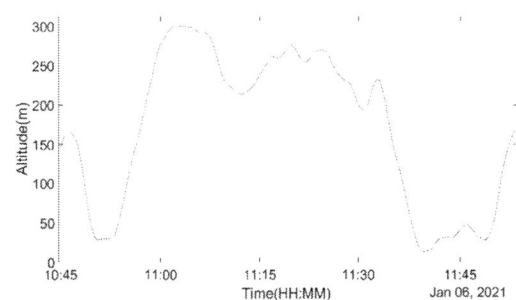

Fig. 5. Altitude on the route

978-1-6654-3236-8/21 $31.00 © 2021 IEEE

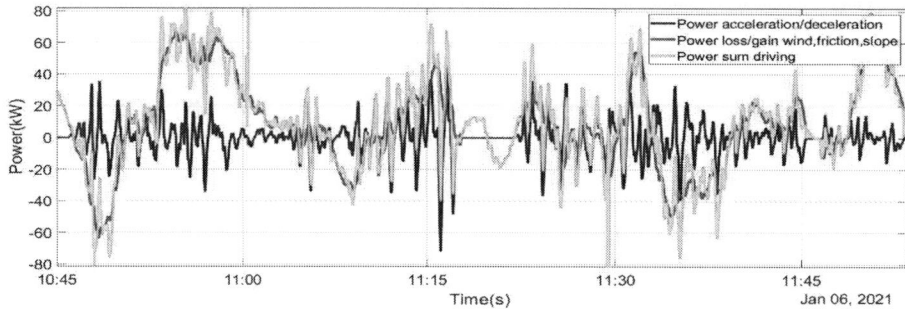

Fig. 6. Power graph

The power consumption/regeneration on the vehicle's wheels is shown in the figure above.

Positive values represent the power that was consumed to achieve movement, where the negative values represent the potential energy that can be used for regenerative braking.

It can be clearly seen that the green line (Total Power) is made of the blue line (Power for acceleration/deceleration) added with the red line (Power to overcome the road load). The power required to overcome the road load has the highest energy requirement, where the highest current will be required.

The energy consumption waveforms of the battery pack can be seen in the following figure.

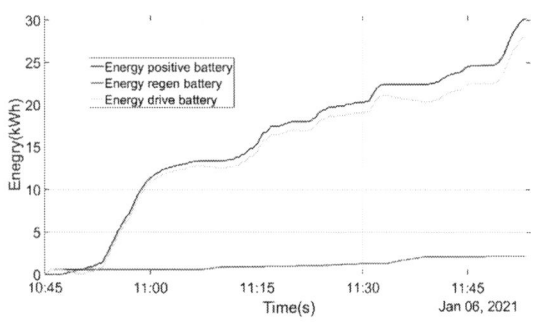

Fig. 7. Energy cumulative sum

The regenerative component (red line) is 3,31 kWh, the drive and auxiliary devices component (blue line) are 32,69 kWh and the total (green line) 29,38 kWh.

The average energy consumption of the drive component is 1,13 kWh/km, regenerative 0,06 kWh/km, auxiliary 0,16 kWh/km, and the total 1,22 kWh/km.

V. CONCLUSION

The purpose of this paper is to present an accurate method of calculating the electrical energy consumption of the vehicle. Instead of using the conventional procedure with the homologian drive cycles and simulated elevation, this paper uses a method based on the measurement of the vehicle's dynamics on a preset route. The measured parameters were used as inputs in a vehicle dynamics model, to calculate the energy consumption.

The myth of the highly efficient electric vehicle has been broken with the efficiency from the wheels to the battery pack

at 60%, and to the grid at about 50%, which is not significantly different from an ICE vehicle efficiency.

Also the regenerative component has been calculated at about 10% of the total energy consumed by the drive and auxiliary devices. So this means that only 10% of the energy that is invested can be regained to charge the battery pack.

This method has provided information to calculate the battery pack's size, and can be used on any electric vehicle. It has achieved an averaging of the route terrain, driving profile and vehicle's specification into one variable (average energy consumption), and it can be scaled to any range that the electric vehicle needs to be able to drive trough.

REFERENCES

[1] '4 Emerging Ways to Pair Electric Vehicles and Renewable Energy', Accessed on: May 13, 2021. [Online]. Available: https://www.wri.org/insights/4-emerging-ways-pair-electric-vehicles-and-renewable-energy

[2] 'Greenhouse gas emissions from transport in Europe', Accessed on: January 22, 2021. [Online]. Available: https://www.eea.europa.eu/data-and-maps/indicators/transport-emissions-of-greenhouse-gases/transport-emissions-of-greenhouse-gases-12

[3] Gaines, Linda L. (2014). Lithium-Ion Batteries || Lithium-Ion Battery Environmental Impacts. , (), 483–508. doi:10.1016/B978-0-444-59513-3.00021-2

[4] Hofman, T.; Dai, C.H. (2010). [IEEE 2010 IEEE Vehicle Power and Propulsion Conference (VPPC) - Lille, France (2010.09.1-2010.09.3)] 2010 IEEE Vehicle Power and Propulsion Conference - Energy efficiency analysis and comparison of transmission technologies for an electric vehicle. , (), 1–6. doi:10.1109/VPPC.2010.5729082

[5] Ali Emadi, Florence Berthold, "Advanced electric drive vehicles". US, CRC Press, 2017. pp. 254

[6] G. Du, W. Cao, S. Hu, Z. Lin and T. Yuan, "Design and Assessment of an Electric Vehicle Powertrain Model Based on Real-World Driving and Charging Cycles," in IEEE Transactions on Vehicular Technology, vol. 68, no. 2, pp. 1178-1187, Feb. 2019, doi: 10.1109/TVT.2018.2884812.

[7] Teoh, Jia Xian; Stella, Morris; Chew, Kuew Wai (2019). [IEEE 2019 IEEE 9th International Conference on System Engineering and Technology (ICSET) - Shah Alam, Malaysia (2019.10.7-2019.10.7)] 2019 IEEE 9th International Conference on System Engineering and Technology (ICSET) - Performance Analysis of Electric Vehicle in Worldwide Harmonized Light Vehicles Test Procedure via Vehicle Simulation Models in ADVISOR. , (), 215–220. doi:10.1109/ICSEngT.2019.8906356

[8] Kroeze, Ryan C.; Krein, Philip T. (2008). [IEEE 2008 IEEE Power Electronics Specialists Conference - PESC 2008 - Rhodes, Greece (2008.06.15-2008.06.19)] 2008 IEEE Power Electronics Specialists Conference - Electrical battery model for use in dynamic electric vehicle simulations. , (0), 1336–1342. doi:10.1109/pesc.2008.4592119

[9] Tom Denton, Automobile Electrical and Electronic Systems, Third edition. Elsevier Butterworth-Heinemann, 2004, page 129.

Sensorless speed control of brushed DC machine

Michal Vidlak
Department of Power Systems and Electric Drives
University of Zilina
Zilina, Slovakia
michal.vidlak@feit.uniza.sk

Lukas Gorel
Department of Power Systems and Electric Drives
University of Zilina
Zilina, Slovakia
lukas.gorel@feit.uniza.sk

Vladimir Vavrus
Department of Power Systems and Electric Drives
University of Zilina
Zilina, Slovakia
vladimir.vavrus@feit.uniza.sk

Pavol Makys
Department of Power Systems and Electric Drives
University of Zilina
Zilina, Slovakia
pavol.makys@feit.uniza.sk

Abstract— The automotive industry is ramping in volume. New models of battery-powered vehicles are introducing to the mass market customer on a monthly based. Upcoming automobiles are using more than 60 rotating electric machines. The dominant part of this amount consists of brushed DC motors dedicated to the car body segment. Cost reduction and simple control are the main reasons why manufacturers prefer using DC machines in applications like window lift, seat positioning, or sunroof control. In addition, safety in the automotive segment pushes control strategies of DC machines toward sensorless operation. In this article, the current ripple caused by commutation is used for accurate sensorless speed control. The new method is analyzed and compared with the traditional encoder-based speed control cascade structure. The real-time control system was developed at dSPACE 1103 and tested using an automotive-qualified H-bridge.

Keywords—sensorless, brushed DC machine, current ripple, bandpass filter

I. INTRODUCTION

A signal processing algorithm of DC machine current, for accurate speed calculation, will be suggested in this article. The proposed design is targeting the well-known phenomenon ripple current caused by DC machines commutator. The commutator, as well known as a mechanical rectifier, is causing ripples in DC machine current during the whole machine working cycle. The ripple component of current is proportional to speed. Hence there is a speed-dependent phenomenon leading towards direct speed observation. The number of ripples depends on the number of pole pairs, commutator segments, and their parity. These parameters are constant and depend on the DC machine construction.

Strategy based on the counting of these ripples can sufficiently deliver accurate information about actual DC machine speed. The current ripple counter is designed in this paper together with a bandpass filter with variable bandwidth. The proposed algorithm combination leads to the separation of specific frequencies typical for actual DC machine speed. Frequency can be further transferred to speed information

while the number of commutator segments is a known parameter.

The automotive body segment is using DC machines for driver comfort applications. Window lifter, sunroof, or other applications which require knowledge about the position creates the main application field for the proposed method. Currently most frequently used methodology for DC machine speed estimation in the automotive industry is revolution counter based on Hall effect probe. The proposed method can replace the existing Hall effect probe and bring more accurate actual speed estimation. The advantages and limitations of this approach are present in this paper.

The sensorless algorithms for brushed DC motor can be divide into two groups. To the first group belongs the algorithms based on the dynamic model of the DC motor. The second group is composed of the algorithms based on the ripple component of the brushed DC motor current. Hence, the sensorless methods lying in the second group are independent of the motor parameters and their variations with temperature and so on.

The methods based on the dynamic model of the DC motor depend on the parameters of the brushed DC motor, such as the resistance, inductance, and back-emf constant. However, these parameters are not constant, and they vary under different operational conditions, for example, with temperature. Varying parameters then lead to uncertainty in speed estimation. Nevertheless, these parameters can be estimated dynamically [1], but this approach usually leads to a nonlinear model, which is more complex, and increased computational time is required. Some techniques indirectly model a motor with the neural networks [2] or the Kalman filter [3]. The combination of neural networks and sliding mode control is used in [4]. Sensorless speed control for very small brushed DC motors with an adaptive estimator is present in [5].

The methods based on the ripple component of motor current do not require any knowledge of motor parameters to obtain information about the actual rotor speed. The motor current already contains this information. The measured motor current is mainly composed of two components. The first component is a DC component, which is responsible for providing torque to the motor. Its amplitude depends on a motor load. The second one is an AC component, also known as the ripple component. The AC component is created by

---

Slovak Scientific Grant Agency VEGA 1/0795/21 and the European Regional Development Fund and the Operational Program Integrated Infrastructure 2014–2020 of the project: Innovative Solutions for Propulsion, Power and Safety Components of Transport Vehicles, code ITMS 313011V334.

978-1-6654-3236-8/21 $31.00 © 2021 IEEE

converting the DC current supplied by a stationary source into an AC current in an armature coil by the commutator and brushes. It should be noted that the AC component contains a ripple, which is directly proportional to the rotor speed. Since the number of ripples in the measured motor current per one rotation is constant, the rotor speed can be estimated. Some methods use the ripple component of the brushed DC motor to estimate the rotor speed.

An adaptive filter was used to estimate the rotational speed in [6], specifically Adaptive Line Enhancer (ALE). Simulation studies of the ALE are present within the paper but without experimental verification. The authors stated in the article that they would like to examine the range and the accuracy of the estimated speed as well as the load effect in practice.

The support vector machines were applied in [7]. The rotor speed was estimated using the inverse distance between the detected pulses, and the position was estimated by counting all detected pulses. The paper shows good results, but there is no explanation for why the results were not present for the low speeds, specifically under 501 rpm. In the article, a response of the proposed method to the applied load is not available either. The paper presents an interesting sensorless method with speed detection from ripple current but without controlling the speed. The complexity of the proposed method also requires a microcontroller with higher computational power due to significant amount of mathematical operations.

The method based on measuring the inductive spikes generated when the motor is turned off was proposed in [8]. The estimated speed was used as a feedback value to the speed controller, while the speed was measured in a steady-state.

Another approach [9] used the spectral components of the motor current. This method is also convenient for the brushed DC motors with a large number of coils where the ripple component is almost negligible. The proposed method was tested within a small speed range. The proposed approach contains a lot of mathematical operations, which could lead to higher computational requirements.

Texas Instruments published an application note [10] where an analog bandpass filter was used to count the current ripples in the form of pulses, which requires the use of additional hardware. This approach used the pulse counting technique without controlling the motor speed or position.

Microchip made a similar application note [11], where a ripple counting technique was proposed. The software implementation of the proposed sensorless method to the microcontroller unit (MCU) is provided in the application note. The main goal of this counting implementation is to count all the ripples until the number of expected ripples is reached. This approach is very suitable to control the motor position by counting pulses. However, a way to control the rotor speed was not the scope and purpose of the application note, so it was not present.

In [12], the comparison between the model-based and non-model-based methods is present. The model-based methods experimentally verified in the article are specifically the pseudo-sliding mode observer and the observer with a PI controller. However, limited accuracy was detected for these methods. According to this fact, they were compared to the non-model based method.

In this article, the proposed method is implemented and tested on the dSPACE system. This non-model-based sensorless approach uses the bandpass filter with variable bandwidth. The bandwidth variation is established using a look-up table to provide sufficient current filtering. Onwards, a filtered current is compared to a reference value, which is set to zero. The measured time between two consecutive zero crossings holds information about the actual speed of the brushed DC motor. The dSPACE DS1103 platform and automotive-qualified H-bridge MC33931 integrated circuit dedicated to the car body segment is used to offer experimental results.

II. BRUSHED DC MACHINE

A. Dynamic model of DC machine

To be able to analyze the behavior of a brushed DC machine, it is necessary to create a mathematical model and equivalent circuit. The equivalent circuit of the DC machine is in Fig. 1.

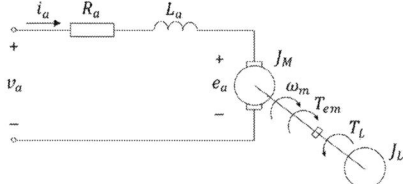

Fig. 1. Equivalent circuit of DC machine.

The mathematical model consists of the following equations (1), (2), (3), (4):

$$v_a = R_a i_a + L_a \frac{di_a}{dt} + e_a \qquad (1)$$

$$e_a = k_E \omega_m \qquad (2)$$

$$T_{em} - T_L = J_{eq} \frac{d\omega_m}{dt} \qquad (3)$$

$$T_{em} = k_T i_a \qquad (4)$$

where R_a represents armature resistance, L_a armature inductance, i_a armature current, u_a terminal voltage, k_E is back-emf constant, ω_m is shaft speed, e_a is back-emf voltage, T_{em} electromagnetic torque, T_L load torque, J_{eq} is the total effective value of the combined inertia of the motor J_M and the inertia of the mechanical load J_L and last but not least constant k_T represents torque constant. The following block diagram (Fig. 2) was created based on (1 - 4).

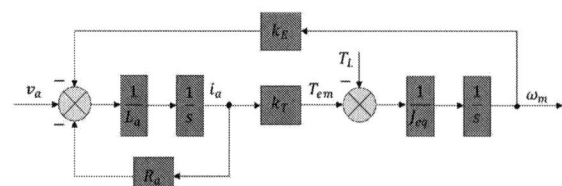

Fig. 2. Block diagram of DC machine.

The standard dynamic model does not include the commutator effect, which is fundamental for proper observer design. The following section briefly describes the principle of current ripple generation.

B. Ripple component of brushed DC machine

The commutation is an essential part of the brushed DC machine working principle. Due to the interaction between the poles and commutator segments, the current ripple is present in the brushed DC motor current. The number of ripples in the motor current is contingent on the three constructional parameters given by (5):

$$number\ of\ ripples_{rot} = c \cdot k \cdot p \qquad (5)$$

where c is the coefficient contingent to the parity of the number of commutator segments k, with $c = 1$ when k is even and $c = 2$ when k is odd and p is the number of pole pairs. Fig. 3 exhibits the experimentally measured brushed DC machine current.

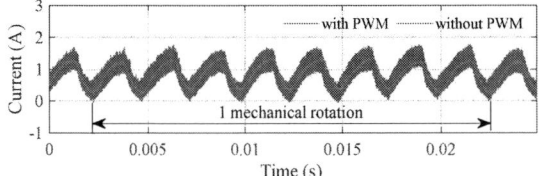

Fig. 3. Brushed DC machine current.

This current was measured on the motor used in the automotive industry as a window lifter. The same one is used to provide experimental results. Constructional parameters of this machine related to the commutation are introduced within TABLE I.

TABLE I.   PARAMETERS OF THE BRUSHED DC MACHINE RELATED TO THE COMMUTATION

Parameter	Symbol	Value
Coefficient of the parity	c	1
Number of commutator segments	k	8
Number of pole pairs	p	1

As is prompt in Fig. 3, one mechanical rotation is given by eight ripples in the motor current. The range between the first and the last ripple is about 0.02 s. The facts associating the commutation with the current ripple are proven by introducing this number to (6):

$$n_m = \frac{60}{T} = \frac{60}{0.02} = 3000\ rpm \qquad (6)$$

The frequency of the current ripple f_r can be calculated concerning the parameters introduced in TABLE I. and actual rotational speed n_m as follows [13]:

$$f_r = \frac{c \cdot k \cdot p \cdot n_m}{60} \qquad (7)$$

Equation (7) shows that for determining an accurate value of the current ripple frequency, it is necessary to know the number of commutator segments and the number of pole pairs. The crucial fact is that these parameters are constant. Hence, they are not changing during motor operation.

Concerning the parameters of the experimental brushed DC motor (TABLE I.) and the actual rotational speed from (6), the current ripple frequency can be acquired as:

$$f_r = \frac{c \cdot k \cdot p \cdot n_m}{60} = \frac{1 \cdot 8 \cdot 1 \cdot 3000}{60} = 400\ Hz \qquad (8)$$

The validity of (8) can be demonstrated by performing Fast Fourier Transform (FFT). The spectral components of the measured motor current after FFT are present in the frequency domain by Fig. 4:

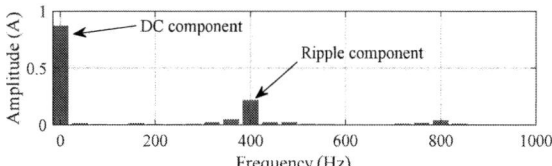

Fig. 4. DC component & ripple component in the frequency domain.

Moreover, the PWM influence on the current ripple is not negligible. The PWM component is part of the spectrum with a frequency of 10 kHz.

III. PROPOSED SENSORLESS METHOD

A. Ripple current signal processing

In this section, a technique of the ripple component extraction from the frequency domain is present. The frequency of the current ripple is proportional to the rotor speed. This fact is an outcome of the brushed DC motor commutation theory and offers a speed or position estimation. This component can be extracted with the bandpass filter, as can be seen in Fig. 5.

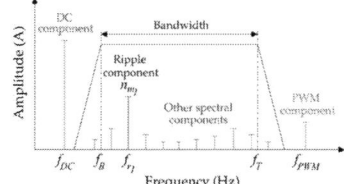

Fig. 5. Frequency domain of the bandpass filter at a lower rotational speed.

The purpose of the bandpass filter is to eliminate the components, which are not corresponding to the actual motor speed. Basically, to these components belong the DC component and the PWM component. However, the filter bandwidth must be large enough to estimate and control the speed in a wider speed range. One can observe that changes in the rotor speed evoke changes in the ripple component in the frequency domain. It is evident from Fig. 6.

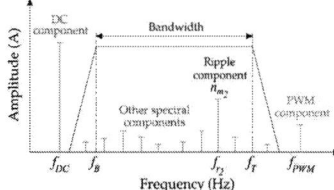

Fig. 6. Frequency domain of the bandpass filter at a higher rotational speed.

The requirement for a wide speed range operation cannot be satisfied with this filter configuration. The reason is that the ripple component is influenced by the other spectral components. These components are the outcome of the brushed DC motor current nonlinearities, for example, cogging torque and so on. Due to these variables, speed can be estimated in a limited range chosen by the user. Beyond this range, the estimated speed becomes inaccurate and wavy.

Nevertheless, most of the other spectral components can be eliminated from the frequency domain with a modified approach. A way to ensure the stable wide speed range operation is to apply a bandpass filter with variable bandwidth present in Fig. 7 and Fig. 8.

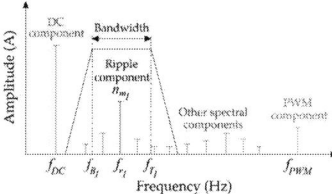

Fig. 7. Frequency domain of the bandpass filter with variable bandwidth at a lower rotational speed.

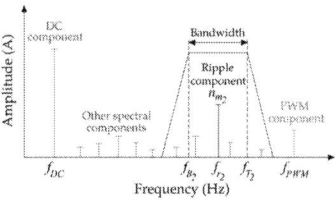

Fig. 8. Frequency domain of the bandpass filter with variable bandwidth at a higher rotational speed.

This modified version of the bandpass filter encloses an area of the actual current ripple frequency. The bandwidth variation is provided by changing the values of the bandpass filter coefficients. These coefficients can be calculated from the transfer function of the bandpass filter. For discrete filters, the transfer function is expressed in terms of b and a coefficients as:

$$H(z) = \frac{B(z)}{A(z)} = \frac{b(1) + b(2)z^{-1} + \cdots + b(n+1)z^{-n}}{a(1) + a(2)z^{-1} + \cdots + a(n+1)z^{-n}} \quad (9)$$

Then the following MATLAB function can be used to get the coefficients offline for desired cut-off frequencies:

$$[b, a] = butter\big(n, W_n, f_{type}\big) \quad (10)$$

which returns the transfer function coefficients for the f_{type} filter, for example, lowpass, highpass, bandpass, or bandstop. The resulting bandpass and bandstop designs are of order $2n$ with normalized cut-off frequency W_n. In terms of the bandpass filter, (10) gives the five coefficients (b_0, b_1, b_2, a_1, a_2), which are maintaining the bandwidth and its position. There are three types of approaches in which the coefficients can be stored and implemented:

- if/else statements,

- polynomial approximations,

- LUTs.

These methods have to be implemented concerning the required rotor speed. However, the required speed cannot be changed with a step if the response of the actual speed is slow or the filter bandwidth is small enough. The best case occurs when the required speed in transient is close as possible to the actual rotor speed. When this condition is satisfied, the ripple component is always in the bandwidth center, as shown in Fig. 7 and Fig. 8. This condition can be partially fulfilled, for example, with a first order input or a ramp. Nevertheless, it is necessary to choose a proper bandwidth. The smaller the bandwidth (e.g., < 10 Hz), the better speed accuracy, particularly at the low speeds. A disadvantage of the small bandwidth lies in the motor transient. In the transient, it is hard to accomplish the ripple component within the borders of the bandwidth. It is causing inaccurate and wavy speed estimation. The first solution to solve this issue is to lower the motor dynamics so the actual rotor speed will be very close to the required speed in the motor transient. The second solution is to expand the bandwidth (e.g., > 50 Hz), so the ripple component is always within the bandwidth borders during the transient. The advantage of the second solution is the step change possibility and improved dynamics. On the contrary, the estimated speed will be more wavy but still accurate. It should be noted that the bandwidth frequency selection depends on (7). It means that the appropriate bandwidth selection depends on the motor parameters related to the commutation. The final decision about the filter bandwidth depends on the target application requirements, e.g.:

- smaller bandwidth:

  – better speed accuracy, more stable low speed operation, limited dynamics,

- larger bandwidth:

  – better dynamics, unstable low speed operation, wavy speed.

In this paper, the 1-D LUTs are used to store the values of the bandpass filter coefficients for a corresponding required speed. Fig. 9 reveals the discrete bandpass filter implementation in Simulink, which is updated with real-time interface (RTI) dSPACE control blocks. 1-D LUTs are present by red blocks.

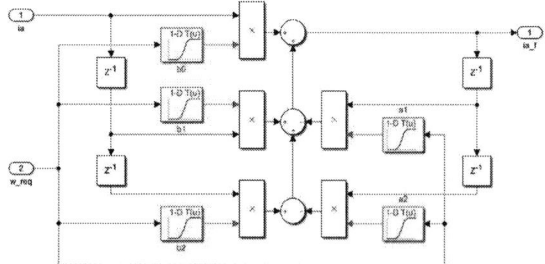

Fig. 9. Discrete bandpass filter with variable bandwidth.

The inputs to the filter are the measured motor current and the required rotor speed, while the output is the filtered motor current. The filtered current has to be as close as possible to the sine wave with eliminated DC component to provide

978-1-6654-3236-8/21 $31.00 © 2021 IEEE

precise speed estimation. Afterward, the filtered current is compared to zero by a discrete comparator (Fig. 10) to produce pulses.

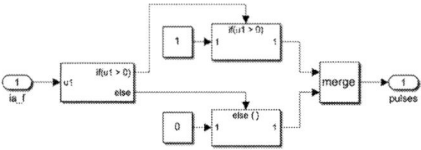

Fig. 10. Discrete comparator.

The time between the pulses is measured by a timer unit (Fig. 11). This time holds information about the actual speed of the brushed DC motor.

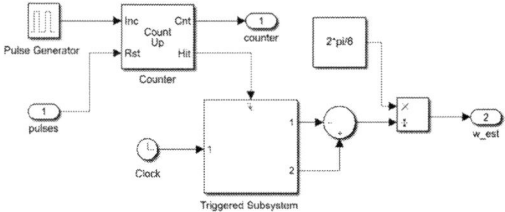

Fig. 11. Timer unit.

The output of the timer has to be divided by the number of ripples per one mechanical rotation to get the estimated speed. The complete block diagram of the current signal processing is shown in Fig. 12.

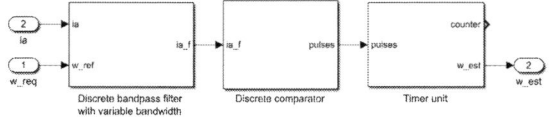

Fig. 12. Current signal processing.

B. Implemenation of the proposed algorithm on dSPACE platform

In this paper, the proposed sensorless algorithm for the brushed DC motor is implemented on the dSPACE platform. The complete block diagram of the proposed sensorless method with a cascade control structure is present in Fig. 13.

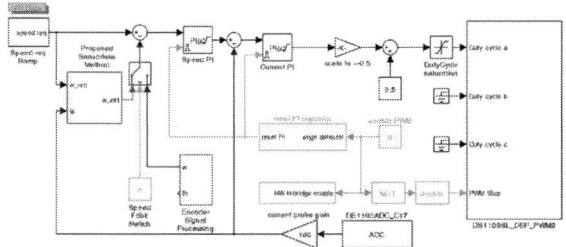

Fig. 13. The proposed control structure implemented on dSPACE platform.

The block with the red border represents the proposed sensorless method discussed and thoroughly described in Section III. The first input to the block is the required rotor speed with a ramp function. The ramp was used as a reference signal because it is a most popular input signal for this type of application. The other reason is that the ramp is very suitable for the proposed sensorless method due to the reasons described in Section III. The second input to the

block is the motor current measured by the current probe. The measured motor current is sampled with an analog to digital converter (ADC). A sampling frequency of the ADC is 10kHz and gained with a constant number of 100.

Blue highlighted blocks in Fig. 13 represent feedback signals. Current up to 5A and incremental encoder was used in practical experiments. Orange highlighted blocks are dedicated for experiment processing like enabling PWM outputs or resetting accumulators in integral parts of PI controllers.

IV. EXPERIMENTAL VERIFICATION

A. Experimental setup

The experimental results were achieved on the setup present in Fig. 14.

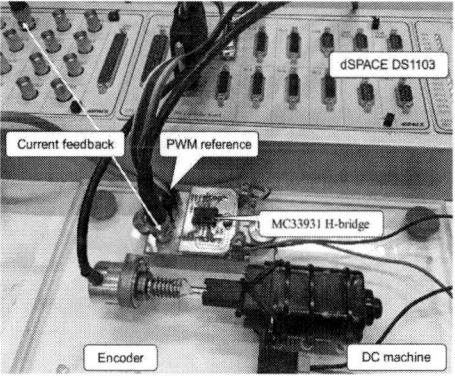

Fig. 14. The photo of the experimental setup.

The dSPACE 1103 platform with the automotive-qualified H-bridge MC33931 integrated circuit dedicated to the car body segment was used to provide experiments. The MC33931 H-bridge is designed primarily for automotive electronic throttle control but is applicable to any low-voltage DC servo motor control application within the current and voltage limits. The maximum PWM frequency of MC33931 is up to 11 kHz. The motor current was measured with a current probe Agilent 1146B connected to the dSPACE input. An incremental encoder was attached to the shaft to provide reliable feedback value about the actual rotor speed. The brushed DC motor with permanent magnets is used as an experimental motor. The parameters of the motor are listed in TABLE II.

TABLE II.   PARAMETERS OF THE BRUSHED DC MACHINE

Parameter	Symbol	Value
Nominal voltage	V_{aN}	12 V
No load current	I_0	1.2 A
Armature resistance	R_a	0.697 Ω
Armature inductance	L_a	1.523 mH
Back-emf constant	k_E	0.0173 Vs/rad
Moment of inertia	J_M	$1.970 \cdot 10^{-6}$ kg·m²

The control strategy in Fig. 13 is using three different execution times. The inner current loop is running at a 10 kHz (100 µs) window. Whole loop execution is including direct path as well as feedback sensing. The outer speed loop is

978-1-6654-3236-8/21 $31.00 © 2021 IEEE   173

running at 1 kHz (1 ms). The same frequency is used for speed feedback sampling. The proposed sensorless method highlighted in red is using a different frequency 40 kHz (25 μs). It is necessary to implement such a short execution time for the timer unit in Fig. 11. In MCU implementation of the proposed algorithm, the core independent timer can replace the block diagram in Fig. 11.

B. Experimental results

In this section, the proposed sensorless method is experimentally verified and compared to the conventional control with the encoder on the rotor shaft. The filter bandwidth is 20 Hz in the whole speed range. A sampling period used to calculate coefficients for the 1D-LUT with the size 80 x 1 is 25 μs.

TABLE III.   PARAMETERS USED IN THE PROPOSED SENSORLESS METHOD

Parameter	Symbol	Value
Bandwidth	Δf_{BPF}	20 Hz
Sampling period of the BPF	T_{BPF}	25 μs
LUT size	-	80 x 1

Fig. 15, Fig. 17, and Fig. 19 show the system performance at steady-state, specifically at 1000 rpm, 3000 rpm, and 6000 rpm. The figures marked with a) are showing the comparison between required speed, actual speed (encoder), and estimated speed. In this case, a feedback value to a speed controller is actual rotor speed to analyze the estimated speed. The figures marked with b) are showing the comparison between the same speeds, however, the feedback value to a speed controller is estimated speed.

The accuracy of the proposed sensorless method is clarified based on Fig. 16, Fig. 18, and Fig. 20. In these figures, the error between the actual speed and estimated speed in sensorless control is present.

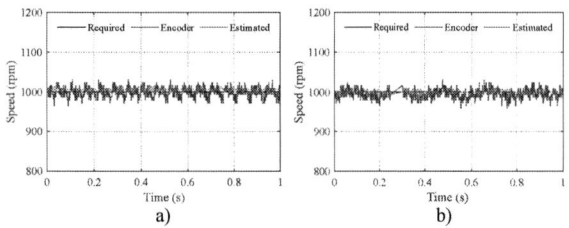

a)   b)

Fig. 15. The steady-state accuracy experimental test at the required speed of 1000 rpm: a) sensor-based control, b) sensorless control.

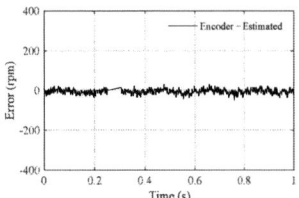

Fig. 16. The error between the actual and estimated speed in sensorless control at the required speed of 1000 rpm.

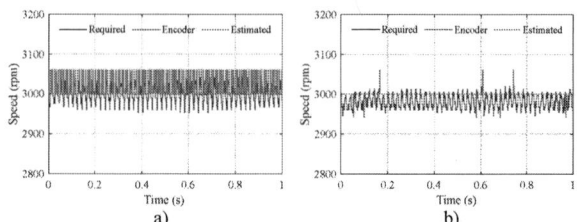

a)   b)

Fig. 17. The steady-state accuracy experimental test at the required speed of 3000 rpm: a) sensor-based control, b) sensorless control.

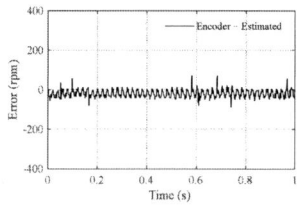

Fig. 18. The error between the actual and estimated speed in sensorless control at the required speed of 3000 rpm.

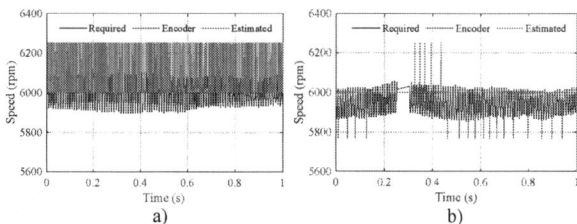

a)   b)

Fig. 19. The steady-state accuracy experimental test at the required speed of 6000 rpm: a) sensor-based control, b) sensorless control.

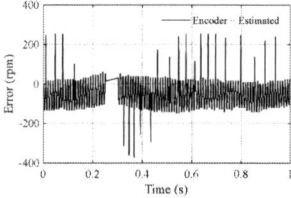

Fig. 20. The error between the actual and estimated speed in sensorless control at the required speed of 6000 rpm.

Fig. 16 and Fig. 18 show almost negligible error proving the sensorless method accuracy at steady-state. Fig. 20 shows the error around 50 rpm, which is upon 1%. This result is acceptable in the high-speed region.

Fig. 21 shows the performance of the proposed sensorless method in a wide speed range as well as in the motor transient. In the figure, the comparison between the speeds in the sensor-based control is present. The required speed was changing with a ramp function. The estimated speed is copying the actual speed from the mechanical sensor. It is proving the validity of the proposed method to use it as a reliable feedback value. It could be useful, for example, in the application where this motor is originally used. In this application, the proposed method could be used for identifying the actual window position.

978-1-6654-3236-8/21 $31.00 © 2021 IEEE

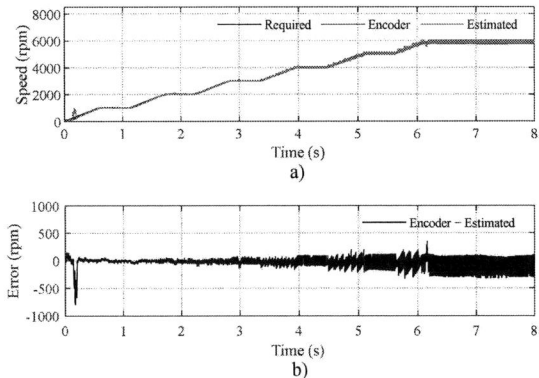

Fig. 21. Sensor-based control: a) the ramp-up speed changes of the required speed, b) the error between the actual and the estimated speed.

However, one can observe a peak, which is present during motor start-up at low speeds. This peak in the estimated speed is the outcome of the motor current transient.

Fig. 22 shows a similar experimental test, where the estimated speed is maintained as a feedback value to the speed controller. Due to the peak present in the previous figure, the motor start-up could not be provided. The motor start-up was provided in the open-loop. The proposed sensorless method was set as a feedback value at 500 rpm. This figure confirms the possibility to even control the rotor speed in a wide speed range by employing the proposed sensorless method.

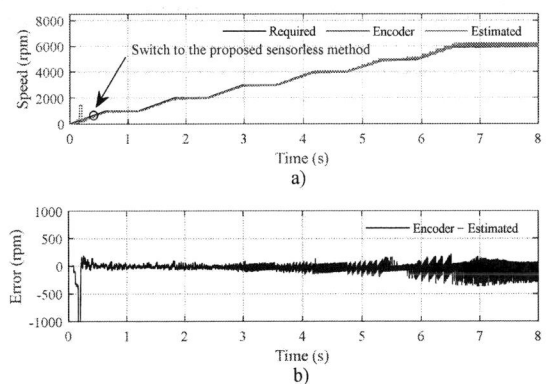

Fig. 22. Sensorless control: a) the ramp-up speed changes of the required speed, b) the error between the actual and the estimated speed.

The operation at lower speeds could be obtained by shrinking the filter bandwidth. However, it is leading to issues regarding motor dynamics. Another limit to control the speed at the low speed region lies in the motor construction. The motors with a small number of ripples per one mechanical rotation cannot be sufficiently controlled at low speeds. The reason is that the motor current has not enough information about the rotor speed to provide stable control. In this case, 16 zero crossings per one mechanical revolution are present. It is a sufficient number for controlling the speed around 500 rpm. However, also the filter bandwidth must be set appropriately due to the spectral components, which are not corresponding to the rotor speed in the low speed region.

V. CONCLUSIONS

The article describes the analysis and practical implementation of a new DC machine sensorless speed control strategy. The proposed method shows the ability to estimate speed based on the ripple current component. The phenomenon caused by the commutator is detected and processed in several stages. All of these stages consist of well-known functions frequently used in real-time control applications. Sensorless speed estimator is using DC machine construction specifications for speed calculation. Estimator showing robustness against parameter change over the whole DC machine working cycle. Using this strategy in a speed loop is quite challenging since the proposed method needs a ripple current component. Inner current loop PI regulators can push ripple component to a negligible, non-detectable level. Hence, high current loop bandwidth cannot cooperate with this type of estimator. The automotive body segment is the perfect application field for the proposed method. Manufacturers preference for this segment is low power DC machines with lower dynamic requirements. The proposed method was experimentally tested on the window lifter brushed DC machine. The experimental results proved accurate speed estimation as well as the appropriate speed control. The proposed method can be used as a redundant and robust sensorless alternative to Hall effect position probes.

ACKNOWLEDGMENT

The authors with to thank the project funding from Slovak Scientific Grant Agency VEGA for project support 1/0795/21 and to projects cofunded from EU sources and the European Regional Development Fund and by the Operational Program Integrated Infrastructure 2014–2020 of the project: Innovative Solutions for Propulsion, Power and Safety Components of Transport Vehicles, code ITMS 313011V334, co-financed by the European Regional Development Fund.

REFERENCES

[1] S. R. Bowes, A. Sevinc, and D. Holliday, "New natural observer applied to speed-sensorless DC servo and Induction motors," IEEE Transactions on Industrial Electronics, vol. 51, pp. 1025–1032, October 2004.

[2] S. Weerasooriya, and M. A. El-Sharkawi, "Identification and control of a DC motor using back-propagation neural networks," IEEE Transactions on Energy Conversion, vol. 6, pp. 663–669, December 1991.

[3] R. Razi, and M. Monfared, "Simple control scheme for single-phase uninterruptible power supply inverters with Kalman filter-based estimation of the output voltage," IET Power Electronics, vol. 8, pp. 1817–1824, May 2015.

[4] E. C. Castaneda, G. A. Loukianov, N. E. Sanchez, and B. Castillo-Toledo, "Discrete-time neural sliding-mode block control for a DC motor with controlled flux," IEEE Transactions on Industrial Electronics, vol. 59, pp. 1194–1207, February 2012.

[5] J. Scott,. J. McLeish, and H. Round, "Speed control with low armature loss for very small sensorless brushed DC motors," IEEE Transactions on Industrial Electronics, vol. 56, pp. 1223–1229, April 2009.

[6] R. M. Ramli, N. Mikami, and H. Takahashi, "Adaptive filters for rotational speed estimation of a sensorless DC motor with brushes," In Proceedings of the 10th International Conference on Information Science, Signal Processing and their Applications (ISSPA 2010), Kuala Lumpur, Malaysia, 10–13 May 2010.

[7] E. Vazquez-Sanchez, J. Gomez-Gil, and J. C. Gamazo-Real, "A new method for sensorless estimation of the speed and position in brushed DC motors using support vector machines," IEEE Transactions on Industrial Electronics, vol. 59, pp. 1397–1408, March 2012.

[8] P. Radcliffe, and D. Kumar, "Sensorless speed measurement for brushed DC motors," IET Power Electronics, vol. 8, pp. 2223–2228, September 2015.

[9] E. Vazquez-Sanchez, J. Sottile, and J. Gomez-Gil, "A novel method for sensorless speed detection of brushed DC motors," Applied Sciences, vol. 7, pp. 14, December 2016.

[10] Automotive brushed-motor ripple counter reference design for sensorless position measurement, Texas Instruments, 2018, accessed 14 June 2018. [Online]. Available: https://www.ti.com/lit/ug/tidud30a/tidud30a.pdf

[11] Sensorless position control of brushed DC motor using ripple counting technique, Microchip Technology Inc, 2019, accesed 17 June 2019. [Online]. Available online: http://ww1.microchip.com/downloads/en/AppNotes/Sensorless-Position-Control-of-Brushed-DC-Motor-Using-Ripple-Counting-Technique-00003049A.pdf

[12] M. Vidlak, P. Makys, and M. Stano, "Comparison between model based and non-model based sensorless methods of brushed DC motor," Transportation Research Procedia, vol. 55, pp. 911–918, July 2021.

[13] B. Yuan, Z. Hu, and Z. Zhou, "Expression of sensorless speed estimation in direct current motor with simplex lap winding," In Proceedings of the International Conference on Mechatronics and Automation, Harbin, China, 5–8 August 2007.

Testing predictive vehicle dynamics control algorithms using a scaled remote controlled car and a roadway simulator

P. Makarun, G. Josipović, M. Švec, Š. Ileš

University of Zagreb Faculty of Electrical Engineering and Computing, Zagreb, Croatia
Email: {petar.makarun, goran.josipovic, marko.svec, sandor.iles}@fer.hr

Abstract—In this paper, a testbed for studying longitudinal and lateral vehicle dynamics is proposed. The testbed consists of a treadmill and a scaled remote-controlled car with four independently controlled brushless DC motors with front steering. The car is equipped with four power converters suitable for field-oriented torque control. The parameters of the scaled vehicle are chosen using the Buckingham PI theorem to achieve dynamic similitude to a full-size vehicle. Such a testbed allows testing of different control algorithms for vehicle dynamics. In this paper, a model predictive steering controller is implemented. Simulations are performed on both the full-scale vehicle in IPG CarMaker and the scaled vehicle in Matlab/Simulink environment. In addition, preliminary experimental results on the scaled vehicle are included to show the real-time suitability of the proposed control algorithm.

Keywords—model predictive control, vehicle dynamics, Buckingham PI theorem, roadway simulator, scaled car

I. INTRODUCTION

It is impossible to imagine a modern car without advanced driver assistance systems. These include cruise control, traction control, Anti-lock Braking System (ABS), Electronic Stability Control (ESC), active steering, Direct Yaw-Moment Control (DYC), etc. All these systems require advanced control algorithms. Most academic research is limited to testing such control algorithms in simulations, since testing on a real car is both expensive and dangerous. It requires a professional driver, booking a private track or closing public roads, especially when the car is pushed to its limits, as there is a possibility of jeopardizing other traffic participants.

For experimental validation of control algorithms, many research groups use scaled vehicle models [1]–[6]. In order to control road conditions without booking the track for scaled vehicles, some research groups have introduced a roadway simulator as a laboratory setup for vehicle dynamics control [7]–[10]. Such a laboratory setup consists of an appropriately sized treadmill that allows testing of control algorithms at high speeds in a confined space. It must be equipped with appropriate sensors for measuring the position and orientation of the vehicle on the treadmill. The treadmill can be equipped with additional degrees of freedom to allow variable track inclinations. In the aforementioned papers, the authors show that by carefully choosing the parameters of the scaled vehicle, a dynamic similitude to the full-size vehicle can be achieved. They also developed active steering algorithms that increase

the maneuverability of the vehicle through front and rear steering using different linear controllers.

In recent years, research in the field of Model Predictive Control (MPC) of vehicle dynamics has been particularly active. Such a control approach uses a model of the system to predict its future behaviour over a finite prediction horizon. By solving an optimization problem at each step the optimal control sequence over the finite horizon is obtained. The first element is used as the controller output and the whole procedure is repeated [11]. The advantage of this approach is the systematic consideration of the dynamical model of the system, as well as its constraints. MPC is used to control the longitudinal dynamics of a vehicle through an adaptive cruise control system to increase passenger comfort, reduce energy consumption, and minimize vehicle-to-vehicle distances while reducing the probability of a collision and reducing driver intervention [12]–[14]. MPC is also used for lateral vehicle dynamics control to improve vehicle handling through predictive active steering [15] active differential [16] or torque vectoring [17]–[19]. The advantages of MPC over linear controllers are presented in [20], [21].

Although MPC has many advantages over simple linear controllers, it requires solving an optimization problem at each sampling time. Therefore, MPC requires more computational power compared to simple linear controllers. It is important to develop computationally efficient algorithms that can be implemented and executed in real time. Before the algorithm is used in a real vehicle, it must be ensured that the maximum time required to solve the optimization problem is less than the selected sampling time.

In this paper, a laboratory setup for testing the real-time capability of predictive algorithms for vehicle dynamics control is presented. It consists of a commercial vehicle dynamics simulation environment (IPG CarMaker), which is used to test the performance of the developed control algorithms on a suitable simulation platform. After this step, the algorithm is tested on a scaled model of a car running on a roadway simulator. The driver can be included in the loop and try out the algorithms using the steering wheel and pedals. An active steering algorithm based on MPC is implemented both in the simulation and on the experimental platform.

The paper is structured as follows: In Section II, a control-oriented mathematical model of the vehicle is presented. In Section III, a roadway simulator is presented. In Section IV, a

978-1-6654-3236-8/21 $31.00 © 2021 IEEE

brief description of the MPC-based active steering algorithm is given. In Section V we present and discuss the simulation and experimental results. Finally, some concluding thoughts are given in Section VI.

II. MATHEMATICAL MODEL

To investigate the fidelity of using the scaled cars for predictive controller evaluation the Buckingham PI theorem [22] is used. The analysis is performed using the simplified bicycle model (Fig 1) presented in [23]. Such a model has two degrees of freedom: the lateral position y and the yaw angle ψ, while the change in longitudinal velocity v_x is treated as a change in the system parameters. The y is measured from the center of rotation of the vehicle O. The front steering angle is denoted by δ and the desired yaw is denoted by ψ_{des}. The l_f and l_r denote the distance of the front and rear wheels, respectively, from the center of gravity CM. The mass of the vehicle is denoted by m and the moment of inertia around the z-axis is denoted by I_z.

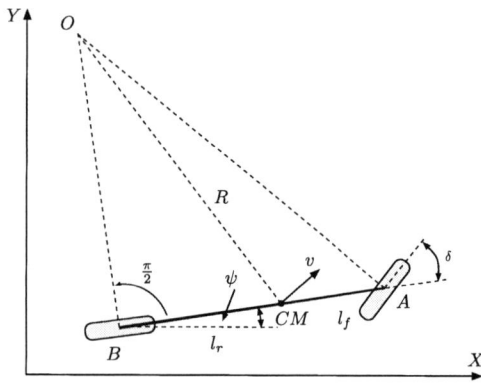

Fig. 1. Bicycle model of a vehicle.

Using the Dugoff tire model [24] and assuming small slip angles, the bicycle model is linearized and the state space model is written in terms of control errors:

$$\dot{e} = A(v_x)e + B\delta + B_{ref}(v_x)\dot{\psi}_{des}, \quad (1)$$

where $e = [e_1\ \dot{e}_1\ e_2\ \dot{e}_2]^T$, $\ddot{e}_1 = \ddot{y} + v_x(\dot{\psi} - \dot{\psi}_{des})$, $e_2 = \psi - \psi_{des}$ and matrices $A(v_x)$, B and B_{ref} are given as follows:

$$A(v_x) = \begin{bmatrix} 0 & 1 & 0 & 0 \\ 0 & -\frac{2C_{\alpha f} + 2C_{\alpha r}}{mv_x} & \frac{2C_{\alpha f} + 2C_{\alpha r}}{m} & -\frac{2C_{\alpha f}l_f - 2C_{\alpha r}l_r}{mv_x} \\ 0 & 0 & 0 & 1 \\ 0 & -\frac{2C_{\alpha f}l_f - 2C_{\alpha r}l_r}{I_z v_x} & \frac{2C_{\alpha f}l_f - 2C_{\alpha r}l_r}{I_z} & -\frac{2C_{\alpha f}l_f^2 + 2C_{\alpha r}l_r^2}{I_z v_x} \end{bmatrix}, \quad (2)$$

$$B = \begin{bmatrix} 0 \\ \frac{2C_{\alpha f}}{m} \\ 0 \\ \frac{2C_{\alpha f}l_f}{I_z} \end{bmatrix}, \quad B_{ref}(v_x) = \begin{bmatrix} 0 \\ -\frac{2C_{\alpha f}l_f - 2C_{\alpha r}l_r}{mv_x} - v_x \\ 0 \\ -\frac{2C_{\alpha f}l_f^2 + 2C_{\alpha r}l_r^2}{I_z v_x} \end{bmatrix}. \quad (3)$$

In the state space model, $C_{\alpha f}$ and $C_{\alpha r}$ denote the cornering stiffness of the front and rear tire, respectively.

Fig. 2. The experiment in the IPG CarMaker simulation software.

The model has eight parameters which can be used to create five PI groups as follows [25]:

$$\Pi_1 = \frac{l_f}{L}, \ \Pi_2 = \frac{l_r}{L}, \ \Pi_3 = \frac{C_{\alpha f}L}{mv_x^2},$$
$$\Pi_4 = \frac{C_{\alpha r}L}{mv_x^2}, \ \Pi_5 = \frac{I_z}{mL^2}, \quad (4)$$

where $L = l_f + l_r$. To achieve dynamic similitude, it is necessary to match the PI groups of the scaled car with the PI groups of the full-size car for the bicycle model.

To create a non-dimensional model of the vehicle, the following transformation is used:

$$e = M\tilde{e}, \quad (5)$$

with

$$M = diag(\begin{bmatrix} L & v_x & 1 & \frac{v_x}{L} \end{bmatrix}), \quad (6)$$

while the time is scaled as follows:

$$t = \frac{L}{v_x}\tilde{t}, \quad (7)$$

where $\tilde{e}$, and $\tilde{t}$ represents non-dimensional error and time, respectively.

III. ROADWAY SIMULATOR FOR TESTING THE CONTROL ALGORITHMS

A. Full size vehicle

Since the full-size vehicle is not available, a generic high-fidelity nonlinear simulation model of the vehicle is simulated in IPG CarMaker (Fig. 2) while the control algorithm is implemented in Matlab/Simulink. Such a simulation model includes a number of nonlinear effects that were neglected in the controller design, such as Ackermann steering geometry, nonlinear tire model, roll dynamics, rolling resistance, etc. The parameters of the linearized model are identified and presented in TABLE I.

B. Experimental setup

The experimental setup shown in Fig. 3 consists of a roadway simulator, a workstation with the CarMaker vehicle dynamics simulation software from IPG Automotive GmbH, dSPACE MicroLabBox and two remote-controlled cars. Each car consists of 4 Maxon 200142 BLDC motors with ESCON

978-1-6654-3236-8/21 $31.00 © 2021 IEEE

TABLE I
PARAMETERS OF THE FULL-SIZE CAR

Parameter	Value	Unit	Description
m	1599.98	kg	mass of the scaled vehicle
l_f	1.311	m	front axle to CM distance
l_r	1.311	m	rear axle to CM distance
I_z	2393.665	$kg \cdot m^2$	moment of inertia around yaw axis
$C_{\alpha f}$	117950	N	lateral front tire stiffness
$C_{\alpha r}$	143700	N/rad	lateral rear tire stiffness

24/2 frequency converters suitable for field-oriented torque control. In addition, each vehicle has a BLDC servo motor for active steering. To be able to involve the driver in experiments, a steering wheel with pedals is included.

Fig. 3. Experimental setup.

The position and orientation of the vehicle is determined by a series of sensors. Attached to the car itself are 4 VL53L1X Time-of-Flight (ToF), laser range sensors that measure distances $d0 - d3$, as shown in Fig. 4. There are also 4 ultrasonic sensors mounted on the treadmill that measure the distance of the car from the "bottom" of the treadmill.

To design the control algorithms for the scaled car model, it is necessary to know a number of physical parameters. The parameters are determined experimentally on the scaled vehicle (Fig. 5) and are given in TABLE II.

IV. MODEL PREDICTIVE CONTROL ALGORITHM

In this section MPC-based active steering is proposed, assuming a low-level longitudinal vehicle dynamics controller. After the brief description of the MPC-based control algorithm, the reference generation is discussed.

TABLE II
ORIGINAL PARAMETERS OF THE SCALED CAR

Parameter	Value	Unit	Description
m	1.173	kg	mass of the scaled vehicle
l_f	0.115	m	front axle to CM distance
l_r	0.141	m	rear axle to CM distance
I_z	0.03373	$kg \cdot m^2$	moment of inertia around yaw axis
$C_{\alpha f}$	8.25	N/rad	lateral front tire stiffness
$C_{\alpha r}$	8.25	N/rad	lateral rear tire stiffness

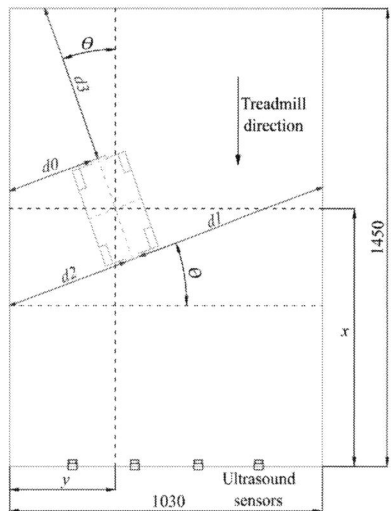

Fig. 4. Treadmill dimensions and sensors setup.

Fig. 5. Experiments for obtaining the center of mass and moment of inertia around yaw axis.

A. Control algorithm

To investigate the applicability of the MPC control algorithms developed for scaled vehicles to the full size vehicles, an MPC algorithm for active steering is implemented. The main objective of the developed control algorithm is to minimize the deviation from the desired yaw angle ψ_{ref} and lateral Y_{ref} position. In addition to the MPC active steering controller, a low-level longitudinal velocity controller is implemented. The main objective of this controller is to maintain the velocity at the set-point by controlling the motor torques. Therefore, the velocity is considered as a varying parameter for the bicycle model of the vehicle.

To achieve this goal, the following objective function is chosen for the active steering algorithm:

$$J_N = e_{k+N}^T P e_{k+N} + \sum_{j=0}^{N-1} \left(e_{k+j}^T Q e_{k+j} + \delta_{k+j}^T R \delta_{k+j} \right) \quad (8)$$

The matrix Q is a diagonal matrix with positive coefficients on its diagonal:

$$Q = diag(\begin{bmatrix} q_1 & q_2 & q_3 & q_4 \end{bmatrix}) \quad (9)$$

where the diagonal terms penalize e_1, e_2 and its derivatives $\dot{e}_1$, $\dot{e}_2$, respectively. The matrix R is a positive matrix penalizing the steering angle, while the matrix P is a positive definite matrix representing the terminal cost.

The only constraint used for lateral vehicle dynamics is the control input constraint:

$$\delta \in [\delta_{min}, \delta_{max}]. \quad (10)$$

This algorithm is implemented for both the full-size car and the scaled model of the car. First, the matrices Q and R for the full-size car are selected and then transformed to be used with the scaled vehicle.

The matrices for the full-size vehicle and the scaled vehicle are related as follows:

$$Q_{scaled} = M^T Q_{full} M, \quad (11)$$

where Q_{scaled}, and Q_{full} are the corresponding matrices for the scaled and the full-size vehicle, respectively.

B. Reference generation

In this paper, the reference generation proposed in [26] is used for the full-size vehicle:

$$\psi_{ref} = \frac{d_{y_1}}{2} \left(1 + \tanh(z_1)\right) - \frac{d_{y_2}}{2} \left(1 + \tanh(z_2)\right),$$
$$Y_{ref} = \arctan(\phi),$$
$$\phi = d_{y_1} \left(\frac{1}{\cosh(z_1)}\right)^2 \frac{1.2}{d_{x_1}} - d_{y_2} \left(\frac{1}{\cosh(z_2)}\right)^2 \frac{1.2}{d_{x_2}}, \quad (12)$$
$$z_1 = \frac{shape}{d_{x_1}} (X - X_{s_1}) - \frac{shape}{2},$$
$$z_2 = \frac{shape}{d_{x_2}} (X - X_{s_2}) - \frac{shape}{2},$$

where $shape = 2.4$, $d_{x_1} = 25$, $d_{x_2} = 21.95$, $d_{y_1} = 4.05$, $d_{y_2} = 5.7$, $X_{s_1} = 27.19$, and $X_{s_2} = 56.46$. The same reference is converted to non-dimensional reference and used for both the full-size vehicle and the scaled vehicle.

At each sampling instant the following optimization problem is solved:

$$\delta_k^\star = \underset{\delta_k}{argmin} \ J_N(e_k, \delta_k)$$
$$\text{s.t } e_{k+1+i} = A(v_{k+i})e_{k+i} + B\delta_{k+i} + B_{ref}(v_{k+i})\dot{\psi}_{des}, \quad (13)$$
$$\delta_{k+i} \in [\delta_{min}, \delta_{max}]^\circ, \ i = 1, \ldots, N-1,$$

where N is the length of the prediction horizon. The result of the optimization is the optimal control sequence $\delta_k^\star = [\delta_k^\star, \delta_{k+1}^\star, \delta_{k+2}^\star, ..., \delta_{k+N-1}^\star]$. However, only the first element of the sequence is applied to the plant and the whole procedure is repeated.

V. RESULTS

First, a vehicle is selected in CarMaker as a test vehicle to test the control algorithms. Some physical parameters such as mass, moment of inertia, and distances of the front and rear axles from the center of mass are directly available in CarMaker. However, the high-fidelity model of the vehicle contains a number of nonlinear effects. For simplicity, it is assumed that the relationship between steering angle and δ is linear. The cornering stiffness of the front and rear tires is estimated from the data. The parameters used for controller synthesis are shown in TABLE I and the controller is tested on both the linear model (Fig. 6) and the full nonlinear model of the vehicle in CarMaker (Fig. 9).

According to the Buckingham PI theorem, the full-size vehicle and the scaled vehicle are dynamically similar if they have the same values of the PI groups. Using the identified parameters of the full-size model and a speed of 80 km/h (22.22 m/s), the PI groups are calculated and given in TABLE V. The same procedure is performed for the scaled vehicle with the original parameters at the speed of 8 km/h (2.22 m/s). As shown in TABLE V, the differences are in the groups Π_3 - Π_5, with the largest difference being in the group Π_5. Since Π_3 and Π_4 depend on the cornering stiffness, a different set of tires is needed for the front and rear wheels to match the parameters accurately. Moreover, to reduce the parameter Π_5, the moment of inertia should be reduced or the mass should be increased. The ideally scaled vehicle is calculated to exactly match the parameters of the full-size vehicle. For this case, the cornering stiffness of the front tires should be decreased by 16 %, while the cornering stiffness of the rear tires should be increased by 2.25 %. Furthermore, the mass of the vehicle should be increased by 1 kg and the experiments should be conducted at 5.2 km/h (1.44 m/s), while the moment of inertia should be decreased by 8 %. The ideal parameters of the scaled vehicle are given in TABLE III. As shown in TABLE V, these parameters correspond to the PI groups of the full-size vehicle

Since the cornering stiffness cannot be changed arbitrarily, the only feasible course of action is to place the additional mass as close as possible to the center of mass of the vehicle. The velocity is chosen to be $v_x = 1.5$ m/s, which gives the feasible parameters of the vehicle (Tab IV).

The PI groups for the modified vehicle are given in TABLE V. It can be seen that the first two parameters match exactly, while the parameters Π_3 - Π_5 show a discrepancy below 11 %.

TABLE III
IDEAL PARAMETERS OF THE SCALED CAR

Parameter	Value	Unit	Description
m	2.173	kg	mass of the scaled vehicle
l_f	0.128	m	front axle to CM distance
l_r	0.128	m	rear axle to CM distance
I_z	0.031	$kg \cdot m^2$	moment of inertia around yaw axis
$C_{\alpha f}$	6.92	N/rad	lateral front tire stiffness
$C_{\alpha r}$	8.43	N/rad	lateral rear tire stiffness

978-1-6654-3236-8/21 $31.00 © 2021 IEEE

TABLE IV
FEASIBLE PARAMETERS OF THE SCALED CAR

Parameter	Value	Unit	Description
m	2.173	kg	mass of the scaled vehicle
l_f	0.115	m	front axle to CM distance
l_r	0.141	m	rear axle to CM distance
I_z	0.03373	$kg \cdot m^2$	moment of inertia around yaw axis
$C_{\alpha f}$	8.25	N/rad	lateral front tire stiffness
$C_{\alpha r}$	8.25	N/rad	lateral rear tire stiffness

TABLE V
PI GROUPS OF THE FULL-SIZE CAR AND THE SCALED CAR

Car	v_x [m/s]	Π_1	Π_2	Π_3	Π_4	Π_5
Full-sized car	22.22	0.5	0.5	0.3914	0.4769	0.2176
Original scaled car	2.22	0.4492	0.5508	0.3646	0.3646	0.4388
Ideal scaled car	1.44	0.5	0.5	0.3914	0.4769	0.2176
Modified scaled car	1.5	0.4492	0.5508	0.4320	0.4320	0.2369

The proposed control algorithm is tested on all the afore-mentioned vehicles. The comparison of the results in Fig. 6 and 7 shows that the scaling is done correctly and the simulation results are exactly the same. Testing the algorithm on the full-size vehicle in CarMaker (Fig. 9) tests the performance of the algorithm in a high-fidelity simulation environment. It can be seen that there is a discrepancy between the linear model used for controller synthesis and the full nonlinear model. This discrepancy reduces the performance of the controller. The same controller is applied to the modified scaled vehicle (Fig. 8). The discrepancy between the ideal model is approximately the same as the discrepancy between the ideal model and the full nonlinear model of the vehicle.

Preliminary experimental results were performed and the results are shown in Fig. 10. The results prove that the proposed experimental setup is capable of simulating vehicle dynamics control algorithms. However, there is a steady-state error which is not present in the simulations due to a mismatch between the experimental setup and the model used for controller synthesis.

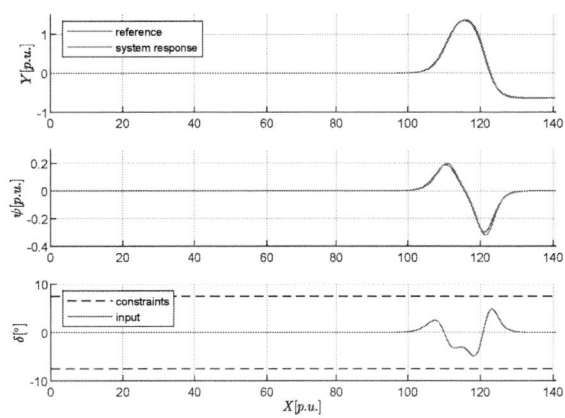

Fig. 7. Simulation results using the linear model: scaled car.

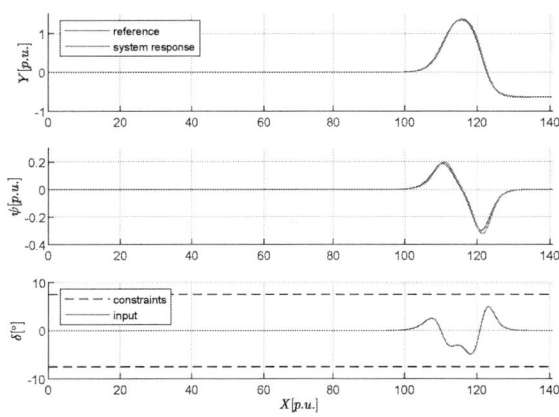

Fig. 8. Simulation results using the linear model: modified scaled car.

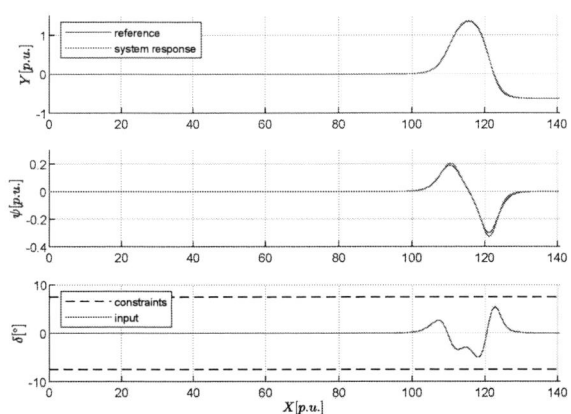

Fig. 6. Simulation results using the linear model: full-size car.

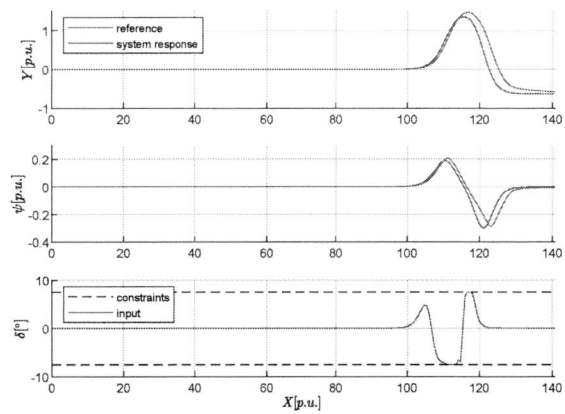

Fig. 9. Simulation results using the CarMaker: full-size car.

978-1-6654-3236-8/21 $31.00 © 2021 IEEE

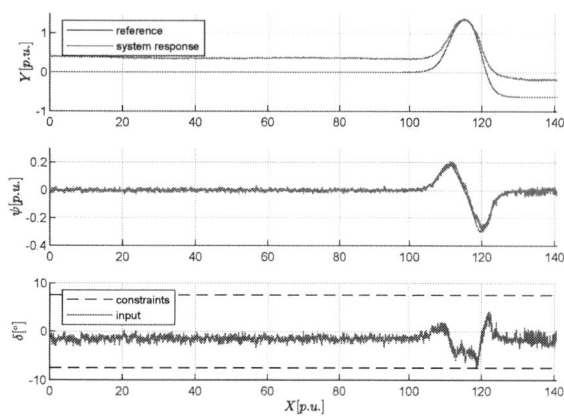

Fig. 10. Experimental results: scaled vehicle.

VI. CONCLUSION

In this paper, a novel experimental setup for the investigation of vehicle dynamics control algorithms is described. Using the high-fidelity simulation software, it is possible to investigate the performance of the developed control algorithms in a relevant simulation environment. For the scaled car, by using the Buckingham PI theorem, it is possible to carefully modify the parameters such as center of mass, moment of inertia, and vehicle speed, given the tire cornering stiffness, to achieve the dynamical similitude with a full-size vehicle at a given speed. When performing the scaling, it is necessary to scale the time and adjust the sampling time accordingly. An active steering algorithm based on model predictive control is implemented in the relevant simulation environment as well as on the proposed experimental setup. The results indicate that the proposed experimental setup can be used as an intermediate step for testing the model predictive control algorithms before they are deployed on the real vehicle.

ACKNOWLEDGEMENT

This work has been supported in part by Croatian Science Foundation under the project UIP-2019-04-6487.

REFERENCES

[1] R. Verma, D. Del Vecchio, and H. K. Fathy, "Longitudinal vehicle dynamics scaling and implementation on a hil setup," in *Dynamic Systems and Control Conference*, vol. 43352, 2008, pp. 979–986.

[2] W. Witaya, W. Parinya, and C. Krissada, "Scaled vehicle for interactive dynamic simulation (sis)," in *2008 IEEE International Conference on Robotics and Biomimetics*. IEEE, 2009, pp. 554–559.

[3] A. Mehra, W.-L. Ma, F. Berg, P. Tabuada, J. W. Grizzle, and A. D. Ames, "Adaptive cruise control: Experimental validation of advanced controllers on scale-model cars," in *2015 American Control Conference (ACC)*. IEEE, 2015, pp. 1411–1418.

[4] Z. Xu, M. Wang, F. Zhang, S. Jin, J. Zhang, and X. Zhao, "Patavtt: A hardware-in-the-loop scaled platform for testing autonomous vehicle trajectory tracking," *Journal of Advanced Transportation*, vol. 2017, 2017.

[5] K. Berntorp, T. Hoang, R. Quirynen, and S. Di Cairano, "Control architecture design for autonomous vehicles," in *2018 IEEE Conference on Control Technology and Applications (CCTA)*. IEEE, 2018, pp. 404–411.

[6] A. Verma, S. Bagkar, N. V. S. Allam, A. Raman, M. Schmid, and V. N. Krovi, "Implementation and validation of behavior cloning using scaled vehicles," SAE Technical Paper, Tech. Rep., 2021.

[7] S. Brennan and A. Alleyne, "The illinois roadway simulator: A mechatronic testbed for vehicle dynamics and control," *IEEE/ASME transactions on mechatronics*, vol. 5, no. 4, pp. 349–359, 2000.

[8] ——, "Using a scale testbed: Controller design and evaluation," *IEEE Control Systems Magazine*, vol. 21, no. 3, pp. 15–26, 2001.

[9] ——, "Dimensionless robust control with application to vehicles," *IEEE Transactions on Control Systems Technology*, vol. 13, no. 4, pp. 624–630, 2005.

[10] S. Lapapong, V. Gupta, E. Callejas, and S. Brennan, "Fidelity of using scaled vehicles for chassis dynamic studies," *Vehicle System Dynamics*, vol. 47, no. 11, pp. 1401–1437, 2009.

[11] D. Q. Mayne, "Model predictive control: Recent developments and future promise," *Automatica*, vol. 50, no. 12, pp. 2967–2986, 2014.

[12] L.-h. Luo, H. Liu, P. Li, and H. Wang, "Model predictive control for adaptive cruise control with multi-objectives: comfort, fuel-economy, safety and car-following," *Journal of Zhejiang University SCIENCE A*, vol. 11, no. 3, pp. 191–201, 2010.

[13] S. E. Li, Z. Jia, K. Li, and B. Cheng, "Fast online computation of a model predictive controller and its application to fuel economy–oriented adaptive cruise control," *IEEE Transactions on Intelligent Transportation Systems*, vol. 16, no. 3, pp. 1199–1209, 2014.

[14] P. Shakouri and A. Ordys, "Nonlinear model predictive control approach in design of adaptive cruise control with automated switching to cruise control," *Control Engineering Practice*, vol. 26, pp. 160–177, 2014.

[15] P. Falcone, F. Borrelli, J. Asgari, H. E. Tseng, and D. Hrovat, "Predictive active steering control for autonomous vehicle systems," *IEEE Transactions on control systems technology*, vol. 15, no. 3, pp. 566–580, 2007.

[16] G. Palmieri, O. Barbarisi, S. Scala, and L. Glielmo, "A preliminary study to integrate ltv-mpc lateral vehicle dynamics control with a slip control," in *Proceedings of the 48h IEEE Conference on Decision and Control (CDC) held jointly with 2009 28th Chinese Control Conference*. IEEE, 2009, pp. 4625–4630.

[17] D. Kasinathan, A. Kasaiezadeh, A. Wong, A. Khajepour, S.-K. Chen, and B. Litkouhi, "An optimal torque vectoring control for vehicle applications via real-time constraints," *IEEE Transactions on Vehicular Technology*, vol. 65, no. 6, pp. 4368–4378, 2015.

[18] E. Siampis, E. Velenis, S. Gariuolo, and S. Longo, "A real-time nonlinear model predictive control strategy for stabilization of an electric vehicle at the limits of handling," *IEEE Transactions on Control Systems Technology*, vol. 26, no. 6, pp. 1982–1994, 2017.

[19] E. Siampis, E. Velenis, and S. Longo, "Rear wheel torque vectoring model predictive control with velocity regulation for electric vehicles," *Vehicle System Dynamics*, vol. 53, no. 11, pp. 1555–1579, 2015.

[20] D. Meola, G. Gambino, G. Palmieri, and L. Glielmo, "A comparison between ltv-mpc and lqr yaw rate-side slip controller," *IFAC Proceedings Volumes*, vol. 42, no. 26, pp. 154–159, 2009.

[21] F. Yakub, S. Lee, and Y. Mori, "Comparative study of mpc and lqc with disturbance rejection control for heavy vehicle rollover prevention in an inclement environment," *Journal of Mechanical Science and Technology*, vol. 30, no. 8, pp. 3835–3845, 2016.

[22] E. Buckingham, "On physically similar systems; illustrations of the use of dimensional equations," *Physical review*, vol. 4, no. 4, p. 345, 1914.

[23] R. Rajamani, *Vehicle dynamics and control*. Springer Science & Business Media, 2011.

[24] H. Dugoff, P. S. Fancher, and L. Segel, "An analysis of tire traction properties and their influence on vehicle dynamic performance," *SAE transactions*, pp. 1219–1243, 1970.

[25] S. Brennan and A. Alleyne, "Robust scalable vehicle control via nondimensional vehicle dynamics," *Vehicle System Dynamics*, vol. 36, no. 4-5, pp. 255–277, 2001.

[26] F. Borrelli, P. Falcone, T. Keviczky, J. Asgari, and D. Hrovat, "Mpc-based approach to active steering for autonomous vehicle systems," *International journal of vehicle autonomous systems*, vol. 3, no. 2-4, pp. 265–291, 2005.

978-1-6654-3236-8/21 $31.00 © 2021 IEEE

Kalman Filter Based Sensor Fusion for Omnidirectional Mechatronic System

Blaž Korotaj, Branimir Novoselnik, Mato Baotić
Department of Control and Computer Engineering
Faculty of Electrical Engineering and Computing, University of Zagreb
Zagreb, Croatia
blaz.korotaj@fer.hr, branimir.novoselnik@fer.hr, mato.baotic@fer.hr

Abstract—The paper describes the sensor fusion for the newly developed omnidirectional mechatronic system. To that end, the kinematic model of the platform and the chosen configuration of omnidirectional Mecanum wheels is described, as well as the principle of operation of all system sensors. The expressions are given for a discrete linear Kalman filter that fuses measurements of a magnetometer and gyroscope, and for a discrete extended Kalman filter that estimates position and orientation of the platform using additional accelerometer measurements. To be able to express the measurement equation four additional states are added to the system model. The developed sensor fusion algorithm was implemented in MATLAB/Simulink programming environment, and very accurate simulation results are reported for estimation of position and orientation of the system. Finally, the real time experimental results are reported for a prototype of the omnidirectional mobile mechatronic system.

Index Terms—Mobile mechatronic system, Omnidirectional drive, Sensor fusion, Discrete extended Kalman filter, State estimation

I. Introduction

Omnidirectional mechatronic systems find applications in diverse areas, from warehouse automation [1] to healthcare system [2]. A design and development of such systems is a complex task because one needs to deal with issues related to energy efficiency, multi-robot coordination, navigation in complex environments by executing optimal trajectories and avoiding obstacles [3]. To achieve all of this, the mobile mechatronic system needs to be able to compute its motions and location based on imprecise measurements from different sensors.

A design of a custom omnidirectional mechatronic system is described in this paper. It is developed from a scratch, using off-the-shelf components (sensors, actuators and embedded computer system). The omnidirectional movement is achieved by Mecanum wheels [4]. The heart of the system is myRIO-1900 [5], an embedded device which can be programmed in LabVIEW [6] but also in C programming language. The main features of the developed mechatronic system are modularity and accessibility so that it can be easily reconfigured and/or expanded as needed. The end goal is to have a fleet of several mobile mechatronic systems which can be used in laboratory

The authors are members of the Laboratory for Renewable Energy Systems at UNIZG-FER (www.lares.fer.hr). This work was been supported in part by the European Regional Development Fund under Grant KK.01.1.1.01.0009 (DATACROSS).

to test different advanced (distributed) control algorithms. This paper describes a basic hardware configuration of one such system and focus on its state estimation (position and orientation) achieved by fusion of measurements from different sensors by using a Kalman filter [12].

The paper is structured as follows: Section II describes the hardware configuration of the developed omnidirectional mobile mechatronic system; system kinematics is analysed in Section III; Kalman filter based sensor fusion, i.e. state estimation, for the system is made in Section IV; Section V reports simulation results in Matlab/Simulink for motion of the mechatronic system; and Section VI reports real-time experimental results obtained with LabVIEW on a system prototype.

II. Mobile Mechatronic System

The platform depicted in Fig. 1, which is constructed with the help of solid modeling computer-aided design tools, is the backbone of the mobile mechatronic system to which all other components are attached. The platform comprises two plates ($240 \times 240 \times 10$ mm), which are connected with four threaded rods so that the height of the platform can be adjusted as needed, up to 300 mm.

Fig. 1. Platform – the backbone of the mobile mechatronic system.

A. Embedded input/output device myRIO-1900

The mobile mechatronic system utilizes an encased myRIO-1900, a portable, reconfigurable input/output device that can

978-1-6654-3236-8/21 $31.00 © 2021 IEEE

connect to a computer via a USB port and Wi-Fi. It is a versatile embedded device that is used for system management in various applications in robotics and mechatronic systems. It can be programmed in LabVIEW and in C programming language. It contains 10 analog inputs (which measure signals ranging from 0 V to 5 V), 6 analog outputs, 40 digital inputs/outputs, LED indicator, accelerometer, Xilinx FPGA and dual-core ARM Cortex-A9 processor that operates at 667 MHz, see [5] for more details.

B. Sensors

The system is equipped with a set of Micro-ElectroMechanical System (MEMS) sensors: 3-axis accelerometer, 3-axis gyroscope, and a magnetometer.

The accelerometer, which is an internal element of the myRIO-1900 device, measures cumulative acceleration, G, of the device in local (x, y, z) coordinates (see Fig. 2)

$$G := \begin{bmatrix} G_x \\ G_y \\ G_z \end{bmatrix} = R_{xyz}(\phi, \psi, \theta) \cdot \begin{bmatrix} 0 \\ 0 \\ g \end{bmatrix} - a, \tag{1}$$

where: $a := [a_x\ a_y\ a_z]^T = [\ddot{x}\ \ddot{y}\ \ddot{z}]^T$ is acceleration of the device in local coordinates, g is gravitational acceleration, and

$$R_{xyz} = \begin{bmatrix} 1 & 0 & 0 \\ 0 & \cos\phi & \sin\phi \\ 0 & -\sin\phi & \cos\phi \end{bmatrix} \cdot \begin{bmatrix} \cos\psi & 0 & -\sin\psi \\ 0 & 1 & 0 \\ \sin\psi & 0 & \cos\psi \end{bmatrix} \cdot$$
$$\cdot \begin{bmatrix} \cos\theta & \sin\theta & 0 \\ -\sin\theta & \cos\theta & 0 \\ 0 & 0 & 1 \end{bmatrix}, \tag{2}$$

is transformation matrix that encapsulates device's rotation around z-axis (the yaw angle, θ), y-axis (the pitch angle, ψ), and x-axis (the roll angle, ϕ), in that order, c.f. [11].

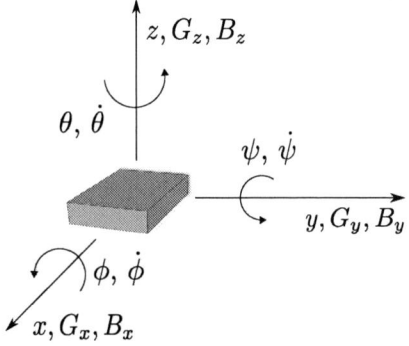

Fig. 2. Local coordinate system of myRIO (and mobile mechatronic system). The rotational angles ϕ, ψ, and θ, are measured with respect to the initial/global coordinate system, which is assumed without loss of generality to have x_I, y_I, z_I axes matching the East-North-Up directions.

In principle, the relative movement of the mechatronic system (odometry) can be calculated from (1) by double integration of $a(t)$. Furthermore, if the mechatronic system does not accelerate, i.e. when $a = [0\ 0\ 0]^T$, we can also determine the roll angle ϕ and the pitch angle ψ, since:

$$\tan\phi = \frac{G_y}{G_z}, \quad \tan\psi = -\frac{G_x}{\sqrt{G_y^2 + G_z^2}}. \tag{3}$$

The mechatronic system is also equiped with a three-axis MEMS digital gyroscope PmodGYRO – STMicroelectronics L3G4200D chip with SPI or I²C communication protocol [8]. It works on the principle of measuring the Coriolis force and provides measurements of angular velocity, ω, of the device:

$$\omega := \begin{bmatrix} \omega_x \\ \omega_y \\ \omega_z \end{bmatrix} = \begin{bmatrix} \dot{\phi} \\ \dot{\psi} \\ \dot{\theta} \end{bmatrix}. \tag{4}$$

In principle, the relative rotation of the system around the x, y, and z axis can be obtained by integration of $\omega(t)$.

To determine the yaw angle θ we use the measurement of the magnetic field, B, at the device's location

$$B := \begin{bmatrix} B_x \\ B_y \\ B_z \end{bmatrix} = R_{xyz}(\phi, \psi, \theta) \cdot \begin{bmatrix} 0 \\ B_E \cos\delta \\ -B_E \sin\delta \end{bmatrix}, \tag{5}$$

where: B_E is the strength of the Earth's magnetic field, δ is magnetic inclination (the angle made with the horizontal by the Earth's magnetic field lines). We measure B with a Digilent's Pmod CMPS2 triple magnetometer, which has a magnetic field resolution of 0.5 mG and connects to the embedded system using the I²C communication protocol with data rates up to 400 kHz [7]. With some algebraic manipulation of (2) and (5), we can calculate the yaw angle θ as

$$\theta = \arctan \frac{B_x \cos\psi + B_y \sin\phi \sin\psi + B_z \cos\phi \sin\psi}{B_y \cos\phi - B_z \sin\phi}. \tag{6}$$

Note that the obtained expression (6) does not depend on (the knowledge of) magnetic inclination δ. Consequently, (6) is valid even if the system is exposed to additional magnetic field sources that are "aligned" with the Earth's magnetic field.

C. Mecanum wheels, DC motor and driver

Two pairs of omnidirectional, Mecanum wheels (left and right) are placed at the bottom of the platform, with each wheel being powered by a DC motor. This wheel selection enables arbitrary translation and rotation of the mechatronic system since Mecanum wheels have three degrees of freedom: rotation around the wheel axis, rotation around the contact point, and rotation of the rollers around their axis. The rollers are actually small wheels placed on the circumference of the main wheel, at an angle of 45° to the wheel axis, see Fig. 3. The wheel radius in our system is $r = 0.03$ m, and the distance between the center of the system and the center of the wheel is $l = 0.17$ m.

The DC motors driving the wheels are controlled with MDD10A, a Dual Channel 10A DC Motor Driver of Cytron Technologies, [9]. It is a two-channel driver designed to run two DC motors with currents up to 10 A continuously. It supports locked-antiphase and sign-magnitude PWM signal.

978-1-6654-3236-8/21 $31.00 ©2021 IEEE

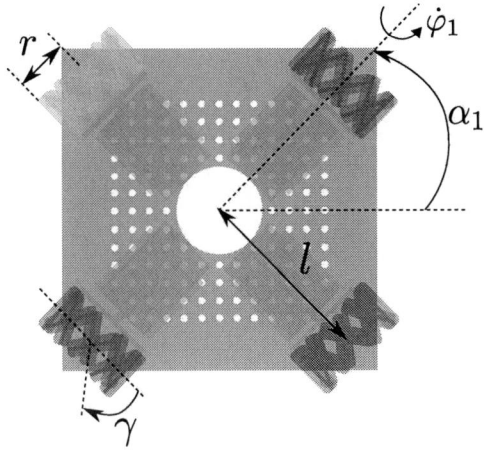

Fig. 3. Placement of DC motors and wheels (top view) on the system.

The construction is built of components that result in faster response times and eliminate mechanical relay wear. It also allows control of the motor in both directions and the speed can be controlled by PWM frequency up to 20 kHz. We use DC motors of nominal voltage 12 V, maximum current 7 A, maximum torque 1.7654 Nm, [10]. The nominal speed is 204 $\frac{1}{\min}$, rated torque 0.3178 Nm, speed without load is 251 $\frac{1}{\min}$ while current without load is 0.35 A. The DC drive is quite powerful, with a gear ratio of 43.7:1, shielded with a metal case that has an integrated quadrature encoder which provides a resolution of 64 pulses per revolution of the main shaft corresponding to 2797 pulses per revolution of the output shaft from the case. This ensures accurate measurement of the angular velocity, $\dot{\varphi}_i$, of each wheel of the system.

III. KINEMATICS OF THE MOBILE MECHATRONIC SYSTEM

For a horizontally placed myRIO (and attached sensors) and the mechatronic system moving on a flat surface one gets simplified version of expressions (1)–(6), because $\phi = \psi = 0$ under those conditions. Henceforth we assume this is the case and refer to the yaw angle θ as *the orientation* of the mobile mechatronic system.

We derive the kinematic model of the mobile mechatronic system under the following assumptions: i) there is only one point of contact between the wheel (i.e. corresponding roller) and the ground, ii) the wheel plane is vertical relative to the ground, and iii) there is no roller slipping.

Since we are not interested in the kinematics of the rollers, the kinematic model of the system can be determined using (only) the rolling equation for all wheels, [11],

$$\begin{bmatrix} \sin\left(\alpha_i + \beta + \gamma\right) \\ -\cos\left(\alpha_i + \beta + \gamma\right) \\ -l\cos\left(\beta + \gamma\right) \end{bmatrix}^{\mathrm{T}} \begin{bmatrix} \dot{x}(t) \\ \dot{y}(t) \\ \dot{\theta}(t) \end{bmatrix} = r\cos(\gamma)\dot{\varphi}_i(t), \quad (7)$$

where: $[\dot{x}, \dot{y}]^{\mathrm{T}}$ is the (translational) velocity of the mechatronic system in local coordinates, $\dot{\theta}$ is the angular velocity of the mechatronic system around z-axis in local coordinates (i.e. it

is the speed of change of system's orientation, which has the same value in local and global coordinates!), β is the angle between the wheel axis and the line connecting the center of the wheel with the center of the system, γ is the angle between the roller axis and the main wheel plane, and $\dot{\varphi}_i$ is the angular velocity of i-th wheel (with positive values for anticlockwise rotation, looking from the wheel towards the center of the system), α_i is the angle of the i-th wheel axis relative to a chosen reference direction, $i = 1, 2, 3, 4$. For our system: $\alpha_1 = 45°$, $\alpha_2 = 135°$, $\alpha_3 = 225°$, $\alpha_4 = 315°$, $\beta = 0$, and $\gamma = 45°$. Now we can derive both the inverse kinematics

$$\begin{bmatrix} \dot{\varphi}_1(t) \\ \dot{\varphi}_2(t) \\ \dot{\varphi}_3(t) \\ \dot{\varphi}_4(t) \end{bmatrix} = \frac{1}{r} \begin{bmatrix} \sqrt{2}\dot{x}(t) - l\dot{\theta}(t) \\ \sqrt{2}\dot{y}(t) - l\dot{\theta}(t) \\ -\sqrt{2}\dot{x}(t) - l\dot{\theta}(t) \\ -\sqrt{2}\dot{y}(t) - l\dot{\theta}(t) \end{bmatrix}, \quad (8)$$

and kinematics of the system in local coordinates

$$\begin{bmatrix} \dot{x}(t) \\ \dot{y}(t) \\ \dot{\theta}(t) \end{bmatrix} = \frac{r}{2\sqrt{2}} \begin{bmatrix} \dot{\varphi}_1(t) - \dot{\varphi}_3(t) \\ \dot{\varphi}_2(t) - \dot{\varphi}_4(t) \\ \dfrac{-\dot{\varphi}_1(t) - \dot{\varphi}_2(t) - \dot{\varphi}_3(t) - \dot{\varphi}_4(t)}{l\sqrt{2}} \end{bmatrix}. \quad (9)$$

Note that by taking into account the connection between system's velocity in global and local coordinates,

$$\begin{bmatrix} \dot{x}_{\mathrm{I}}(t) \\ \dot{y}_{\mathrm{I}}(t) \end{bmatrix} = \begin{bmatrix} \cos\theta(t) & -\sin\theta(t) \\ \sin\theta(t) & \cos\theta(t) \end{bmatrix} \begin{bmatrix} \dot{x}(t) \\ \dot{y}(t) \end{bmatrix}, \quad (10)$$

system kinematics in global coordinates is readily available from (9).

IV. STATE ESTIMATION

In the following we give details of the recursive, discrete-time, state estimators for the orientation of the system θ and position of the system in global coordinates x_{I}, y_{I} by using the knowledge of the measurement model, disturbances, and past values of system states.

With any real sensors we expect to see some level of measurement noise. Measurements obtained from sensors enter the state estimator for which there is an a-priori state knowledge. In our case the initial position of the system and its orientation are (assumed to be) known. At each time instant we obtain estimation of the system state and associated covariance matrix, i.e. the uncertainty of the estimation. Process noise is present due to the imperfection of the actuator in the system (DC motor in this case) and the fact that we do not have the exact dynamical mathematical model of the system.

A. Magnetometer and gyroscope fusion

A discrete linear Kalman filter (DKF, [12]) is used to improve estimation of the orientation of the system, θ, by the fusion of the angle measurement from the magnetometer, here denoted as θ^{m} (see (6)) and the angular velocity around z-axis measurement from the gyroscope, here denoted as ω^{m} (see (4)).

978-1-6654-3236-8/21 $31.00 © 2021 IEEE

For a discrete-time linear system

$$\xi_k = \Phi_{k-1}\xi_{k-1} + \Gamma_{k-1}\mathbf{u}_{k-1} + \mathbf{w}_{k-1}, \quad (11)$$

$$\mathbf{z}_k = H_k\xi_k + \mathbf{v}_k, \quad (12)$$

where ξ_k is the state vector, $\mathbf{u}_k$ is the input vector, $\mathbf{z}_k$ is the output vector, and $\mathbf{w}_k$ and $\mathbf{v}_k$ are the process and measurement noise which are assumed to be Gaussian with zero mean and covariance Q_k and R_k respectively

$$\mathbf{w}_k \sim N(0, Q_k), \quad \mathbf{v}_k \sim N(0, R_k), \quad (13)$$

we can design DKF as follows (cf. [12] for more details):

$$
\begin{aligned}
P_k^- &= \Phi_{k-1}P_{k-1}^+\Phi_{k-1}^{\mathrm{T}} + Q_{k-1}, \\
K_k &= P_k^- H_k^{\mathrm{T}}(H_k P_k^- H_k^{\mathrm{T}} + R_k)^{-1}, \\
\hat{\xi}_k^- &= \Phi_{k-1}\hat{\xi}_{k-1}^+ + \Gamma_{k-1}\mathbf{u}_{k-1}, \\
\hat{\xi}_k^+ &= \hat{\xi}_k^- + K_k(\mathbf{z}_k - H_k\hat{\xi}_k^-), \\
P_k^+ &= (I - K_k H_k)P_k^-(I - K_k H_k)^{\mathrm{T}} + K_k R_k K_k^{\mathrm{T}},
\end{aligned}
\quad (14)
$$

where $\hat{\xi}_k^+$ and P_k^+ are the *a posteriory* state estimate and its covariance matrix (with known initial values $\hat{\xi}_0^+$ and P_0^+).

To estimate the mechatronic system orientation θ we use the gyroscope measurement, ω^{m}, as an input in the state update equation (11) and the magnetometer measurement, θ^{m}, as an observation in the output equation (12). Hence we execute DKF algorithm (11)–(14) with:

$$
\begin{aligned}
&\xi_k := \theta(kT), \ \mathbf{u}_{k-1} := \omega^{\mathrm{m}}(kT), \ \mathbf{z}_k := \theta^{\mathrm{m}}(kT), \\
&\Phi_k = 1, \ \Gamma_k = T, \ H_k = 1,
\end{aligned}
\quad (15)
$$

where $T = 0.01\,\mathrm{ms}$ is the sampling time of our system.

To avoid (bias) accumulation of the error (due to the integration of the angular velocity from the gyroscope), we can extend system model to include bias of the gyroscope. In such case we execute DKF algorithm (11)–(14) with:

$$
\begin{aligned}
&\xi_k := [\theta(kT) \ b_k]^{\mathrm{T}}, \ \mathbf{u}_{k-1} := \omega^{\mathrm{m}}(kT), \ \mathbf{z}_k := \theta^{\mathrm{m}}(kT), \\
&\Phi_k = \begin{bmatrix} 1 & -T \\ 0 & 1 \end{bmatrix}, \Gamma_k = \begin{bmatrix} T \\ 0 \end{bmatrix}, H_k = \begin{bmatrix} 1 & 0 \end{bmatrix}.
\end{aligned}
\quad (16)
$$

The estimated orientation, $\xi_{k,1}^+$, henceforth denoted as θ^{f} is used as an improved "observation" (instead of a measurement $\theta^{\mathrm{m}}(kT)$) during the estimation of the system position, which is treated next.

B. Estimation of system position (and orientation) by discrete extended Kalman filter

Since kinematic model of the platform in global coordinates (9)–(10) is nonlinear we use a Discrete Extended Kalman Filter (DEKF, [12]) to estimate global position: x_{I} and y_{I}, and orientation θ of the mechatronic system.

For a discrete-time non-linear system

$$\xi_k = \mathbf{f}_k(\xi_{k-1}, \mathbf{u}_{k-1}, \mathbf{w}_{k-1}), \quad (17)$$

$$\mathbf{z}_k = \mathbf{h}_k(\xi_k, \mathbf{v}_k), \quad (18)$$

where ξ_k is the state vector, $\mathbf{u}_k$ is the input vector, $\mathbf{z}_k$ is the output vector, and $\mathbf{w}_k$ and $\mathbf{v}_k$ are the process and measurement noise which are assumed to be Gaussian with zero mean and covariance Q_k and R_k respectively

$$\mathbf{w}_k \sim N(0, Q_k), \quad \mathbf{v}_k \sim N(0, R_k), \quad (19)$$

we can design DEKF as follows (cf. [12] for more details):

$$
\begin{aligned}
\Phi_{k-1} &= \left.\frac{\partial \mathbf{f}_{k-1}}{\partial \xi}\right|_{\hat{\xi}_{k-1}^+}, \ \Lambda_{k-1} = \left.\frac{\partial \mathbf{f}_{k-1}}{\partial \mathbf{w}}\right|_{\hat{\xi}_{k-1}^+}, \\
P_k^- &= \Phi_{k-1}P_{k-1}^+\Phi_{k-1}^{\mathrm{T}} + \Lambda_{k-1}Q_{k-1}\Lambda_{k-1}^{\mathrm{T}}, \\
\hat{\xi}_k^- &= \mathbf{f}_{k-1}(\hat{\xi}_{k-1}^+, \mathbf{u}_{k-1}, 0), \\
H_k &= \left.\frac{\partial \mathbf{h}_k}{\partial \xi}\right|_{\hat{\xi}_k^-}, \ M_k = \left.\frac{\partial \mathbf{h}_k}{\partial \mathbf{v}}\right|_{\hat{\xi}_k^-}, \\
K_k &= P_k^- H_k^{\mathrm{T}}(H_k P_k^- H_k^{\mathrm{T}} + M_k R_k M_k^{\mathrm{T}})^{-1}, \\
\hat{\xi}_k^+ &= \hat{\xi}_k^- + K_k(\mathbf{z}_k - \mathbf{h}_k(\hat{\xi}_k^-, 0)), \\
P_k^+ &= (I - K_k H_k)P_k^-,
\end{aligned}
\quad (20)
$$

where $\hat{\xi}_k^+$ and P_k^+ are the *a posteriory* state estimate and its covariance matrix (with known initial values $\hat{\xi}_0^+$ and P_0^+). We chose velocities of each wheel $\dot{\varphi}_i$ as the inputs signals in the nonlinear state update equation. As measurement signals in the output equation we use accelerations measured by the accelerometer in local coordinates, denoted here as $a_{\mathrm{x}}^{\mathrm{m}}$ and $a_{\mathrm{y}}^{\mathrm{m}}$ (see (1)), and orientation obtained by the fusion of the magnetometer and gyroscope in the previous subsection, denoted here as θ^{f}

$$
\xi_k := \begin{bmatrix} x_{\mathrm{I}}(kT) \\ y_{\mathrm{I}}(kT) \\ \theta(kT) \\ x_{\mathrm{I}}(kT - T) \\ x_{\mathrm{I}}(kT - 2T) \\ y_{\mathrm{I}}(kT - T) \\ y_{\mathrm{I}}(kT - 2T) \end{bmatrix}, \ \mathbf{u}_k := \begin{bmatrix} \dot{\varphi}_1(kT) \\ \dot{\varphi}_2(kT) \\ \dot{\varphi}_3(kT) \\ \dot{\varphi}_4(kT) \end{bmatrix}, \ \mathbf{z}_k := \begin{bmatrix} a_{\mathrm{x}}^{\mathrm{m}}(kT) \\ a_{\mathrm{y}}^{\mathrm{m}}(kT) \\ \theta^{\mathrm{f}}(kT) \end{bmatrix}.
\quad (21)
$$

From (8)–(10), with disretization, we obtain the state update equations (i.e. $\mathbf{f}_k$ in (17))

$$
\begin{aligned}
\xi_{k,1} =\ &\xi_{k-1,1} + \\
&+ (\Delta x + \mathbf{w}_{k-1,1})\cos(\xi_{k-1,3} + \Delta\theta + \mathbf{w}_{k-1,3}) + \\
&- (\Delta y + \mathbf{w}_{k-1,2})\sin(\xi_{k-1,3} + \Delta\theta + \mathbf{w}_{k-1,3}), \\
\xi_{k,2} =\ &\xi_{k-1,2} + \\
&+ (\Delta x + \mathbf{w}_{k-1,1})\sin(\xi_{k-1,3} + \Delta\theta + \mathbf{w}_{k-1,3}) + \\
&+ (\Delta y + \mathbf{w}_{k-1,2})\cos(\xi_{k-1,3} + \Delta\theta + \mathbf{w}_{k-1,3}), \\
\xi_{k,3} =\ &\xi_{k-1,3} + \Delta\theta + \mathbf{w}_{k-1,3}, \\
\xi_{k,4} =\ &\xi_{k-1,1} + \mathbf{w}_{k-1,4}, \\
\xi_{k,5} =\ &\xi_{k-1,4} + \mathbf{w}_{k-1,5}, \\
\xi_{k,6} =\ &\xi_{k-1,2} + \mathbf{w}_{k-1,6}, \\
\xi_{k,7} =\ &\xi_{k-1,6} + \mathbf{w}_{k-1,7},
\end{aligned}
\quad (22)
$$

where we use shorthand notation

$$
\begin{aligned}
\Delta x &:= \frac{Tr}{2\sqrt{2}}(\mathbf{u}_{k,1} - \mathbf{u}_{k,3}), \\
\Delta y &:= \frac{Tr}{2\sqrt{2}}(\mathbf{u}_{k,2} - \mathbf{u}_{k,4}), \\
\Delta\theta &:= -\frac{4T}{l}(\mathbf{u}_{k,1} + \mathbf{u}_{k,2} + \mathbf{u}_{k,3} + \mathbf{u}_{k,4}).
\end{aligned}
\quad (23)
$$

978-1-6654-3236-8/21 $31.00 © 2021 IEEE

The output equations (i.e. $\mathbf{h}_k$ in (24)) look like

$$
\begin{aligned}
\mathbf{z}_{k,1} =\ & \frac{\xi_{k,1} - 2\xi_{k,4} + \xi_{k,5}}{T^2} \cos\left(\xi_{k,3}\right) + \\
& + \frac{\xi_{k,2} - 2\xi_{k,6} + \xi_{k,7}}{T^2} \sin\left(\xi_{k,3}\right) + \mathbf{v}_{k,1}, \\
\mathbf{z}_{k,2} =\ & -\frac{\xi_{k,1} - 2\xi_{k,4} + \xi_{k,5}}{T^2} \sin\left(\xi_{k,3}\right) + \\
& + \frac{\xi_{k,2} - 2\xi_{k,6} + \xi_{k,7}}{T^2} \cos\left(\xi_{k,3}\right) + \mathbf{v}_{k,2}, \\
\mathbf{z}_{k,3} =\ & \xi_{k,3} + \mathbf{v}_{k,3}.
\end{aligned}
\tag{24}
$$

As a final result of the DEKF computation (17)–(24) we obtain (as the first 3 components of the state vector $\hat{\xi}^+$) the *a posteriory* estimate of the mobile mechatronic system position in global coordinates $\hat{x}_I^+$, $\hat{y}_I^+$, and its orientation $\hat{\theta}^+$.

V. SIMULATION RESULTS IN MATLAB/SIMULINK

Position and orientation estimation with the DEKF was implemented and simulated in Matlab/Simulink programming environment. The process and measurement noise covariance matrices and the initial state vector are set as

$$
\begin{aligned}
Q_k &= \operatorname{diag}(0.4, 0.4, 0.004, 0.4, 0.4, 0.4, 0.4), \\
R_k &= \operatorname{diag}(0.001, 0.01, 0.01), \\
\xi_0 &= [0\ 0\ 0\ 0\ 0\ 0\ 0]^{\mathrm{T}}.
\end{aligned}
\tag{25}
$$

In testing scenario we set the reference values of position and orientation $x_{\mathrm{ref}} = -10$ m, $y_{\mathrm{ref}} = 30$ m, $\theta_{\mathrm{ref}} = 180°$. The obtained position and orientation estimation responses are shown in Fig. 4 and Fig. 5. In Fig. 5 the orange line shows the angle measurements, which was generated by artificially adding white noise to the angle θ from the process output, the black line shows the estimated angle value, and the blue line shows the estimation error.

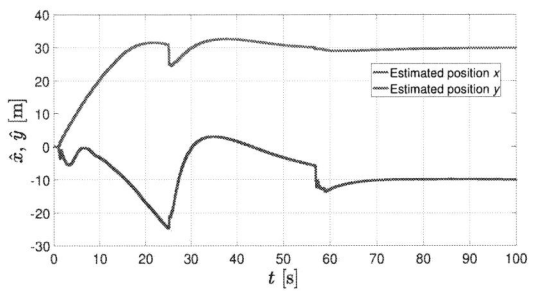

Fig. 4. Simulation in Matlab of mechatronic system position estimation with DEKF.

From the responses shown in Fig. 4 and Fig. 5 it can be seen that DEKF works very accurately for this case and that the correct position and orientation value is achieved fairly quickly. The accuracy of the state estimation can also be detected in the response of the trace of the error covariance matrix shown on Fig. 6. Namely, it can be seen that the trace of the covariance error matrix has a sharp drop from the initial

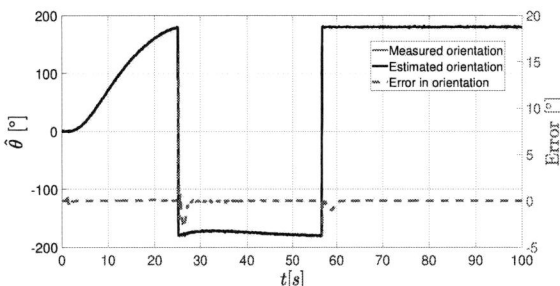

Fig. 5. Simulation in Matlab of mechatronic system orientation estimation with DEKF.

value of $40,000$ and then converges to a reduced value, which confirms that the estimation works very well.

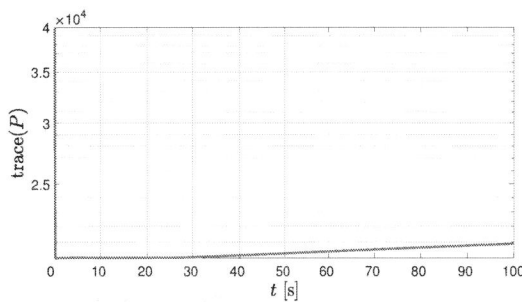

Fig. 6. Trace of the state estimation covariance matrix for DEKF implementation in Matlab.

VI. REAL-TIME OPERATION OF THE MOBILE MECHATRONIC SYSTEM WITH LABVIEW

The prototype of the mobile mechatronic system, shown in Fig. 7. was tested using the LabVIEW programming environment. To obtain the fastest possible response without amplitude overshoot, a PI controller is used, for which the process variable is the speed in revolutions per minute, and the setpoint is the speed obtained by the inverse kinematics. Motor speed is PWM controlled. The parameters of the PI controller used in the testing are as follows: $K_c = 0.3$, $T_I = 0.001$ s, and the frequency of the PWM signal is $1\,\mathrm{kHz}$.

Functionality of the sensor fusion was tested by rotating the system around its center of mass, and the obtained orientation response is shown in Fig. 8. From the response it can be seen that the magnetometer has a substantive noise, while the orientation obtained by DKF has almost no noise, which is an indicator of a good estimation. The obtained response confirms the expectation – combination of all available measurements can always give a more accurate estimate than the one obtained by omitting some measurements (e.g. the least reliable).

We note that the discrete extended Kalman filter estimation was tested on the actual system but with limitations on the available motion due to the setup of the testing environment. Namely, the length of the wire of the power supply and the

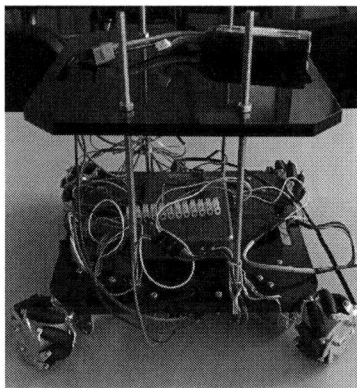

Fig. 7. Prototype of the omnidirectional mobile mechatronic system.

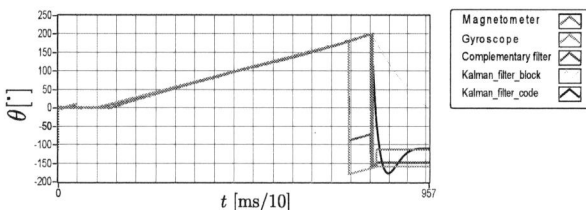

Fig. 8. Experimental results (with LabVIEW) of sensor fusion.

size of the table on which the test are performed (chosen due to the greater adhesion of the wheels to the ground) limits the span of the movement of the system. An additional problem that affects the experimental results is presence of a modeling error due to the inability to use four completely identical wheels, which affects inverse and direct kinematics of the system. Finally, on a prototype system one wheel is not perfectly connected to the motor shaft, and sometimes there is a slip, which introduces additional error in the movement of the system. In Fig. 9 we report the estimated values of the platform position for the reference values: $x_{\text{ref}} = 0.2\,\text{m}$, $y_{\text{ref}} = 0\,\text{m}$, $\theta_{\text{ref}} = 60°$. The response shows that DEKF

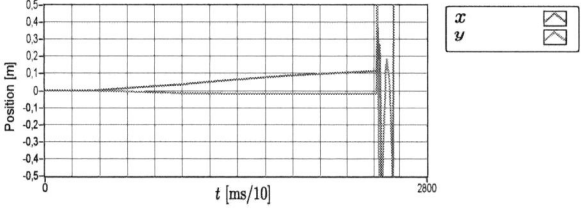

Fig. 9. Experimental results (with LabVIEW) of position estimation of the system for reference values: $x_{\text{ref}} = 0.2\,\text{m}$, $y_{\text{ref}} = 0\,\text{m}$, $\theta_{\text{ref}} = 60°$.

estimates the x component of the position to $0.1\,\text{m}$ due to the slightly larger dimensions of the platform, so the wheel tips quickly reach the end of the work table and it is not possible to let the platform complete the movement. The disturbance after 24 seconds that appears on the response shown in figure Fig. 9

is a consequence of stopping the movement and is not part of the position estimation. Also for this wheel arrangement, it is not possible to examine the motion in the y direction with any great accuracy due to the large rotation of the platform.

VII. CONCLUSION

The developed omnidirectional mechatronic system uses the measurements from three sensors: magnetometer, gyroscope and accelerometer. For a successful design of the control system it is crucial to obtain good estimate of system states. To increase the accuracy and precision of estimation of the position and orientation of the system we used a discrete extended Kalman filter. In particular, DKF based fusion of the magnetometer and gyroscope measurement is used to obtain orientation of the platform, which is then used as an input to the DEKF together with the measurements from the accelerometer. The state estimation responses obtained in the simulation illustrate satisfactory accuracy and speed of the selected estimation procedure. Despite the limitations of the used wheel layout and imperfections of the grooves connecting the wheel to the motor shaft, which may lead to the slipping of the wheels in practice, the real-time experimental results on a prototype system also confirm the efficiency of the sensor fusion.

REFERENCES

[1] A. A. Sklyarov, G. E. Veselov, S. A. Sklyarov and T. E. Pohilina, "Synthesis of the synergetic control law of the transport robotic platform," 2017 IEEE II International Conference on Control in Technical Systems (CTS), 2017, pp. 285-288, doi: 10.1109/CTSYS.2017.8109547.

[2] M. Wada, "An omnidirectional 4WD mobile platform for wheelchair applications," Proceedings, 2005 IEEE/ASME International Conference on Advanced Intelligent Mechatronics., 2005, pp. 576-581, doi: 10.1109/AIM.2005.1511044.

[3] W. Kowalczyk, M. Przybyła and K. Kozłowski, "Rapid navigation function control for omnidirectional mobile platform," 2017 22nd International Conference on Methods and Models in Automation and Robotics (MMAR), 2017, pp. 137-140, doi: 10.1109/MMAR.2017.8046812.

[4] B. I. Adamov, "Influence of mecanum wheels construction on accuracy of the omnidirectional platform navigation (on exanple of KUKA youBot robot)," 2018 25th Saint Petersburg International Conference on Integrated Navigation Systems (ICINS), 2018, pp. 1-4, doi: 10.23919/ICINS.2018.8405889.

[5] National Instruments:"User Guide And Specifications, NI myRIO-1900", 2016. Accessed on: Jun. 15, 2021. [Online]. Available: https://www.ni.com/pdf/manuals/376047c.pdf

[6] National Instruments:"LabVIEW", Accessed on: Jun. 15, 2021. [Online]. Available: https://www.ni.com/en-rs/shop/labview.html

[7] Digilent:"Pmod CMPS2 Reference Manual", 2017. Accessed on: Jun. 15, 2021. [Online]. Available: https://reference.digilentinc.com/_media/reference/pmod/pmodcmps2/pmod_cmps2_rm.pdf

[8] Digilent:"Pmod Gyro Reference Manual", 2016. Accessed on: Jun. 15, 2021. [Online]. Available: https://reference.digilentinc.com/_media/reference/pmod/pmodgyro/pmodgyro_rm.pdf

[9] Cytron:"Dual Channel 10 A DC Motor Driver User's Manual", 2013. Accessed on: Jun. 15, 2021. [Online]. Available: https://www.robotshop.com/media/files/pdf/user-manual-mdd10a.pdf

[10] DFRobot:"12V DC Motor 251RPM w/Encoder". Accessed on: Jun. 15, 2021. [Online]. Available: https://www.robotshop.com/en/12v-dc-motor-251rpm-encoder.html

[11] D.Scaramuzza, Introduction to Autonomous Mobile Robots, 2nd ed., Massachusetts Institute of Technology, 2011.

[12] D. Simon, Optimal State Estimation - Kalman, H_∞, and Nonlinear Approaches, John Wiley and Sons, 2006.

Analysis of FPGA Implementation of Set-based Predictive Control for Grid-tied Inverters

Bruno Vilić Belina
Faculty of Electrical Engineering and Computing
University of Zagreb
Zagreb, Croatia
bruno.vilic-belina@fer.hr

Renato Babojelić
Faculty of Electrical Engineering and Computing
University of Zagreb
Zagreb, Croatia
renato.babojelic@fer.hr

Šandor Ileš
Faculty of Electrical Engineering and Computing
University of Zagreb
Zagreb, Croatia
sandor.iles@fer.hr

Jadranko Matuško
Faculty of Electrical Engineering and Computing
University of Zagreb
Zagreb, Croatia
jadranko.matusko@fer.hr

Abstract—This paper analyses an field-programmable gate array (FPGA) implementation of a set-based model predictive control algorithm for controlling a grid-tied inverter with an LCL filter. Parallelizing the MPC algorithm in FPGA hardware led to substantial decrease in computation time. The implementation is show to be synthesizeable on a commercial mid-range FPGA device.

Index Terms—power converters, model predictive control, robust control, fast gradient projection method, field programmable gate array (FPGA).

I. Introduction

Model predictive control (MPC) has seen increased use for controlling power converters in the last two decades, this is accredited mainly to the increasing computational power and the advent of powerful microprocessors [1]–[3]. Using either of two distinct approaches, continuous-control set (CCS-MPC) or finite-control set (FCS-MPC) model predictive control. Fixed number of switching states in a power converter in combination with short prediction horizons (typically 1 or 2) FCS-MPC algorithms use simple exhaustive searches for finding the optimal switching state, this state is then applied to the power converter. On the other hand controllers employing a CCS-MPC solve an optimization problem at every sample time and output a continuous control signal that is applied to the converter using a modulator.

FCS-MPC approach enables the use of simple and fast optimization algorithms, but the discrete control set implies the use of suboptimal control signal will almost always be used. This requires additional care when designing the controller to guarantee stability [4]–[6]. Lengthening the control horizon has also been considered, but it may lead to untractable optimization problems that require advanced optimization algorithms [7], [8].

The ability to work with different power levels and systems [9] and constant switching frequency are some of the advantages of CCS-MPC. Constant switching frequency is especially important when used with LCL filters as the filter design depends on the switching frequency [10].

The main challenge implementing controllers, using either approach, is satisfying strict execution time constraints imposed by fast power converter dynamics. Because of its inherent parallelism the field-programmable gate array (FPGA) present most promise for implementing MPC. Traditionally used as hardware interface logic circuits, in the last decade they have seen dramatic reduction in price with simultaneous increase in density of logic elements on a chip. Furthermore, FPGAs provide advantages over microcontrollers or DSPs in power converter applications shortening the computational time, which results in a lower control delay and better dynamic performance [11].

In this paper we build on our previous work on set-based approach to control of grid-tied power converters with an LCL filter and propose a FPGA implementation of set-based MPC algorithm using the fast gradient projection method for solving the optimization problem. The proposed approach was tested in simulation using Xilinx ISE Design Suite with the results compared to the FGM implemented on a Texas Instruments microcontroller.

The paper is divided in sections as follows: Section II describes the mathematical model of a grid-tied two-level inverter with an LCL filter, Section III briefly presents a set-based model predictive controller, IV presents a fast gradient method for solving the optimisation problem of a set-based MPC. Section V describes the FPGA implementation details of a proposed controller. Section VI shows the analysis of the FPGA implementation of a proposed controller and Section VII concludes the paper.

978-1-6654-3236-8/21 $31.00 © 2021 IEEE

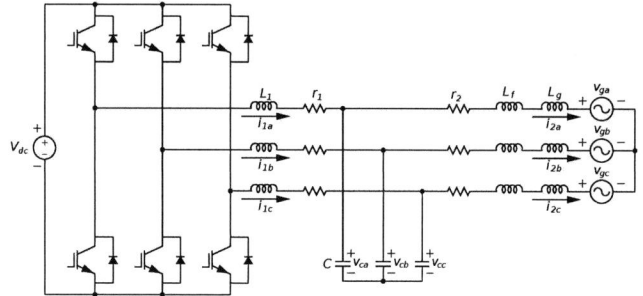

Fig. 1. Two-level grid connected inverter with LCL filter

II. Mathematical Model of a Grid Connected Two Level Converter with an LCL Filter

We consider a grid-tied three-phase inverter with an LCL filter in Fig. 1, to obtain a model in synchronous reference frame, three-phase quantities $i_1 = [i_{1a} \ i_{1b} \ i_{1c}]^T$, $v_c = [v_{ca} \ v_{cb} \ v_{cc}]^T$, $i_2 = [i_{2a} \ i_{2b} \ i_{2c}]^T$ representing converter side current, capacitor voltage and grid side current respectively are transformed as follows:

$$\begin{bmatrix} i_{1d} & i_{1q} \end{bmatrix}^T = T_{dq} \begin{bmatrix} i_{1a} & i_{1b} & i_{1c} \end{bmatrix}^T$$
$$\begin{bmatrix} v_{cd} & v_{cq} \end{bmatrix}^T = T_{dq} \begin{bmatrix} v_{ca} & v_{cb} & v_{cc} \end{bmatrix}^T$$
$$\begin{bmatrix} i_{2d} & i_{2q} \end{bmatrix}^T = T_{dq} \begin{bmatrix} i_{2a} & i_{2b} & i_{2c} \end{bmatrix}^T$$

where T_{dq} stands for well-known Park transformation:

$$T_{dq} = \frac{2}{3} \begin{bmatrix} \sin\theta & \sin\left(\theta - \frac{2}{3}\pi\right) & \sin\left(\theta + \frac{2}{3}\pi\right) \\ \cos\theta & \cos\left(\theta - \frac{2}{3}\pi\right) & \cos\left(\theta + \frac{2}{3}\pi\right) \end{bmatrix}. \quad (1)$$

Let $x(t) = [i_{1d}(t) \ i_{1q}(t) \ v_d(t) \ v_q(t) \ i_{2d}(t) \ i_{2q}(t)]^T$ be the state vector with quantities in stationary frame. Vector $u(t) = [u_d(t) \ u_q(t)]^T$ represents the voltage applied by the inverter and vector $v(t) = [v_d \ v_q]^T$ represents the grid voltage. The controlled output $y(t)$ are the grid currents $[i_{2d}(t) \ i_{2q}(t)]$. The continuous time state-space linear model is given by

$$\dot{x}(t) = Ax(t) + Bu(t) + Dv(t)$$
$$y(t) = Cx(t) \quad (2)$$

where

$$A = \begin{bmatrix} -\frac{r_1}{L_1} & \omega & -\frac{1}{L_1} & 0 & 0 & 0 \\ -\omega & -\frac{r_1}{L_1} & 0 & -\frac{1}{L_1} & 0 & 0 \\ \frac{1}{C} & 0 & 0 & \omega & -\frac{1}{C} & 0 \\ 0 & \frac{1}{C} & -\omega & 0 & 0 & -\frac{1}{C} \\ 0 & 0 & \frac{1}{L_2} & 0 & -\frac{r_2}{L_2} & \omega \\ 0 & 0 & 0 & \frac{1}{L_2} & -\omega & -\frac{r_2}{L_2} \end{bmatrix},$$

$$B = \begin{bmatrix} 1/L_1 & 0 \\ 0 & 1/L_1 \\ 0 & 0 \\ 0 & 0 \\ 0 & 0 \\ 0 & 0 \end{bmatrix}, D = \begin{bmatrix} 0 & 0 \\ 0 & 0 \\ 0 & 0 \\ 0 & 0 \\ -1/L_2 & 0 \\ 0 & -1/L_2 \end{bmatrix}$$

and

$$C = \begin{bmatrix} 0 & 0 & 0 & 0 & 1 & 0 \\ 0 & 0 & 0 & 0 & 0 & 1 \end{bmatrix},$$

where $\omega = 2\pi f$, with f being the grid frequency. Inductance on the grid side L_2 is modeled as

$$L_2 = L_f + L_g \quad (3)$$

with L_f and L_g standing for grid side filter inductance and unknown grid inductance respectively, and $L_g \in [0, L_{\max}]$. To take this uncertainty into consideration we assume parameter L_2 in a range around the given nominal value, ie.

$$L_{2,\text{nom}} - \xi \le L_2 \le L_{2,\text{nom}} + \xi \quad (4)$$

Now the system model (2), considering the said uncertainty, can be written as a polytopic LPV model [12]. Vertices $(A_{c,i}, B_{c,i}, D_{c,i}, C_{c,i})$, $i \in \{1,2\}$ represent 2 possible combinations of extreme values of parameter L_2 and define a polytopic matrix family

$$\mathcal{M} = \left\{ \sum_{i=1}^{2} \alpha_i (A_{c,i}, B_{c,i}, D_{c,i}, C_{c,i}) \ \Big| \ \sum_{i=1}^{2} \alpha_i = 1, \alpha_i \ge 0 \right\}. \quad (5)$$

That is, every system with parameter L_2 inside a predefined interval (4) can be written as a convex combination of vertices $(A_{c,i}, B_{c,i}, D_{c,i}, C_{c,i})$, $i \in \{1,2\}$.

The system can now be written as

$$\dot{x}(t) = A_c(\alpha(t))x(t) + B_c(\alpha(t))u(t) + D_c(\alpha(t))v(t)$$
$$y(t) = C_c(\alpha(t))x(t) \quad (6)$$

where $\alpha : \mathbb{R}_0^+ \to \mathbb{R}^2$ is a function of time that for every $t \ge 0$ provides scalars α_i, $i \in \{1,2\}$ that define the systems as a convex combination of vertices of a family $\mathcal{M}$, obviously $(A_c(\alpha(t)), B_c(\alpha(t)), D_c(\alpha(t)), C_c(\alpha(t))) \in \mathcal{M}$. (For brevity we omit writing the argument t of a function α in the text below.)

To discretize the system with constant sampling rate T we use the forward Euler discretization and obtain

$$x(k+1) = A_d(\alpha)x(k) + B_d(\alpha)u(k) + D_c(\alpha)v(k)$$
$$y(k) = C_c(\alpha)x(k) \quad (7)$$

with

$$A_d(\alpha) = I + TA_c(\alpha),$$
$$B_d(\alpha) = TB_c(\alpha),$$
$$C_d(\alpha) = C_c(\alpha),$$
$$D_d(\alpha) = TD_c(\alpha).$$

A. Equilibrium point

An equilibrium point $[x_d \ u_d]^T$ for system (6), considering a desired reference r of the output $y = [i_{2d} \ i_{2q}]^T$ under constant grid voltage v, is computed as

$$\begin{bmatrix} x_d \\ u_d \end{bmatrix} = \begin{bmatrix} A_c(\alpha) & B_c(\alpha) \\ C_c(\alpha) & 0 \end{bmatrix}^{-1} \left(\begin{bmatrix} 0 \\ r \end{bmatrix} - \begin{bmatrix} D_c(\alpha) \\ 0 \end{bmatrix} v \right). \quad (8)$$

Define $e(k) := x(k) - x_d$ as an error around equilibrium point $[x_d \ u_d]^T$ of a system (7), for a given output reference r under constant grid voltage v, then the error dynamics is obtained

$$e(k+1) = A_d(\alpha)e(k) + B_d(\alpha)u_{\text{err}}(k), \quad (9)$$

where $u_{\text{err}}(k) = u(k) - u_d$.

It is obvious this is also a polytopic family with vertices $(A_{d,i}, B_{d,i})$, $i \in \{1,2\}$ and interior points given by

$$(A_d(\alpha), B_d(\alpha)) = \sum_{i=1}^{2} \alpha_i (A_{d,i}, B_{d,i}) \quad (10)$$

for $\alpha_i \in [0,1]$ for all $i \in \{1,2\}$, $\sum_{i=1}^{2} \alpha_i = 1$.

III. Ellipsoidal MPC

In this section we briefly recall the main results from [13]. A set $\mathcal{I}$ is called a **control invariant** for the polytopic system (9) if for all $e \in \mathcal{I}$ and every $\alpha(t)$ there exists control action $u_{\text{err}} \in \mathcal{U}$ such that $A_d(\alpha)e + B_d(\alpha)u_{\text{err}} \in \mathcal{I}$ holds. Given such set $\mathcal{I}$ it is possible to construct a family of sets $\{\mathcal{I}_i \mid i \in \mathbb{N}\}$ with

$$\begin{aligned} \mathcal{I}_0 &= \mathcal{I} \quad (11) \\ \mathcal{I}_i &= \{e \mid \exists u_{\text{err}} \in \mathcal{U}, \ \forall \alpha(t), \\ & \quad A_d(\alpha)e + B_d(\alpha)u_{\text{err}} \in \mathcal{I}_{i-1}\}. \end{aligned}$$

Sets $\mathcal{I}_i$ contain the states that can be steered to $\mathcal{I}_{i-1}$ in one step. To decrese the computational burden for calculating such sets we adopt the following ellipsoidal inner approximations of sets $\mathcal{I}_i$ proposed in [14]. As shown in [15], given a stabilizing feedback gain K and an nonempty ellipsoidal control invariant set $\mathcal{E} \subset \mathbb{R}^n$, that is:

$$(A_{d,i} - B_{d,i}K)e \in \mathcal{E}, \ \forall e \in \mathcal{E}, \ i \in \{1,2\}, \quad (12)$$

there exists the family of ellipsoidal sets $\{\mathcal{E}_i \mid i \in \mathbb{N}\}$ satisfying the recursion

$$\begin{aligned} \mathcal{E}_0 &= \mathcal{E} \quad (13) \\ \mathcal{E}_i &= \mathbf{In}(\{e \mid \exists u_{err} \in \mathcal{U}, \ \forall \alpha(t), \\ & \quad A_d(\alpha)e + B_d(\alpha)u_{\text{err}} \in \mathcal{E}_{i-1}\}), \end{aligned}$$

where **In** is the operation finding inner ellipsoidal approximation of a given set.

For the recursive construction of ellipsoidal sets a feedback gain K that is stabilizing for the vertices of a system (9)

needs to be obtained. Using the linear matrix inequality (LMI) approach presented in [16] the synthesized feedback gain K is guaranteed to be close loop stable for system (9).

Ellipsoidal sets allow the following representation: let $\mathcal{Q}$ be a positive semidefinite matrix then

$$\mathcal{E}_{\mathcal{Q}} = \{x \in \mathbb{R}^n \mid x^T \mathcal{Q} x \le 1\} \quad (14)$$

is an ellipsoidal set with a centre in origin. Terminal ellipsoidal set $\mathcal{E}_0$ is now obtained solving the optimization problem

$$\begin{aligned} \min_{a \in \mathbb{R}} \quad & a \\ \text{s.t.} \quad & K'K - a\mathcal{P} \le 0, \quad (15) \\ & a \ge 0, \end{aligned}$$

and setting its defining positive semifinite matrix

$$\mathcal{P}_0 = \frac{a}{u_{max}^2}\mathcal{P}, \quad (16)$$

where u_{max} is the maximal magnitude of a control signal u_{err}.

For iterative construction of ellipsoids $\mathcal{E}_i$, $i \in \{1, \ldots n\}$ we use the following equation as proposed in [14]

$$\mathcal{E}_i = \Pi_e \left(\mathbf{In} \left(\tilde{\mathcal{E}}_{i-1}^1 \cap \tilde{\mathcal{E}}_{i-1}^2 \cap (\mathbb{R}^6 \times \mathcal{E}^{\mathcal{U}}) \right) \right) \quad (17)$$

where $\tilde{\mathcal{E}}_{i-1}^j$ are ellipsoids defined in the extended space $[e \ u_{err}]^T \in \mathbb{R}^8$ as

$$\tilde{\mathcal{E}}_{i-1}^j = \{[e \ u_{err}]^T \mid A_{d,j}e + B_{d,j}u_{err} \in \mathcal{E}_{i-1}\}, \quad (18)$$

$\mathcal{E}^{\mathcal{U}}$ is ellipsoidal representation of the set of control inputs and Π_e is projection operation to the state space from the extended space.

Set-based ellipsoidal MPC algorithm is given with Algorithm 1, where J is some appropriate cost function.

Algorithm 1: Set-based Ellipsolidal MPC

$k = 0$
L: Find $i(k) = \min\{i \mid e(k) \in \mathcal{E}_i\}$
if $i(k) == 0$ **then**
   | $u_{err}(k) = -Ke(k)$
end
else

$$\begin{aligned} u(k) &= \min J(e(k), u(k)) \\ \text{s.t.} & \quad (19) \\ A_{d,j}e(k) &+ B_{d,j}u(k) \in \mathcal{E}_{i(k)-1}, j \in \{1,2\} \end{aligned}$$

end
$k = k+1$; jump to **L**

IV. Computationally efficient set-based MPC based on fast gradient projection method

In this section we recall of the computationally efficient method for solving the optimization problem (19), based on fast gradient method presented in our previous work [17].

A. Set-based fast gradient projection method

The optimization problem (19) that need to be solved online at each time step is a quadratic problem with set membership constraints that requires the next system's states to lie in ellipsoidal set $\mathcal{E}_{i(k)-1}$. However, such the formulation is not suitable to solve it using fast gradient method. In order to overcome this problem we redefine the problem (19) to replace a state constraint with corresponding input constraint:

$$u(k) = \min_{} J(e(k), u(k)) \\ \text{s.t.} \ \ u(k) \in \mathcal{U}_k(e(k)) \tag{20}$$

where the set $\mathcal{U}_k(e(k))$ is defined as follows:

$$\mathcal{U}_k(e(k)) = \{u(k) \mid A_{d,j}e(k) + B_{d,j}u(k) \in \mathcal{E}_{i(k)-1}, j \in \{1,2\}\}. \tag{21}$$

In other words, given state $e(k)$ the set $\mathcal{U}_k(e(k))$ represents a set of all inputs that will steer the next state $e(k+1)$ to set $\mathcal{E}_{i(k)-1}$. The set $\mathcal{U}_k(e(k))$ can be easily calculated from the set $\bar{\mathcal{E}}_i = \mathbf{In}\left(\tilde{\mathcal{E}}_i^1 \cap \tilde{\mathcal{E}}_i^2 \cap (\mathbb{R}^6 \times \mathcal{E}^{\mathcal{U}})\right)$ (see equation 18) defined in an extended space by fixing $e(k)$. The obtained set is also ellipsoidal and as such represents a set that is simple enough to allow an efficient calculation of the projection operation onto it. In the extended space the ellipsoid $\bar{\mathcal{E}}_i$ is represented as:

$$\begin{bmatrix} e \\ u \end{bmatrix}^T \begin{bmatrix} P_1 & P_{12} \\ P_{12}^T & P_2 \end{bmatrix} \begin{bmatrix} e \\ u \end{bmatrix} \leq 1 \tag{22}$$

which can be further rewritten, for $e = e(k)$, as:

$$u^T P_2 u + 2u^T P_{12} e(k) + e(k)^T P_1 e(k) \leq 1 \tag{23}$$

Equation (23) represents a scaled and translated ellipsoid $u^T P_2 u \leq 1$ which can be written as:

$$(u-a)^T P_2 (u-a) \leq \gamma, \tag{24}$$

where $a = P_2^{-1} P_{12} e(k)$ and $\gamma = 1 + e(k)^T (P_{12}^T (P_2^{-1})^T P_{12} - P_1) e(k)$.

Once a candidate v^i solution at i-th iteration has been calculated its projection onto set $\mathcal{U}_k(e(k))$ can be easily calculated as:

$$u^i = \Pi_{\mathcal{U}_k(e(k))} v^i = \begin{cases} v^i, & ||v^i - a|| \leq \gamma \\ \frac{v^i - a}{||v^i - a||}\gamma + a, & ||v^i - a|| > \gamma \end{cases} \tag{25}$$

V. FPGA IMPLEMENTATION

In this section we describe the implementation of Set-based MPC in FPGA hardware. Possibility of parallel computing in FPGA is exploited to calculate the values of quadratic forms (each corresponding to an ellipsoidal set) for the given error state vector e, in order to find the smallest ellipsoidal set containing it. FGM (Algorithm 2) is used for solving the MPC optimisation problem (20). MPC algorithm implemented in FPGA calculates control signal for two level grid-tied inverter depending on the current state vector.

So called Half-precision floating-point (16bit) number format [18] is used as a number format in order to reduce the usage of logic cells of the FPGA fabric.

Algorithm 2: Set-based fast gradient method for constrained minimization

Data: Initial point $u^0 \in \mathcal{U}_\mathcal{N}, y^0 = u^0$, number of iterations $i_{\max}$, Lipschitz constant L, scaling factors $\beta^0, \ldots, \beta^{i_{\max}-1}$

Calculate set $\mathcal{U}_k(e(k))$ from the set $\bar{\mathcal{E}}_i$ by fixing $e(k)$ and parameters a and γ;

for $i = 0 \to i_{\max} - 1$ **do**
$\quad v^{i+1} = y^i - \frac{1}{L}\nabla J_N(y^i) = y^i - \frac{1}{L}(2Hy^i + h) = My^i + g$;
$\quad$ Calculate $u^{i+1} = \Pi_{\mathcal{U}_k(e(k))}(v^{i+1})$ using (25);
$\quad y^{i+1} = U^{i+1} + \beta^i(u^{i+1} - u^i)$;
end

Block diagram of main modules is shown in Figure 2 Design consists of four main modules and their functionality is described in this section.

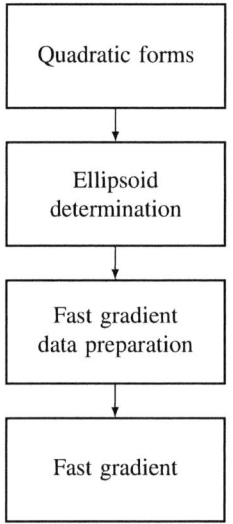

Fig. 2. Block diagram of main FPGA modules

A. Quadratic forms

In the state space every ellipsoid $\mathcal{E}_i$, $i \in \{0, \ldots, N\}$ is represented by a positive definite matrix $\mathcal{P}_i$. To check if the error vector e is contained in the ellipsoid $\mathcal{E}_i$ we evaluate the quadratic form inequality:

$$e^T \mathcal{P}_i e \leq 1. \tag{26}$$

This module calculates the values of quadratic forms in (26) for the current error vector for all ellipsoids 1 to N in parallel. Because the matrices $\mathcal{P}_i$, $i \in \{0, \ldots, N\}$ are positive definite, they can be decomposed as a product of upper triangular matrices U using the well known Cholesky decomposition:

$$\mathcal{P}_i = U^T U \tag{27}$$

This reduces the number of mathematical operations that need to be done in order to evaluate the quadratic form.

978-1-6654-3236-8/21 $31.00 © 2021 IEEE

B. Ellipsiod determination

If the quadratic form inequality (26) is satisfied, error e is contained in the i-th ellipsoid. Depending on the quadratic form values of all ellipsoids, this module finds the index i of a smallest ellipsoid that contains the current error. Of all quadratic forms with value less than 1, form with value closest to 1 is the one with the current error vector e closest to the ellipsoid border, that is, the corresponding ellipsoid is the smallest ellipsoid containing current error vector. The functionality of this module is implemented as sequential search of maximal quadratic form value with parallelised mathematical operations ignoring values greater than 1.

C. Fast gradient data preparation

This module prepares data for fast gradient method (Algorithm 2) based on the current ellipsoid i and error e. Values calculated are the parameters a and γ defined in (23), and used in the projection step of the algorithm; and the parameter $g = He$ needed for gradient descent step in each iteration, where matrix

$$H := -\frac{2}{L} B_{d,j}^T A_{d,j} \qquad (28)$$

is calculated offline.

D. Fast gradient

This module performs fast gradient method described in Algorithm 2 for solving the MPC optimisation problem. First step is potential solution calculation v^{i+1}, i.e. the gradient step. Second step is the projection step; if potential solution lays outside of a feasible control set ($||v^i - a|| > \gamma$) it is projected onto the feasible region using (25). The projection operation involves computing an euclidean norm of a vector, which needs to evaluate a square root, and dividing by the norm value. Both operations are very resource heavy to implement in hardware. Fast inverse square root [19] ocuupies less slices (FPGA resources) than reciprocal unit [20] and square root unit [21] combined. The so called Quake's fast inverse square root algorithm [22] solves this problem as it uses only shift, product and sum operations that can be efficiently mapped in FPGA hardware components. The last step calculates final control signal value in each iteration. This is the step which accelerates algorithm based on function gradient.

VI. RESULTS

For the grid tied inverter with an LCL filter and parameters given in Table I, an MPC controller is designed using the procedure described in [13]. State feedback controller K is synthesized using parameter values of $r = 0.42$, $d = 0.5$ and $\mu = 1000$ with control gains

$$K^T = \begin{bmatrix} 38.3729 & -1.5875 \\ 1.1164 & 38.5296 \\ -0.4380 & -0.1198 \\ 0.0601 & -0.5018 \\ -33.6588 & 0.8758 \\ -0.8565 & -34.2581 \end{bmatrix}. \qquad (29)$$

TABLE I
INVERTER AND GRID PARAMETERS

Symbol	Description	Value	Unit
r_1	Inverter resistance	0.5	Ω
L_1	Filter inductance	1.7	mH
C	Filer capacitance	5	μF
r_2	Grid resistance	0.5	Ω
L_f	Filter inductance	1.7	mH
L_g	Grid inductance	0–1	mH
V_{dc}	DC link voltage	500	V
v_g	Grid peak voltage	180	V
f	Grid frequency	60	Hz
f_{PWM}	PWM frequency	10	kHz

The objective functions are defined as:

$$J_{i(k)} = \\ \max_j (A_{d,j}e(k) + B_{d,j}u(k))^T \mathcal{P}_{i(k)-1} (A_{d,j}e(k) + B_{d,j}u(k)) \qquad (30)$$

where $\mathcal{P}_{i(k)-1}$ is matrix representing $i(k) - 1$ ellipsoid, is chosen to provide a pseudo minimum time control.

Based on the objective function $J_{i(k)}$, and for every $i \in \{0, \dots, N\}$ the Lipschitz constant L, scaling factors $\beta^0, \dots, \beta_{\max}^k$ and matrix M, used in Algorithm 2, are computed offline and used as constants in our FPGA implementation.

Modules described in Section V were synthesized for the Xilinx Artix-7 XC7A200T device using Xilinx's ISE Design Suite development environment. Depending on the maximum number of gradient steps $i_{\max}$ in Algorithm 2, we compared the FPGA resources utilized for $i_{\max} \in \{1, \dots, 7\}$. Results, shown in Table VI, confirm our implementation viable as the utilized logic on said FPGA device ranges between 56% and 87% for $i_{\max} = 1$ and $i_{\max} = 7$ respectively. Static timing tests were also performed for said maximum number of gradient steps. Figure 3 shows the obtained timing results compared to the implementation on Texas Instruments TMS320F28335 DSP presented in our previous work [17].

Execution time directly depends on the number of fast gradient iterations. The results show great improvement in computational speed of FPGA implementation over the DSP one, for any number of iterations.

VII. CONCLUSION

In this paper, we propose the FPGA implementation of a set-based fast gradient projection model predictive control algorithm for control of a two-level grid-tied inverter with an LCL filter. Using the FPGAs inherent potential for parallel computation we parallelized parts of the set-based MPC algorithm. The proposed approach is verified in simulation using Xilinx ISE Design Suite environment, implementation is shown to be synthesizeable on a commercial mid-range FPGA device, and the computation is carried out in the microsecond range.

TABLE II

FPGA RESOURCE UTILIZATION FOR SET-BASED MPC ALGORITHM DEPENDING ON NUMBER OF FAST GRADIENT ITERATIONS

Iterations	Occupied slices (% of 33 650)	Slice LUTs (% of 134 600)	Unused flip flops (% of 60 508)	Fully used LUT-FF pairs (% of 60 508)	DSPs (% of 740)
1	18 987 (56%)	60 508 (44%)	60 505 (99%)	3 (1%)	363(49%)
2	19 820 (58%)	65 135 (48%)	65 192 (99%)	6 (1%)	393(53%)
3	24 773 (73%)	70 349 (52%)	70 340 (99%)	9 (1%)	423(57%)
4	25 801 (76%)	75 155 (55%)	75 143 (99%)	12 (1%)	453(61%)
5	26 951 (80%)	79 995 (59%)	79 980 (99%)	15 (1%)	483(65%)
6	28 270 (84%)	84 834 (63%)	84 816 (99%)	18 (1%)	513(69%)
7	29 409 (87%)	89 669 (66%)	89 648 (99%)	21 (1%)	543(73%)

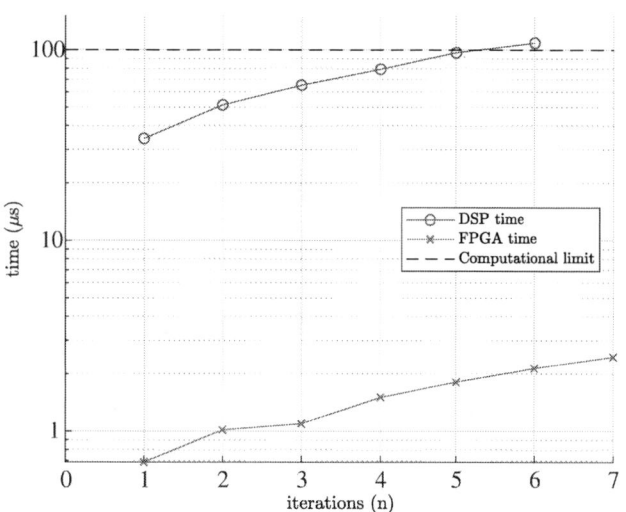

Fig. 3. Comparison of computational times between DSP and FPGA implementations for different values of maximum number of iterations

REFERENCES

[1] S. Kouro, M. A. Perez, J. Rodriguez, A. M. Llor, and H. A. Young, "Model predictive control: Mpc's role in the evolution of power electronics," *IEEE Industrial Electronics Magazine*, vol. 9, no. 4, pp. 8–21, 2015.

[2] S. Vazquez, J. Rodriguez, M. Rivera, L. G. Franquelo, and M. Norambuena, "Model predictive control for power converters and drives: Advances and trends," *IEEE Transactions on Industrial Electronics*, vol. 64, no. 2, pp. 935–947, 2016.

[3] P. Karamanakos, E. Liegmann, T. Geyer, and R. Kennel, "Model predictive control of power electronic systems: Methods, results, and challenges," *IEEE Open Journal of Industry Applications*, vol. 1, pp. 95–114, 2020.

[4] R. P. Aguilera and D. E. Quevedo, "Predictive control of power converters: Designs with guaranteed performance," *IEEE Transactions on Industrial Informatics*, vol. 11, no. 1, pp. 53–63, 2014.

[5] M. Preindl, "Robust control invariant sets and lyapunov-based mpc for ipm synchronous motor drives," *IEEE Transactions on Industrial Electronics*, vol. 63, no. 6, pp. 3925–3933, 2016.

[6] H. A. Young, M. A. Perez, J. Rodriguez, and H. Abu-Rub, "Assessing finite-control-set model predictive control: A comparison with a linear current controller in two-level voltage source inverters," *IEEE Industrial Electronics Magazine*, vol. 8, no. 1, pp. 44–52, 2014.

[7] T. Geyer and D. E. Quevedo, "Multistep finite control set model predictive control for power electronics," *IEEE Transactions on power electronics*, vol. 29, no. 12, pp. 6836–6846, 2014.

[8] P. Falkowski and A. Sikorski, "Finite control set model predictive control for grid-connected ac–dc converters with lcl filter," *IEEE Transactions on Industrial Electronics*, vol. 65, no. 4, pp. 2844–2852, 2017.

[9] M. Preindl and S. Bolognani, "Comparison of direct and pwm model predictive control for power electronic and drive systems," in *2013 Twenty-Eighth Annual IEEE Applied Power Electronics Conference and Exposition (APEC)*, IEEE, 2013, pp. 2526–2533.

[10] A. Reznik, M. G. Simoes, A. Al-Durra, and S. Muyeen, "Lcl filter design and performance analysis for grid-interconnected systems," *IEEE transactions on industry applications*, vol. 50, no. 2, pp. 1225–1232, 2013.

[11] Z. Ma, S. Saeidi, and R. Kennel, "Fpga implementation of model predictive control with constant switching frequency for pmsm drives," *IEEE transactions on industrial informatics*, vol. 10, no. 4, pp. 2055–2063, 2014.

[12] S. Boyd, L. El Ghaoui, E. Feron, and V. Balakrishnan, *Linear matrix inequalities in system and control theory*. Siam, 1994, vol. 15.

[13] R. Babojelić, Š. Ileš, and J. Matuško, "Set-based predictive control of a grid-tied inverter with lcl filter under variable grid inductance conditions," in *2020 IEEE 11th International Symposium on Power Electronics for Distributed Generation Systems (PEDG)*, 2020, pp. 236–241. DOI: 10.1109/PEDG48541.2020.9244331.

[14] D. Angeli, A. Casavola, G. Franzè, and E. Mosca, "An ellipsoidal off-line mpc scheme for uncertain polytopic discrete-time systems," *Automatica*, vol. 44, no. 12, pp. 3113–3119, 2008.

[15] D. Angeli, A. Casavola, and E. Mosca, "Predictive control with partial state information," in *2001 European Control Conference (ECC)*, IEEE, 2001, pp. 962–967.

[16] L. Maccari, J. Massing, L. Schuch, C. Rech, H. Pinheiro, V. Montagner, and R. Oliveira, "Robust h_∞ control for grid connected pwm inverters with lcl filters," in *2012 10th IEEE/IAS International Conference on Industry Applications*, IEEE, 2012, pp. 1–6.

[17] R. Babojelić, Š. Ileš, V. Šunde, and J. Matuško, "Computationally efficient set-based predictive control for grid-tied inverters," in *2021 22nd IEEE International Conference on Industrial Technology (ICIT)*, IEEE, vol. 1, 2021, pp. 1283–1288.

[18] M. S. Committee *et al.*, *754-2019-ieee standard for floating-point arithmetic*, 2019.

[19] S. Zafar and R. Adapa, "Hardware architecture design and mapping of 'fast inverse square root' algorithm," in *2014 International Conference on Advances in Electrical Engineering (ICAEE)*, IEEE, 2014, pp. 1–4.

[20] D. Chen, B. Zhou, Z. Guo, and P. Nilsson, "Design and implementation of reciprocal unit," in *48th Midwest Symposium on Circuits and Systems, 2005.*, IEEE, 2005, pp. 1318–1321.

[21] S. Suresh, S. F. Beldianu, and S. G. Ziavras, "Fpga and asic square root designs for high performance and power efficiency," in *2013 IEEE 24th International Conference on Application-Specific Systems, Architectures and Processors*, IEEE, 2013, pp. 269–272.

[22] C. Lomont, "Fast inverse square root," *Technical Report*, vol. 32, 2003.

Grid-connected and Islanded Control of Energy Storage Converter

Božo Terzić
Department of Power Engineering
FESB, University of Split
Split, Croatia
bterzic@fesb.hr

Ozren Bego
Department of Power Engineering
FESB, University of Split
Split, Croatia
obego@fesb.hr

Marin Despalatović
Department of Power Engineering
FESB, University of Split
Split, Croatia
despi@fesb.hr

Goran Majić
Department of Power Engineering
FESB, University of Split
Split, Croatia
gomajic@fesb.hr

Ante Kriletić
Department of Power Engineering
FESB, University of Split
Split, Croatia
akrileti@fesb.hr

Mislav Blajić
Department of Power Engineering
FESB, University of Split
Split, Croatia
mblaji00@fesb.hr

Abstract—The development of distributed generation systems allows for the widespread application of microgrids in electric power systems. In most cases, renewable energy sources (RES) are used to provide power generation. Use of RES reduces environmental impact over conventional generation units but can lead to power outages in microgrids. In order to alleviate this and other drawbacks of RES, battery energy storage systems (BESS) can be used in coordination with RES. Amongst other functions, BESS can provide microgrids with power during outages. The main parts of BESS are the voltage source converter (VSC) and batteries. VSC allows for both grid-connected and islanded operation of the BESS. This paper presents simple control algorithms of battery storage converter during grid-connected and islanded mode of operation as well as algorithms for transition from one mode to another and vice versa. Experimental verification of presented algorithms is performed using prototype of 100 kVA storage converter and battery string with capacity of 70 kWh.

Keywords—energy storage, lithium-ion battery, PWM converter, grid-connected, islanded mode of operation

I. INTRODUCTION

In the last decade, a high increase of renewable energy sources (RES) can be observed in electric power systems. This is mainly because of their reduced environmental impact, in comparison with conventional generation units, and the fact that different types of RES can be installed in almost every location in the world. One of their advantages is that they can be used in remote microgrids disconnected from the main grid. Nevertheless, renewable energy systems are highly dependent on weather conditions that can lead to power outages in microgrids or use of additional conventional sources in large electrical systems.

Battery energy storage systems (BESS) are interfaced to the grid through voltage source converters (VSC) and can be used for different applications. One of those applications is uninterruptible power supply (UPS) which allows BESS to be implemented with RES in order to address its drawbacks. Microgrids can operate in two modes. In grid connected mode of operation, RES and/or electric power grid provide power to the microgrid. In islanded mode of operation, BESS is the only source in the microgrid. During grid-connected operation, the grid dictates voltage and frequency levels and only active and reactive power of BESS can be controlled. On other hand, during islanded operation, BESS is required to provide stable voltage and frequency levels throughout the microgrid.

Generally, when grid-connected mode of operation is active, batteries are charged, and are discharged during islanded operation. Furthermore, transition between the two modes of operation should be carried out in such a way to avoid transient overcurrents. This requires development of algorithms that can detect different operating modes in microgrids and transition from grid-connected to islanded mode of operation while avoiding the loss of power.

Control techniques of voltage source converters, used in renewable energy systems, are well known and provide stable operation and rapid response with low levels of current harmonic distortion [1]. Two strategies for the transition between grid-connected and islanded mode are proposed in [2]. In both cases, islanding algorithm based on voltage and frequency measurement is used to determine whether the grid-connected mode controller or the islanding mode controller sets the referent values for the PWM algorithm. However, validation of the proposed strategies is carried out using only simulations. The transition algorithm described in [3] requires two phase locked loops (PLL) and three PI controllers whose gains are determined through trial and error. The proposed algorithm is evaluated using simulations and hardware-in-the-loop (HIL) technique. As stated in [3], this evaluation process does not take into account inherent time delays present in the actual system that may affect the resulting waveforms. Riding-through control and virtual inductance concept are used to suppress transient currents in [4] and transition can be achieved without grid-side measurements. Although, [2] – [7] present various algorithms there is a need for a straightforward approach that can be implemented in a low-cost microcontroller.

This paper describes very simple control system during islanded (off-grid) mode of operation as well as algorithms for transition from grid-connected (on-grid) to islanded mode and vice versa. The system description is shown in Section II. It presents descriptions of the power circuit and the proposed control algorithms. In Section III, the laboratory setup is explained and experimental results are given. The final section presents the conclusion.

II. SYSTEM OVERVIEW

A. Power circuit

Fig. 1 shows power circuit and control system of energy storage system for grid-connected and islanded mode of operation. A battery string consisting of lithium-ion batteries is used to store electrical energy. The three-phase IGBT

This paper is a result of research project "Active system for electric energy storage and stabilization of electric grid" that is co-financed from EU structural funds under grant number KK.01.2.1.01.0026.

978-1-6654-3236-8/21 $31.00 © 2021 IEEE

voltage source PWM converter is fed by the battery string and is connected to the grid through LCL filter. The system operates in grid-connected or islanded mode of operation depending on whether the grid circuit breaker (GCB) is switched-on or off, respectively. At the point of common coupling (PCC) local consumers are also connected to the grid (400 V, 50 Hz) and remain connected in islanded mode of operation after grid disconnection.

The basic system parameters are listed in Table I.

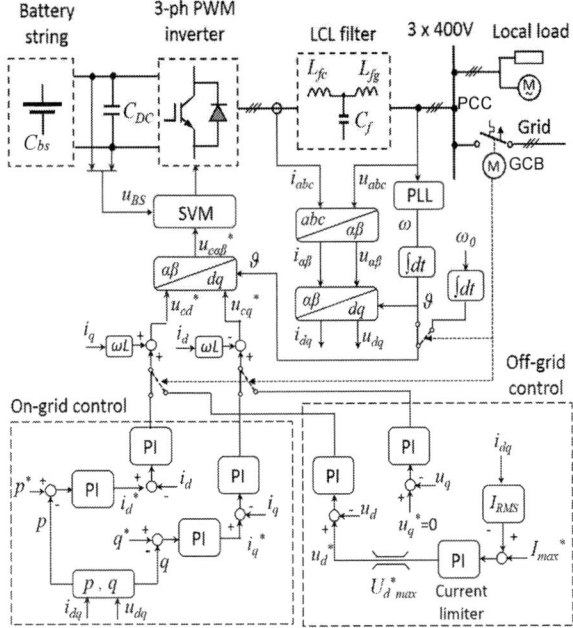

Fig. 1. Block diagram of control system for on- and off-grid operation

TABLE I. PARAMETERS OF SYSTEM SHOWN IN FIG. 1

Symbol	Quantity (Parameter)	Value
U_g	Nominal grid voltage (rms)	400 V
f	Grid frequency	50 Hz
I_g	Nominal converter current on ac side (rms)	150 A
L_{fg}	Inductance of grid-side filter inductors	40 μH
L_{fc}	Inductance of converter-side inductors	120 μH
C_f	Capacitance of filter capacitors	200 μF
C_{DC}	DC link capacitance	4.4 mF
f_s	Converter switching frequency	12 kHz
U_{BS}	Nominal battery string voltage	700 V
C_{BS}	Capacity of battery string	70 kWh

B. On-grid control

The grid-connected control structure is cascaded with the outer loops controlling active (p) and reactive (q) power using PI controllers. Since power is calculated using d- and q-components of the converter current (i_{abc}), the reference reactive power at PCC (Q^*) should be corrected for the

reactive power of the LCL filter. Both active and reactive power can be positive and negative. There are two inner current loops that are implemented in synchronous dq-rotating frame with d-axis aligned to the grid voltage space vector. The reference value of the active current component (i_d^*) is generated by the active power controller. On the other hand, the reactive current reference (i_q^*) is generated by the reactive power controller. To decouple the d- and q-current dynamics, decoupling terms $-i_d \omega L_f$ and $i_q \omega L_f$ are added to the outputs of the current PI controllers. The resultant signals are d- and q-reference of the converter voltages (u_{cd}^*, u_{cq}^*). These signals are transformed to the stationary reference frame, and obtained signals ($u_{c\alpha}^*$, $u_{c\beta}^*$) are used as inputs for space vector modulation (SVM) algorithm of the IGBT converter. Transformation from the synchronous rotating reference frame to the stationary reference frame ($dq{\rightarrow}\alpha\beta$) is performed using angle ϑ obtained by integrating the grid frequency (ω_g) determined by the PLL algorithm. Double second order generalized integrator PLL (DSOGI-PLL) as described in [8] is implemented in the laboratory setup. This algorithm is chosen due to its robust performance during unbalance load conditions. PLL algorithm is used to synchronize the supply system to the grid in grid-connected mode. In addition, PLL algorithm is active in islanded mode of operation in order to calculate angle used for transformation from the stationary to the synchronous rotating reference frame ($\alpha\beta{\rightarrow}dq$).

C. Off-grid control

In islanded mode of operation, transformation between the synchronous rotating reference frame and the stationary reference frame for obtaining reference voltage ($u_{c\alpha}^*$, $u_{c\beta}^*$) is performed using the reference frequency (ω_o). Active and reactive voltage controllers are used to control the converter voltage. The reference value of the reactive voltage controller (u_q^*) is usually set to zero, while the reference value of the active voltage controller (u_d^*) is set by the current limiter. The reference value of the active voltage controller is reduced from value U_{dmax}^* only if the converter current is greater than the maximum allowed value (I_{max}^*). It should be pointed out that initial values of parameters of PI controllers for off-grid control have been determined using simulations. However, optimal system performance was achieved through trial and error after fine tuning of PI controller parameters.

D. Transition algorithm

Active and reactive current controllers are not used to set the SVM reference values during islanded mode of operation but remain active in follow mode. When in follow mode, active and reactive current controllers' outputs are equal to active and reactive voltage controllers' outputs, respectively. This requires setting the controller integrator's output to match the voltage controller's output and using the current feedback as reference values of the current controllers. The same principle applies for active and reactive voltage controllers during grid-connected mode of operation.

Transition to islanded mode of operation can be carried out based on grid circuit breaker status or grid voltage measurement. In this paper, transition between two modes of operation is activated by the GCB's auxiliary output state. When the auxiliary output is in OFF state, microgrid transitions to islanded mode. Angle ϑ, used for transformation between the synchronous (dq) rotating reference frame and the stationary ($\alpha\beta$) reference frame is set by the microcontroller starting from the last value obtained during grid-connected mode.

III. EXPERIMENTAL SETUP AND RESULTS

A. Laboratory setup

Measurements were performed using laboratory setup consisting of 100 kVA storage converter, battery string with capacity of 70 kWh and local load in accordance with Fig. 1. Transition algorithm and (grid-connected and islanded) control of storage converter are implemented in Texas Instruments' TMS320F28335 microcontroller. Battery system includes LiFePO4 cells connected in series (219S1P) and battery management system (BMS). Nominal data of a single battery cell are given in Table II. Storage converter and battery string of the laboratory setup are shown in Fig. 2 and Fig. 3, respectively. Local load consists of adjustable three-phase resistive heaters with maximum power of 56 kW and an induction motor with rated mechanical power of 30 kW. Induction motor is mechanically coupled to a four-quadrant DC drive that serves as an adjustable load.

TABLE II. BATTERY CELL PARAMETERS

Quantity (Parameter)	Value
Rated capacity (0.2C)	100 Ah
Nominal voltage (0.2C)	3.2 V
Max. charging current	$< 0.8C - 1C$
Continuous current	1C
Constant Current	100 A
Power density	> 800 W/kg
Internal resistance	≤ 0.4 mΩ
Operating temperature	0°C~ +40°C
Self discharge rate (month)	≤ 2 %

B. Experimental results

Experiments were carried out for various operating points, as well as for several cases of transition between two operating modes. Initially, experimental results in islanded mode of operation will be given for both steady-state and transient operation. Afterwards, results presenting transition between two modes of operation will be shown. The voltage differential and current probes are used to measure voltages and storage converter's currents at PCC.

Figs. 4 and 5 show steady-state waveforms and average rms spectrum of phase-to-phase voltages at PCC during islanded mode of operation, respectively. Figs. 6 and 7 show steady-state waveforms and average rms spectrum of converter currents for the same operating point, respectively. BESS supplies resistive heaters with 47 kW of output power and additional 26 kW is supplied due to induction motor operation. Both voltages and currents exhibit sinusoidal waveforms with low levels of harmonic distortion as can be seen in Figs. 5 and 7. For this operating point, total harmonic distortion (THD) of voltages at PCC is 1.2%, while the THD of converter current is 1.1 %.

During islanded mode of operation, reference voltage level is kept constant at U_{dmax}^* if the converter current is below the limit. When the current reaches the maximum allowed value (I_{max}^*), the reference voltage value is reduced. Fig. 8 represents limitation of converter current during motor start-up. For this

experiment only, induction motor with nominal power of 5 kW was used. Phase-to-phase voltages at PCC are presented with red and blue lines while the converter currents are shown with green and purple lines. Induction motor with no load is started at $t = 0.04$ s and during start-up, voltage reference value is adjusted.

Fig. 2. Prototype of 100 kVA storage converter

Fig. 3. 70 kWh LiFePO4 battery string of the laboratory setup

Fig. 9 presents phase-to-phase voltages at PCC (red and blue lines) and converter currents (green and purple lines) during transition from islanded to grid-connected mode of operation. While in islanded mode of operation, storage system supplies 73 kW of output power (resistive heaters and induction motor). At $t = 0.042$ s, GCB is switched on and afterwards local load is supplied by the grid. In grid-connected mode of operation, the supply system compensates the reactive power of the LCL filter capacitors. As can be seen, transition to grid-connected mode of operation is almost seamless and does not lead to transient overcurrents due to optimized system performance.

Fig. 10 shows phase-to-phase voltages at PCC (red and blue lines) and converter currents (green and purple lines) during transition from grid-connected to islanded mode of operation. At t = 0.068 s, GCB is switched off. When in islanded mode of operation, the storage system supplies 9 kW of resistive heaters and an induction motor operating at 19 kW of input power. Voltage at PCC is restored after approximately 0.038 s. During transition to islanded mode of operation, transient overcurrents can be observed.

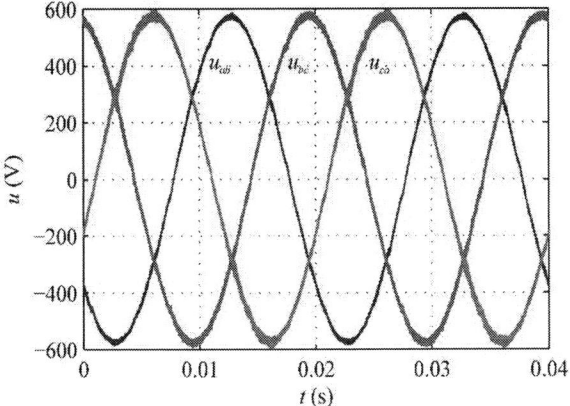

Fig. 4. Measured phase-to-phase voltages at PCC in islanded mode of operation

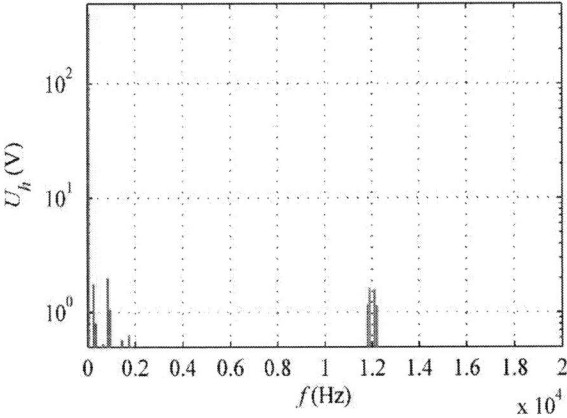

Fig. 5. Average rms spectrum for voltages presented in Fig. 4

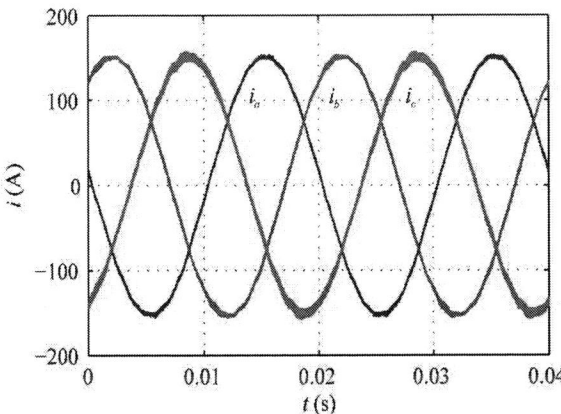

Fig. 6. Measured converter currents in islanded mode of operation

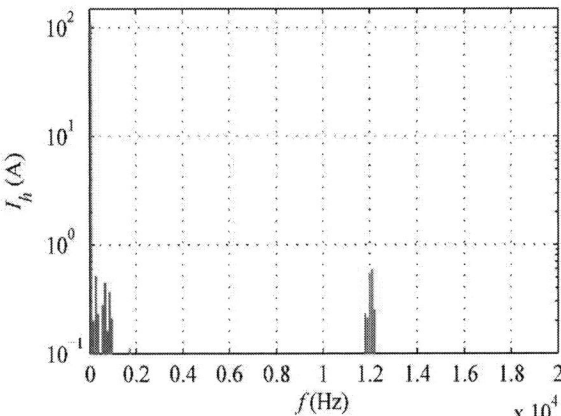

Fig. 7. Average rms spectrum for currents presented in Fig. 6

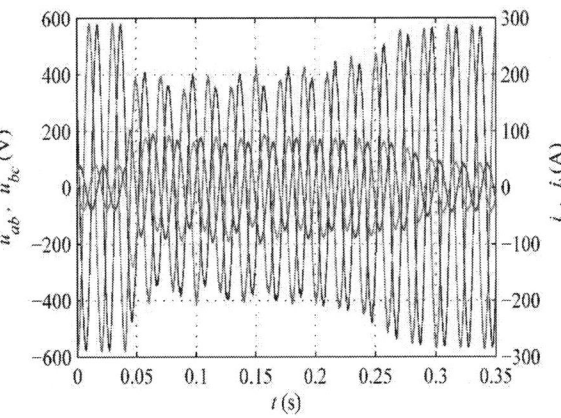

Fig. 8. Measured phase-to-phase voltages voltages at PCC (red and blue lines) and converter currents (green and purple lines) during motor start-up in islanded mode of operation

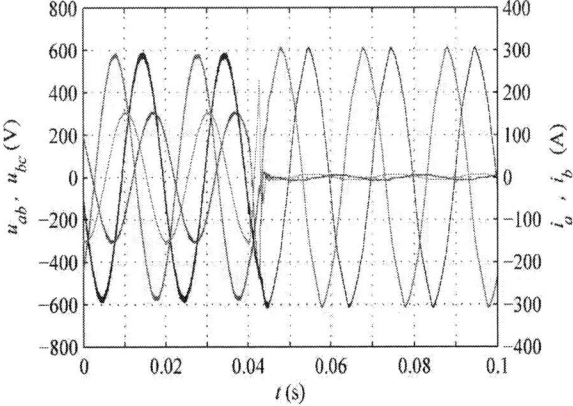

Fig. 9. Measured phase-to-phase voltages at PCC (red and blue lines) and converter currents (green and purple lines) during transition from islanded to grid-connected mode of operation

IV. CONCLUSION

In this paper, BESS system is used to provide power to the microgrid during islanded mode of operation. This paper presents control system of BESS during grid-connected and islanded mode of operation as well as algorithms for transition from grid-connected to islanded mode and vice versa. The proposed approach is straightforward and simple to

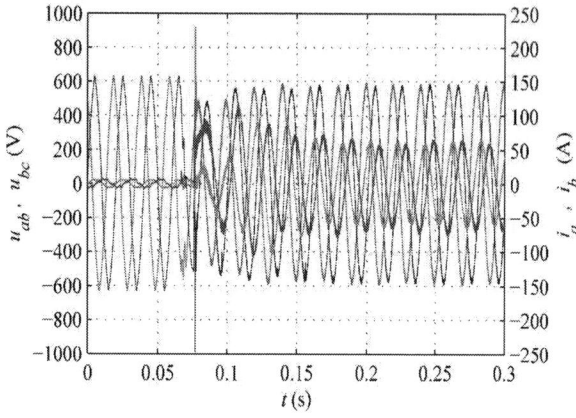

Fig. 10. Measured phase-to-phase voltages at PCC (red and blue lines) and converter currents (green and purple lines) during transition from grid-connected to islanded mode of operation

implement. In grid-connected mode, storage converter provides reference values of active and reactive power to the microgrid. On the other hand, while in islanded mode, BESS system supplies the microgrid and maintains stable voltage and frequency levels. For steady-state operation in islanded mode, experimental results indicate stable operation with low levels of voltage and current distortion. In addition, experimental results show that by using the proposed algorithm, smooth transition between two modes of operation can be achieved with acceptable levels of overcurrent. The future work will involve development of algorithms to reduce transient overcurrents. Furthermore, future work will focus on off-grid control of BESS in microgrids with nonlinear loads.

REFERENCES

[1] S. Tahib, J. Wang, M.H. Baloch, G.S. Kaloi, "Digital Control Techniques Based on Voltage Source Inverters in Renewable Energy Applications,"A Review, Electronics, vol. 7, no. 2: 18, 2018.

[2] D. Das, G. Gurrala, U. J. Shenoy, "Transition between grid-connected mode and islanded mode in VSI-fed microgrids," Sadhana, vol. 42, no. 8, pp. 1239–1250, 2017.

[3] C. N. Papadimitriou, V. A. Kleftakis, N. D. Hatziargyriou, "Control strategy for seamless transition from islanded to interconnected operation mode of microgrids," Journal of Modern Power Systems and Clean Energy, vol. 5, pp. 169–176, 2017.

[4] S.-H. Hu, T.-L. Lee, C.-Y. Kuo, J.M. Guerrero, "A Riding-through Technique for Seamless Transition between Islanded and Grid-Connected Modes of Droop-Controlled Inverters," Energies, vol. 9, no. 9: 732, 2016.

[5] Y. Li, M. Wang, "Control stratories for grid-connected and island dual-mode operated inverter under unbalanced grid voltage conditions," Proceedings of The 7th International Power Electronics and Motion Control Conference, 2012, pp. 2152-2156, 2012.

[6] X. Li, H. Zhang, M. B. Shadmand and R. S. Balog, "Model Predictive Control of a Voltage-Source Inverter With Seamless Transition Between Islanded and Grid-Connected Operations," IEEE Transactions on Industrial Electronics, vol. 64, no. 10, pp. 7906-7918, 2017.

[7] M. N. Arafat, S. Palle, I. Husain and Y. Sozer, "Transition control strategy between standalone and grid connected operation of voltage source inverters," 2011 IEEE Energy Conversion Congress and Exposition, pp. 1994-2000, 2011.

[8] X. Guo, W. Wu, H. Gu, "Phase locked loop and synchronization methods for grid- interfaced converters: a review," Przegląd Elektrotechniczny, vol. 87, no. 4, pp. 182-187, 2011.

978-1-6654-3236-8/21 $31.00 © 2021 IEEE

LCL Filter Design with Amorphous Core Inductor for 100 kVA Energy Storage Converter

Božo Terzić
Department of Power Engineering
FESB, University of Split
Split, Croatia
bterzic@fesb.hr

Ozren Bego
Department of Power Engineering
FESB, University of Split
Split, Croatia
obego@fesb.hr

Marin Despalatović
Department of Power Engineering
FESB, University of Split
Split, Croatia
despi@fesb.hr

Goran Majić
Department of Power Engineering
FESB, University of Split
Split, Croatia
gomajic@fesb.hr

Ante Kriletić
Department of Power Engineering
FESB, University of Split
Split, Croatia
akrileti@fesb.hr

Mislav Blajić
Department of Power Engineering
FESB, University of Split
Split, Croatia
mblaji00@fesb.hr

Abstract—Today, three-phase voltage source converters (VSC) are usually connected to the grid through LCL filters. The design of the LCL filter has a major impact on the overall system performance and, consequently much attention should be given to the selection of basic filter parameters as well as the choice of magnetic materials and dimensions of inductor core. In this paper, the procedure for selecting the basic parameters of LCL filter for a 100 kVA grid-connected converter is described. A simple design procedure for determination of winding turns number and air-gap length of amorphous core inductor is presented. The prototype of three-phase inductor with amorphous core is manufactured in respect to specific constraints. Afterwards, experimental verification of LCL filter performance is carried out using 100 kVA energy storage converter and lithium-ion battery string with capacity of 70 kWh.

Keywords — LCL filter, energy storage converter, PWM converter, amorphous core

I. Introduction

In recent years, the grid connected voltage source converter (VSC) is most commonly used as bidirectional interface between the grid and various dc sources like renewable energy and energy storage. To filter the higher grid current harmonics, the LCL filter is used as they are more cost-effective compared to simple L-filters because smaller inductors can be used to achieve the same damping of the switching harmonics. Fig. 1 shows the block diagram of energy storage system using VSC with LCL filter.

To reduce the dimensions, weight and price of the LCL filter, it is possible to increase the switching frequency of the converter, which allows the selection of filter inductors with smaller inductance. In this case, there may be a problem of increased heating of the filter inductor on the converter side (L_{fc}) due to additional losses resulting from high-frequency current pulsations flowing through that choke. A substantial part of these losses occurs in the inductors core. Comparison of magnetic properties of typical core materials presented in [2] – [4] was used to minimize the core size and losses. Several core materials were analyzed (silicon-steel, ferrite, iron-powder, nanocrystalline, amorphous) and the amorphous core was selected as the most suitable choice for inductor with high-frequency current pulsations.

The first part of this paper describes the procedure for selecting the basic parameters of LCL filter for a 100 kVA

grid-connected converter. In the next section, properties of magnetic materials are compared and reasons for selection of amorphous core are explained. Furthermore, this part of the paper presents a straightforward design procedure for determination of winding turns number and air-gap length of amorphous core inductor. The prototype of three-phase inductor with amorphous core is manufactured and several production stages are also shown in Section III. Section IV presents experimental verification of LCL filter performance carried out using laboratory setup of an energy storage system consisting of 100 kVA converter and battery string with capacity 70 kWh. The final section presents the conclusion.

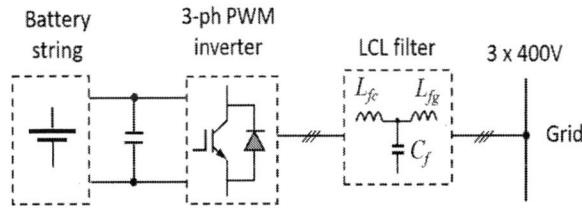

Fig. 1 Three-phase converter with LCL filter for battery energy storage application

II. LCL Filter Design

Many papers deal with LCL filter design for grid-connected converters [4] - [6]. Most of these methods are based on the use of analytical expressions to calculate the reactive power of the filter capacitors, the maximum allowable pulsations of the converter current, the resonant frequency and the Bode diagrams of the filter. The design procedure presented in this paper uses some of these formulas but also uses simulation procedures. It should be pointed out that system is modelled using PLECS software while optimization procedures are performed in MATLAB. Besides determining current ripple, simulations are also used to confirm that the selected LCL filter parameters do not cause resonant instability of the system.

Filter design involves selection of three basic parameters (L_{fc}, L_{fg}, C_f) after the converter switching frequency range has been defined. The first step is to choose capacitance of capacitor. Normally reactive power of LCL filter capacitor (C_f) is kept within 5 % of total system power rating and can be calculated by, [4]:

This paper is a result of research project "Active system for electric energy storage and stabilization of electric grid" that is co-financed from EU structural funds under grant number KK.01.2.1.01.0026.

$$C_f = \frac{0.05 \cdot S_n}{2\pi \cdot f_g \cdot U_g^2} = 100\ \mu\text{F} \qquad (1)$$

where:

$S_n = 100$ kVA, nominal apparent power of system

$U_g = 400$ V, phase-to-phase grid voltage (rms)

$f_g = 50$ Hz, grid frequency

The other two filter parameters (L_{fc}, L_{fg}) and converter switching frequency are selected as follows:

1. For various combinations of L_{fc}, and L_{fg}, the resonant frequencies of the filter are calculated and presented in the corresponding diagrams.

2. Based on the condition that the resonant frequency should be between 25 % and 50 % of the converter switching frequency, using the diagrams from the previous step, the appropriate sets of LCL filter parameters and the switching frequency range are selected [3].

3. For these sets of parameters, the operation of VSC is simulated and based on results, the maximum current pulsation through inductor L_{fc} is determined. These pulsations are then plotted on separate diagrams as a function of the switching frequency for different values of L_{fc}. The parameter L_{fc} and the switching frequency are ultimately selected according to the criterion of minimum current pulsations through inductor L_{fc}.

Based on the described procedure the following LCL filter parameters are selected: $L_{fc} = 150\ \mu$H, $L_{fg} = 50\ \mu$H, $C_f = 100\ \mu$F. For these filter parameter values, optimal converter performance is expected for switching frequencies in the range of 10 - 14 kHz.

III. AMORPHOUS CORE INDUCTOR DESIGN AND PRODUCTION

A. Selection of magnetic material

As aforementioned, due to high switching frequency current ripple, the converter side inductor of the LCL filter is expected to have greater losses and is more likely to overheat than the grid side inductor if the same material is used for both cores. There are several magnetic materials appropriate for construction of inductor cores considering this specific application. Choice of material depends on numerous criteria such as power, voltage and current ratings, fundamental and switching frequencies of the converter, overall efficiency and losses, as well as size and weight of the components [2]. Although, the ideal material would have high saturation, linear permeability and lower power losses, there are some applications where the overall system performance can be improved by selecting nonideal materials [2], [7]. In most cases, materials that are considered for inductor cores are silicon steel, ferrite, iron powder, amorphous and nanocrystalline [2]. For selection of the core material in this paper, the following materials were analyzed: ferrite, nanocrystalline and amorphous. Properties of abovementioned magnetic materials are given in Table I, where symbols +, ++ and +++ indicate low, average and higher specific core losses, respectively [4]. Amongst materials analyzed and given the weight and thermal constraints of the specific application, amorphous core is selected because it has low core losses, high saturation flux density, wider operating frequency and relatively high operating temperature [4].

In [4] detailed amorphous core inductor design procedure is presented. That procedure includes: (*i*) choice of type and dimensions of magnetic core, (*ii*) bobbin and winding design, (*iii*) air-gap length calculation, (*iv*) inductor losses analysis and (*v*) thermal analysis. Amorphous core used in this paper is selected and purchased based on the analysis and results presented in [2] – [4]. To reduce overall cost, the standard core type AFEC 150b manufactured by *Gaotune Technologies* is selected. Fig. 2 shows physical layout (a) and dimensions (b) of three-phase amorphous E-shape core. Dimensions of this core type are (in mm): $a = 48.6$, $b = 50$, $c = 154$, $d = 75$, $e = 250.8$, $f = 256.2$. Based on these parameters, one can obtain average value of magnetic path length equal to:

$$MPL \approx 2 \cdot (a+b) + c + f = 60.2\ \text{cm} \qquad (2)$$

Similarly, cross-sectional area can be calculated as:

$$A_c = a \cdot d = 36.45\ \text{cm}^2 \qquad (3)$$

TABLE I. PROPERTIES OF VARIOUS MAGNETIC MATERIALS

PHYSICAL PROPERTY	MAGNETIC MATERIAL		
	Ferrite	Nanocrystalline	Amorphous
Saturation flux density (T)	0.4	1.2	1.56
Initial relative permeability	1000-4000	15000-150000	3000-6000
Mass density (g/cm³)	5	7.2	7.18
Core loss without air-gap (W/kg)	+	++	+++
Continous operating temperature (°C)	250	120	150

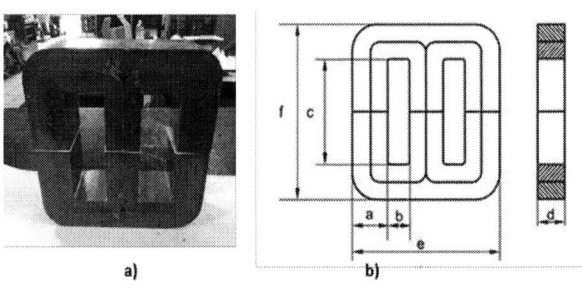

Fig. 2 Physical layout (a) and dimensions (b) of amorphous EE core

B. Number of winding turns and air-gap length

After selecting the amorphous magnetic core, it is necessary to calculate the number of winding turns per inductor phase and the length of the air-gap. Although the rated phase current of the inductor is 160 A, for the calculation of the number of winding turns and the length of the air-gap we will use a 50 % higher current than the rated one. This is because the converter during transients limits current values greater than 150 % of the rated value, i.e. the converter current limit is set to $I_L = 240$ A. In order to calculate the number of winding turns, the inductance of the choke is required. Since this high-frequency choke is intended to be on the converter-side of the LCL filter, based on the analysis in Section II, inductance of $L_{fe} = 150$ μH was chosen.

The voltage drop per inductor phase is calculated according to the following relation:

$$V_L = I_L \cdot \omega \cdot L = 11.3 \, \text{V} \qquad (4)$$

The required number of turns is determined from the law of electromagnetic induction according to:

$$N_L = \frac{V_L \cdot 10^4}{K_f \cdot B_{ac} \cdot f_g \cdot A_c} = 9.3 \qquad (5)$$

where:

$K_f = 4.44$, form factor of voltage waveform,

$B_{ac} = 1.5$ T, maximum magnetic flux density,

$f = 50$ Hz, rated grid frequency,

$A_c = 36.45$ cm², core cross sectional area,

and, for obvious reasons, N_L was rounded to the nearest integer, i.e. $N_L = 9$ turns.

Air-gap length is calculated from:

$$l_g = \frac{0.4 \cdot \pi \cdot N_L^2 \cdot A_c \cdot 10^{-8}}{L} - \left(\frac{MPL}{\mu_m} \right) = 0.23 \, \text{cm} \qquad (6)$$

where:

$MPL = 60.2$ cm, magnetic path length,

$\mu_m = 3000$, permeability of amorphous core.

The air-gap increases fringing flux that results in an increase of inductance by the following factor, [1]:

$$F = 1 + \frac{l_g}{\sqrt{A_c}} \cdot \ln \left(\frac{2 \cdot G}{l_g} \right) = 1.19 \qquad (7)$$

where $G = 15.4$ cm is the core height (designated with symbol c in Fig. 2).

Due to fringing flux, it is necessary to recalculate the number of inductor turns using the factor F according to, [1]:

$$N_L^* = \sqrt{\frac{L \cdot l_g}{0.4 \cdot \pi \cdot A_c \cdot F \cdot 10^{-8}}} = 7.96 \qquad (8)$$

that can be rounded to the nearest integer, i.e. $N_L^* = 8$ turns.

C. Inductor prototype production

Based on winding turns number and air-gap length calculated in the previous part, the three-phase inductor with amorphous core is manufactured. An insulating material, type Nomex 410, with a thickness of 0.42 mm is placed in the air-

gap. Six layers of insulating material are placed for one air-gap, thus air-gap length of 2.52 mm is obtained, which is approximately 10% more than the calculated air-gap length given by (6).

The rated current of the inductor is 160 A, which means that by selecting a current density of 2.5 A / mm², a wire with a cross section of 64 mm² should be used. A wire of such cross-section would be difficult to wind on the bobbin. Thus, a wire of cross-section 14 x 2.3 = 32 mm² and a different winding design was selected. In order to obtain the same current carrying capacity and the same inductance at the ends of the windings, it is necessary to connect in parallel two windings of 8 turns for each phase. During the production of windings, the usual technological procedures used to manufacture the windings of electrical devices were employed, such as polyester resin impregnation and subjecting components to curing at high temperatures up to 110 °C for a period of 10 hours.

Fig. 3 shows inductor after several steps of production, while Fig. 4 shows final version of inductor prototype. During production phase, it was observed that amorphous material is relatively brittle and much attention needs to be given when handling it.

Fig. 3 Amorphous core inductor after several steps of production

Fig. 4 Final prototype of amorphous core inductor

IV. EXPERIMENTAL VERIFICATION

A. Inductance measurement

The laboratory setup for inductance measurements is shown in Fig. 5. Synchronous generator supplies inductors with variable AC voltage. Output frequency is kept constant (50 Hz) using a vector controlled three-phase induction motor. The voltage differential and current probes are used to measure inductor's currents and voltages.

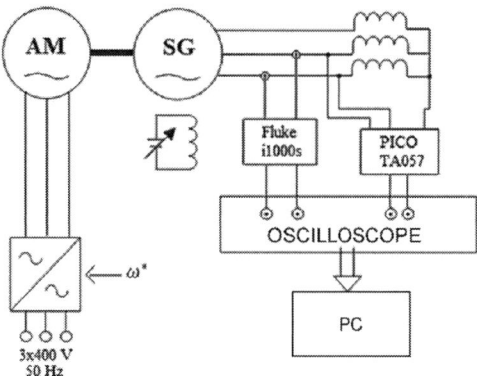

Fig. 5 Laboratory setup for inductance measurement

The inductance of the amorphous core inductor was measured at a rated sinusoidal current of 160 A and frequency of 50 Hz using Ohm's law. Using the same measurement procedure, the linearity of the inductance up to 150 % of the rated current was tested. At the nominal current the measured inductance is 125 μH which is about 16 % less than the calculated value. The reason for this is the slightly larger air-gap in the manufactured inductor in relation to the calculated value. At 150 % of the rated current the inductance is 110 μH, which means that the magnetic flux in the inductor is slightly saturated at currents

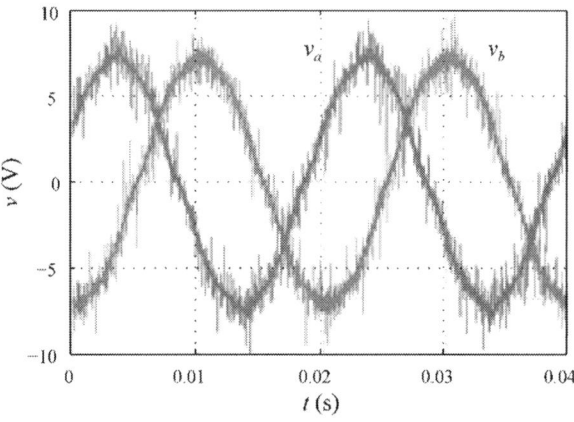

Fig.6 Voltage drop across the inductor at 100% of the inductor's nominal current

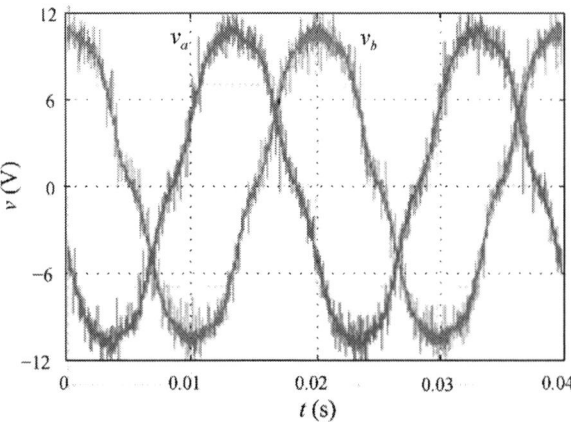

Fig.7 Voltage drop across the inductor at 150% of the inductor's nominal current

Fig.8 LCL filter and storage converter of the laboratory setup used for experimental verification of LCL filter design

above nominal value. Figs. 6 and 7 present voltage drop across the inductor at 100 % and 150 %of the inductor's rated current, respectively. The saturation effect at 150 % of the rated current cannot be detected by current measurement since the stator inductance of the synchronous generator is significantly larger than the inductance of the manufactured inductors. Therefore, the measured current exhibits sinusoidal waveform for both measuring points. Figures presenting almost ideal sinusoidal waveforms are omitted from this paper for the sake of brevity.

B. Energy storage system steady-state operation

The design procedure of the LCL filter inductors with amorphous core is validated using energy storage system shown in Fig. 1. Fig. 8 shows LCL filter and part of storage converter inside the electrical enclosure that is used for experimental verification of the LCL filter design. DC link voltage, phase-to-phase grid voltage, grid current and converter side current at nominal operating point are shown in Fig. 9. Converter side current exhibits high frequency

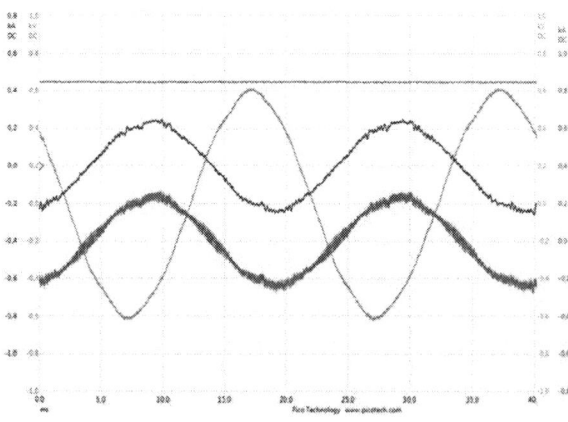

Fig. 9 DC link voltage (green), phase-to-phase grid voltage (orange), grid current (blue) and converter current (red).
Voltage scale: 200 V/div, Current scale: 200 A/div

pulsations at converter switching frequency of 12 kHz, while the grid current is practically sinusoidal without pulsations. Total harmonic distortion (THD) of the grid current is 3.3 %.

V. CONCLUSION

This paper presents the design procedure of converter-side inductors of LCL filter considering a 100 kVA grid-connected converter application. The first step of the design procedure is the choice of basic filter parameters based on well-known relations involving converter rated data (nominal power and current, switching frequency, …), as well as simulations. The selection of magnetic material and dimensions of inductors core is based on results presented in [2] – [4]. Amorphous material is selected because it has relatively low core losses, high saturation flux density, wider operating frequency and relatively high operating temperature when compared to other materials considered for this specific application. Afterwards, a straightforward approach is used to determine the winding turns number and air-gap length of amorphous core. Taking into account constraints specific to the given application, three-phase amorphous core inductor was manufactured. In order to validate the proposed design procedure, experiments were conducted. Inductance measurement was performed in the range from 100 % to 150 % of the rated inductor's current. Results show slight saturation at 150% of the rated current. Verification of LCL filter performance was carried out by steady-state measurements on 100 kVA energy storage converter and lithium-ion battery string with capacity of 70 kWh. Results demonstrate stable operation of the overall system with low harmonic distortion of the grid current. Finally, it can be concluded that the proposed filter design procedure may be used in other applications with different constraints.

REFERENCES

[1] B. Colonel WM. T. McLyman: Transformer and Inductor Design Handbook (4ed), CRC Press Taylor and Francis Group, Boca Raton, FL, 2011.

[2] A.Hilal, B. Cogo, " Optimal inductor design and material selection for high power density inverters used in aircraft applications," IEEE International Conference on lectrical Systems for Aircraft, Railway, Ship Propulsion and Road Vehicles, Toulouse, France, Nov. 2016.

[3] Y.Jiao, F.C.Lee, " LCL Filter Design and Inductor Current Ripple Analysis for a Three-Level NPC Grid Interface Converter," IEEE Trans. On Power Electronics, Vol. 30, No. 9, Sep. 2015.

[4] Y. Liu, K.Y.See, " S. Yin, R. Simanjorang, C.F.Tong: LCL Filter Design of 50 kW 60-kHz SiC Inverter with Size and Thermal Consideration for Aerospace Application," IEEE Trans. On Ind. El., Vol. 64, No. 10, Oct. 2017.

[5] M. Liserre, F. Blaabjerg, and S. Hansen, "Design and Control of an LCL-Filter-Based Three-Phase Active Rectifier," IEEE Transactions on Industry Applications, vol. 41, no. 5, pp. 1281– 1291, Sep. 2005.

[6] M. Liserre, F. Blaabjerg and A. Dell'Aquila, "Step-by-step design procedure for a grid-connected three-phase PWM voltage source converter," Int. J. Electronics, vol. 91, no. 8, pp. 445-460, Aug. 2004.

[7] B. Terzić, G. Majić, A. Slutej, "Stability Analysis of Three-Phase PWM Converter with LCL Filter by Means of Nonlinear Model," Automatika, vol. 51, no. 3, pp. 221-232, 2010.

Power Loss Analysis of Multi-converter System with Single Wire and Wireless Energy Trasfer

Marcin Zygmanowski
Faculty of Electrical Engineering
The Silesian University of Technology
Gliwice, Poland
marcin.zygmanowski@polsl.pl

Marcin Kasprzak
Faculty of Electrical Engineering
The Silesian University of Technology
Gliwice, Poland
marcin.kasprzak@polsl.pl

Kamil Kierepka
Faculty of Electrical Engineering
The Silesian University of Technology
Gliwice, Poland
marcin.kierepka@polsl.pl

Jarosław Michalak
Faculty of Electrical Engineering
The Silesian University of Technology
Gliwice, Poland
jaroslaw.michalak@polsl.pl

Grzegorz Jarek
Faculty of Electrical Engineering
The Silesian University of Technology
Gliwice, Poland
grzegorz.jarek@polsl.pl

Krzysztof Przybyła
Faculty of Electrical Engineering
The Silesian University of Technology
Gliwice, Poland
krzysztof.przybyla@polsl.pl

Abstract—This paper presents power loss analysis of the multi-converter system with single wire and wireless power transfer. This system is intended to be used in underground auxiliary transportation where the need for highly efficient power electronic converter plays a crucial role. Presented in the paper multi-converter system consist of six converters. It is supplied from the three-phase power line and utilizes both the single-wire energy transfer and wireless energy transfer for charging the battery of the machine moving on a suspended rail. The purpose of the single-wire energy transfer use is to limit the cost of the system while the wireless energy transfer increases the safety of energy delivery to the moving machine.

Keywords—power loss analysis, single-wire power transfer, wireless power transfer, SiC MOSFET converters

I. INTRODUCTION

Power losses are one of the most important issues in power electronic converter design. This is particularly true for systems where the application of forced cooling may be difficult or impossible to implement e.g., in underground mining. In underground mines due to the high ambient temperature, it is recommended to use highly efficient electrical equipment. With the high efficiency the generated power losses are small, which limits the heat dissipated to the surrounding of the equipment. For power electronic converters this recommendation means that the exact recognition of converter power losses is of great importance. Furthermore, due to the possibility of operation in the explosive atmosphere the converters are encapsulated inside of flameproof enclosures. This allows to use the air-forced cooling system but the heat exchange from the converter to the surrounding of the enclosure is only through the convection to the air and the radiation. This issue additionally limits the permitted value of power losses generated in converters.

Major challenges in underground transportation are high cost of system components and the risk of electrical shock associated with electrical wiring. To overcome these challenges single-wire energy transfer and wireless energy transfer are proposed [1], [2]. Single-wire energy transfer systems are economically justified due to the decrease of number of conductors employed to deliver energy to the load [3]. Wireless energy transfer reduces the problems with the risk of electrical shock and is also cost effective solution compared to wire based supply systems [4]–[5].

Scientific paper published as part of an international project co-financed by the European Commission Research Fund for Coal and Steel (RFCS) in the years 2020-2023; grant agreement no: 899469.

Scientific paper published as part of an international project co-financed by the Ministry of Science and Higher Education's program "PMW" in the years 2020-2023; contract no. 5122/FBWiS/2020/2

The paper is organized as follows. In section II the description of the multi-converter system is presented. Section III shows the analysis of power losses in all converters of the system. In section IV the laboratory setup with experimental results is presented. Section V will draw the conclusions.

II. DESCRIPTION OF THE MULTICONVERTER SYSTEM

The analyzed multi-converter system is connected to three-phase power grid with rms line-to-line voltage equal to 500 V and feeding the ac output with the rated power of 5 kW and the battery-operated suspended tractor with the rated power of 2 kW. The power delivered to the tractor is for charging its battery and is transferred through wireless energy transfer (WET). The rated output voltage supplying the tractor battery is 48 V. The schematic of the multi-converter system is shown in Fig. 1.

The multi-converter system consists of six converters among which there are:

- REC1 converter which is a MOSFET based three-phase PWM boost rectifier responsible for transfer the energy from the grid at a unity power factor,

- INV1 converter is an inverter connected between the dc link of REC1 converter with rated voltage of 800 V and the step-up transformer TR1 which supplies the single-wire energy transfer (SWET),

- REC2 is diode rectifier receiving the energy from the SWET through the step-down transformer TR2 and delivers it to the dc-link circuit. The rated voltage of the dc-link is 800 V and the energy from this circuit is transferred to the AC output and to the wireless energy transfer WET,

- INV2 is a three-phase MOSFET based inverter,

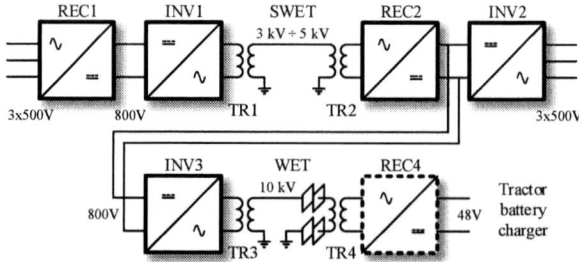

Fig. 1. Schematic of multi-converter system with the single wire energy transfer (SWET) and wireless energy transfer (WET).

978-1-6654-3236-8/21 $31.00 © 2021 IEEE

- INV3 is an inverter supplying the WET through the step-up transformer TR3,

- REC4 is the diode rectifier which transfers the energy from the step-down transformer TR4 to the voltage level required for the tractor battery charger.

In the proposed multi-converter system, the converters are divided into three groups which are in separate enclosures. In the first group there are REC1 and INV1, in the second group there are three converters REC2, INV2 and INV3, and in the last there is REC4. Power losses of two first converter groups are investigated in this paper. The switching frequency used in all converters is set close to 50 kHz, while REC1 and INV2 are controlled with PWM technique with the fundamental frequency set to 50 Hz. Inverters INV1 and INV3 are resonant inverters including resonant circuits. The converters REC1, INV1, INV2 and INV3 are based on SiC MOSFET half-bridge modules FF23MR12W1M1P_B11 while rectifier REC2 is build using MSC030SDA170B SiC Schottky diodes.

III. POWER LOSS ANALYSIS

The analysis of power losses which is performed in this section is based on the manufacturer data provided to all semiconductor devices. None of the six converters is operating with the same conditions therefore power losses generated inside of these converters will be different and need to be recognized separately [6]-[7] .

A. Power losses of REC1 converter

The rated power of this converter is P_n = 7 kW. This converter is connected to the three-phase ac grid with the rms value of the line-to-line voltage equal to V_n = 500 V. The maximum amplitude of the phase current is given as:

$$I_m = \sqrt{2}\frac{P_n}{\sqrt{3}V_n} = 11.4\,\text{A} \qquad (1)$$

The fundamental frequency of the grid voltage is 50 Hz and the switching frequency f_S = 50 kHz. It is assumed that in the REC1 converter the modulation index is close to unity, $m_a \approx 1$. This assumption is done because the converter has to generate the dc voltage equal to V_{dc} = 800 V. It should be noted here that the modulation index m_a has no significant effect on power losses when the converter utilizes MOSFETs. The analyzed three-phase REC1 converter is based on two-level converter topology, which the single phase-leg is shown in Fig. 2.

The conduction power losses of a MOSFET T_1 are calculated from (2).

$$
\begin{aligned}
P_{conT1} &= \frac{1}{2\pi}\int_0^{2\pi}\left\{-i_A(t)\cdot v_T(t)\cdot s_{MT}(\omega t)\right\}d\omega t \\
&= \frac{1}{2\pi}\int_0^{2\pi}\left\{r_T\left[I_m\sin(\omega t)\right]^2\cdot\left[\frac{1+m_a\sin(\omega t)}{2}\right]\right\}d\omega t \quad (2) \\
&= \frac{I_m^2 r_T}{4}
\end{aligned}
$$

where $s_{MT}(\omega t)$ is the modulation function dependent on the modulation index m_a, v_T is the transistor drain-source voltage during the conduction $v_T(t) = i_A(t)r_T$, and r_T is the on-state

resistance of the MOSFET. The same conduction losses, as from (2), are generated in transistor T_2 and in transistors of other converter phase legs.

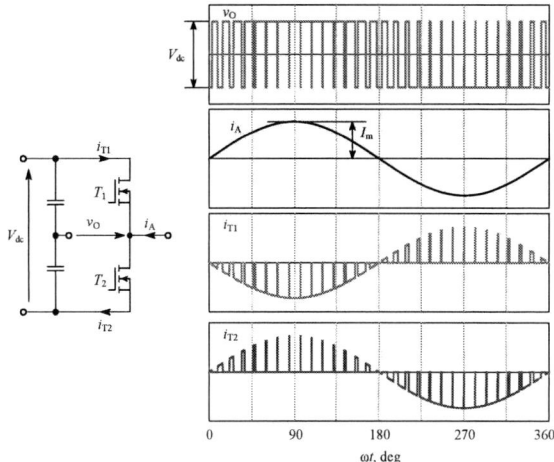

Fig. 2. Single leg of SiC MOSFET converter operated under pulse-width modulation with currents and voltage waveforms.

The conduction power losses in a single MOSFET for the maximum amplitude of the phase current I_m, given by (1), and for transistor resistance r_T = 32 mΩ, are P_{conT1} = 1.05 W.

The second components of power losses are switching power losses (3). These losses are calculated by time integration of switching energies e_{on} and e_{off} given as functions of transistor current $i_T(\omega t)$ (Fig. 3.a), which for simplification is written as i_T. The switching energy loss functions $e_{on}(i_T)$, and $e_{off}(i_T)$ are approximated by cubic equations (4), which polynomial coefficients are calculated by using fitting curve tool basing on the MOSFET datasheet data. The polynomial coefficients from (4) are listed in Table I.

$$P_{swT1} = k_{Vdc}\frac{f_S}{2\pi}\int_\pi^{2\pi}\left[k_{RGon}e_{on}(i_T)+k_{RGoff}e_{off}(i_T)\right]d\omega t \quad (3)$$

$$
\begin{aligned}
e_{on}(i_T) &= a_{on}i_T^3 + b_{on}i_T^2 + c_{on}i_T + d_{on} \\
e_{off}(i_T) &= a_{off}i_T^3 + b_{off}i_T^2 + c_{off}i_T + d_{off}
\end{aligned} \quad (4)
$$

The switching losses (3) are linearly dependent on the switching frequency f_S and dc voltage which occur at transistor during the off-state. This linear dependence is provided by using a factor k_{Vdc} which is the ratio of the dc voltage to the voltage used in manufacturer measurement data (V_{DS} = 600 V) thus $k_{Vdc} = V_{dc}/V_{DS}$ = 800 V/ 600 V = 1.33. The power switching losses (3) are also influenced by the gate resistance of the driver circuit. In the analyzed system transistors are switched by gate drivers with different gate resistors (R_{Gon} = 7.4 Ω and R_{Goff} = 3.7 Ω) compared to the manufacturer data obtained for R_G = 1 Ω. The change of the losses due to the change of gate resistance is modeled by introduction of two coefficients k_{RGon} and k_{RGoff} for switching-on and switching-off energies respectively. These two coefficients are obtained from functions $e_{on}(R_G)$ and $e_{off}(R_G)$ provided by the manufacturer for transistor current I_D = 50 A as shown in Fig. 3 with using (5).

978-1-6654-3236-8/21 $31.00 ©2021 IEEE

$$k_{\text{RGon}} = \frac{e_{\text{on}}\left(R_{\text{Gon}} = 7.4\,\Omega\right)}{e_{\text{on}}\left(R_{\text{G}} = 1\,\Omega\right)}; k_{\text{RGoff}} = \frac{e_{\text{off}}\left(R_{\text{Goff}} = 3.7\,\Omega\right)}{e_{\text{off}}\left(R_{\text{G}} = 1\,\Omega\right)} \quad (5)$$

TABLE I.   POLYNOMIAL COEFFICIENTS FOR SWITCHING ENERGY
LOSSES OF FF23MR12W1M1P_B11 MODULE

Symbol	Value	Symbol	Value	Unit
a_{on}	$5.26 \cdot 10^{-10}$	a_{off}	$5.46 \cdot 10^{-10}$	$\text{J/(A}^3)$
b_{on}	$-1.26 \cdot 10^{-7}$	b_{off}	$-2.68 \cdot 10^{-8}$	$\text{J/(A}^2)$
c_{on}	$1.33 \cdot 10^{-5}$	c_{off}	$7.36 \cdot 10^{-7}$	J/A
d_{on}	$1.34 \cdot 10^{-4}$	d_{off}	$5.68 \cdot 10^{-5}$	J

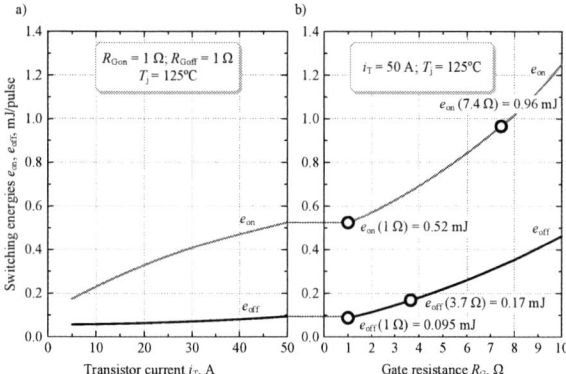

Fig. 3. Switching energy losses of FF23MR12W1M1P_B11 module: a) as functions of the transistor current i_T for the gate resistance $R_G = 1\,\Omega$ and b) as functions of the gate resistance R_G for the transistor current $i_T = 50$ A.

By using characteristics from Fig. 3.a and Fig. 3.b the gate resistance coefficients are $k_{\text{RGon}} = 1.74$ and $k_{\text{RGoff}} = 1.79$ respectively. By substituting (4) and (5) into (3) the switching power losses are given as (6).

$$P_{\text{swT1}} = k_{\text{Vdc}} f_S \left[\begin{array}{l} k_{\text{RGon}} \left(\dfrac{2a_{\text{on}}}{3\pi} I_m{}^3 + \dfrac{b_{\text{on}}}{4} I_m{}^2 + \dfrac{c_{\text{on}}}{\pi} I_m + d_{\text{on}} \right) \\[2ex] k_{\text{RGoff}} \left(\dfrac{2a_{\text{off}}}{3\pi} I_m{}^3 + \dfrac{b_{\text{off}}}{4} I_m{}^2 + \dfrac{c_{\text{off}}}{\pi} I_m + d_{\text{off}} \right) \end{array} \right] \quad (6)$$

For $I_m = 11.4$ A, $k_{\text{Vdc}} = 1.33$ and $f_S = 50$ kHz the switching power losses in the single transistor are equal to $P_{\text{swT1}} = 16.6$ W. This is evident that switching losses in REC1 converter are dominant over conduction losses.

B. Power losses of INV1 converter

The INV1 converter operates as a resonant inverter with the rectangular output voltage and the quasi-sinusoidal current. It is assumed that the converter is based on half-bridge converter topology with capacitor voltage divider (Fig. 4).

The switching frequency is slightly higher than the resonant frequency of the resonant circuit. This condition allows to operate the converter in ZVS operation which is desirable due to the reduction of switching power losses.

The analyzed converter is operating with the switching frequency $f_{S1} = 42$ kHz. The rated power of the INV1 converter is $P_n = 7$ kW. The amplitude of the output current I_{m1}, which is treated as sinusoidal (Fig. 4), is calculated from (7).

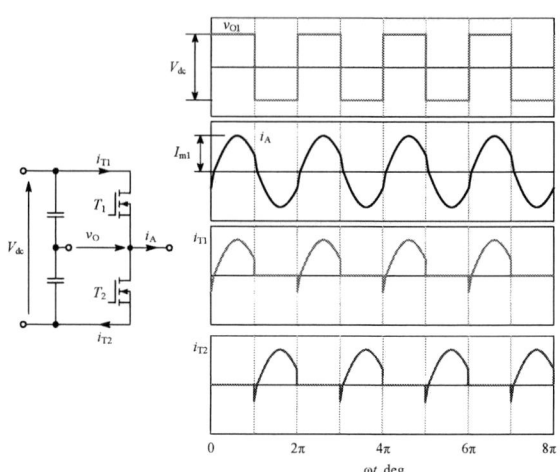

Fig. 4. Current and voltage waveforms in the resonant inverter leg representing INV1 converter.

$$I_{m1} = \sqrt{2}\frac{P_n}{V_{O1\text{rms}}} = \sqrt{2}\frac{P_n}{\frac{1}{\sqrt{2}}\frac{4}{\pi}\frac{V_{\text{dc}}}{2}} = \pi\frac{P_n}{V_{\text{dc}}} = 27.5\,\text{A} \quad (7)$$

The conduction power losses of a single MOSFET are given by (8).

$$P_{\text{conT1}} = \frac{1}{2\pi}\int_0^\pi \left\{ r_T \left(i_A(t) \right)^2 \right\} d\omega t = \frac{I_{m1}{}^2 r_T}{4} \quad (8)$$

As is seen from (8) the conduction losses of a single MOSFET in the resonant inverter are described by the same formula as for transistors in REC1 converter. Taking into account the on-state resistance of the MOSFET $r_T = 32$ mΩ and current amplitude $I_{m1} = 27.5$ A, the conduction power losses (8) are equal to $P_{\text{conT1}} = 6.05$ W.

Due to ZVS condition in the INV1 converter, the switching power losses occur only when MOSFETs are switched-off. The output current at the time of switching is smaller than the amplitude I_{m1} and depends on the quality factor of the load. In this paper it is assumed that this current can change in the range $0.5\,I_{m1}$ to 0 and for further analysis the average value equal to $0.25\,I_{m1}$ has been chosen. The switching power losses in a single MOSFET are given by (9).

$$P_{\text{swT1}} = k_{\text{Vdc}} f_{S1} k_{\text{RGoff}} \left[\frac{a_{\text{off}}}{16} I_{m1}{}^3 + \frac{b_{\text{off}}}{8} I_{m1}{}^2 + \frac{c_{\text{off}}}{4} I_{m1} + d_{\text{off}} \right] \quad (9)$$

Where $k_{\text{Vdc}} = V_{\text{dc}}/V_{\text{DS}} = 800$ V/600 V = 1.33 and $f_{S1} = 42$ kHz. The switching losses are calculated from (9) for the maximum output power and are equal to $P_{\text{swT1}} = 6.09$ W. One can see that in the resonant converter INV1 the switching losses are comparable to the conduction power losses.

C. Power losses of REC2 rectifier

The REC2 rectifier is a based on a full-wave SiC Schottky diode rectifier topology. It is assumed that power losses in this converter are generated due to the conduction. This assumption is done due to the application of SiC Schottky diodes MSC030SDA170B, which have no reverse recovery. The diode currents can be approximated as half wave sinusoidal waveforms with the frequency of f_{S1} (the same as the switching frequency in INV1 converter). The conduction power losses of a single diode in REC2 are calculated from (11) where v_D is the voltage across a diode when it is forward-biased. Each diode output characteristic can be approximated by the threshold voltage $V_{F0} = 0.8$ V and the on-state resistance $r_D = 36$ mΩ. These values are taken from the datasheet output characteristic for the junction temperature $T_j = 125°C$. The REC2 rectifier is supplied from step-down transformer connected to the SWET line (Fig. 1). The transformer secondary side voltage is rectangular with the maximum value of 800 V. Because the REC2 converter rated power is the same as in REC1 and INV1 converters the rated amplitude of the current I_{m2} is calculated from (10).

$$I_{m2} = \sqrt{2}\frac{P_n}{V_{O2rms}} = \sqrt{2}\frac{P_n}{\frac{1}{\sqrt{2}}\frac{4}{\pi}V_{dc}} = \pi\frac{P_n}{2V_{dc}} = 13.75\,\text{A} \quad (10)$$

$$P_{conD1} = \frac{1}{2\pi}\int_0^\pi \left\{ i_A(\omega t)\cdot v_D(\omega t) \right\} d\omega t$$
$$= \frac{1}{2\pi}\int_0^\pi \left\{ I_{m2}\sin(\omega t)\cdot\left[V_{F0} + r_D I_{m2}\sin(\omega t) \right] \right\} d\omega t \quad (11)$$
$$= \frac{V_{F0}I_{m2}}{\pi} + \frac{I_{m2}^2 r_D}{4}$$

For the resonant current with the rated amplitude I_{m2} the REC2 diode conduction power losses are $P_{D1con} = 5.2$ W.

D. Power losses of INV2 converter

The INV2 converter is the three-phase two-level converter constructed from the same MOSFETs as REC1 and INV1 converters. The INV2 converter operates with sinusoidal PWM technique generating three-phase output voltages with the line-to-line rms value equal to $V_{ll} = 500$ V. The switching frequency is set to 50 kHz, therefore the power losses generated in the INV2 converter can be calculated from the same expressions for power losses which are applied to the REC1 converter. There is one difference in the value of the current amplitude because the rated power at the ac output of INV2 converter is equal to $P_{n3} = 5$ kW. With the assumption that the load has a unity power factor the current amplitude is

$$I_{m3} = \sqrt{2}\frac{P_{n3}}{\sqrt{3}V_n} = 8.1\,\text{A}. \quad (12)$$

For such current amplitude the conduction losses (2) in the single MOSFET are equal to $P_{T1con} = 0.53$ W.

The switching losses generated in the single transistor of INV2 converter, which are calculated from (6), are equal to $P_{T1sw} = 15.1$ W. The switching power losses generated in INV2 converter are dominant over conduction power losses which is similar to the losses in the REC1 converter.

E. Power losses of INV3 converter

The rated power which the INV3 converter transfers to the output is equal to $P_{n4} = 2$ kW. It is assumed that the INV3 converter is operating similarly to the INV1 converter as a resonant inverter with the resonant frequency lower than the switching frequency. The resonant frequency f_{r4} is set close to

42 kHz. The amplitude of the resonant current I_{4m} is given by (13). It was assumed that the INV3 converter is based on the H-bridge converter topology thus the output voltage maximum value is equal to $V_{dc} = 800$ V. By substituting (13) to (8) and (9) it is possible to calculate the power losses in converter MOSFETs, which are equal to $P_{conT1} = 0.12$ W and $P_{swT1} = 5.82$ W respectively.

$$I_{m4} = 2\frac{P_{n4}}{\frac{4}{\pi}V_{dc}} = 3.9\,\text{A} \quad (13)$$

F. Power losses of REC4 rectifier

The REC4 converter in the proposed multi-converter system is mounted on the suspended tractor and its power losses are not investigated in this paper.

G. Conclusion on power loss analysis

In this section total power losses P_{tot} generated in five converters of the system are collected in the Table II. These losses consider the number of switching devices of selected converters. Total power losses are divided into conduction and switching power losses. The losses generated outside of the converters i.e., in the SWET or WET lines are not shown because in this paper their analysis is out of the scope.

TABLE II. CALCULATED POWER LOSSES IN WATTS

Converter	REC1	INV1	REC2	INV2	INV3
P_n	7000	7000	7000	5000	2000
P_{con}	6.27	12.09	20.80	3.15	0.49
P_{sw}	99.55	12.68	0	90.73	22.91
P_{tot}	105.82	24.27	20.80	93.88	23.40
Total losses	259.24				

As is shown in Table II the highest power losses occur in hard switching converters i.e., in REC1 and INV2. Power losses in these converters are at least four times higher than in other converters. The next section provides the laboratory verification of power losses.

IV. LABORATORY SETUP

A. Converter models

The component converters of the multi-converter system have been constructed and tested in the laboratory separately. The laboratory model of converters REC1 and INV2 with the controller is depicted in Fig. 5.

Fig. 5. Laboratory model of REC1 and INV2 converters.

978-1-6654-3236-8/21 $31.00 © 2021 IEEE

The REC1 and INV2 converters are composed of three half bridge SiC-MOSFET modules. The controller is based on digital signal microcontroller. Similar converter model, which is based on the same SiC-MOSFET modules as for REC1 and INV2 converter, has been constructed for INV1 and INV3 converters with using two half-bridge modules (Fig. 6). The REC2 converter is based on SiC Schottky diodes is shown in Fig. 7. All constructed converter models are designed for dc-link voltage of $V_{dc} = 800$ V.

Fig. 6. Laboratory model of INV1 and INV3 converters.

Fig. 7. Laboratory model of REC2 converter.

B. Experimental results

The aim of the experimental test is to verify the theoretical analysis of power losses presented in this paper. The converter models are tested at the output power close to their rated powers by using precision power analyzer WT5000. The exemplary results obtained for the INV2 converter at the output $P_O = 6.442$ kW are shown in Fig. 8.

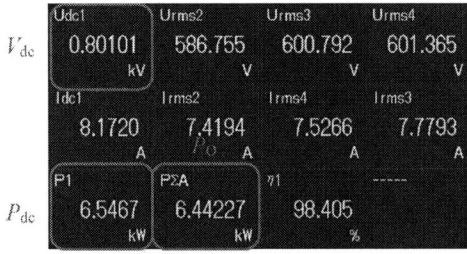

Fig. 8. Experimental results of power measurements obtained with power analyzer for INV2 converter.

From Fig. 8 it is seen that the input power, measured at the dc side of the converter, is $P_{in} = 6.546$ kW, resulting the total power losses are equal to $P_{tot} = P_{in}\text{-}P_O = 104.43$ W. From analytical formulae at the output power $P_O = 6.442$ kW the power losses are 102.46 W. If parasitic resistances of terminals and pcb traces, which have been measured at 10 mΩ, are considered the calculated total power losses increases to 104.12 W, which fits measurements very well.

The tests performed on resonant converters INV1 and INV3 reveals some discrepancies which are not present in PWM modulated converters (REC1 and INV2). In Fig. 9 one can see the output voltage with output current of the converter at the switching frequency $f_S = 42.2$ kHz. The output power is 7.207 kW and the measured power losses are 42.7 W.

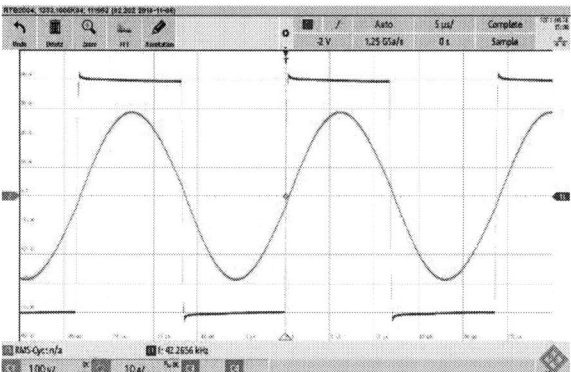

Fig. 9. Output voltage and current waveforms of INV1 converter operating at $P_O = 7.207$ kW.

The analytical total power losses for the INV1 converter are equal to 23.6 W. However, when additional resistance equal to 30 mΩ are considered the total power losses increase to 31.57 W. These resistances occur in the converter due to PCB tracks, dc-link reactor and converter terminals. The still existing discrepancy is because the additional resistance is measured for dc component. It is expected that for high frequency components with frequency in the range of the resonant frequency f_r the additional resistance would be even larger.

Power losses of the REC2 converter are measured for the output power equal to 6.4 kW and are equal to 22.3 W. The power losses obtained from analytical formulae for the same output power is 18.48 W. Similar measurements of additional resistances have revealed that these resistances are equal to 10 mΩ placed in series with each diode. After considering these resistances the analytical total power losses in REC2 converter are equal to 20.06 W.

V. CONCLUSIONS

The paper presents an analysis of power losses of five converters included in the multi-converter system. In this analysis mainly power losses of semiconductor devices are considered. As a result of experimental tests, it was shown that in all converters there are additional losses occurring at parasitic resistances on pcb tracks, converter terminals and inductive filters. The exact recognition of power losses is very important for presented multi-converter system because converters operate in encapsulated enclosures with limited heat exchange to the environment. The analysis also reveals for different converters the division of power losses among switching and conduction power losses. All these results can be useful during next design stages of the multi-converter system.

REFERENCES

[1] R. Song, S. Lu, T. Sirojan, B. T. Phung and E. Ambikairajah, "Power quality monitoring of single-wire-earth-return distribution feeders," 2017 International Conference on High Voltage Engineering and Power Systems (ICHVEPS), 2017, pp. 404-409, doi: 10.1109/ICHVEPS.2017.8225879.

[2] B. Nkom, A. P. R. Taylor and C. Baguley, "Narrowband Modeling of Single-Wire Earth Return Distribution Lines," in IEEE Transactions on Power Delivery, vol. 33, no. 4, pp. 1565-1575, Aug. 2018, doi: 10.1109/TPWRD.2017.2775189.

[3] B. Regensburger, S. Sinha, A. Kumar, J. Vance, Z. Popovic and K. K. Afridi, "Kilowatt-scale large air-gap multi-modular capacitive wireless power transfer system for electric vehicle charging," 2018 IEEE Applied Power Electronics Conference and Exposition (APEC), 2018, pp. 666-671, doi: 10.1109/APEC.2018.8341083.

[4] Z. Zhang, H. Pang, A. Georgiadis and C. Cecati, "Wireless Power Transfer—An Overview," in IEEE Transactions on Industrial Electronics, vol. 66, no. 2, pp. 1044-1058, Feb. 2019, doi: 10.1109/TIE.2018.2835378.

[5] S. K. Mishra R. Adda S. Sekhar A. Joshi and A. K. Rathore "Power transfer using portable surfaces in capacitively coupled power transfer technology" IET Power Electron. vol. 9 no. 5 pp. 997-1008 2016, doi.org/10.1049/iet-pel.2015.0332.

[6] Z. Li et al., "Loss Analysis and Efficiency Test of a 3 MW 10 kV AC to ±750 V DC Power Electronic Transformer," 2020 IEEE 9th International Power Electronics and Motion Control Conference (IPEMC2020-ECCE Asia), 2020, pp. 2068-2072, doi: 10.1109/IPEMC-ECCEAsia48364.2020.9368216.

[7] E. S. Glitz and M. Ordonez, "MOSFET Power Loss Estimation in LLC Resonant Converters: Time Interval Analysis," in IEEE Transactions on Power Electronics, vol. 34, no. 12, pp. 11964-11980, Dec. 2019, doi: 10.1109/TPEL.2019.2909903.

Analysis of Regenerative Cycles and Energy Efficiency of Regenerative Elevators

Dora Erica, Damjan Godec, Martina Kutija, Luka Pravica and Ivana Pavlić

University of Zagreb Faculty of Electrical Engineering and Computing

Zagreb, Croatia

dora.erica@fer.hr, damjan.godec@fer.hr, martina.kutija@fer.hr, luka.pravica@fer.hr, ivana.pavlic@fer.hr

Abstract—This paper analyzes the regenerative elevator operation and energy savings of the faculty elevator. The study is based on measurements taken during a week of high travel demand. The results are compared to previous weekly measurements taken with the same elevator during lower travel demand. The effects of travel demand on regenerative potential and energy efficiency are analyzed. Energy savings are calculated using two different approaches and the impact of standby energy is highlighted. In addition, the energy efficiency, total annual energy consumption and annual energy savings are calculated according to the energy efficiency classification standards VDI 4707 and ISO 25745 and compared with the results of measurements performed at low and high travel demand. The regenerative cycles are studied in detail to determine the requirements for a future energy storage system. The results have shown that increased travel demand increases the energy efficiency of the elevator system. This effect is not obvious when considering only the results obtained according to the elevator energy efficiency classification standards. Moreover, the overall energy efficiency of the elevator system can be further improved by reducing the energy consumption in standby mode.

Index Terms—elevator, regenerative energy, energy efficiency, energy savings, standby energy, VDI 4707, ISO 25745

I. INTRODUCTION

Traction sheave elevators convert gravitational potential energy into electrical energy due to the difference in weight between the counterweight and the car, and thus can serve as a source of electrical energy in certain modes of operation. When the elevator decelerates, mechanical braking occurs, and the energy is usually dissipated at the braking resistor in the form of heat. Instead, the regenerative energy can be fed back into the grid or stored for later use. Due to the increasing use of elevators and thus their high energy demand (they consume 3 to 8% of a building's total electricity [1]), but also the tendency to reduce their negative impact on the environment, the use of regenerative elevator energy is currently the most researched approach in the field of increasing the energy efficiency of elevators and buildings. Regenerative energy technologies are still rarely used today and can only be found in 2% of the total installed elevators, mainly due to the high price of bidirectional energy converters [2].

This work has been fully supported by the European Regional Development Fund under the project EULIFT - Development of a Smart Modular Elevator Drive System for Increasing the Energy Efficiency of a Building (EFRR-IRI-II-KK.01.2.1.02.0077).

In recent years, various ways of utilizing the generated energy have been developed to increase the overall efficiency of the system, such as optimizing the elevator control algorithm [3]–[5], optimizing the design of the counterweight to achieve higher energy efficiency and reduce the losses [6], [7], storing the regenerated energy in batteries and supercapacitors [2], [8]–[10], and returning the energy to the power grid.

To improve the efficiency of the existing elevator system without installing new expensive components, several methods have been proposed based on the improvement of the elevator control algorithm. The authors in [3] analysed the energy consumption of an elevator to develop an energy consumption model for testing a multiobjective genetic algorithm (MOGA), which, according to the simulation results, could reduce the average energy consumption by 23.6%. In [5], the authors proposed an energy saving method based on optimising the schedule of an elevator group that can reduce the total consumption during peak hours. Both approaches resulted in longer waiting times for passengers. On the other hand, in [4], the authors designed an energy-saving elevator system that changes its speed during travel, which reduces travel time and increases energy savings by 12% compared to a constant-speed elevator. According to the present results, significant energy saving obtained by improving the control algorithm usually results in longer waiting time for passengers. Therefore, in order to significantly increase the efficiency of the system, the proposed methods should be combined with some of the following solutions [2], [8]–[13] that allow the use of generated elevator energy.

One of the most effective ways to improve the overall efficiency of the elevator system is to use batteries and supercapacitors to store the energy generated by the elevator for later use. The authors in [8] have proposed a simulation model and prototype of an energy recovery system with super-capacitor bank to store the regenerated energy and concluded that such a system can reduce the overall energy consumption by about 20%. In order to design the supercapacitor bank capacity to reduce the cost of the elevator system with energy storage, the authors in [2] analyzed the difference in the traffic flow of the elevator using the neural network method. In [10], the authors compared the energy efficiency of two regenerative braking methods: the grid inverter method and the supercapacitor method using Simulink to simulate typical

978-1-6654-3236-8/21 $31.00 © 2021 IEEE

elevator usage. The results showed that the grid inverter method is more efficient in situations with longer regenerative braking, while the supercapacitors were found to be suitable for shorter average travel distances and fast decelerations.

The solutions that use supercapacitors to increase energy efficiency achieve favorable percentages of energy savings, but to properly design such solutions, one must analyze the regenerative cycles in detail. Such an analysis has not yet been reported in the literature. Moreover, there are only two studies in the literature based on actual measurements that estimate energy savings from regenerative elevator operation. According to [11], the average energy savings after replacing non-regenerative elevator drives with regenerative drives in different building types in Taiwan was 23.1%. The second study was conducted on a faculty elevator where the regenerative potential of the elevator was analyzed to estimate the annual energy consumption and energy savings based on weekly measurements, VDI 4707 and ISO 25745 [12]. The measurements were performed under pandemic conditions, where elevator traffic was about 30% lower than usual. The results showed that under these conditions 15.9% of the total energy can be regenerated and fed back to the grid on an annual basis.

This paper presents a measurement study of regenerative cycles and elevator energy savings during regenerative operation to determine energy storage system requirements. The study is based on measurements during a typical work week with high travel demand. The results are compared with those obtained for the same elevator with lower travel demand presented in [12]. The impact of travel demand on regenerative potential and energy savings is analyzed, and the impact of standby energy on energy savings is highlighted. The energy efficiency of the elvator system, the total annual energy consumption and the annual energy savings at higher travel demand are calculated according to the energy efficiency classification standards VDI 4707 and ISO 25745 and compared with the previously obtained measurements at lower travel demand. The distribution of regenerative and motor starts, peak power demand, and peak and average energy flow during regenerative operation are analyzed. The influence of the ride type distribution on the regenerative potential of the elevator is also presented.

This paper is organized as follows. Section II describes the working principle of an elevator and the regenerative elevator system studied. Section III analyzes the experimental results of measurements made during a typical work week with high travel demand and compares them with the results obtained for the same elevator under conditions with lower travel demand. It also calculates annual energy consumption and energy savings according to energy efficiency classification standards and energy savings using two different approaches, highlighting the impact of standby energy on energy efficiency. In section IV, the regenerative operating cycles of elevators are analyzed in detail. The conclusion is given in section V.

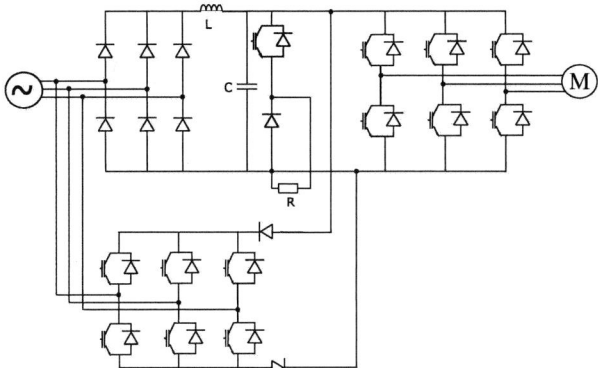

Fig. 1. Power converter with regenerative unit.

II. ELEVATOR SYSTEM DESCRIPTION

The main components of a traction elevator are the car, the counterweight, the electric drive, the controls, and the guide rails [14]. The mass of the counterweight is generally equal to the sum of the mass of the empty car and 40-50% of the rated elevator load. Depending on the direction of motion and the difference in weight between the car and the counterweight, the elevator can operate in four quadrants. Regenerative operation occurs when the elevator is moving up with a light load or down with a heavy load. The energy saving potential of the elevator system increases with the greater weight difference between car and counterweight.

Nowadays, elevators are controlled by variable frequency drives, and the most commonly used are the three-phase power converters with diode bridge rectifiers as the front-end topology. This type of drive can use regenerative energy only if it is supplemented by a regenerative unit, as shown in Fig. 1, or a storage system is added, which is rarely the case. Usually, the regenerative energy is dissipated at the braking resistor.

The considered faculty elevator system consists of the mechanical system, the gearless drive with a permanent magnet motor and the frequency converter with a diode bridge rectifier, and the regenerative unit. The regenerative unit feeds the regenerated energy back into the mains during regenerative elevator operation. It is connected between the DC-link and the mains, as shown in Fig. 1. The technical data of the studied elevator are shown in Table I.

The nominal data of the permanent magnet motor, the power converter and the regenerative unit are as follows:

Motor: P_n = 12.1/8.4 kW, f_n = 32 Hz, V_n = 360 V, I_n = 32/22 A, n_n = 192 rpm, T_n = 600/420 Nm.

Power converter: V_n = 400 V, f_n = 50 Hz, I_n = 32 A, P_n = 12,1 kW.

Regenerative unit: V_n = 400 V, f_n = 50 Hz, P_n = 5 kW.

The measurement of energy consumption of a faculty elevator was performed for a typical working day under real conditions. The energy consumption of the elevator was measured and analyzed at 1 second intervals. The measurements were performed using a METREL MI2892 power analyzer,

978-1-6654-3236-8/21 $31.00 © 2021 IEEE

TABLE I
ELEVATOR DATA

Parameter	Value
Nominal load, Q [kg]	630
Nominal speed, v_N [m/s]	2.5
Car weight [kg]	800
Counterweight compensation [%]	50
Operating days per year, d_{op}	365
Lifting height, F_H [m]	45.6
Number of floors	13
Maximum number of passengers	8

Fig. 2. Measurement setup.

which belongs to class A according to IEC 61000-4-30. The meter was connected behind the main switch of the elevator, as shown in Fig. 2.

III. ENERGY EFFICIENCY ANALYSIS

The energy efficiency analysis was performed on a faculty elevator under high travel demand during one week. The results are compared with the analysis reported in [12], which was performed on the same elevator during a week of lower travel demand due to pandemic conditions. Three-phase currents and voltages and power and energy components were measured, and the number of starts, standby and running energy, and energy savings were calculated from the measured results. The influence of the travel demand on the regenerative energy potential and energy savings was analyzed.

The results of the measured weekly energy consumption at the faculty elevator and the comparison with the results obtained at the same elevator with lower travel demand are given in the subsection III-A. Energy efficiency calculations

and annual energy estimates based on ISO 25745, VDI 4707 and measurement results are given in subsection III-B. In the subsection III-C, the energy savings are analyzed and calculated in two different ways. It has been observed how the increase in travel demand affects the energy savings and energy efficiency of the regenerative elevator. The results have shown that greater energy savings can be achieved during periods of increased travel demand.

A. Energy consumption under different travel demand

The results of the weekly measurement at higher travel demand are analysed and compared with the results obtained at lower travel demand. The measured regenerative energy (E_{reg}) and total energy consumption (E_{tot}) of the studied regenerative elevator, the calculated consumed energy (E_{con}) of a non-regenerative elevator, the standby energy (E_{st}), the running energy (E_{run}), the average standby time (t_{st}) and the average running time (t_{run}) are reported. The regenerative energy E_{reg} represents the energy generated by an elevator in regenerative mode and fed back to the grid. The consumed energy E_{con} represents the energy that the regenerative elevator takes from the grid, which is also the total energy consumed by the non-regenerative elevator during its operation. The total energy E_{tot} is the total energy consumed by the regenerative elevator, which is calculated by subtracting the regenerative energy (E_{reg}) from the consumed energy (E_{con}). The consumed energy is composed of the standby energy E_{st} and the running energy E_{run}. Running energy is the energy consumed when the elevator is running, and standby energy is the energy consumed when the elevator is not running.

Table II shows the results of weekly measurement when travel demand is higher and lower. For both cases, the average and the best day of the week are analysed, where the best day is the day when the travel demand is the highest and the amount of regenerated energy is the largest. The energy savings ES were also calculated for each case as the ratio of regenerated energy E_{reg} to consumed energy E_{con} and presented in Table II. The energy savings achieved with higher travel demand are about 2.3% higher on a weekly basis, indicating that the regenerative potential of elevators increases with more starts.

Fig. 3 shows the standby energy (E_{st}) and running energy (E_{run}) of the elevator for each day of the observed week. The running energy of the elevator is slightly higher than the standby energy on each day of the week, while the situation is reversed on weekends when the elevator starts are reduced. In contrast, the standby energy was higher than the running energy for each day of the observed week when the travel demand was lower.

Fig. 4 shows the average number of starts per hour during the measured week. Due to the rather rough and unpredictable distribution of starts during the day, the faculty elevator does not belong to any of the typical elevator categories (residential buildings, office buildings, hospitals, hotels) and therefore it is very difficult to determine the average distribution of starts

TABLE II
COMPARISON OF THE ENERGY CONSUMPTION UNDER DIFFERENT TRAVEL DEMAND

Parameter	Value			
	High Demand		Low Demand	
	average day	best day	average day	best day
E_{reg}, [kWh]	2.0	3.1	1.8	2.7
E_{con}, [kWh]	11.2	14.7	11.3	14.2
E_{tot}, [kWh]	9.2	11.7	9.5	11.5
E_{run}, [kWh]	4.6	7.2	3.9	6.0
E_{st}, [kWh]	4.5	4.5	5.6	5.5
t_{run}, [h]	2.5	3.8	2.1	2.6
t_{st}, [h]	21.5	20.2	21.9	21.4
Number of starts	457	703	372	582
ES, [%]	18.2	21.7	15.9	18.4

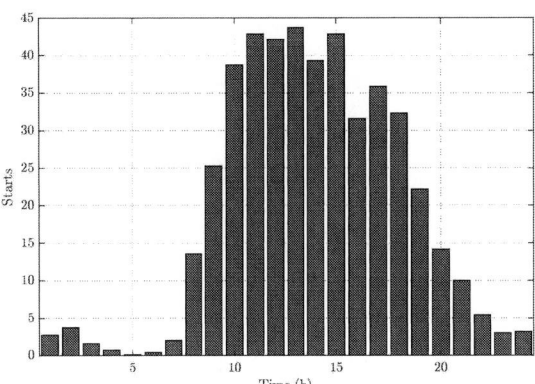

Fig. 4. Average number of starts per hour during the week.

TABLE III
ISO 25745, VDI 4707 AND MEASUREMENT BASED ESTIMATION

Parameter		E_{run} [kWh]	E_{st} [kWh]	E_d [kWh]	E_y [kWh]	ES [%]
ISO	Non-Reg.	8.4	4.5	12.9	4720	24.0
	Reg.	5.3	4.5	9.8	3585	
VDI	Non-Reg.	12.4	4.8	17.1	6259	26.7
	Reg.	7.8	4.8	12.6	4588	
Meas.	Non-Reg.	6.7	4.5	11.2	4088	17.9
	Reg.	4.6	4.5	9.2	3358	

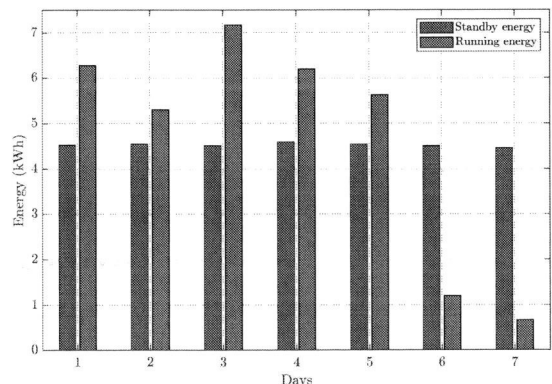

Fig. 3. Elevator standby and running energy during the week.

as well as the peak hours for this example of an elevator installation.

B. Efficiency calculation and annual energy estimation

The energy efficiency of the studied elevator system and the total annual energy consumption for the regenerative and non-regenerative elevator are calculated based on energy efficiency classification standards VDI 4707 and ISO 25745, and weekly measurement results. The daily running energy consumption (E_{run}), daily standby energy consumption (E_{st}), total daily energy consumption (E_d), estimated yearly energy consumption (E_y) and energy savings (ES) are presented in Table III. Energy saving (ES) is calculated using the expression:

$$ES = \frac{regenerated\ energy}{consumed\ energy\ of\ a\ non\text{-}regenerative\ elevator} \quad (1)$$

The standards for classifying the energy efficiency of elevators ISO 25745 and VDI 4707 assume elevator usage determined by various factors such as the number of starts per day, elevator speed, building type, building height, and

the ratio of running to standby time. Therefore, despite a 30% increase in traffic, the elevator where the measurements were taken belongs to the same usage categories under high and low travel demand and the same results are obtained for the estimation of annual energy consumption and efficiency. On the other hand, the results of energy saving (ES) based on the previously performed measurements showed that the regenerative operation provided an annual saving of about 16% for the considered system (Table II). Based on the new measurement results obtained at high travel demand, the increase in traffic volume increased the savings by about 2%.

C. Energy savings

There are several ways to calculate the energy savings of elevators, and depending on the method of calculation, the percentages achieved can vary significantly. Several studies have examined the efficiency of elevator systems and the potential energy savings that can be achieved through the use of advanced technologies [1], [15]–[17]. The consumption of an elevator generally depends on many factors, such as the type of drive (hydraulic, geared traction, gearless traction) and the efficiency of the individual components (motor, gearbox, fans, brakes) [1], [16], the control and operating equipment and the load - the frequency of use, the number of passengers and the number of starts [18]. Since most elevators are idle most of the time, standby energy accounts for a significant portion of total elevator energy consumption, and the trend to reduce it is becoming increasingly important [19].

978-1-6654-3236-8/21 $31.00 © 2021 IEEE

TABLE IV
ENERGY SAVINGS CALCULATION

Method	Expression	Savings	
		Absolute[kWh]	Relative [%]
E_{s1}	E_{reg}/E_{con}	2.0	18.2
E_{s2}	E_{reg}/E_{run}	2.0	49.8

The share of standby energy in elevators varies between 5% and 95% and depends mainly on the frequency of use of the elevator where running energy of the elevator increases with a larger number of starts [20]. The energy consumption in running and standby mode strongly depends on the technology used and its energy efficiency. Solutions proposed in [1] use the best components available on the market and put the installation in sleep mode when not in use, reducing the standby energy by 80%.

In this section, the elevator energy savings are calculated in two different ways (E_{s1}, E_{s2}), based on measurements during a typical workday with high travel demand, to illustrate the differences in the results depending on the approach chosen. Given the large contribution of standby energy to total energy consumption [12] and its negative impact on system efficiency, the calculation of energy savings is sometimes approached in a way that ignores its impact. The first approach presented in this paper considers standby energy throughout the day when calculating energy savings, while the second approach calculates energy savings based only on energy consumed in running mode.

The first approach calculates the energy saving (E_{s1}) as the ratio of regenerated energy (E_{reg}) to the energy consumed by the non-regenerative drive (E_{con}):

$$E_{s1} = \frac{E_{reg}}{E_{con}} \qquad (2)$$

This calculation method was used to calculate energy savings in the [12] article and in the III-A and III-B subsections.

The second approach calculates the energy savings (E_{s2}) while the elevator is running as the ratio between the regenerative energy E_{reg} and the total energy consumed in the running mode (E_{run}):

$$E_{s2} = \frac{E_{reg}}{E_{run}} \qquad (3)$$

The energy savings for an average day of high travel demand were calculated using these approaches and presented in Table IV. When the non-regenerative elevator is replaced with a regenerative elevator, the energy savings (E_{s1}) is 18.2%. When the standby energy of the elevator is neglected (E_{s2}), the energy saving obtained with the regenerative solution is 49.8%. The results confirm the importance of reducing standby energy for elevator systems and that the savings can vary greatly depending on the calculation method. In each of the above cases, the regenerative potential increases with the increase in travel demand.

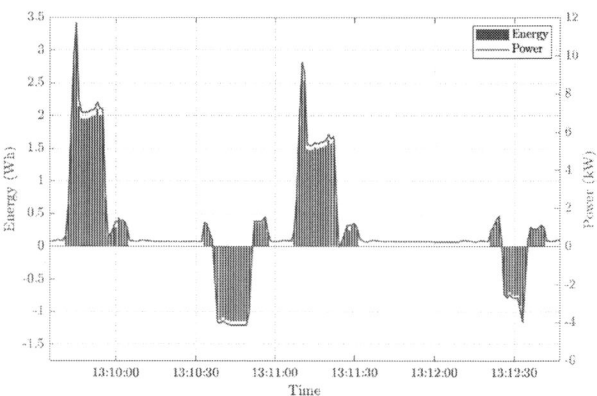

Fig. 5. High power demand and energy consumption cycles of the observed elevator.

IV. REGENERATIVE CYCLES ANALYSIS

A detailed analysis of the regenerative elevator cycles and the distribution of elevator ride types was performed to determine the energy storage system requirements. In the IV-A subsection, peak power demand, peak, and average energy flow during regenerative operation, and distribution of regenerative and motor starts are analyzed. Regenerative elevator cycles are analyzed in the IV-B subsection. In the subsection IV-C the distribution of elevator ride types and their impact on the regenerative potential of the elevator are analyzed.

A. Power and energy demand

A typical elevator ride consists of the opening and closing of the car doors, a steady acceleration, a constant speed travel, a steady deceleration, and the opening and closing of the doors. Fig. 5 shows the energy consumption and power demand for four consecutive trips with high power and energy demand of the studied elevator. The successive trips shown have high energy demand in motor mode and high energy return in regenerative mode.

The flow of energy during elevator travel can be divided into three phases: the acceleration of the system during start-up, the constant velocity during steady state, and the deceleration of the system during braking. At the beginning of the regenerative trip, the motor consumes energy from the grid while the brake is released. During the constant speed travel phase, the elevator generates energy in regenerative mode and consumes it in motor mode. During the deceleration phase, before reaching landing speed, the drive operates in regenerative mode during regenerative trip but can sometimes operate in regenerative mode during motor trip. At the end of deceleration, the elevator travels at a low landing speed and no energy is generated during deceleration and the drive operates in motor mode. Therefore, the drive can operate in motor mode for part of the elevator ride and in regenerative mode for another part. The

978-1-6654-3236-8/21 $31.00 © 2021 IEEE

TABLE V
ELEVATOR ENERGY CONSUMPTION DURING A WEEK

Day	1	2	3	4	5	6	7
E_{con} [kWh]	12.9	11.8	14.1	13.0	12.2	6.1	5.4
E_{reg} [kWh]	2.7	2.5	3.1	2.8	2.5	0.5	0.3
$E_{reg_{avg}}$ [Wh]	112.5	104.2	129.2	116.7	104.2	20.8	12.5
$E_{reg_{max}}$ [Wh]	289.7	322.5	320.5	306.6	296.1	56.5	54.3
Regenerative starts	304	258	339	309	264	61	31
Motor starts	315	270	365	305	293	59	32

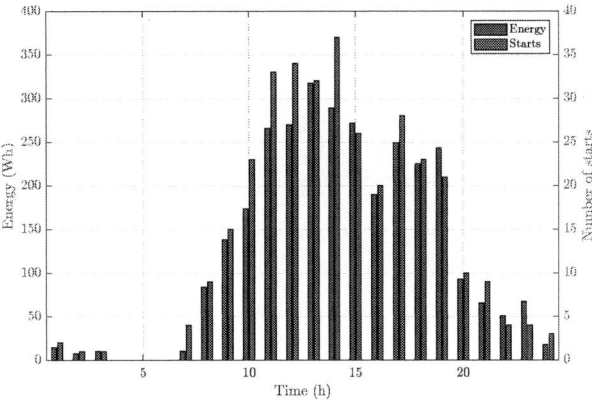

Fig. 7. Number of regenerative starts and total energy per hour of the best day.

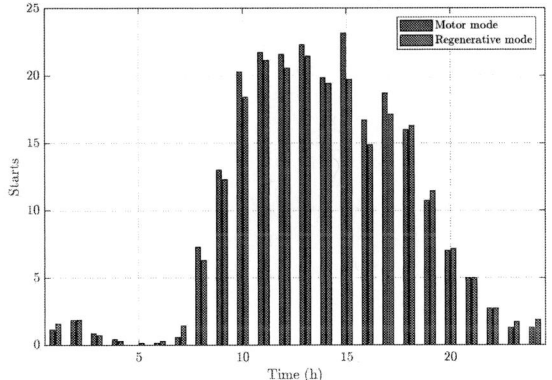

Fig. 6. Number of regenerative and motor starts per hour during the day.

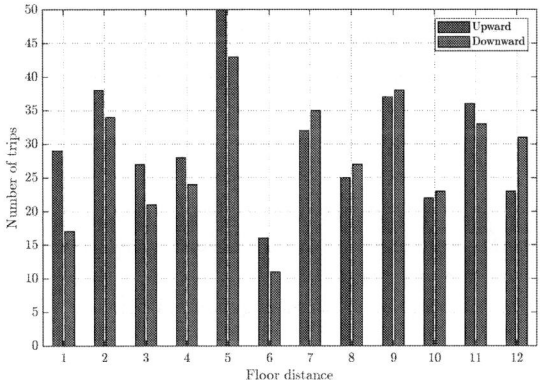

Fig. 8. Number of upward and downward elevator starts during the best day.

highest power demand is in motor mode during acceleration of the elevator car.

B. Analysis of regenerative elevator cycles

Fig. 6 shows the average number of elevator starts per hour in regenerative and motor modes during a measured week. Elevator runs where the elevator generates the energy only during deceleration are categorized as motor starts. The number of motor and regenerative starts is approximately equal over the course of a day.

Table V contains the daily consumed energy (E_{con}), daily regenerated energy (E_{reg}), average regenerative energy per hour ($E_{reg_{avg}}$), maximum regenerative energy per hour ($E_{reg_{max}}$), and the number of regenerative and motor starts for each day of the week. To optimise the energy storage system, the instances of consecutive regenerative starts were counted. During the weekly measurement, two consecutive regeneration starts (two regenerative trips in a row) occurred 151 times, while three consecutive regeneration starts occurred 22 times during the observed week.

Fig. 7 shows the total regenerative energy per hour and the number of regenerative starts per hour for the best day. The average regenerative energy per hour is 127 Wh. During peak hours, from 10 am to 7 pm, the average regenerative energy per hour is 249 Wh. A larger number of starts results in a better ratio between consumed and regenerative energy. While an average regenerative ride during the week generates 8 Wh, the amount of energy varies from ride to ride and depends on

many factors such as the weight of the car, the direction of travel and the duration of the ride.

C. Elevator ride type distribution

During the observed week, the exact position of the elevator car was also measured. The exact position, expressed in number of floors, was used to determine the distribution of elevator ride types. Fig. 8 shows the number of starts during the best day for each distance, expressed in number of floors, with upward rides in blue and downward rides in red. For each distance, there are approximately the same number of upward and downward rides. The absolute number of elevator rides depends on the distribution of people, labs, offices, etc. in a building, which is not uniformly distributed. Table VI shows the number of motor and regenerative starts for downward or upward rides during the week. The table shows that about 98% of the regenerative energy is generated during upward rides.

Fig. 9 shows the average and maximum regenerative energy per floor distance during the week. As the distance increases, the maximum and average regenerative energy increases linearly. Table VII contains the total amount of regenerative

TABLE VI
NUMBER OF MOTOR AND REGENERATIVE RIDES IN CORRELATION WITH UPWARD OR DOWNWARD MOVEMENT.

	Motor mode	Regenerative mode
Upward	243	1540
Downward	1392	25

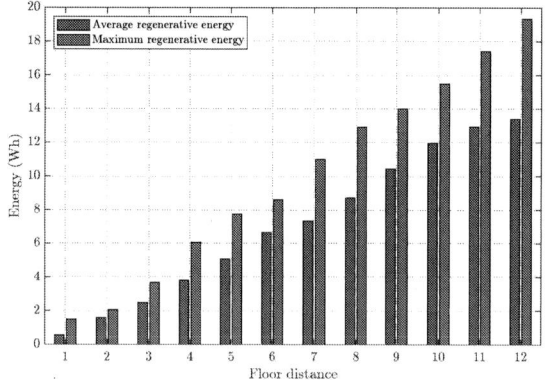

Fig. 9. Average and maximum energy per regenerative trip during the week.

energy and the number of starts for each floor distance during the week.

V. CONCLUSION

In this paper, the analysis of the regenerative cycles and energy efficiency of the faculty elevator under high travel demand is performed. The study is based on measurements made during one week. An analysis of energy efficiency is performed in comparison to a previous study using the same elevator at a lower travel demand. Energy savings are calculated using two different approaches. Annual energy consumption and annual energy savings were estimated according to the elevator energy efficiency classification standards ISO

TABLE VII
NUMBER OF STARTS AND TOTAL REGENERATIVE ENERGY AND PER FLOOR DISTANCE.

Floor distance	Number of starts	Total regenerative energy [kWh]
1	48	26.7
2	156	247.2
3	147	364.1
4	132	502.5
5	177	897.2
6	81	539.1
7	154	1131.7
8	105	915.0
9	176	1837.3
10	87	1040.5
11	152	1963.9
12	150	2008.3

25745 and VDI 4707 and based on actual weekly measurement results. The regenerative cycles are analyzed to define the energy storage solution requirements. The power demand and energy consumption, the number of regenerative starts and the total regenerative energy per hour and per start are observed.

The results have shown that the daily distribution of regenerative and motor starts for the faculty elevator is approximately the same. Comparison with previous measurements made at lower travel demand has shown that the higher travel demand increases the regenerative potential of the elevator and improves the energy efficiency. It also shows that calculating the energy efficiency solely on the basis of the energy efficiency classification standards for elevators ISO 25745 and VDI 4707, the results may differ significantly from the actual situation if the use of the elevator does not correspond to the categories indicated there. Therefore, estimation should be based on actual measurements whenever possible. The large impact of standby energy on energy efficiency is also shown by the comparison. The results confirm the importance of reducing standby energy for elevator systems and that the savings can vary greatly depending on the calculation method.

REFERENCES

[1] Aníbal De Almeida, Simon Hirzel, Carlos Patrão, João Fong, and Elisabeth Dütschke. Energy-efficient elevators and escalators in europe: An analysis of energy efficiency potentials and policy measures. *Energy and buildings*, 47:151–158, 2012.

[2] Hong-peng Liu, Kuan Liu, and Bai-nan Sun. Analysis of energy management strategy for energy-storage type elevator based on supercapacitor. In *2017 11th IEEE International Conference on Compatibility, Power Electronics and Power Engineering (CPE-POWERENG)*, pages 175–180. IEEE, 2017.

[3] Zhangyong Hu, Yaowu Liu, Qiang Su, and Jiazhen Huo. A multi-objective genetic algorithm designed for energy saving of the elevator system with complete information. In *2010 IEEE International Energy Conference*, pages 126–130, 2010.

[4] Muhammad Z Hasan, Rainer Fink, Muthuvel Raj Suyambu, and Manoj Kumar Baskaran. Assessment and improvement of elevator controllers for energy efficiency. In *2012 IEEE 16th International Symposium on Consumer Electronics*, pages 1–8. IEEE, 2012.

[5] Jinglong Zhang and Qun Zong. Energy-saving scheduling optimization under up-peak traffic for group elevator system in building. *Energy and Buildings*, 66:495–504, 2013.

[6] Toni Tukia, Semen Uimonen, Marja-Liisa Siikonen, Harri Hakala, and Matti Lehtonen. A study for improving the energy efficiency of lifts with adjustable counterweighting. *Building Services Engineering Research and Technology*, 38(4):421–435, 2017.

[7] Toni Tukia, Semen Uimonen, Matti Lehtonen, Marja-Liisa Siikonen, and Claudio Donghi. Evaluating and improving the energy efficiency of counterbalanced elevators based on passenger traffic. In *2016 IEEE 16th International Conference on Environment and Electrical Engineering (EEEIC)*, pages 1–6, 2016.

[8] Seema Mathew, Pooja Mogre, Rohan Chouthai, PB Karandikar, and NR Kulkarni. Supercapacitor based energy recovery system for an elevator. In *2017 International Conference on Advances in Computing, Communication and Control (ICAC3)*, pages 1–5. IEEE, 2017.

[9] Nikolaos Jabbour and Christos Mademlis. Supercapacitor-based energy recovery system with improved power control and energy management for elevator applications. *IEEE Transactions on Power Electronics*, 32(12):9389–9399, 2017.

[10] Jaesung Kim, Zongyou Han, Pintian Huang, Donovan O'Donnell, and Narayan C Kar. Comparative analysis of the utilization of supercapacitor versus grid-tie inverter regenerative braking methods for elevator systems. In *IECON 2019-45th Annual Conference of the IEEE Industrial Electronics Society*, volume 1, pages 2535–2540. IEEE, 2019.

[11] Kun-Yu Lin and Kuang-Yow Lian. Actual measurement on regenerative elevator drive and energy saving benefits. In *2017 International Automatic Control Conference (CACS)*, pages 1–5. IEEE, 2017.

[12] Martina Kutija, Luka Pravica, Damjan Godec, and Dora Erica. Regenerative Energy Potential of Roped Elevator Systems-A Case Study. In *2021 IEEE 19th International Power Electronics and Motion Control Conference (PEMC)*, pages 284–291. IEEE, 2021.

[13] G Nobile, AG Sciacca, M Cacciato, C Cavallaro, A Raciti, G Scarcella, and G Scelba. Energy harvesting in roped elevators. In *2014 International Symposium on Power Electronics, Electrical Drives, Automation and Motion*, pages 533–540. IEEE, 2014.

[14] Boonyang Plangklang and Sittichai Kantawong. Study of power generation for permanent magnet motor elevator by energy regenerative unit (eeru). *Energy Procedia*, 56:591–597, 2014. 11th Eco-Energy and Materials Science and Engineering (11th EMSES).

[15] Harri Hakala, Marja-Liisa Siikonen, Tapio Tyni, and Jari Ylinen. Energy-efficient elevators for tall buildings. In *6th World Congress on Tall Buildings and Urban Habitat*, 2001.

[16] Harvey M Sachs. Opportunities for elevator energy efficiency improvements. American Council for an Energy-Efficient Economy Washington, DC, 2005.

[17] Rohan Sirsi and Danesh Kamath. Energy efficient elevators and technologies. In *2017 14th IEEE India Council International Conference (INDICON)*, pages 1–4. IEEE, 2017.

[18] A Sharif. Lift and escalator energy consumption. *Transportation Systems in Buildings*, 13:102–108, 2000.

[19] Harvey Sachs, Harry Misuriello, and Sameer Kwatra. Advancing elevator energy efficiency. *Report A1501*, 2015.

[20] Carlos Patrão, Aníbal De Almeida, João Fong, and Fernando Ferreira. Elevators and escalators energy performance analysis. In *ACEEE Summer Study on Energy Efficiency in Buildings*, 2010.

2021 International Conference on Electrical Drives & Power Electronics (EDPE)   Dubrovnik, 22-24 Sept. 2021

DC/DC Converter Topologies for Elevator Energy Storage Systems Based on Supercapacitors

Martin Makar, Martina Kutija, Luka Pravica, Filip Jukić
University of Zagreb Faculty of Electrical Engineering and Computing
Zagreb, Croatia
martin.makar@fer.hr, martina.kutija@fer.hr, luka.pravica@fer.hr, filip.jukic@fer.hr

Abstract—To increase the energy efficiency of traction elevators, the regenerative energy must be stored or fed back into the grid. The regenerative energy can be stored in batteries or supercapacitors using the appropriate DC/DC converter. In this paper, the DC/DC converter topologies typically used in supercapacitor-based energy storage systems for elevator applications are investigated. The requirements for the DC/DC converters are analyzed from two perspectives: elevator drive and supercapacitor bank. The performance of the most commonly used topologies is compared using simulation tests, and their advantages and disadvantages are discussed.

Index Terms—DC/DC converter, supercapacitor, energy storage, elevator

I. INTRODUCTION

Increasing the energy efficiency of the elevator system has received much attention recently [1], [2]. Traction elevators can achieve energy savings of up to about 25% by utilizing their regenerative energy potential [3], [4]. The traction elevator is shown in Fig. 1 and typically consists of an electric motor (connected to the traction sheave), a car, and a counterweight whose weight is typically equal to the weight of an empty car plus 50% of the rated load. The drive can operate in generator or motor mode, depending on the weight of the load and the direction of motion. The elevator is typically controlled by a variable frequency drive with a diode bridge rectifier as the front-end topology. In this case, the energy is dissipated through a braking resistor during regenerative operation. Several solutions have been proposed in the literature to improve the energy efficiency of the elevator. These solutions mainly consider feeding regenerative energy back into the grid or storing regenerative energy in batteries or supercapacitor-based storage systems and using it for motor operation and other purposes (e.g., evacuation in case of power failure, smoothing high and short-time power, solicitations).

Elevators have a highly modulated power demand with high starting peak power in motor mode and high continuous power during regenerative operation. For this reason, supercapacitors represent an attractive solution in energy storage systems for elevator applications [5], [6]. Compared to classical batteries, supercapacitors contain a higher power density and can be

This work has been fully supported by the European Regional Development Fund under the project EULIFT - Development of a Smart Modular Elevator Drive System for Increasing the Energy Efficiency of a Building (EFRR-IRI-II-KK.01.2.1.02.0077).

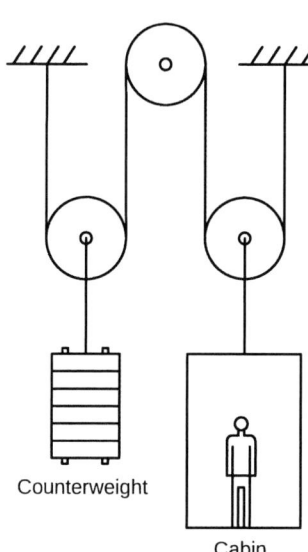

Fig. 1. Schematic of traction sheave elevator

charged and discharged very quickly. Furthermore, a high life expectancy, considering a large number of charging and discharging cycles, makes them suitable for elevator applications [7]. One of the energy storage systems main components is the bidirectional DC/DC converter that transfers the energy between the supercapacitor bank and the elevator drive.

In general, DC/DC converters can be divided into two main categories: isolated and non-isolated. The non-isolated DC/DC topology is well suited for applications where isolation is not required and where the voltage transformer ratio (between primary and secondary) is four or less [8]. On the other side, the advantages of the isolated DC/DC topology are that it can provide isolation and a larger voltage transformer ratio [9]. In elevator energy storage systems, non-isolated topology is typically used, such as basic buck/boost [10]–[15] and interleaved buck/boost [16]. The transformer-based topologies such as Dual Active Bridge (DAB) and resonant are also reported [17].

978-1-6654-3236-8/21 $31.00 © 2021 IEEE   220

However, research on elevator energy storage systems has mainly focused on the control and design of the energy storage device, with less attention paid to the design of a DC/DC converter. In [11], a control scheme for elevators with a supercapacitor-based energy storage system is proposed in which a non-isolated bidirectional DC/DC converter is chosen due to the efficiency, weight, size, and cost of this type of DC converter compared to an isolated converter. In [16], the main design aspects for an energy storage system are analyzed, focusing on the supercapacitor bank. A non-isolated DC/DC converter was chosen for the DC/DC converter topology. In [15], the use of power factor correction (PFC) in combination with a supercapacitor energy storage system is proposed. Here, a non-isolated buck/boost topology is also proposed for supercapacitor energy storage. However, the comparison of different topologies of DC/DC converters has not yet been reported in the literature.

This paper gives an overview of the typically used DC/DC converters in elevator storage systems and analyses their advantages and disadvantages by simulation tests. Furthermore, the design requirements of the DC/DC converter given by the elevator system and the supercapacitor bank are analysed in detail, and potential problems are highlighted. The main objective of this paper is to summarise the requirements of the DC/DC converter in an elevator energy storage system, show the advantages and disadvantages of the typically used topologies, and give the guidelines for designing such a system.

The paper is organized as follows. Section II specifies the requirements for the DC/DC converter used in the elevator energy storage system. Section III gives an overview of the typically used DC/DC topologies, which are compared by the simulation tests in section IV. In section V the simulation results are analyzed. A conclusion is given in section VI.

II. REQUIREMENTS FOR DC/DC CONVERTER IN ENERGY STORAGE SYSTEM

A. Elevator system requirements

Induction or synchronous motors with permanent magnets, typically used in traction sheave elevators, are controlled by AC/DC/AC frequency converters shown in Fig. 2. In most cases, a diode bridge rectifier is used for AC/DC conversion, and in regenerative mode, the energy is dissipated at the braking resistor. Typically, the elevator drives are powered from the three-phase 400 V mains, and the DC link voltage is about 540-580 V. Generally, the DC-link voltage can range between about -10/+40 % of $\sqrt{3} \cdot \sqrt{2} \cdot U_{phase}$ DC. The lower limit is defined by the undervoltage protection of the frequency converter and is reached when the power supply fails. The upper limit depends on the voltage level at which the braking chopper is activated and is reached during regenerative operation. The upper limit cannot be changed for commercially available drives. Therefore the DC/DC converter used for energy storage must be activated below this limit when the drive is operating in regenerative mode. The overvoltage protection is typically activated around 800 V, and the

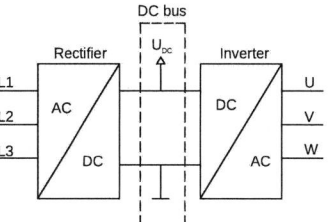

Fig. 2. Frequency converter basic schematic

frequency converter is switched off. If the energy storage is used for emergency operation in case of mains failure, the DC/DC converter must detect the DC-link voltage drop and supply the drive instead of the mains. DC/DC converter can generate common mode and differential mode noise which can produce electromagnetic interference (EMI) problems. This EMI problems can affect on work of frequency converter and other devices in system, such as non-predictable behavior of frequency converter. To prevent this, there must be a filter between DC/DC converter and frequency converter.

Therefore, the elevator drive system determines the voltage range in which the DC/DC converter must operate. It also affects the voltage level at which the DC/DC converter starts storing energy in the energy storage and the voltage level at which it uses energy storage. The peak power in regenerative mode is defined by the elevator drive system and depends on the load, the speed of the elevator, and the system efficiency.

B. Supercapacitor bank requirements

A supercapacitor is characterized by a high capacitance (several hundred farads or more), so it is used in energy storage systems. The expression determines the energy that can be stored in a supercapacitor:

$$E = \frac{U_C^2 \cdot C}{2}, \qquad (1)$$

where E is the energy, U_C is the voltage across the supercapacitor, and C is the capacitance of the supercapacitor.

To make the best use of the available energy of the capacitor, the voltage across the capacitor is usually controlled between U_{max} and $U_{max}/2$, where U_{max} is the voltage of the fully charged capacitor. In this case, 75% of the total energy stored in the supercapacitor is used [16]. The supercapacitor current is determined by the regenerative mode power and the supercapacitor voltage. In [11] and [16], it is pointed out that to reduce the supercapacitor heating, the ripple of the charge/discharge current must be reduced. To keep the losses as low as possible, this paper focuses on keeping the value of the current ripple close to the average value. A lower current ripple means that the RMS value of the current is close to the average value. Therefore, from the supercapacitor bank point of view, the main requirements of a DC/DC converter depend on the voltage range of the supercapacitor bank and the peak power required. Unlike batteries, the voltage across

2021 International Conference on Electrical Drives & Power Electronics (EDPE)

the discharged capacitor is 0 V, so the DC/DC converter must be able to charge the specified package from 0 V to $U_{max}/2$ DC.

III. OVERVIEW OF DC/DC CONVERTER TOPOLOGIES

In this section, a brief overview of DC/DC topologies used in elevator energy storage systems is given.

A. Basic buck/boost topology

The basic buck/boost topology is shown in Fig. 3. The main components are transistors (T1, T2), forming a half-bridge configuration and inductor L. The capacitor SC represents the equivalent capacitance of the supercapacitor bank, and the resistor ESR represents the equivalent series resistance of the supercapacitor bank. DC bus is represented as the voltage source connected to the terminals U_{DC} and GND, and C1 is the filter capacitor. In this topology, the supercapacitor bank voltage must be lower than the voltage source U_{DC}. In buck mode, the energy is transferred from DC bus U_{DC} to the supercapacitor bank by controlling the PWM duty cycle of transistor T1. In boost mode, the energy is transferred from the supercapacitor bank to the DC bus U_{DC} by controlling the PWM duty cycle of transistor T2. In this way, bidirectional energy transfer is achieved. In [18], it is pointed out that zero-voltage switching (ZVS) is not possible in this type of topology in continuous conduction mode, which means that the switching losses are higher at a higher frequency.

The expected basic parameters of the buck/boost topology when used in an elevator energy storage system can be obtained from [10], and [14]. In [10], simulation tests are performed using a supercapacitor bank of 48 V / 83 F with a DC bus voltage of 308 V. According to their calculations, the required inductance of the coil was 0.5 mH for a frequency of 4 kHz with a current ripple of 20 %. In [14], a DC bus voltage in the range of 530-680 V, a capacitor operating voltage of 240-400 V with a capacitance of 17 F, and an inductance of 500 µH were used for the simulation. During charging operation, the current can rise to 35 A (the ripple is between 25 and 35 A), while the discharging current can reach up to 50 A peak.

The basic buck/boost topology is also used in other types of energy storage systems, such as metro vehicles [19] and microgrid systems [20].

B. Interleaved buck/boost topology

The interleaved buck/boost topology is shown in Fig. 4. Compared to the basic buck/boost topology (Fig. 3), it consists of at least two half-bridges (transistors T1, T2, and T3, T4) and two inductors (L1, L2). The minimum number of half-bridges is two since it forms the basic buck/boost topology with one half-bridge. The main difference between basic and interleaved converters is the control method. In the interleaved converter, there is a phase shift of $2\pi/N$ between the PWM signals controlling each half-bridge/phase [21], where N is the number of half-bridges/phases. In this way, it is achieved that each half-bridge can behave as a buck/boost converter in the discontinuous operation mode, which then eliminates turn-on

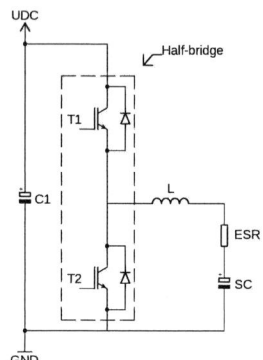

Fig. 3. Basic buck/boost topology

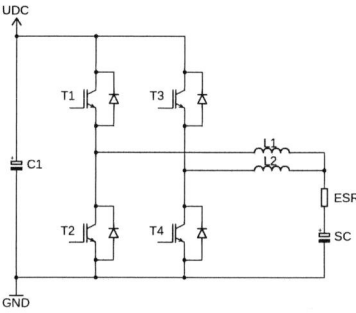

Fig. 4. Interleaved buck/boost topology

losses and increases the overall efficiency of the converter. The total current (current through the supercapacitor bank) is equal to the sum of the currents through the inductors. In addition to the above benefits, [21] also points out that interleaving helps reduce the size and losses of the input and output filters.

In [16], the design process of a supercapacitor-based storage system for elevators is presented. The 5-kW supercapacitor-based storage system with the interleaved DC / DC converter described in [22] has been built, and the test results obtained under laboratory conditions and on an actual elevator are presented. It has been shown that the efficiency in discontinuous operation is higher than in continuous operation. The required inductor volume is also reduced in discontinuous operation compared to continuous operation.

The interleaved buck/boost topology is also used in electric vehicle applications. [23].

C. Dual active bridge topology

Dual Active Bridge (DAB) DC/DC converter uses an inductor to transfer power from one source to another. It can operate as an isolated or non-isolated DC/DC converter. In this section, an overview of isolated topology is given. Fig. 5 shows an isolated bidirectional DAB DC/DC converter. It consists of two full-bridges (one full-bridge is composed of two half-bridges) formed by transistors T1-T8 connected through transformer Tr1 and inductor L1. DAB operation is

978-1-6654-3236-8/21 $31.00 © 2021 IEEE

2021 International Conference on Electrical Drives & Power Electronics (EDPE)   Dubrovnik, 22-24 Sept. 2021

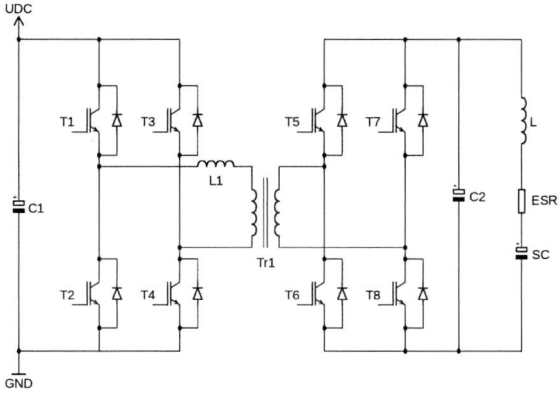

Fig. 5.   Isolated DAB topology

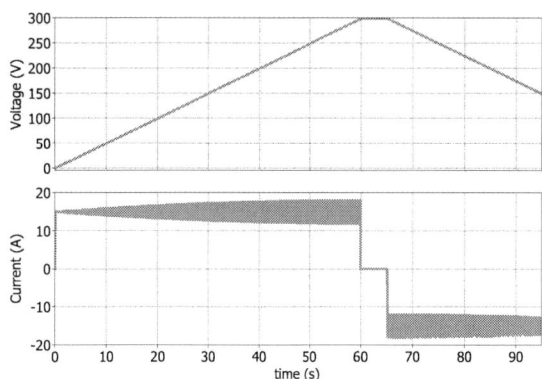

Fig. 6.   Voltage and current of supercapacitor bank for basic buck/boost topology

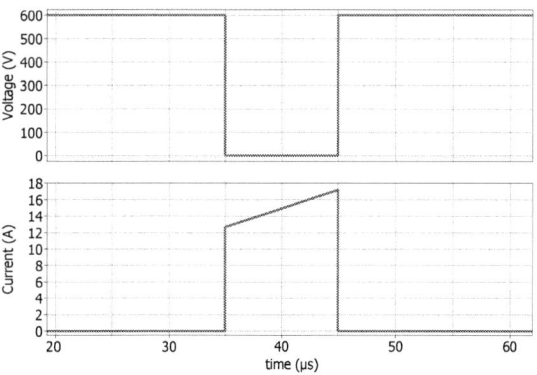

Fig. 7.   Voltage and current of transistor for basic buck/boost topology

based on two phase-shifted (each full-bridge a one AC signal) voltage AC square wave signals, which are then brought to the transformer and inductor. If there is no phase difference between the primary and secondary voltages, there is no power transfer. However, depending on the phase difference between the primary and secondary voltages, power transfer occurs between the primary and secondary sources. As mentioned in [18], the inductor L1 behaves as a power link between the two sides (half-bridges). To reduce the current ripple through the supercapacitor bank, the inductor L is combined with the capacitor C2. In [18], two advantages of DAB are highlighted, namely the natural power transfer due to phase shift and the elimination of synchronous rectification. Some technical difficulties of DAB topology are [9]: the possibility of transformer saturation, high reactive power, and problems in achieving good efficiency at low load. In [17], which deals with retrofitting existing elevators to increase efficiency, it is pointed out that a transformer-coupled DC /DC converter may be a better choice for ultrasonic operation (either resonant or DAB topology).

The DAB converter topology is also used as a battery charger in some applications [24], [25].

IV. SIMULATION RESULTS

The simulation tests are performed in PLECS to analyze the current and voltage waveforms of the studied topologies. In the simulations, it is assumed that the DC bus voltage of the frequency converter is constant at 600 V, the supercapacitor bank has a capacitance of 3 F and the maximum voltage of 300 V. The expected peak power that can be generated by the elevator is 4.5 kW, which is calculated for the existing elevator system.

The simulation starts with fully discharged supercapacitors and a charging current of 15 A. At $t = 60$ s, the supercapacitors are charged to a value of 300 V, and the current is set to 0 A. At $t = 65$ s, the current is set to -15 A, and the supercapacitors are discharged to 150 V. Specific voltage or current waveforms are also analyzed as a function of the converter topology.

A. Basic buck/boost topology

In all simulation tests, the operating frequency of the DC/DC converter is 25 kHz and the inductance of the coil is 1 mH. In Fig. 6, the voltage and current of the supercapacitor bank are shown. The average value of charge and discharge current is 15 A, while the value of RMS is approximately the average value, with a current ripple of +/- 3 A in the worst case in both buck and boost modes. Fig. 7 shows voltage and current of the transistor. It can be seen that in buck mode, the upper transistor (transistor T1 in Fig. 3) has turn-on and turn-off losses as it turns on and off at higher current and voltage.

B. Interleaved buck/boost topology

For the simulation verification of the interleaved buck/boost topology, a topology with six phases is chosen, where the operating frequency of each phase is 25 kHz, and the inductances are 350 μH. In Fig. 8, the supercapacitor bank current and voltage are shown for one cycle of charging from 0 V to 300

978-1-6654-3236-8/21 $31.00 © 2021 IEEE   223

2021 International Conference on Electrical Drives & Power Electronics (EDPE)   Dubrovnik, 22-24 Sept. 2021

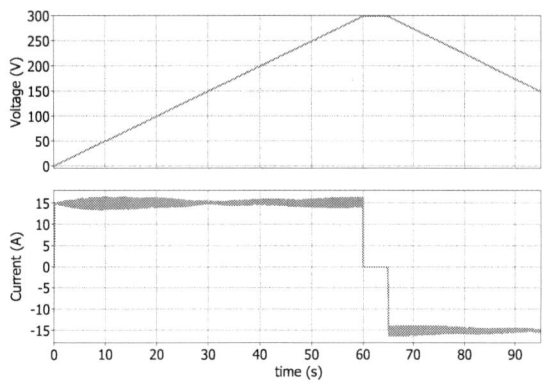

Fig. 8. Voltage and current of supercapacitor bank for interleaved buck/boost topology

Fig. 10. Voltage and current of transistor for interleaved buck/boost topology

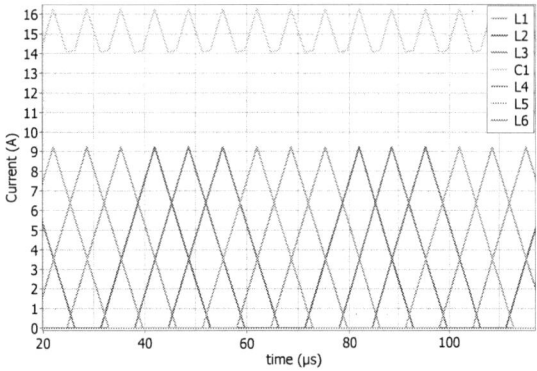

Fig. 9. Inductor currents and total current (supercapacitor bank current) in interleaved buck/boost topology

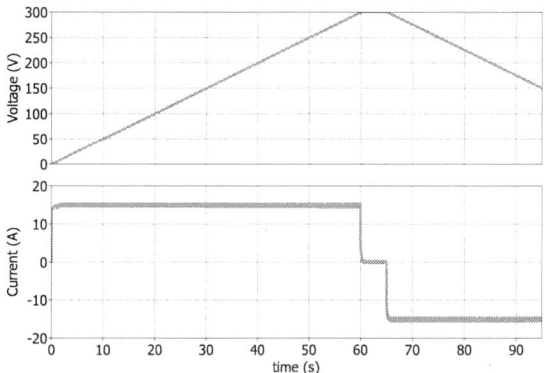

Fig. 11. Voltage and current of supercapacitor bank for DAB topology

V and then discharging to 150 V. The average value of the charge and discharge current is 15 A with a current ripple of +/- 1.5 A in the worst case. In Fig. 9, the currents of each inductor and the current of the supercapacitor bank are shown. The RMS value of the inductor current is 4 A, and the peak value is 9 A.

The switching waveform of the upper transistor in half-bridge configuration for buck mode is shown in Fig. 10. It can be seen that the current value is 0 A when the transistor is turning on, so there are no turn-on losses, only turn-off losses.

C. Dual active bridge topology

In all simulations, the primary inductance is 30 μH, the secondary inductance is 65 μH, the transformer mutual inductance is 10 mH, and the transformation ratio is 2:1 (chosen to correspond to a 600 V to 300 V transformation). DAB is controlled by a single-phase shift (SPS) at a frequency of

25 kHz. In Fig. 11, the simulation results for the converter topology DAB are shown. The current ripple is about +/- 0.4 A.

In Fig. 12, the currents of the primary and secondary sides of the transformer for charging the supercapacitor bank with 15 A at 300 V are shown. Fig. 13 shows the primary and secondary currents of the transformer at 150 V on the supercapacitor bank. It can be seen that there are high current peaks on the primary and secondary sides of the transformer in both cases.

Fig. 14 shows the waveforms of voltage and current of the transistor in full bridge on the primary side, while the DAB converter transfers energy from the DC bus to the supercapacitor bank. It can be seen that the current is negative when the transistor is turned on, but due to dead time, the diode connected in parallel with the transistor starts to conduct. In this way, there is a negligible voltage drop when the transistor is turned on, so there is no turn-on loss.

978-1-6654-3236-8/21 $31.00 © 2021 IEEE   224

2021 International Conference on Electrical Drives & Power Electronics (EDPE) Dubrovnik, 22-24 Sept. 2021

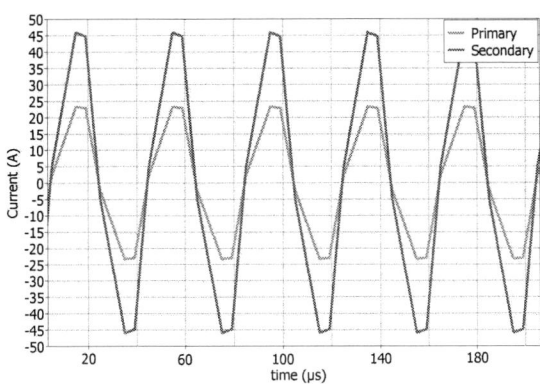

Fig. 12. Primary and secondary currents of transformer in DAB topology at 300 V on supercapacitor bank

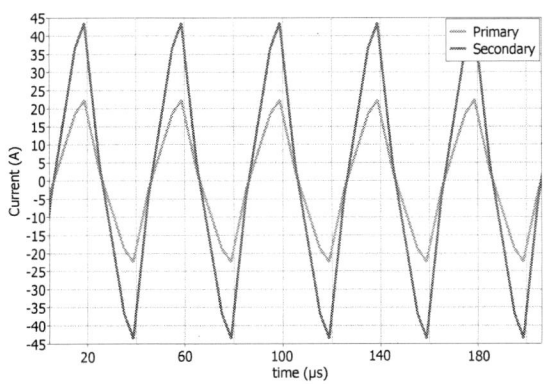

Fig. 13. Primary and secondary currents of transformer in DAB topology at 150 V on supercapacitor bank

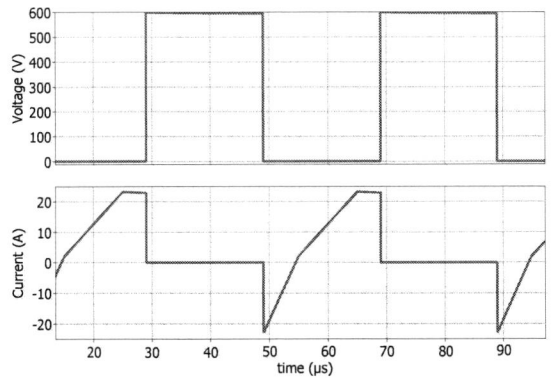

Fig. 14. Voltage and current of transistor in DAB topology

TABLE I
EFFICIENCY IN BUCK MODE

	Efficiency		
	Basic	*Interleaved*	*DAB*
η / %	98	97	89

TABLE II
EFFICIENCY IN BOOST MODE

	Efficiency		
	Basic	*Interleaved*	*DAB*
η / %	98	97	88

D. Simulation of efficiency's

To compare efficiency of chosen DC/DC converter topology, all simulations of topologies are conducted at same heat sink temperature and with same semiconductor switch model. Efficiency calculation takes only semiconductor losses in count. For semiconductor switch it is used IGBT with parallel diode. Comparison of efficiency is conducted at maximum power (4.5 kW), that is at 300 V on supercapacitor bank with charge/discharge current of 15 A. DC bus voltage is kept at 600 V during simulations.

In table I are shown efficiency's in buck mode (charging supercapacitor bank), while in table II are shown efficiency's in boost mode (discharging supercapacitor bank).

V. ANALYSIS OF SIMULATIONS RESULTS

The simulation results have shown that the current ripple in the basic buck/boost topology increases with the voltage increase across the supercapacitor bank. This current ripple can be reduced by increasing either the switching frequency or the inductance of the inductor. However, these changes result in larger switching losses or larger dimensions of the DC / DC converter.

Interleaved DC / DC converters have lower current ripple compared to the basic buck/boost topology. Also, the current values of inductors such as peak, RMS and average are smaller than the basic buck/boost topology. A major advantage of the interleaved DC / DC converter is that it can be operated in discontinuous mode while achieving low output current ripple. When operating in discontinuous mode, semiconductor switching losses are lower.

Comparing the basic and interleaved buck/boost topologies, it can be seen that for the same switching frequency (25 kHz), the inductance of the inductor in the interleaved topology is about three times smaller, and the peak current value and the value RMS are also smaller, which allows a smaller inductor size. In addition, there is no turn-on loss in the interleaved buck/boost topology compared to the basic buck/boost topology. However efficiency of basic buck/boost topology is slight better in comparison with interleaved for conducted simulation point.

When considering the basic and interleaved buck/boost topologies, it should be noted that if the drive is shut down for

978-1-6654-3236-8/21 $31.00 © 2021 IEEE   225

TABLE III
SUMMARIZED COMPARISON OF TOPOLOGIES

	Topologies		
	Basic	**Interleaved**	**DAB**
Isolation	no	no	yes
Turn on losses	yes	no*	no*
Turn off losses	yes	yes	yes
Main magnetic component size	big	small	medium
Good for voltage transform ration range	≤4	≤4	all
Efficiency** at max power	>95 %	>95 %	<95 %

*Special case's
**At same heat sink temperature and using same semiconductor switch

any reason, there is a possibility that the DC bus of the frequency converter will be at the potential of the supercapacitor bank, which must be taken into account.

For the DAB converter, the current ripple in the supercapacitor bank is the lowest. However, there are higher current peaks and RMS values in the primary and secondary sides of the transformer. As mentioned in [26], in addition to SPS control, there is also dual phase shift (DPS) control which can reduce the switch stress and peak inductor current. DAB converter can operate in the zero voltage switching (ZVS) region depending on the operating conditions (load and voltages). The problem of operating in ZVS region, high peak current and reactive power is addressed in [27]–[30]. By increasing the ZVS range, it is possible to achieve higher efficiency at a higher working voltage range. As expected, comparing efficiency of DAB topology to basic and interleaved buck/boost topologies, DAB topology has lower efficiency.

Final comparison of analyzed topologies by taking in account simulations results is given in table III.

VI. CONCLUSION

This paper analyses typical DC/DC converter topologies for elevator energy storage systems based on supercapacitors. The elevator system and supercapacitor bank requirements for the DC/DC converter are analysed, and the performance of three different topologies: basic buck/boost, interleaved buck/boost, and Dual Active Bridge (DAB) are verified by simulation tests. The simulation tests have shown that low current ripple is easier to achieve with the interleaved buck/boost converter. The efficiency's of basic and interleaved buck/boost topology are similar although interleaved topology doesn't have turn on losses. Comparing the interleaved buck/boost converter topology and the DAB topology, it can be seen that the DAB topology has high current peaks and high RMS values of transformer primary and secondary currents but lower charging current ripple. However, with some techniques, higher efficiency (smaller current ripple and wider ZVS range) of the DAB converter can be achieved. If the application does not require isolation, an basic or interleaved buck/boost converter might be a good choice. However, if there is a larger difference between DC bus voltage and fully charged supercapacitor

bank, a topology with transformer as DAB topology could be a good choice.

REFERENCES

[1] Harvey Sachs, Harry Misuriello, and Sameer Kwatra. Advancing elevator energy efficiency. *Report A1501*, 2015.

[2] Toni Tukia, Semen Uimonen, Marja-Liisa Siikonen, Claudio Donghi, and Matti Lehtonen. High-resolution modeling of elevator power consumption. *Journal of Building Engineering*, 18:210–219, 2018.

[3] Martina Kutija, Luka Pravica, Damjan Godec, and Dora Erica. Regenerative energy potential of roped elevator systems-a case study. In *2021 IEEE 19th International Power Electronics and Motion Control Conference (PEMC)*, pages 284–291. IEEE, 2021.

[4] Kun-Yu Lin and Kuang-Yow Lian. Actual measurement on regenerative elevator drive and energy saving benefits. In *2017 International Automatic Control Conference (CACS)*, pages 1–5. IEEE, 2017.

[5] Alfred Rufer and Philippe Barrade. A supercapacitor-based energy-storage system for elevators with soft commutated interface. *IEEE Transactions on industry applications*, 38(5):1151–1159, 2002.

[6] Nikolaos Jabbour and Christos Mademlis. Supercapacitor-based energy recovery system with improved power control and energy management for elevator applications. *IEEE Transactions on Power Electronics*, 32(12):9389–9399, 2017.

[7] P Barrade and A Rufer. Supercapacitors as energy buffers: a solution for elevators and for electric busses supply. In *Proceedings of the Power Conversion Conference-Osaka 2002 (Cat. No. 02TH8579)*, volume 3, pages 1160–1165. IEEE, 2002.

[8] Seref Soylu. *Electric vehicles: modelling and simulations*. BoD–Books on Demand, 2011.

[9] Ding Wu. *Control of a super-capacitor based energy storage system*. The University of Manchester (United Kingdom), 2014.

[10] Seema Mathew, Pooja Mogre, Rohan Chouthai, PB Karandikar, and NR Kulkarni. Supercapacitor based energy recovery system for an elevator. In *2017 International Conference on Advances in Computing, Communication and Control (ICAC3)*, pages 1–5. IEEE, 2017.

[11] Nikolaos Jabbour, Christos Mademlis, and Iordanis Kioskeridis. Improved performance in a supercapacitor-based energy storage control system with bidirectional dc-dc converter for elevator motor drives. 2014.

[12] Nikolaos Jabbour and Christos Mademlis. Improved control strategy of a supercapacitor-based energy recovery system for elevator applications. *IEEE Transactions on Power Electronics*, 31(12):8398–8408, 2016.

[13] C Attaianese, V Nardi, and G Tomasso. Virtual testing of high speed elevators using supercapacitor recovery system. In *2004 IEEE International Conference on Industrial Technology, 2004. IEEE ICIT'04.*, volume 2, pages 728–733. IEEE, 2004.

[14] Li Bin, Wan Jianru, Li Mingshui, and Ge Ang. Research on elevator drive device with super capacitor for energy storage. In *2011 4th International Conference on Power Electronics Systems and Applications*, pages 1–5. IEEE, 2011.

[15] C Attaianese, V Nardi, and G Tomasso. A high efficiency conversion system for elevators. In *2007 International Conference on Clean Electrical Power*, pages 236–242. IEEE, 2007.

[16] Sergio Luri, Ion Etxeberria-Otadui, Alejandro Rujas, Endika Bilbao, and Antonio González. Design of a supercapacitor based storage system for improved elevator applications. In *2010 IEEE Energy Conversion Congress and Exposition*, pages 4534–4539. IEEE, 2010.

[17] Estanis Oyarbide, Ivan Elizondo, Abelardo Martínez-Iturbe, Carlos Bernal, and Javier Irisarri. Ultracapacitor-based plug & play energy-recovery system for elevator retrofit. In *2011 IEEE International Symposium on Industrial Electronics*, pages 462–467. IEEE, 2011.

[18] Deshang Sha and Guo Xu. *High-Frequency Isolated Bidirectional Dual Active Bridge DC–DC Converters with Wide Voltage Gain*. Springer, 2018.

[19] Yi-cheng Zhang, Lu-lu Wu, Xue-jun Zhu, and Hai-quan Liang. Design of supercapacitor-based energy storage system for metro vehicles and its control rapid implementation. In *2008 IEEE Vehicle Power and Propulsion Conference*, pages 1–4. IEEE, 2008.

[20] Jin Li-Jun, Yang Guang-Yao, Jiang Miao-Miao, Cheng Yi-Fan, Zhu Hai-Peng, and Zhou Ke. Study of bi-directional dc-dc converter of micro-grid hybrid energy storage system. In *2015 IEEE 10th Conference on Industrial Electronics and Applications (ICIEA)*, pages 1166–1169. IEEE, 2015.

[21] Petar J Grbovic. *Ultra-Capacitors in Power Conversion Systems: applications, analysis, and design from theory to practice.* John Wiley & Sons, 2013.

[22] Blaise Destraz, Yannick Louvrier, and Alfred Rufer. High efficient interleaved multi-channel dc/dc converter dedicated to mobile applications. In *Conference Record of the 2006 IEEE Industry Applications Conference Forty-First IAS Annual Meeting*, volume 5, pages 2518–2523. IEEE, 2006.

[23] Ahmed M Omara and M Sleptsov. Bidirectional interleaved dc/dc converter for electric vehicle application. In *2016 11th International Forum on Strategic Technology (IFOST)*, pages 100–104. IEEE, 2016.

[24] Yuto Takayama and Hiroaki Yamada. Experimental verification of dual active bridge converter based battery charger in a stand-alone wind power generation system. In *2020 23rd International Conference on Electrical Machines and Systems (ICEMS)*, pages 1022–1026. IEEE, 2020.

[25] Dehao Qin, QiuYe Sun, Dazhong Ma, and Jiazheng Sun. Model predictive control of dual-active-bridge based fast battery charger for plug-in hybrid electric vehicle in the future grid. In *2019 IEEE Innovative Smart Grid Technologies-Asia (ISGT Asia)*, pages 2162–2166. IEEE, 2019.

[26] Bhimisetty Manoj Kumar, Anupam Kumar, AH Bhat, and Pramod Agarwal. Comparative study of dual active bridge isolated dc to dc converter with single phase shift and dual phase shift control techniques. In *2017 Recent Developments in Control, Automation & Power Engineering (RDCAPE)*, pages 453–458. IEEE, 2017.

[27] Giuseppe Guidi, Atsuo Kawamura, Yuji Sasaki, and Tomofumi Imakubo. Dual active bridge modulation with complete zero voltage switching taking resonant transitions into account. *EPE Journal*, 22(1):5–12, 2012.

[28] M Yaqoob, Ka Hong Loo, and YM Lai. Extension of soft-switching region of dual-active-bridge converter by a tunable resonant tank. *IEEE Transactions on Power Electronics*, 32(12):9093–9104, 2017.

[29] Masood Soleimanifard and Ali Yazdian Varjani. A bidirectional buck-boost converter in cascade with a dual active bridge converter to increase the maximum input and output currents and extend zero voltage switching range. In *2020 11th Power Electronics, Drive Systems, and Technologies Conference (PEDSTC)*, pages 1–6. IEEE, 2020.

[30] Song Chi, Peng Liu, Xue Li, Mochen Xu, and Shanhu Li. A novel dual phase shift modulation for dual-active-bridge converter. In *2019 IEEE Energy Conversion Congress and Exposition (ECCE)*, pages 1556–1561. IEEE, 2019.

Loss Minimization Control and Energy Consumption Improvements in PMSM drive

1st Martin Novak
Department of Instrumentation and Control Engineering
Faculty of Mechanical Engineering
Czech Technical University in Prague
Prague, Czech Republic
Martin.Novak@fs.cvut.cz

2nd Jaroslav Novak
Department of Instrumentation and Control Engineering
Faculty of Mechanical Engineering
Czech Technical University in Prague
Prague, Czech Republic
Jaroslav.Novak@fs.cvut.cz

Abstract—**This paper focuses on loss minimization control (LMC) for an electric vehicle (EV) driven by a permanent magnet synchronous machine (PMSM). The goal is to minimize the consumed electrical energy for a given driving cycle. First a theoretical description of LMC will be provided. Then a simulation of the machine (PMSM 100 kW, 5000 min-1) will show the feasibility and expected improvements of the method. The expectation is around 0.5 to 1.5 percent improvement. Experimental validation of the PMSM model is then performed. The EV simulation model is then build with the PMSM model, the EV is driven in a driving cycle and the total energy consumption is calculated. The results show that when the motor parameter are suitable, the efficiency increase can be in the order of few percent. However not all motors can profit of the LMC, the method is strongly dependent on the motor parameters.**

Index Terms—**Permanent magnet synchronous machine; Loss minimization control; Maximum efficiency control**

I. Introduction

There is a constant push to improve efficiency of devices. This is true also in the area of electric vehicles (EV) such as electric cars or buses. With an improved efficiency the EV can drive for a longer distance. The limiting factor for today's EVs is the limited battery capacity. Therefore any improvement in the efficiency of the drivertain is important. Especially when it can be achieved only by a different control algorithm and hence only software changes are required. Among others there are several vector control "optimization" techniques, for example Maximum Torque Per Ampere (MTPA), Maximum Torque Per Motor Loss (MTPML), and Maximum Torque Per System Loss (MTPSL) [1]. Some of them, such as MTPA are suited for high dynamic events, such as rapid acceleration [2] while others can be used to optimize the efficiency (minimize losses) [2].

With loss minimization control (LMC) or maximum efficiency control there are basically two possible approaches. Offline or online. In the offline approach, for each operating point

This research has been partially funded from project Josef Bozek National Center of Competence for Surface Vehicles, TN01000026, Technology Agency of the Czech Republic

of the machine an optimal setting has to be known in advance and typically stored in a look-up table [2]. The LMC can also be based on a three dimensional table as shown in [3]. The offline approach however represents several disadvantages. The offline approach cannot inherently cover changes in machine parameters such as with varying temperature, saturation etc. On the other hand, the algorithm is simple. The online approaches search for the optimal settings of the parameters continuously when the machine is in operation and can therefore adapt to the inherent nonlinear nature of the machine. It can for example perform minimum-copper-loss control in a interior permanent-magnet synchronous machine (PMSM) under the current and voltage limit of the drive system as shown in [4]. Although the online approaches can cover various non-linearities, they also take more time to find the optimal point than the offline methods where the optimal point is known in advance. Some methods present various disadvantages, e.g. torque ripple, high measurement effort, heavy assumptions and convergence issues [5]. It has been shown [6] [7] that the LMC can improve the efficiency of an PMSM. The efficiency optimization control of IPMSM can also consider varying machine parameters such as inductance and equivalent circuit resistance [8].

The efficiency optimization methods can be further extended not only to the electric machine but the the inverter as well as shown in [9], [10]. Here the power losses are considered include switching and conduction losses in the inverter and fundamental and harmonic losses in the stator winding, core and rotor magnets. The maximum efficiency control can optimize the copper loss, iron loss and power converter loss together as shown in [11].

All the presented methods have one in common. In order to improve the efficiency they do not require hardware changes, only software changes in the control algorithm are required. In this paper LMC will be applied to a PMSM control used in an EV (bus) with the goal to minimize the consumed electrical energy for a given driving cycle.

II. Theoretical description

A. Machine model

The approach is based on a standard PMSM model in the rotor dq reference frame. The electrical equations (neglecting the iron losses that will be added later) are as follows

$$v_d = R_s i_d + \frac{d\lambda_d}{dt} - \omega_r \lambda_q \tag{1}$$

$$v_q = R_s i_q + \frac{d\lambda_q}{dt} + \omega_r \lambda_d \tag{2}$$

where v_d and v_q are d- and q-axis stator voltages, R_s is stator resistance, i_d and i_q are d- and q-axis stator currents, λ_d and λ_q are d- and q-axis flux linkages, ω_r is angular velocity.

The d- and q-axis flux linkages components are

$$\lambda_d = L_d i_d + \lambda_m \tag{3}$$

$$\lambda_q = L_q i_q \tag{4}$$

where L_d and L_q are d- and q-axis inductance, respectively and λ_m is the flux linkage of the permanent magnets.

The machine model then becomes

$$v_d = R_s i_d + L_d \frac{di_d}{dt} - \omega_r L_q i_q \tag{5}$$

$$v_q = R_s i_q + L_q \frac{di_q}{dt} + \omega_r L_d i_d + \omega_r \lambda_m \tag{6}$$

The equivalent circuit model for equations (1) - (6) is shown in figure 1. The iron losses, modeled with resistor R_c, will be explained later.

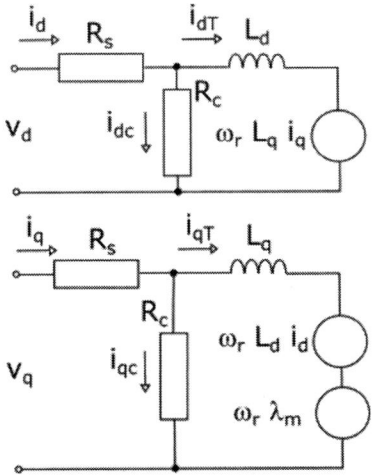

Fig. 1. Equivalent circuit diagram for PMSM considering iron loss

The torque of the machine is

$$T_e = (3/2)P_p(\lambda_m + (L_d - L_q)i_d)i_q \tag{7}$$

where P_p is number of pole-pairs.

The magnet torque is the part in equation (7) with λ_m, the reluctance torque is the part with $(L_d - L_q)i_d$. For surface mounted PMSM typically $L_d \doteq L_q$, hence the reluctance torque is zero. This is also the case of this paper that will use a surface mount PMSM.

The mechanical components are modeled with the following differential equation

$$\frac{d\omega_r}{dt} = (1/J)(T_e - B\omega_r - T_L) \tag{8}$$

where J is the moment of inertia, B is damping and T_L is load torque.

B. Loss model

There are several components of the losses in a PMSM. In order to understand theirs influence on the loss minimization and the influence of other parameters such as load variations, they are described in this section. Electrical losses P_L are the sum of copper losses P_{Cu} and iron losses P_{Fe}. The copper loss P_{Cu} is calculated as follows

$$P_{Cu} = (3/2)R_s(i_q{}^2 + i_d{}^2) \tag{9}$$

Iron losses are [2]

$$P_{Fe} = (3/2)R_C(i_{dc}{}^2 + i_{qc}{}^2) \tag{10}$$

where

$$i_{dc} = -(\omega_r \rho L_d i_{qT})/R_C \tag{11}$$

$$i_{dc} = ((L_d i_{dT} + \lambda_m)\omega_r)/R_C \tag{12}$$

where ρ is saliency ratio. For a non-salient PMSM, $L_d = L_q = L_S$, hence $\rho = 1$

The mechanical loss in the PMSM is modeled as being directly proportional to angular velocity

$$P_{mech} = k_{mech} \cdot \omega_m \tag{13}$$

where k_{mech} is the constant of proportionality for mechanical losses (found out experimentally) and ω_m is mechanical speed $\omega_m = \omega_r/P_p$.

As can be seen from the equivalent circuit diagram in figure 1 the currents i_d and i_q are split into the iron loss component i_{dc} and i_{qc} and a component used to create torque i_{dT} and i_{qT}. The components can be calculated as follows

$$i_{dT} = i_d - i_{dc} \tag{14}$$

$$i_{qT} = i_q - i_{qc} \tag{15}$$

The optimal current i_{dT} to obtain minimal losses can be calculated as [2]

$$i_{dT} = \frac{\omega_r^2 L_s (R_s + R_C) \lambda_m}{R_s + R_C{}^2 + \omega_r L_s{}^2 (R_s + R_C)} \quad (16)$$

The optimal current i_{qT} is found from equation (7) by substituting the just found value of i_{dT} and the desired value of torque T_e the machine should produce

$$i_{qT} = \frac{T_e}{(3/2)p_p[\lambda_m + (L_d - l_q)i_{dT}]} \quad (17)$$

The reference currents used as inputs into the current controllers are then calculated from equations (11) and (12) and passed to the control algorithm.

III. SIMULATION RESULTS

The goal of this analysis is to find an optimal value of currents i_d and i_q such that the losses in the machine are minimal. The method was first verified with simulation. The used parameters are summarized in table I. The simulation model in Simulink is shown in figure 2. The block "PMSM model" contains the machine model described with equations (1)-(7). The decoupling block decouples influence from i_d to v_q and from i_q to v_d. The decoupling equations are generally known and hence not presented in this paper. The rest of the structure is formed by two PI controllers for currents i_d and i_q and one speed controller. The calculation of optimal currents is based on the shown equations and is performed in block "calculate losses".

In order to show the influence of different values of current i_d the following simulation was performed. With a speed controller the machine first starts to nominal speed 5000 min-1. The machine is loaded with nominal torque 191 Nm. The startup takes 5 seconds. When steady state is reached, the search for the optimal value of i_d and i_q begins. The current i_d is changed gradually and changes in efficiency and losses are observed. As can be seen in figure 3 as current i_d is decreased, the efficiency increases as the iron losses decrease. There certainly exists an optimal point. There are however several limitations. The current i_d cannot decrease indefinitely. It is limited by the maximal possible motor current and the requirement that the produced machine torque should not decrease (the requested speed and torque of the machine should not change). The simulation takes this into account, the maximal value of current i_d is limited, the limit was not reached in figure 3.

Figure 4 shows the losses in the machine. After the initial start up of the machine to the nominal operating point a steady state is reached and the decrease of i_d begins. It can be seen that as current i_d is decreased also iron losses are decreasing. On the other hand, as copper losses are $P_{Cu} = (3/2)R_s(i_q{}^2 + i_d{}^2)$ copper losses are increasing with decreasing current (increasing magnitude) of i_d. At this operating point the limit of minimal

TABLE I
PMSM PARAMETERS

parameter	value
Nominal torque [Nm]	191
Nominal speed [min-1]	5000
Nominal power [kW]	100
Nominal current [A]	200
Nominal voltage [V]	3x350
damping coefficient B [Nm.s/rad]	8.10^{-4}
moment of inertia J [kgm2]	0.345
torque constant KT [Nm/A]	0.955
voltage constant Kv [V/kRPM]	60
Permanent magnet flux linkage λ_m [Wb]	0.22
stator inductance Ls [H]	$3.7.10^{-4}$
number of pole pairs Pp [-]	2
iron resistance RFe [Ω]	20
Saliency ratio ρ [-]	1
Stator resistance Rs [mΩ]	6.6

current i_d is given by the maximal possible motor current. At this operating point the machine efficiency can be increased by 0.94 percent. At other operating points this increase might be different.

IV. EXPERIMENTAL VALIDATION

The presented PMSM model was experimentally verified on a dynamometer. The output power (mechanical) was calculated from torque and speed, the input power was measured. The PMSM was loaded with a given torque and the value of current id has been changed by changing the setpoint of the id current controller. The effects of i_d changes on efficiency were observed.

In order to validate the simulation with an experiment, essentially two types of tests were done. An unloaded measurement and measurement with load.

In the unloaded measurement the speed of PMSM was changed and the input power was measured. The PMSM was still connected to the dynamometer since it is complicated to decouple, but with load torque zero. The unloaded test allowed to make an estimate of mechanical and iron losses. The fit of the simulation and the experiment is shown in figure 5. It can be seen that the match between the model and experiment is sufficient. The model underestimates the losses in the lower speeds and overestimates the losses for higher speeds. In the middle of the speed range, the match is much closer. The accuracy of the match could probably be improved by fine tuning the model coefficients such as damping coefficient and iron loss resistance. However the match for the purpose of this calculation was sufficient.

The second performed test was the loaded test. The PMSM was loaded with the dynamometer and the input and output power was measured and the efficiency was calculated. Due to the parameters of the available dynamometer the PMSM could only be loaded up to 70Nm, 5000 min-1. It was not possible

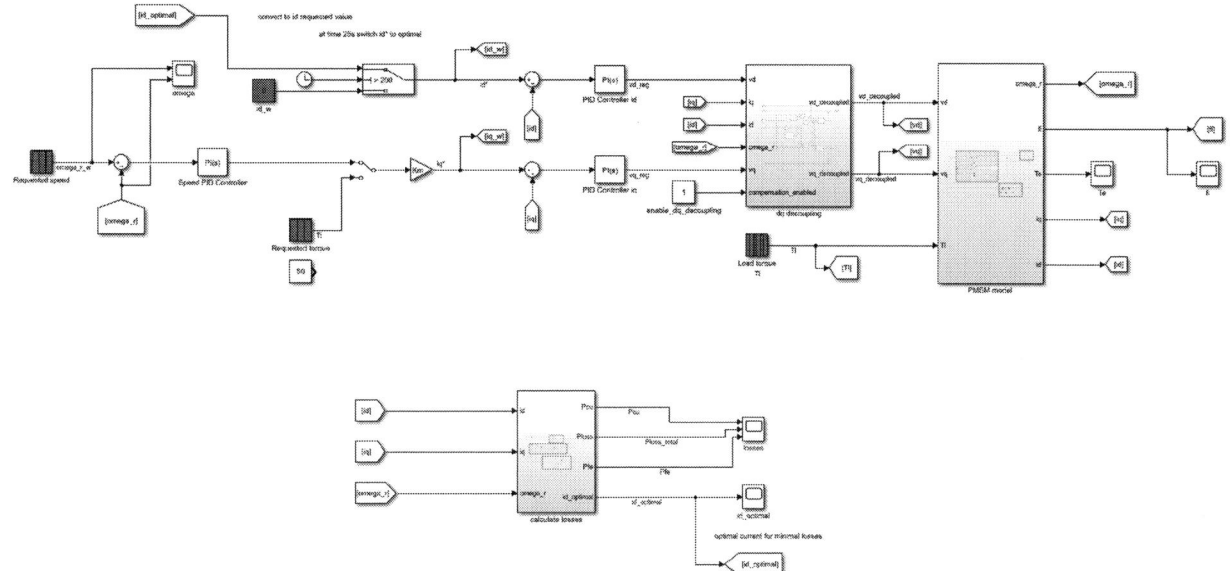

Fig. 2. Simplified Simulink block diagram for simulations

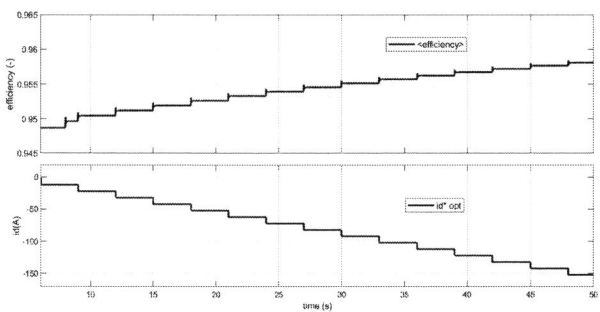

Fig. 3. Machine efficiency with changes of requested current id

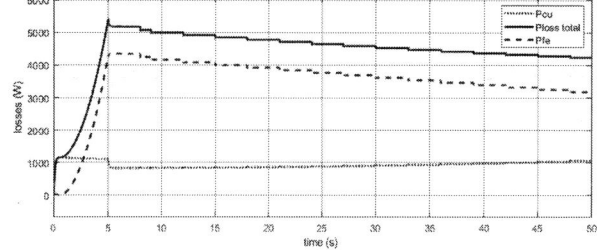

Fig. 4. Machine copper, iron and total losses with changes of requested current id

to load the PMSM up to its nominal point. Example results from loaded test are shown in figure 6. It can be seen that the model calculates correctly the PMSM efficiency. For other speeds and torques within the measured range of 1000 - 5000 min^{-1} and 12 - 70 Nm, the match of efficiency is in general within approximately 0.1 percent.

Both in simulation and experiment the same trend could be observed: the efficiency changes with current i_d, a certain value can be found that maximizes the efficiency. However it is possible that this optimal value of current i_d is not reachable in reality. This can be for example due to the maximal current limitation of the inverter, current request to create a given torque etc.

V. RESULTS AND DISCUSSION

In order to calculate the total energy consumption of the bus for a given driving cycle, the presented model was extended with a gearbox model and bus motion model. The bus motion was then simulated in a simple driving cycle and the total consumed energy was calculated. The driving cycle consists from a simple ramp to get up to the desired speed, then holds the set speed for 2 minutes and then slow down with a defined ramp.

The driving cycle was first simulated with Id = 0 and then with the optimal i_d. The consumed energy was calculated with a time integral over the constant speed area of the driving cycle in order to eliminate the effects of acceleration and deceleration. The used bus parameters are summarized in table II.

978-1-6654-3236-8/21 $31.00 © 2021 IEEE

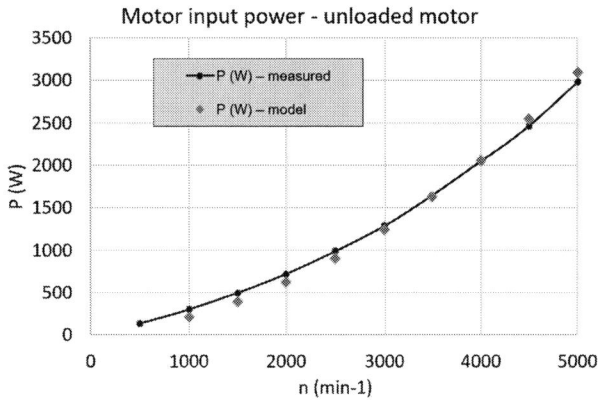

Fig. 5. Input power of unloaded motor vs. speed - simulation vs. experiment

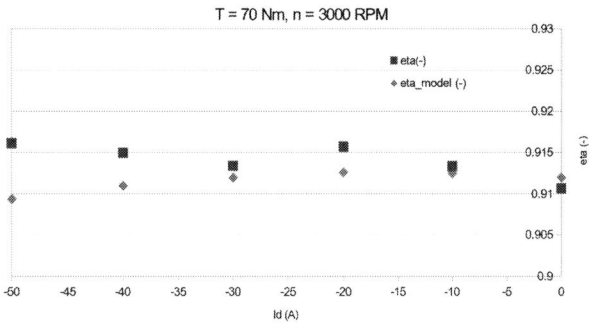

Fig. 6. Loaded test, 70 Nm, 3000 min^{-1} - simulation vs. experiment

The results are summarized in table III. It can be seen that for the used PMSM the improvement is very small. In reality there would probably be no advantage in using this method at all since this simulation does not consider inverter losses. The losses in the inverter are a function of current, with high value of current i_d inverter losses will increase and negate the very small effect of efficiency increase. It can even be expected that the total efficiency would drop compared to $i_d = 0$.

It can be concluded, that this particular PMSM does not

TABLE II
BUS PARAMETERS

parameter	value
vehicle mass m [kg]	14000
aerodynamic drag coefficient Cd [-]	0.55
front equivalent area A [m2]	5
wheel radius r_{wheel} [m]	0.43
rolling resistance coefficient Cr [-]	0.013
moment of inertia - front axle J_{front} [kgm2]	15
moment of inertia - rear axle J_{rear} [kgm2]	45
moment of inertia - Cardan J_{Cardan} [kgm2]	1.99

TABLE III
CALCULATED ENERGY CONSUMPTION AND CONTRIBUTIONS OF USING OPTIMAL VALUE OF i_d^*

vehicle speed (km/h)	E [kWh] per 1 km, id = 0	optimal id [A]	E [kWh] per 1 km, for optimal id	improv. (%)
10	0.744	0	0.744	-0.03
20	0.708	0	0.708	-0.01
30	0.715	-5	0.715	0.01
40	0.738	-8	0.738	0.02
50	0.773	-13	0.772	0.02
60	0.816	-18	0.816	0.03
70	0.869	-24	0.868	0.03
80	0.929	-32	0.929	0.04
90	0.998	-40	0.997	0.04
100	1.074	-50	1.074	0.04

have suitable parameters for loss minimization control. It has a dominant copper loss over the iron loss. This results in increasing the copper losses when currents are increased. Even when the iron loss decreases with the optimal value of current i_d, the coppers loss increases and negates the improvement.

VI. SIMULATION WITH ANOTHER MOTOR

Since the efficiency improvement with the presented PMSM was not satisfactory, another PMSM, with different parameters was simulated. This motor is not available for experimental testing, therefore only simulation with the previously validated model could be done. The PMSM parameters are summarized in table IV. This PMSM differs in the ratio between copper and iron losses. PMSM 2 has a dominant iron loss, whereas the previous PMSM had a dominant copper loss. Therefore, for PMSM 2 the efficiency optimization will yield significantly better results.

TABLE IV
PARAMETERS OF PMSM 2

parameter	motor 2
nominal torque [Nm]	917
nominal speed [min-1]	1684
maximal speed [min-1]	3368
nominal power [kW]	162
nominal current [A]	471
nominal voltage [V]	235
moment of inertia - motor J_M [kgm2]	0.1
torque constant KT [Nm/A]	1.38
permanent magnet flux linkage λ_m [Wb]	0.173
stator inductance Ls [μH]	200
number of pole pairs Pp [-]	8
iron resistance R_c [Ω]	18
saliency ratio ρ [-]	1
stator resistance Rs [mΩ]	2.6
coefficient of mechanical resistance k_{mech} [-]	500e-3

The same driving cycle simulations have been performed for PMSM 2. The results are summarized in table V. It can be seen that PMSM 2 gives significantly better results. By changing the

TABLE V

CALCULATED ENERGY CONSUMPTION AND CONTRIBUTIONS OF USING OPTIMAL VALUE OF i_d^* - PMSM 2

vehicle speed (km/h)	E [kWh] per 1 km, id = 0	optimal id [A]	E [kWh] per 1 km, for optimal id	improv. (%)
10	0.579	-60	0.577	0.4
20	0.613	-200	0.600	2.1
30	0.654	-250*	0.624	4.5
40	0.701	-250*	0.657	6.3
50	0.754	-250*	0.696	7.7
60	0.814	-250*	0.742	8.8
70	0.880	-250*	0.795	9.6
80	0.952	-250*	0.854	10.3
90	1.030	-250*	0.919	10.8
100	1.115	-250*	0.991	11.1

value of current i_d, the total energy consumption is significantly smaller. The improvement goes up to 11 percent. However it has to be noted that the simulation model does not consider inverter losses. With inverter losses included it is expected that the improvement would be smaller and the optimal point would likely be different. Nevertheless PMSM 2 is suitable for loss optimization, the improvement in the energy consumption is in the order of few percent.

VII. ADVANTAGES / DISADVANTAGES

Advantages:

- Simple - the presented method is simple, allows to calculate in advance the optimal value of current id for a given speed and load, the values can be stored in a table and accessed by the control algorithm
- Can improve energy consumption - but only when motor parameters are suitable

Disadvantages:

- In the presented form it does not consider variable motor parameters. For example the stator resistance varies with temperature, inductance varies with saturation etc. Those variations are not considered in this work.
- Motor parameters have to be known in advance, the control algorithm in its current state cannot search for the optimal values. This is an implementation limit, not the limit of the loss minimization control itself

VIII. CONCLUSIONS

As it was shown in this paper it is possible to improve the efficiency and decrease the energy consumption of an electric bus by changing the set point of current id. However the effect depends very strongly on the parameters of the used motor. When the motor has an unsuitable ratio of copper and iron losses, this method will not bring anything at all. It may even increase the energy consumption and decrease efficiency.

When the motor parameters are suitable, the loss minimization control may bring an improvement in the order of few percent as it was shown in this paper.

The shown calculations did not include the inverter losses, that are a function of current and switching frequency. The inverter and battery losses will be included in the future. It is expected that with the inverter losses included, the improvement will be smaller, however the expected improvement remains still in the order of few percent. Considering the fact that the presented method is essentially free (it just has to be programmed, no hardware changes are required in the existing system), the loss minimization control can bring an interesting energy consumption reduction.

REFERENCES

[1] B. Gallert, G. Choi, K. Lee, X. Jing, and Y. Son, "Maximum efficiency control strategy of pm traction machine drives in gm hybrid and electric vehicles," in *2017 IEEE Energy Conversion Congress and Exposition (ECCE)*, 2017, pp. 566–571.

[2] S. Vaez-Zadeh, *Control of Permanent Magnet Synchronous Motors*. Oxford University Press, 2018.

[3] G.-S. Li, J.-F. Xie, and L.-Y. Xu, "Maximum efficiency control method of permanent magnet synchronous motor based on three-dimensional table," in *2015 International Conference on Advanced Mechatronic Systems (ICAMechS)*, 2015, pp. 392–396.

[4] Y. Jeong, S. Sul, S. Hiti, and K. Rahman, "Online minimum-copper-loss control of an interior permanent-magnet synchronous machine for automotive applications," *IEEE Transactions on Industry Applications*, vol. 42, no. 5, pp. 1222–1229, 2006.

[5] J. Bonifacio and R. Kennel, "Online maximum torque per ampere control of interior permanent magnet synchronous machines (ipmsm) for automotive applications," in *8th IET International Conference on Power Electronics, Machines and Drives (PEMD 2016)*, 2016, pp. 1–5.

[6] T. Bariša, D. Sumina, and M. Kutija, "Comparison of maximum torque per ampere and loss minimization control for the interior permanent magnet synchronous generator," in *2015 International Conference on Electrical Drives and Power Electronics (EDPE)*, 2015, pp. 497–502.

[7] R. Ni, D. Xu, G. Wang, L. Ding, G. Zhang, and L. Qu, "Maximum efficiency per ampere control of permanent-magnet synchronous machines," *IEEE Transactions on Industrial Electronics*, vol. 62, no. 4, pp. 2135–2143, 2015.

[8] S. Yang, K. Liu, Y. Hu, L. Chu, and S. Chen, "Efficiency optimization control of ipmsm considering varying machine parameters," in *2018 IEEE Student Conference on Electric Machines and Systems*, 2018, pp. 1–6.

[9] A. Balamurali, G. Feng, C. Lai, V. Loukanov, and N. C. Kar, "Investigation into variation of permanent magnet synchronous motor-drive losses for system level efficiency improvement," in *IECON 2017 - 43rd Annual Conference of the IEEE Industrial Electronics Society*, 2017, pp. 2014–2019.

[10] A. Balamurali, C. Lai, H. Dhulipati, V. Loukanov, and N. C. Kar, "Improved analytical eddy current loss modelling considering carrier harmonics towards maximum efficiency control of pmsm-drive systems." in *2018 IEEE International Magnetics Conference (INTERMAG)*, 2018, pp. 1–1.

[11] Q. Guo, C. Zhang, L. Li, M. Wang, L. Pei, and T. Wang, "Maximum efficiency control of permanent magnet synchronous motor system with sic mosfets for flywheel energy storage," in *2016 19th International Conference on Electrical Machines and Systems (ICEMS)*, 2016, pp. 1–5.

State of Health and Aging Estimation Using Kalman Filter in Combination with ARX Model for Prediction of Lifetime Period of Li-Ion

Lukáš Krčmář
Faculty of Mechatronics, Informatics and Interdisciplinary Studies
Technical University of Liberec
Liberec, Czech Republic
lukas.krcmar@tul.cz

Pavel Rydlo
Faculty of Mechatronics, Informatics and Interdisciplinary Studies
Technical University of Liberec
Liberec, Czech Republic
pavel.rydlo@tul.cz

Aleš Richter
Faculty of Mechatronics, Informatics and Interdisciplinary Studies
Technical University of Liberec
Liberec, Czech Republic
ales.richter@tul.cz

Jakub Eichler
Faculty of Mechatronics, Informatics and Interdisciplinary Studies
Technical University of Liberec
Liberec, Czech Republic
jakub.eichler@tul.cz

Pavel Jandura
Faculty of Mechatronics, Informatics and Interdisciplinary Studies
Technical University of Liberec
Liberec, Czech Republic
pavel.jandura@tul.cz

Abstract — **The state of health (SOH) is a critical factor to guarantee that a battery system will operate in a safe and reliable manner for the whole lifetime period. The estimation of the lifetime period is important for effective and failure - free working. Many factors affect the rate of the degradation of batteries. Every cell has a typical dissimilar mechanism of degradation. A method of the state of health (SOH) is discussed in this contribution. A hybrid method takes advantage of combining the autoregressive exogenous battery model (ARX) and the Kalman filter.**

Keywords — battery estimation; Kalman filter; state of health SOH; autoregressive exogenous battery model ARX; state of charge SOC.

I. INTRODUCTION (HEADING 1)

The state of health (SOH) of a battery is a measure of the battery ability to meet its defined performance. Note that all the parts of lithium-ion batteries age during their lifetime. The rate and degree of battery degradation is affected by many factors. The SOH parameter is a measure of the reduction in battery performance due to the aging process. The SOH determination is the subject of this article. A number of methods are currently used to estimate the SOH parameter. Most methods require the knowledge of the battery model. Because each battery component has a different degradation mechanism, the exact battery model is usually complex. If a complex model is chosen, it is very difficult to estimate the parameter of this model. Therefore, the article focuses on the use of the most simplified model and assessing whether it is suitable for ensuring a sufficiently accurate estimation of the SOH parameter. The article uses a hybrid method that consists in combining the ARX model and Kalman filter. The block diagram of the hybrid structure is shown in Fig.1

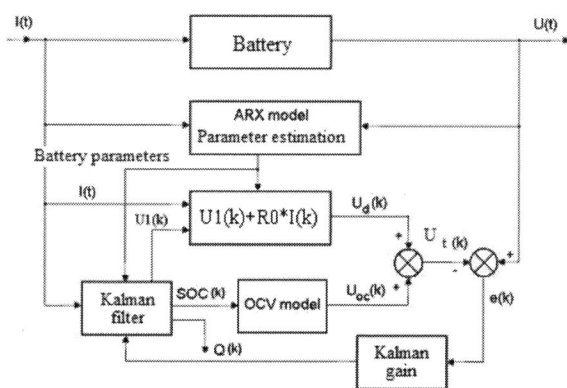

Fig. 1. Block diagram of battery estimation of SOH and SOC parameters

II. ESTIMATION OF BATTERY PARAMETERS AND CONSTRUCTION OF BATTERY MODEL

Note that the battery parameters change during the battery lifetime. Therefore, it is necessary to estimate these parameters during the lifetime of the battery. Note that the battery parameters change slowly as the battery ages. It is therefore not necessary to estimate these parameters in each cycle of the battery operation. For continuous estimation of the battery parameters, parameter identification can be used using the "ARX model" (Auto Regressive with External Input), the method of least squares. In this case, it is the experimental identification, in which we use measurements performed on an existing system. Based on the input (current) and output (voltage) data, discrete transmission of the monitoring system can be determined, even if we do not know the internal structure of the system [1], [2], [3]. Based on the knowledge of this transmission it is possible to use the

978-1-6654-3236-8/21 $31.00 © 2021 IEEE

dynamic behavior of a li-ion battery. We use the model shown in Fig.2 according to [1].

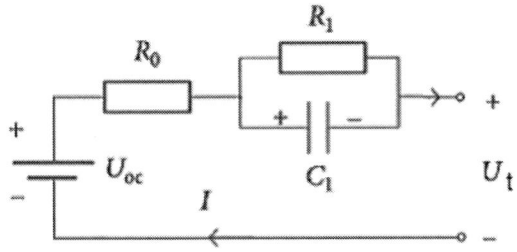

Fig. 2. Substitution diagram of li-ion baterie

U_{OC} Open Circuit Voltage

R_0 Internal Resistance

R_1 Diffusion Resistance

C_1 Diffusion Capacity

U_t Terminal Voltage of Battery

The transfer function of the battery model in Fig.2 is:

$$G(s) = \frac{U(s)}{I(s)} = R_0 + \frac{R_1}{1 + R_1 C_1 s} \quad (2)$$

The discrete transfer function is defined using bilinear transformation:

$$s = \frac{2}{T} \cdot \frac{1 - Z^{-1}}{1 + Z^{-1}} \quad (3)$$

$$G_d(z) = \frac{2R_0\tau(1 - z^{-1}) + (R_0 + R_1)(1 + z^{-1})}{(1 + 2\tau) + z^{-1}(1 - 2\tau)} \quad (4)$$

Now it is possible to obtain the estimated transfer function from the measured data using ARX model

$$G(z) = \frac{b_0 + b_1 z^{-1}}{1 + a_1 z^{-1}} \quad (5)$$

The parameter comparison of the model transfer function and estimated transfer function allows calculate the components of the substitution diagram (see Fig.2):

$$R_0 = \frac{b_0 - b_1}{1 - a_1} \qquad \tau = \frac{1 - a_1}{2(1 + a_1)}$$

$$R_1 = \frac{b_0 + b_1}{1 + a_1} - \frac{b_0 - b_1}{1 - a_1} \qquad C_1 = \frac{\tau}{R_1} \quad (6)$$

III. ESTIMATION OF SOC (STATE OF CHARGE) PARAMETER

In order to estimate the current value of the battery capacity Q and the SOC parameter, we use a model developed using the Kalman filter [5], [6], [7], [8]. In this model, we will use the above estimated parameters of the monitored battery. The parameters of the discrete state equation used to design the Kalman filter are as follows:

$$\mathbf{A} = [1/T-1/(R_1*C_1)\ 0;\ 0\ 1/T\ 0];$$

$$\mathbf{B} = [1/C_1;\ 1/Q];$$

$$\mathbf{C} = [-1\ 0.01];$$

IV. OCV MODEL (OPEN CIRCUIT VOLTAGE)

Various equations are used for the mathematical description of the dependence UOC = f (SOC). In our case, we will use a simplified description of this equation

$$U_{OC} = K_1 + K_2 * SOC$$

Where: $K_1 = 3.7$ and $K_2 = 0.052$

V. EVALUATION OF MEASUREMENT PERFORMED ON NEW BATTERY

The aging tests were realized for the INR18650-25R battery produced by SAMSUNG Co. This cell is a typical lithium-ion energy accumulator. These batteries are characterized by comparatively low numbers of working cycles (decrease of battery capacity by 20 % after 500 cycles).

The battery tester Chroma 11710 was used in the experiments. The charging as well as discharging current was chosen 4 A, it is the maximum value of the SAMSUNG datasheet. The battery was discharged by the constant current (CC - 4 A). The charging was in the CC - CV mode (constant current-constant voltage). The charging starts with the constant current 4 A. When the voltage reaches 4.2 V, the battery charging continues in the constant voltage mode. When the current drops to 20 mA, the battery charging is finished. The temperature was constant 25 ºC for all experiments.

The measurement was performed on a Chroma 11710 measuring apparatus together with a CTS t 40 100 temperature chamber. The devices are photographed in Figure 3.

Fig. 3. Chroma test equipment and CTS chamber

The estimated discrete battery transfer obtained using the ARX method is as follows

$$G(z) = \frac{0.023165 - 0.023015z^{-1}}{1 - 0.999996z^{-1}} = \frac{b_0 + b_1 z^{-1}}{1 + a_1 z^{-1}} \quad (7)$$

Calculated parameters of the new battery:

$R_0 = 0.02\ [\Omega];$ $R_1 = 0.01\ [\Omega];$ $C_1 = 10\ 000\ [F];$

Comparison of the measured and estimated voltage profiles on the battery is shown in Fig.4.

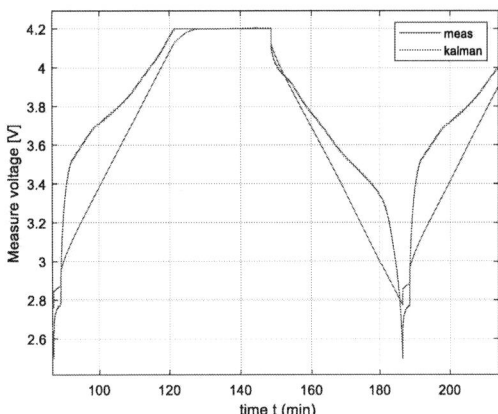

Fig. 4. Comparison of measured and estimated voltage profiles (new battery)

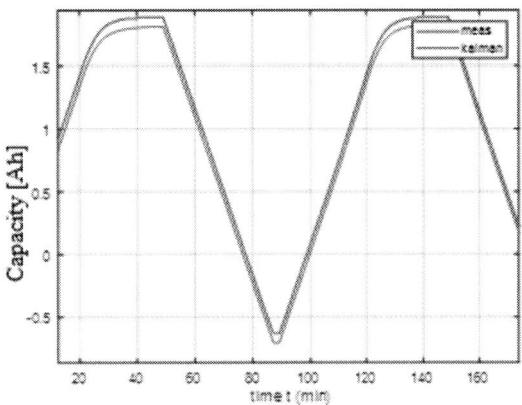

Fig. 5. Comparison of measured and estimated Q capacity profiles (new battery)

VI. EVALUATION OF MEASUREMENTS PERFORMED AFTER 498 CYCLES

The measurement was performed 500 times, the first and last measurement was discarded, which is the number 498.

The voltage and current waveforms that were measured during charging and discharging the battery after 498 cycles are shown in Fig.6.

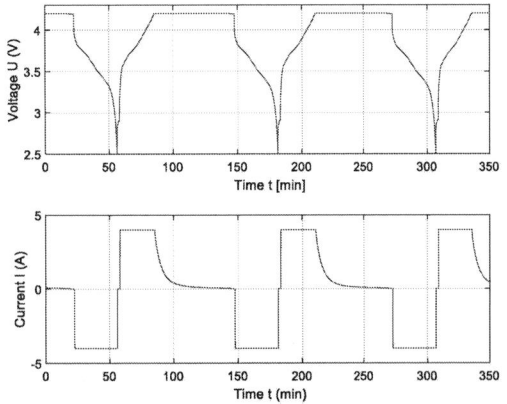

Fig. 6. Shape of voltage and current waveforms, battery after 498 cycles

The estimated transfer function (using the ARX method) is as follows:

$$G(z) = \frac{0.051705 - 0.05151z^{-1}}{1 - 0.999995z^{-1}} \tag{8}$$

Calculated parameters of the battery after 498 cycles:

R0 = 0.05 [Ω]   R1 = 0.05 [Ω]   C1=30 000 [F]

Comparison of the measured and estimated voltage profiles on the battery after 498 cycles is shown in Fig.7.

Fig. 8 shows the shape of measured and estimated Q capacity of the battery after 498 cycles.

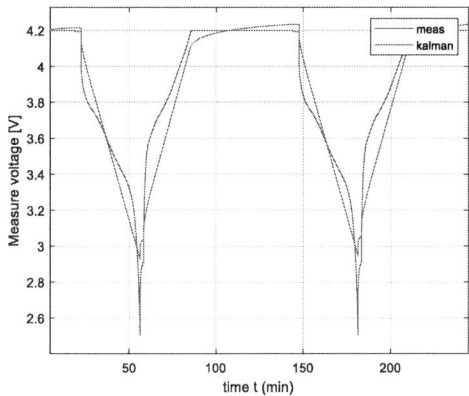

Fig. 7. Comparison of measured and estimated voltage profiles after 498 cycles

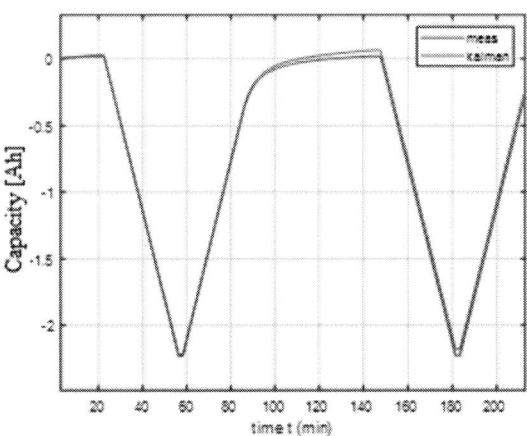

Fig. 8. Comparison of measured and estimated Q capacity after 498 cycles

VII. CALCULATION OF SOH PARAMETER

The SOH parameter is an indicator of the rate of reduction in battery performance due to the aging process [3]. The following relation is used in practice to calculate the SOH parameter:

$$SOH = \frac{Q}{Q_n} \tag{9}$$

Q – current capacity of battery during working [Ah]

Q_n – maximum capacity of new battery [Ah]

Q is the maximum capacity value in the given cycle from the full charging to discharging (Qn = 2.45 Ah). The SOH parameter for different number of cycles determined by the above method is given in Table 1.

TABLE 1. CALCULATED VALUES OF SOH PARAMETER IN RELATION TO NUMBER OF AGING CYCLES

No. of cycles	1	125	188	373	498
Q[Ah]	2.45	2.31	2.25	2.1	1.97
SOH	1	0.942	0.918	0.857	0.8

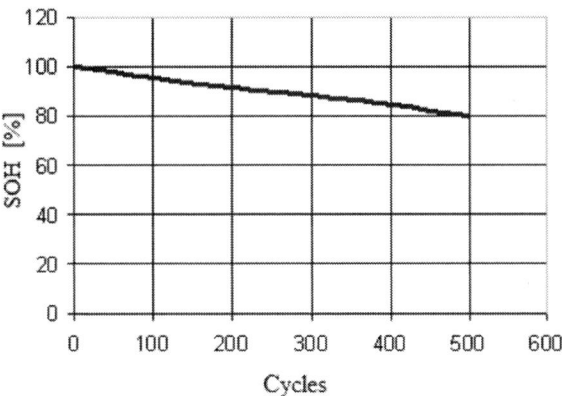

Fig. 9.   Course of SOH parameters related to aging cycle number

The course of the SOH parameter depending on the number of cycles is shown in Fig.9.

VIII. CONCLUSION

The article describes a method for determining the SOH parameter that is a measure of the battery performance reduction due to the aging process. The article lists the measurements that were performed during the battery aging. Using the "ARX model" method, the battery parameters were estimated based on the measured values that changed as expected. When comparing the measured and estimated voltage profile, we can observe a larger error in discharging the battery. The error is caused by using a simplified model. The maximum deviation between the estimated and measured course of the current Q capacity is 5 %. The estimated value of the parameter SOH after 498 cycles corresponds to the catalogue data of the battery type INR 18650-25R. The performed experiments show that even the use of the maximum simplified model of the battery allows a sufficiently accurate estimation of the SOH parameter.

ACKNOWLEDGEMENT

This work was partially supported by the Ministry of Education, Youth and Sports of the Czech Republic and the European Union (European Structural and Investment Funds – Operational Programme Research, Development and Education) in the frames of the project "Modular platform for autonomous chassis of specialized electric vehicles for freight and equipment transportation", Reg. No. CZ.02.1.01/0.0/0.0/ /16_025/0007293.

This work was supported by the programme *H2020-WIDESPREAD* under the project „ Networking For Research And Development Of Human Interactive And Sensitive Robotics Taking Advantage Of Additive Manufacturing " (R2P2, ID: 857061).

This work was partly supported by the Student Grant Competition of the Technical University of Liberec under the project No. SGS-2020-3042.

REFERENCES

[1] Bizhong Xia, Zizhou Lao, Ruifeng Zhang , Yong Tian, Guanghao Chen, Zhen Sun, WeiWang, Wei Sun, Yongzhi Lai, Mingwang Wang and Huawen Wang : Online Parameter Identification and State of Charge Estimation of Lithium-Ion Batteries Based on Forgetting Factor Recursive Least Squares and Nonlinear Kalman Filter,

[2] S.J. Lee, J.H. Kim, J.M. Lee, and Bo H. Cho: The State and Parameter Estimation of an Li-Ion Battery Using a New OCV-SOC Concept, https://www.researchgate.net/publication/224289849

[3] Carlo Taborelli and Simona Onori: Advanced battery management system design for SOC/SOH estimation for e-bikes applications,

[4] Carlo Taborelli; Simona Onori: State of charge estimation using extended Kalman filters for battery management systém,

[5] Dae-Won Chung, Seung-Hak Yang: SOC Estimation of Lithium-Ion Battery Based on Kalman Filter Algorithm for Energy Storage System in Microgrids,

[6] Agus Hasan, Martin Skriver, and Tor Arne Johansen: eXogenous Kalman Filter for State-of-Charge Estimation in Lithium-ion Batteries,

[7] Yu Ding-xuan , Gao Yan-xia: SOC estimation of Lithium-ion battery based on Kalman filter algorithm, Proceedings of the 2nd International Conference on Computer Science and Electronics Engineering (ICCSEE 2013)

[8] Shifei Yuan , Hongjie Wu and Chengliang Yin: State of Charge Estimation Using the Extended Kalman Filter for Battery Management Systems Based on the ARX Battery Model, Energies 2013, 6, 444-470; doi:10.3390/en6010444

[9] Habiballah Rahimi Eichi, Mo-Yuen Chow: Adaptive Parameter Identification and State-of-Charge Estimation of Lithium-Ion Batteries, The 38th Annual Conference of the IEEE Industrial Electronics Socie

Online Optimization of Firing Angles for Switched Reluctance Motor Control

Peter Bober
Faculty of Electrical Engineering and Informatics
Technical University of Košice
Košice, Slovakia
peter.bober@tuke.sk
http://orcid.org/0000-0002-8743-5220

Želmíra Ferková
Faculty of Electrical Engineering and Informatics
Technical University of Košice
Košice, Slovakia
zelmira.ferkova@tuke.sk
http://orcid.org/0000-0002-3846-2660

Abstract—**The switched reluctance motor can run on high efficiency even if it is driven by the constant current with simple firing angles adjustment according to the speed and the current. The paper presents the procedure for online and model-based optimization of firing angles for high-efficiency variable speed drive. The measurement of motor efficiency shows that the objective function is continuous and has one extreme. Therefore, it is possible to use simple gradient method width lower computational time requirements. The firing angles adjustment function can be a second-order polynomial or lookup table. Therefore, the drive control can be implemented by a low-cost microcontroller.**

Keywords—*Switched reluctance motor, optimization, efficiency, current control, FEA*

INTRODUCTION

Today, energy saving is becoming more and more urgent for the future of humanity. At the same time, manufacturing companies are constantly looking for new ways to reduce production costs, which ultimately means less burden on the environment. The switched reluctance machines (SRM) offer the opportunity to increase the efficiency of converting electrical energy into mechanical [1], reduce production costs, and increase reliability due to simple construction. The SRM has windings only on the stator, and the torque is created by changing the magnetic reluctance while the rotor is moving. The torque ripple is naturally larger than with an induction motor or permanent magnet motor, which is the most significant disadvantage of the SRM.

Various control schemes have been developed for SRM variable speed drives. These can, in principle, be divided into voltage pulse control, by maintaining a constant current or by maintaining a constant torque [2]. The voltage pulse control is the simplest controller for high-speed control and for systems that do not require high performance. The current control with firing angles modulation can achieve high efficiency as well as torque ripple reduction [3],[4],[5]. However, a further reduction in torque oscillation can only be achieved by using advanced torque control methods such as Average torque control (ATC), Direct torque control (DTC), or Torque sharing function (TSF) based control [6]. More complex control methods require a powerful microcontroller along width high-speed ASIC (Application Specific Integrated Circuit) or FPGA (Field Programmable Gate Array). In addition, efforts to reduce torque ripple lead to reduced efficiency, as mentioned, for example, in [7] and [8].

Therefore, this paper focuses on the current control method that a low-cost microcontroller can implement. We will show how to determine the optimal turn-on and turn-off angles for each SRM operating point for this method. The optimization objective is maximum efficiency. The measurement of efficiency showed that the objective function is continuous and has one extreme. Therefore, it is possible to use one of the versions of the gradient method. Otherwise, more complex optimization methods that require longer computational time must be used [3],[4],[5].

Although the presented SRM control has a relatively large torque ripple compared to more advanced methods, it can still be used by many applications in the growing SRM market, where motor efficiency is essential, e.g., in automotive applications to prolong the battery life.

SWITCHED RELUCTANCE MOTOR

The shape and construction of the SRM on which the presented control method will be verified are in Fig. 1. The motor has a salient pole stator and rotor, $2p_1/2p_2 = 12/8$. The rated power of the motor is 200W, maximum phase current $I_{max} = 6A$, and DC-link voltage $U_{DC} = 120V$. Each phase winding has two parallel branches.

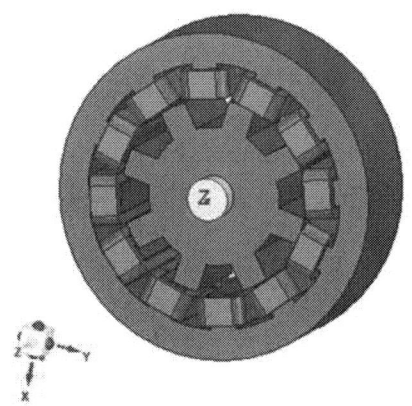

Fig. 1.   3D model of used SRM.

The SRM was modeled in the program Ansys Maxwell. The distribution of magnetic field density in the XY plane passing through the center of the motor is in Fig. 2. The FEA (Finite Element Analysis) model was verified against several measurements and is sufficiently accurate. Fig. 3 shows the 3D view of measured static torque characteristics used for verification.

978-1-6654-3236-8/21 $31.00 © 2021 IEEE   238

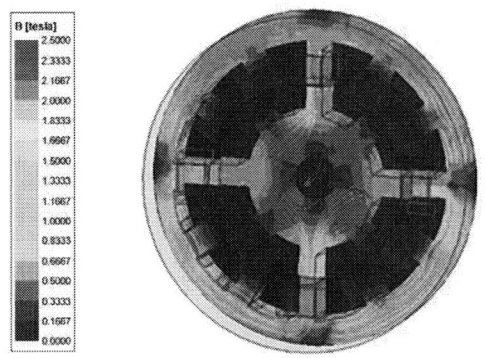

Fig. 2. Magnetic field density in the XY plane passing through the center of the motor

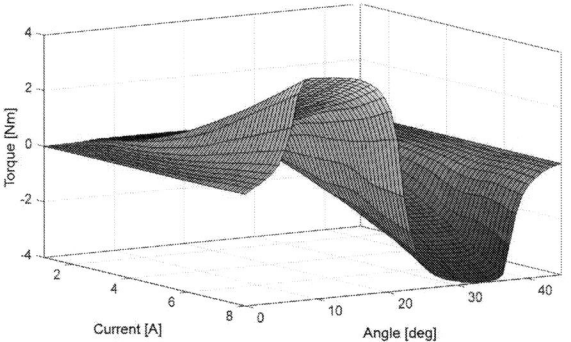

Fig. 3. 3D view of measured static torque of SRM

OPTIMIZATION OF FIRING ANGLES

A. Current control with angle modulation

The block diagram of the SRM current control method is shown in Fig. 4. The hysteresis current controller controls phase currents I_A, I_B, and I_C. The speed controller provides torque reference T_{ref} that is translated to the reference current command I_{ref}. Angles θ_{on} and θ_{off} determine a region where phase A is energized. Firing angles for phases B and C are shifted by 15° and 30°, respectively, for SRM geometry in Fig. 1. The phase current waveform in Fig. 5 is controlled by reference current I_{ref} and by the position of firing angles. The phase winding inductance and the rotor speed define the rising and falling slope of the current.

The efficiency is dependent on firing angles. The task of the block *Current and Firing Angles Calculation* in Fig. 4 is to calculate the reference current $I_{ref} = f_l(T_{ref},\omega)$, angles $\theta_{on} = f_{on}(T_{ref},\omega)$ and $\theta_{off} = f_{off}(T_{ref},\omega)$ so that the motor efficiency is highest for each torque and speed value.

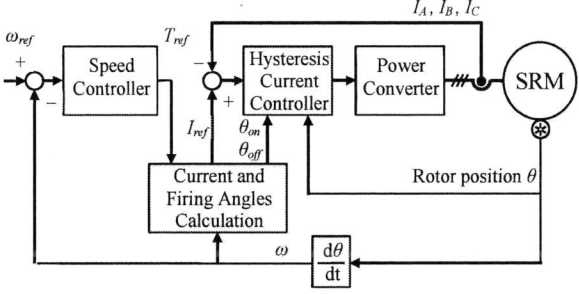

Fig. 4. Block diagram of the SRM current control with reference current and firing angle adjustment.

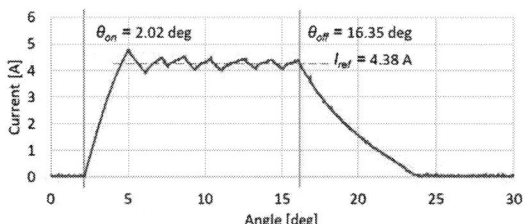

Fig. 5. Measured phase current maintained by hysteresis current controller at angle interval $\langle \theta_{on}, \theta_{off} \rangle$ (1200 rpm, torque 1 Nm).

B. Current and Firing Angles Calculation

Functions $f_l(T_{ref},\omega)$, $f_{on}(T_{ref},\omega)$, and $f_{off}(T_{ref},\omega)$ are calculated numerically. The optimization procedure determines optimal values of I_{ref}, θ_{on}, and θ_{off} for a finite set of operating points defined by speed and load torque combination. The regression uses obtained values to calculate the formulas for the f_l, f_{on}, and f_{off} functions. The most commonly used is the second-order polynomial. If the regression model is not accurate enough, an alternative is a lookup table. Section C and D describe the measured objective function and calculated one, respectively.

The time required for optimization depends on the shape of the objective function and the chosen optimization method. In the case of one extreme, it is possible to use a gradient method that requires relatively little evaluation of the objective function. The evaluation of the function itself can be performed online by direct measurement in real-time or by calculation with the FEA model.

C. Measured Objective Function

The measurement setup in Fig. 6 allows automatic online measurement of motor efficiency, drive efficiency, power factor, current ripples, and torque ripples for low motor speed. A personal computer controls the measurement. The list of instruments is in TABLE I. The objective function is the motor efficiency in this research. Fig. 7 shows a measured motor efficiency map, and Fig. 8 shows the corresponding current reference map for one operating point. The white spots on the maps mean that the SRM cannot maintain the operating point (T,ω) for the particular combination of firing angles and the phase current limitation of 6.5 A.

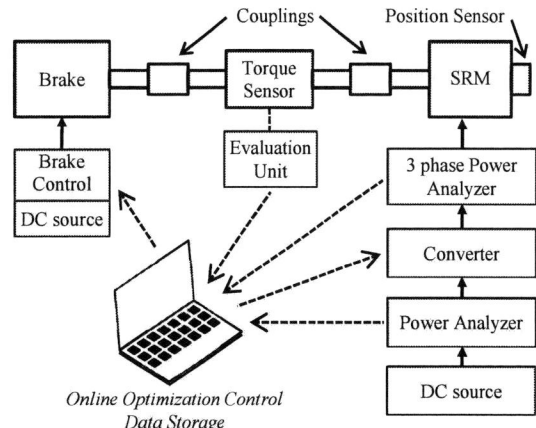

Fig. 6. Structure of the measurement system.

978-1-6654-3236-8/21 $31.00 © 2021 IEEE

TABLE I.	LIST OF MEASUREMENT INSTRUMENTS

Instrument	Type / Description
Torque Sensor	KISTLER 4520A
Evaluation Unit	CoMo torque evaluation instrument 4700BP0UA
3 phase Power Analyzer	Fluke NORMA 4000 Power Analyzer
Converter	Three-phase asymmetric H-bridge, prototype with STM32F303RET6 microcontroller
Power Analyzer	Infratek 106A Power Analyzer

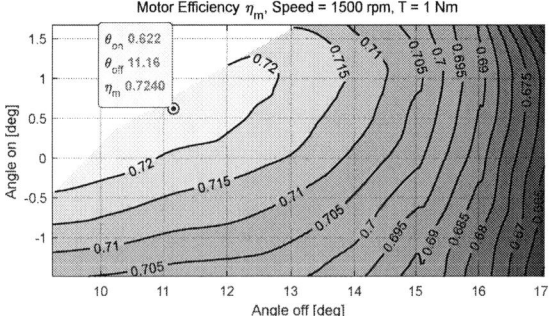

Fig. 7. Measured motor efficiency map for speed 1500 rpm / 1 Nm.

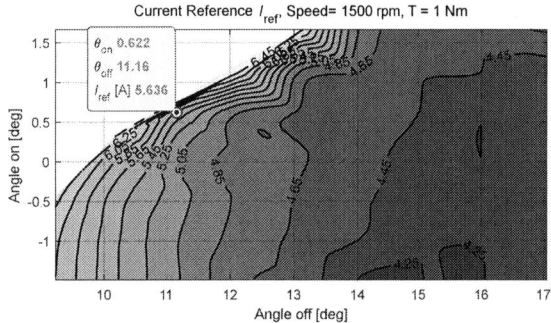

Fig. 8. Measured current reference map for speed 1500 rpm / 1 Nm.

The motor efficiency measurement for one point in the map has the following steps: set the firing angles, wait for SRM to reach the operating point (10 sec), measure efficiency (2 minutes). There were two reasons to construct the efficiency map:

- To see the shape of the selected objective function.
- To find the extreme to tune-up and to verify the optimization procedure.

According to the efficiency map in Fig. 7 and the current reference map in Fig. 8, there is only one combination (I_{ref}, θ_{on}, θ_{off}) for the operating point (1500 rpm, 1 Nm) where the motor efficiency is highest.

D. Calculated Objective Function Using the FEA Model

Model-based offline optimization of firing angles uses the scheme of the SRM drive in Fig. 11. The model of the controller and the converter were created in Ansys Twin Builder. The FEA SRM model is a part of the scheme. The calculated motor efficiency map and current reference map are shown in Fig. 9 and Fig. 10, respectively, for one operating point. There are two differences between measured and calculated maps. First, the area of maximal efficiency on the

calculated map is shifted toward the highest angles. The reason identified is a later increase in inductance profile in the FEA model than in the SRM. This is probably due to a local change in the magnetic material parameters due to sheet metal cutting. The second difference is in the efficiency values, which are lower for the SRM, especially in the area of low efficiencies. Due to the change in efficiency, the temperature of the motor and subsequently the resistance of the winding changed. In the FEA model, the winding resistance was constant. It is a small machine where joule losses are dominant.

The Optimetrics block of Twin Builder searches for minimum by default. Therefore, the objective function (1) has minus sing at the beginning:

$$FEA\ objective\ function = -\left(\eta - w\left|T - T_{ref}\right|\right) \quad (1)$$

where η, w, and T are efficiency, weight, and torque, respectively. The term $w\left|T_{ref} - T_{ref}\right|$ guarantees that the torque is as close as possible to the operating point torque T_{ref}.

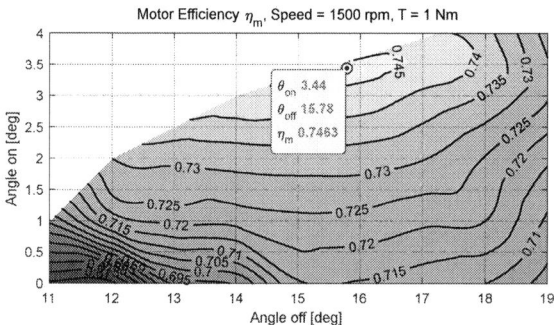

Fig. 9. Calculated motor efficiency map for speed 1500 rpm / 1 Nm.

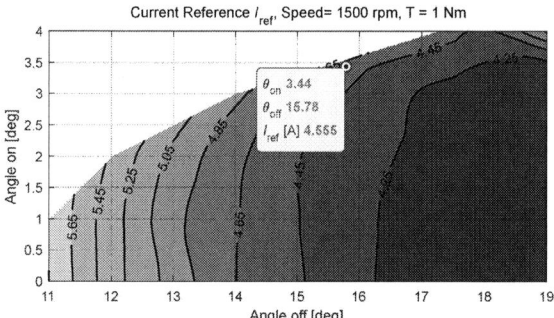

Fig. 10. Calculated current reference map for speed 1500 rpm / 1 Nm.

RESULTS

A. Online Optimization

This section describes the results of the online optimization procedure. The procedure finds optimal values of I_{ref}, θ_{on}, and θ_{off} for one operating point (1500 rpm, 1 Nm). Fig. 12 shows paths of the gradient optimization starting in five different points. All paths finish in the area of high efficiency. However, paths 1 and 2 stops farther from the efficiency maximum than paths 3, 4, and 5. The distance is not so big if we compare the endpoint efficiency deviations in TABLE II. The measurement precision plays a substantial role here as the values are below 0.5%, and the efficiency map is flat in this region.

978-1-6654-3236-8/21 $31.00 © 2021 IEEE   240

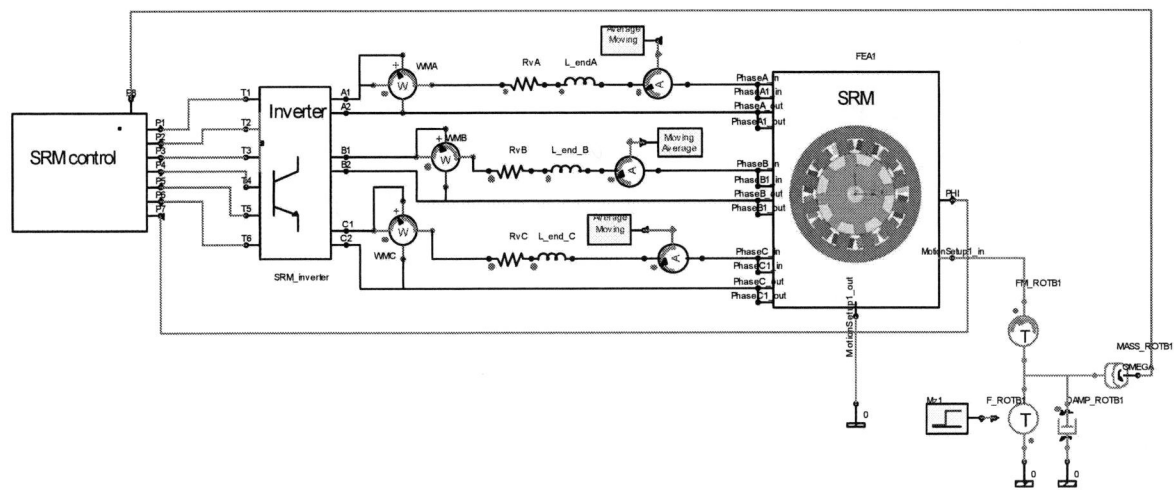

Fig. 11. The scheme of the SRM control in Ansys Twin Builder.

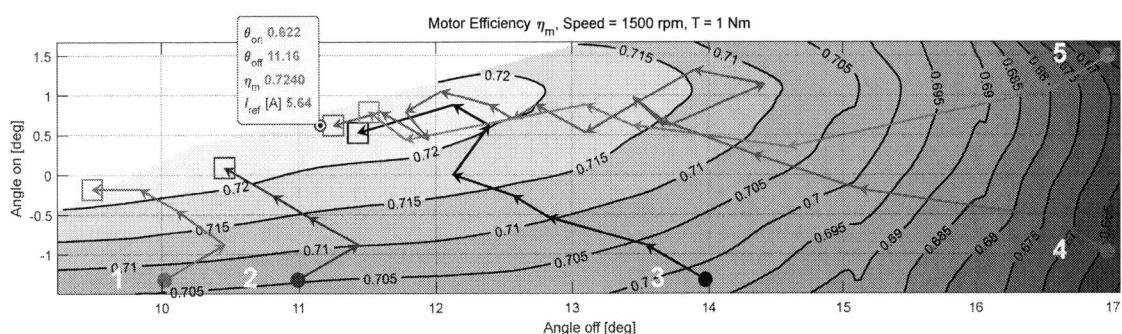

Fig. 12. Online optimization paths from different starting points.

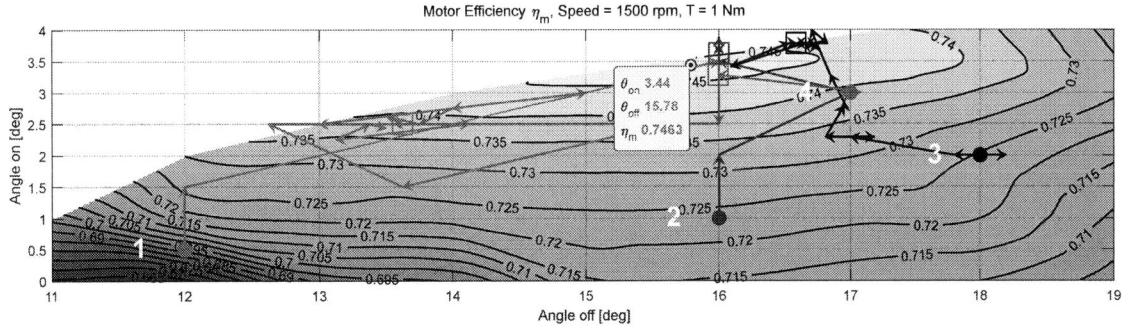

Fig. 13. FEA optimization paths from different starting points.

TABLE II. ONLINE OPTIMIZATION PATHS RESULTS

Path No	Angle on (deg)	Angle off (deg)	η_m	η_m deviation from max
max	0.62	11.16	0.724	-
1	-0.18	9.49	0.7196	0.6%
2	0.09	10.46	0.7235	0.1%
3	0.53	11.43	0.7202	0.5%
4	0.62	11.25	0.7204	0.5%
5	0.79	11.51	0.7238	0.0%

The measurement confirms the presumption that the simple gradient optimization method can be used to find I_{ref}, θ_{on}, and θ_{off} for one operating point. However, the automatic finding of optimal values for all operating points requires several optimization runs to ensure a sufficiently reliable result. Once the optimization procedure is tuned, the automatic measurement of optimal values for construction of functions $f_I(T_{ref},\omega)$, $f_{on}(T_{ref},\omega)$, and $f_{off}(T_{ref},\omega)$ can be scheduled.

B. FEA Optimization

Model-based offline optimization (Fig. 13) exhibits similar results as the online one. The FEA objective function is slightly different than in the previous case. Therefore, various optimization methods available in the Optimetrics block of Twin Builder were tested. Tracks 1 and 2 in Fig. 13 use Pattern search, tracks 3 and 4 use Quasi-Newton (gradient). These two methods give reliable results. The TABLE III. summarizes numerical values.

TABLE III.   FEM OPTIMIZATION PATHS RESULTS

Path No	Angle on (deg)	Angle off (deg)	η_m	η_m deviation from max
max	3.44	15.78	0.7463	-
1	2.5	13.56	0.74	0.8%
2	3.625	16	0.7427	0.5%
3	3.78	16.59	0.745	0.2%
4	3.73	17.45	0.7425	0.5%

METHOD MODIFICATION FOR DIFFERENT OBJECTIVES

The described procedure can be used to find functions $f_I(T_{ref},\omega)$, $f_{on}(T_{ref},\omega)$, and $f_{off}(T_{ref},\omega)$ that are optimal according to another objective. In each case, it is necessary to determine the shape of the objective function and find a suitable optimization method. After tuning the optimization, it is possible to start a batch of experiments to obtain the optimal values of parameters for the entire range of motor operating points, from which the functions f_I, f_{on}, and f_{off} are determined.

CONCLUSION

This paper describes the procedure to find the optimal firing angles for the current control method to run the SRM at maximum efficiency and do not demand a high-performance microcontroller. First, the optimization procedure that finds the optimal values of reference current and firing angles for each operating point is selected and tuned. Then the automated batch of experiments is run to obtain the set of optimal values for the regression model or lookup table. The optimization can run on a real motor, where the online motor measurement evaluates the objective function, or the FEA model calculation is used instead.

The results show that the optimization procedure finds values that are near enough to the optimum for real motor measurement as well as for FEA model calculation. However, the optimization paths tended to end in two different areas in bought cases. Additional analysis of the objective functions showed that there is another local maximum located in the region of small firing angle values in addition to the global maxima. Nevertheless, every path that started at more reliable values of angles in the middle or the right side of the efficiency map finished at the global extreme.

Optimization with the online measurement as well as FEA calculation has its field of application. The online measurement requires a hundred times less time than FEA calculation; however, the measurement is subject to error and can only be performed for an existing motor.

Further work after tuning the optimization will be performing the automated batch of experiments.

ACKNOWLEDGMENT

This work was supported by the Slovak Research and Development Agency under contract No. APVV-18-0436 (90%) and APVV-15-0750 (10%).

REFERENCES

[1] A. Chiba and K. Kiyota, "Review of research and development of switched reluctance motor for hybrid electrical vehicle," 2015 IEEE Workshop on Electrical Machines Design, Control and Diagnosis (WEMDCD), 2015, pp. 127–131, https://doi.org/10.1109/WEMDCD.2015.7194520.

[2] M. F. S. Pereira, A. Mamede, R. E. Araújo, "Switched Reluctance Motor Drives: Fundamental Control Methods," In: Rui Esteves Araújo, José Roberto Camacho (Eds.), Modelling and Control of Switched Reluctance Machines, IntechOpen, 2020, https://doi.org/10.5772/intechopen.90476.

[3] J. H. Fisch, Yun Li, P. C. Kjaer, J. J. Gribble and T. J. E. Miller, "Pareto-optimal firing angles for switched reluctance motor control," Second International Conference On Genetic Algorithms In Engineering Systems: Innovations And Applications, 1997, pp. 90–96, https://doi.org/10.1049/cp:19971161.

[4] M. Debouza, A. Al-Durra, H. M. Hasanien, S. Leng and W. Taha, "Optimization of Switched Reluctance Motor Drive Firing Angles Using Grey Wolf Optimizer for Torque Ripples Minimization," IECON 2018 - 44th Annual Conference of the IEEE Industrial Electronics Society, 2018, pp. 619–624, https://doi.org/10.1109/IECON.2018.8591473.

[5] B. Anvari, M. Kaya, S. Englebretson, S. Hajimirza and H. A. Toliyat, "Surrogate-Based Optimization of Firing Angles for Switched Reluctance Motor," 2018 IEEE Transportation Electrification Conference and Expo (ITEC), 2018, pp. 1023–1028, https://doi.org/10.1109/ITEC.2018.8450092.

[6] C. Gan, J. Wu, Q. Sun, W. Kong, H. Li and Y. Hu, "A Review on Machine Topologies and Control Techniques for Low-Noise Switched Reluctance Motors in Electric Vehicle Applications," IEEE Access 6:31430-31443. https://doi.org/10.1109/ACCESS.2018.2837111.

[7] S. Song, G. Fang, R. Hei, J. Jiang, R. Ma and W. Liu, "Torque Ripple and Efficiency Online Optimization of Switched Reluctance Machine Based on Torque per Ampere Characteristics," in IEEE Transactions on Power Electronics, vol. 35, no. 9, pp. 9608-9616, Sept. 2020, https://doi.org/10.1109/TPEL.2020.2974662.

[8] M. Balaji, V. Kamaraj, "Evolutionary computation based multi-objective pole shape optimization of switched reluctance machine," International Journal of Electrical Power & Energy Systems, Volume 43, Issue 1, 2012, pp. 63–69, ISSN 0142-0615, https://doi.org/10.1016/j.ijepes.2012.05.011.

Influence of Rotor Slot Number on Magnetic Noise in a Squirrel-cage Induction Motor for Traction Applications

Ivan Milažar
KONČAR – Generators and Motors Inc.
Zagreb, Croatia
imilazar@koncar-gim.hr

Damir Žarko
University of Zagreb Faculty of Electrical Engineering and Computing
Department of Electrical Machines, Drives and Automation
Zagreb, Croatia
damir.zarko@fer.hr

Abstract— In this paper, the influence of rotor slot number on the magnetic noise of a squirrel cage motor for traction applications is analyzed. The introductory part explains the importance of noise analysis and measurement and gives an overview of the types of noise that occur in this type of motor. The dominance of a particular type of noise depending on the design of the induction motor (number of poles, type of cooling) and the operating range of the motor is also commented. The cause and occurrence of magnetic noise are explained by analytical expressions. Methods are listed that can be used to effectively influence the magnetic noise level at the design stage. The difference between sound pressure level and sound power level is explained as noise level is expressed in both forms. The dominant stress components due to the radial forces acting on the stator teeth and the resulting sound power maxima are calculated for three combinations of stator and rotor slot numbers using the software package MANATEE.

Keywords— magnetic noise, induction motor, slot number, simulation

I. INTRODUCTION

Noise is an undesirable phenomenon in electrical machines, so it is important to measure the noise level and analyze the noise source. The maximum permissible noise level of motors is defined by International Electrotechnical Commission (IEC) [1]. An additional incentive to minimize motor noise sometimes comes from the motor purchaser itself, who sets stricter criteria than those prescribed by IEC. Induction motor noise is classified into three groups: fan noise, bearing noise and magnetic noise. Fan noise is dominant for motors with 2 or 4 poles, since the speed of the induction motor is directly dependent on the number of poles

$$n_s = \frac{60 \cdot f}{p} \qquad (1)$$

where n_s is the synchronous speed, f is the frequency, and p is the number of pole pairs. When the fan diameter is decreased, the noise level of the fan is decreased at the same time. However, it must be considered that with the reduction of the fan diameter, the air flow for motor cooling is also reduced.

The bearing noise is not dominant in relation to the fan noise and the magnetic noise if the bearing is not worn. When the bearing is worn and overheated, the bearing noise increases noticeably.

The magnetic noise level increases as the number of poles of the motor increases, since in this case the yoke becomes thinner (Table I) and the stator then becomes more susceptible to vibrational excitation.

TABLE I. RATIO OF STATOR OUTER DIAMETER AND STATOR INNER DIAMETER AS A FUNCTION OF THE NUMBER OF POLES

$2p$	2	4	6	8	10	12
D_o/D	1.65-1.8	1.5-1.6	1.4-1.45	1.35-1.4	1.3-1.35	1.3-1.35

For motors with more than 6 poles, the magnetic noise is more dominant than the fan noise. Induction motors for electric vehicles are connected to a power converter that supplies a voltage with high harmonic content, which causes additional magnetic noise. The contribution of higher order harmonics is very noticeable at low speeds of motors that have a fan mechanically connected to the rotor (e.g. cooling method IC0A1), since at low motor speeds the fan speed and its contribution to the noise is also low. In this paper, the magnetic noise is analysed on a V6AOJ 205-04 induction traction motor (Fig. 1) for a low-floor tram manufactured by Končar Generators and Motors Inc. The basic motor data are listed in Table II.

TABLE II. TECHNICAL DATA FOR MOTOR V6AOJ 205-04

$U,$ V	$P,$ kW	$M,$ Nm	$n,$ min^{-1}	$I,$ A	Power factor	$f,$ Hz
320	65	365	1700	151	0,84	58

Fig. 1. Induction motor V6AOJ 205-04 [2]

II. MAGNETIC NOISE OF INDUCTION MOTORS

A. Causes of Magnetic Noise

As mentioned in the introduction, magnetic noise is generated by vibrations or elastic deformations of the stator core of the motor. The rotor core is much stiffer than the stator core, so rotor vibrations are usually ignored. The cause of core vibrations is electromagnetic forces generated by the radial component of the rotating magnetic field. Since most of the magnetomotive force (MMF) of the winding is spent on the magnetization of the air gap, it can be written

$$\Theta \approx H_\delta 2\delta \qquad (2)$$

978-1-6654-3236-8/21 $31.00 © 2021 IEEE

where Θ is the magnetomotive force, H_δ is the magnetic field strength, and δ is the air gap width. Magnetic field strength and magnetic flux density are related according to

$$B = \mu_0 H_\delta \approx \mu_0 \frac{\theta}{2\delta} \qquad (3)$$

The rotating MMF contains, in addition to the fundamental, higher harmonics and thus, according to (3), also the flux density in the air gap [3]

$$B(x,t) = \sum_{\nu=1}^{\infty} B_\nu \cos(\nu \frac{\pi}{\tau_p} x - \omega_\nu t - \varphi_\nu) \qquad (4)$$

where B_ν is the peak value of flux density (ν^{th} harmonic), τ_p is the pole pitch, x is the circumferential position, ω_ν is angular frequency of the νth harmonic, and φ_ν is the phase shift of the νth harmonic. Higher harmonics in the air gap flux density result in the presence of higher harmonics in the radial force acting on the stator core teeth

$$F(x,t) = \frac{[B(x,t)]^2 \cdot S}{2\mu_0} \qquad (5)$$

where S is the tooth surface, and μ_0 is the permeability of vacuum. The resulting stress is given by

$$\sigma(x,t) = \frac{F(x,t)}{S} = \frac{[B(x,t)]^2}{2\mu_0} \qquad (6)$$

The distribution of stresses along the circumference of the stator bore can lead to different vibration modes [3], which are shown in Fig. 3. For a motor with a pole number of $2p$ and an integer-slot winding, the dominant mode of vibration $r = 2p$ can be noticed. The vibration amplitude is largest when the force frequency coincides with the natural frequency of the core stack due to resonance. The natural frequency for zero mode can be calculated according to [3]

$$f_0 = \frac{1}{2\pi R_m} \sqrt{\frac{E}{\Delta \rho}} \qquad (7)$$

where R_m is the stator mean radius (Fig. 2), E is the Young's modulus of elasticity of the core material, ρ is the density of the core material and Δ is defined as follows

$$\Delta = 1 + \frac{G_z + G_w}{G_y} \qquad (8)$$

where G_z is the total weight of the teeth, G_w is the weight of the winding and G_y is the weight of the yoke (weight of the core excluding the teeth and excluding the winding). Natural frequencies for vibration modes greater than or equal to 2 can be calculated according to [3]

$$f_r = f_0 \cdot i \cdot \frac{r(r^2-1)}{\sqrt{r^2+1}} \cdot \frac{1}{\sqrt{1+i^2\left(\frac{r^2-1}{r^2+1}\right)\cdot[3+r^2\left(4+\frac{\Delta_m}{\Delta}\right)]}} \qquad (9)$$

where r is the vibration mode. The parameter i is defined according to

$$i = \frac{h}{2\sqrt{3}\cdot R_m} \qquad (10)$$

where h is the thickness of the stator yoke (Fig. 2). The parameter Δ_m is given by

$$\Delta_m = 1 + \frac{N_1 \theta_z}{2\pi I R_m} \qquad (11)$$

where I is the section moment of inertia for the yoke defined according to

$$I = \frac{Lh^3}{12} \qquad (12)$$

where L is stator axial length. The parameter θ_z is defined according to

$$\theta_z = b_z L h_s^3 \cdot \left[\frac{1}{3} + \frac{h}{2h_s} + \left(\frac{h}{2h_s}\right)^2\right] \cdot \frac{G_z + G_w}{G_z} \qquad (13)$$

where b_z is the mean tooth width, and h_s is the tooth height. Table III contains the parameters of the motor V6AOJ 205-04 required for the calculation of the natural frequencies using equations (7) to (13).

It should be noted that stator core contains as many natural frequencies as there are vibration modes.

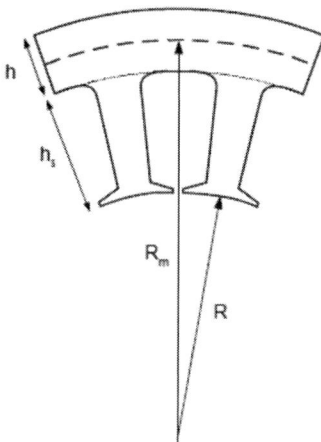

Fig. 2. Geometry of the stator yoke and teeth [3]

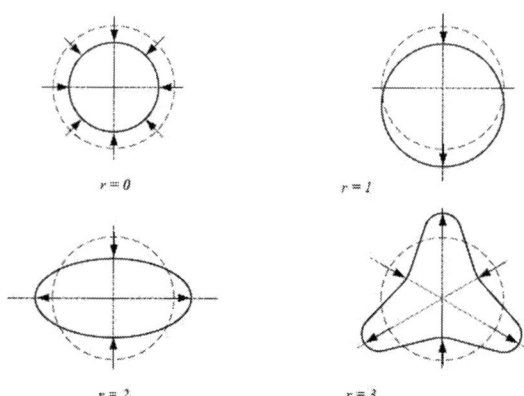

Fig. 3. Vibration modes [3]

TABLE III. MOTOR PARAMETERS FOR CALCULATION OF THE NATURAL STATOR FREQUENCIES

Parameter	Value
E	$20,6\cdot10^{10}$ N/m^2
ρ	7650 kg/m^3
R	0,1325 m
R_m	0,14875 m
h	0,0325 m
h_s	0,03 m
b_z	0,00839 m
L	0,25 m
G_z	170 N
G_w	225,6 N
G_y	581 N

The noise level is defined by the sound pressure level [4]

$$L_r = 20\log(9{,}05 \cdot 10^4 f_r\, y_r) + 10\log(P_{rel}) \qquad (14)$$

where f_r is the frequency of the force harmonic, y_r is the amplitude of deformation of the stator core, and P_{rel} is the correction in sound pressure level that considers the cylindrical surface of the motor

B. Methods for Magnetic Noise Reduction

The magnetic noise can be affected in several ways:

- combinations of stator and rotor slot numbers [5],
- dimensions of the slots,
- stator and rotor slot opening widths [6],
- lamination stack fastening,
- air gap width [7],
- air gap flux density,
- slot skewing [8],
- magnetic wedges in the slots [9].

According to [10], some recommended slot combinations for 4-pole motors are listed in Table IV.

TABLE IV. RECOMMENDED SLOT COMBINATIONS [10]

| 2p | stator | rotor | |
	N_1	N_2 – without skew	N_2 – with skew
4	24	32, 34	16, 20
	36	26, 28, 44	27, 28, 30
	42	34, 50, 52	33, 34, 38
	48	36, 38, 52, 56	36, 39, 40
	60	46, 50, 52, 68, 76	49, 51, 56
	72	58, 62, 64, 80, 82	61, 63, 68

According to [4], when choosing the number of slots, it is important to follow the rule

$$N_1 - N_2 \neq 2p \pm r \qquad (15)$$

where N_1 is the number of stator slots, N_2 is the number of rotor slots, and r is the vibration mode to be avoided.

Noticeable noise is present if resonance occurs in the stator core, i.e. when the frequency of a particular force harmonic and the natural frequency of the stator coincide. Any change in the slot dimensions simultaneously affects the dimensions b_z and h_s in (13) and thus the value of the stator natural frequency. Additional longitudinal construction elements for fastening the lamination stack have an influence on its natural frequency as well (Fig. 4).

Fig. 4. Longitudinal elements for fastening the stator package

If voltage drops on the stator resistance and leakage reactance are neglected, the basic voltage equation of an induction motor is

$$U \approx 4{,}44\, \Phi f w_s k_w \qquad (16)$$

where U is the phase voltage, Φ is the air gap magnetic flux per pole, f is the frequency, w_s is the number of turns per phase connected in series, and k_w is the winding factor. According to (16), the increase of the number of turns per coil reduces the air gap flux per pole, i.e. it decreases the air gap flux density and, according to (6), also decreases the core stress and the magnetic noise level. The air-gap flux density can also be reduced if the core is axially extended because flux Φ is proportional to the product of the flux density and the stack length and at the same time it is determined by the applied voltage and frequency.

In addition, as the air gap width increases, the level of magnetic noise decreases [7].

C. Results of simulations for different slot combinations

The magnetic noise in the traction motor V6AOJ 205-04 is analysed using the simulation software Magnetic Noise Acoustic Analysis Tool for Electrical Engineering (MANATEE). Three combinations of slot numbers listed in Table V are used.

TABLE V. SLOT NUMBER COMBINATIONS

	N_1	N_2
Combination 1	36	28
Combination 2	36	27
Combination 3	36	30

For a better understanding, it is important to clarify the difference between sound pressure level (L_p) and sound power level (L_w). The sound pressure level L_p depends on the distance to the source, the power of the source and the acoustic properties of the surrounding space

$$L_p = 20\log\frac{p}{p_0} \qquad (17)$$

where p_0 is the referent pressure which equals 20 µPa. The sound power level represents the measure of the sound power P radiated from the source into the surrounding space

$$L_w = 10\log\frac{P}{P_0}. \qquad (18)$$

where P_0 is the referent power in the amount of 1 pW.

Table VI contains sound pressure levels obtained by simulation in MANATEE for the V6AOJ 205-04 motor with 36 stator slots and 28 rotor slots, at no-load and supplied with sinusoidal voltage. This is a slot number combination of the actual motor shown in Fig. 1. The ratio of voltage and frequency was kept constant for all speeds, assuming constant flux operation. The influence of stator resistance on the voltage to frequency ratio is neglected since the motor is at no load and the minimum speed is 300 min^{-1} where the effect of stator resistance is still negligible.

Table VII contains sound pressure levels of the unloaded motor fed with sinusoidal voltage in steady-state operation, but for the combination with 36 slots on the stator and 27 slots on the rotor.

Table VIII contains results for the combination with 36 slots on the stator and 30 slots on the rotor.

Tables IX, X and XI contain the sound pressure and sound power levels of a loaded motor V6AOJ 205-04 with stator/rotor slot number combinations 36/28, 36/27 and 36/30 respectively, with sinusoidal voltage supplied at four speeds: 300 min^{-1}, 600 min^{-1}, 1200 min^{-1} and 1700 min^{-1}.

TABLE VI. SOUND PRESSURE LEVELS FOR SLOT NUMBER RATIO 36/28

U, V	f, Hz	n, min^{-1}	L_p, dBA
446,1	140	4200	66,2
424,3	133,2	3994,7	65,3
380,7	119,5	3584,2	63,8
358,9	112,6	3378,9	63,4
315,3	98,9	2968,4	63,9
271,7	85,3	2557,9	73,8
249,8	78,4	2352,6	71,7
206,3	64,7	1942	75,6
184,5	57,9	1736,8	67
140,9	44,2	1326,3	59,2
119,1	37,4	1121,1	56,3
75,5	23,7	710,5	47,7
31,9	10	300	48,6

TABLE VII. SOUND PRESSURE LEVELS FOR SLOT NUMBER RATIO 36/27

U, V	f, Hz	n, min^{-1}	L_p, dBA
446,1	140	4200	63,1
424,3	133,2	3994,7	65,7
380,7	119,5	3584,2	61,2
358,9	112,6	3378,9	61,9
315,3	98,9	2968,4	66
271,7	85,3	2557,9	79,9
249,8	78,4	2352,6	76,5
206,3	64,7	1942	71,2
184,5	57,9	1736,8	66,5
140,9	44,2	1326,3	61,9
119,1	37,4	1121,1	68,7
75,5	23,7	710,5	56
31,9	10	300	48,8

TABLE VIII. SOUND PRESSURE LEVELS FOR SLOT NUMBER RATIO 36/30

U, V	f, Hz	n, min^{-1}	L_p, dBA
446,1	140	4200	75,9
424,3	133,2	3994,7	76,5
380,7	119,5	3584,2	78
358,9	112,6	3378,9	78,8
315,3	98,9	2968,4	80,7
271,7	85,3	2557,9	83,5
249,8	78,4	2352,6	85,4
206,3	64,7	1942	92,6
184,5	57,9	1736,8	105,8
140,9	44,2	1326,3	82,3
119,1	37,4	1121,1	73,7
75,5	23,7	710,5	62
31,9	10	300	50,3

TABLE IX. SOUND PRESSURE AND SOUND POWER LEVELS FOR A LOADED MOTOR WITH THE SLOT NUMBER RATIO 36/28

M, Nm	n, min^{-1}	U, V	f, Hz	L_p, dBA	L_w, dBA
339	1700	170,52	58,1	69,6	78,9
339	1200	121,21	41,5	70,7	80
339	600	63,07	21,76	72,7	81,5
340	300	36,4	12,8	61,7	71

TABLE X. SOUND PRESSURE AND SOUND POWER LEVELS FOR A LOADED MOTOR WITH THE SLOT NUMBER RATIO 36/27

M, Nm	n, min^{-1}	U, V	f, Hz	L_p, dBA	L_w, dBA
339	1700	170,7	58,16	70,1	79,4
339	1200	121,42	41,57	74,2	83,5
340	600	63,33	21,85	71,8	81,1
340	300	36,83	12,95	63,8	73,1

TABLE XI. SOUND PRESSURE AND SOUND POWER LEVELS FOR A LOADED MOTOR WITH THE SLOT NUMBER RATIO 36/30

M, Nm	n, min^{-1}	U, V	f, Hz	L_p, dBA	L_w, dBA
341	1700	170,23	58	109,2	118,5
341	1200	120,92	41,4	80,6	89,9
340	600	62,7	21,63	74,4	83,7
341	300	35,78	12,58	55,1	64,4

The results in the tables show that the reduction of the rotor slot number from 28 to 27 resulted in an increase of the sound pressure level in 8 out of 13 operating points at no load and in 3 out of 4 points under load (comparison of the last columns in Tables VI and VII, and in Tables IX and X). The increase in the number of rotor slots from 28 to 30 has led to an increase in the sound pressure level in all no-load operating points and in 3 out of 4 points under load (comparison of the last columns in Tables VI and VIII, and Tables IX and XI). For a loaded motor with 30 slots on the rotor at the speed of 1700 min^{-1}, the noise level exceeds the IEC limit (Fig. 5), so the slot number combination 36/30 is not applicable.

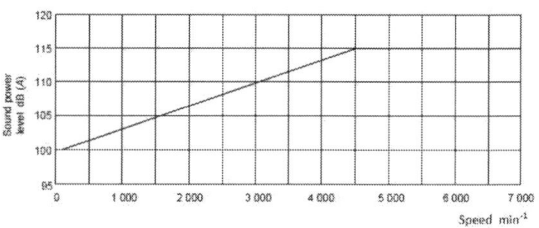

Fig. 5. Limiting mean sound power level for traction motors [1]

Figs. 6, 7 and 8 show the contributions of several vibration modes to the sound power level in the motor V6AOJ 205-04 with stator/rotor slot number combinations 36/28, 36/27 and 36/30 respectively, loaded with 339 Nm at 1700 min^{-1}, fed with sinusoidal voltage. According to (7) and (9), the stator natural frequency is 4282 Hz for the zero mode, 688 Hz for the second mode, 1784 Hz for the third mode, 3088 Hz for the fourth mode, 4486 Hz for the fifth mode, 5918 Hz for the sixth mode, 7355 Hz for the seventh mode, 8786 Hz for the eighth mode and 10206 Hz for the ninth mode. Considering that the number of stator slots was not changed in this simulation, the given natural frequencies are valid for all three combinations of stator/rotor slot numbers.

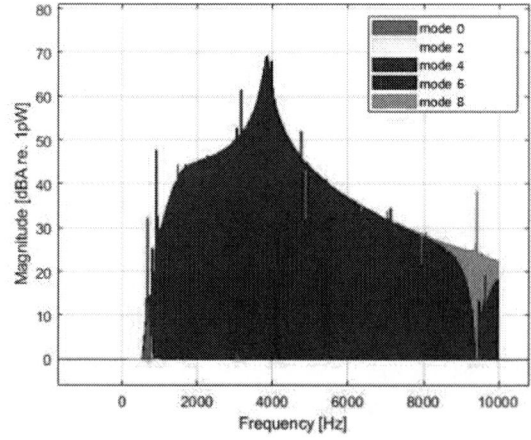

Fig. 6. Contributions of several vibration modes (28 slots on the rotor) to the sound power level

978-1-6654-3236-8/21 $31.00 © 2021 IEEE

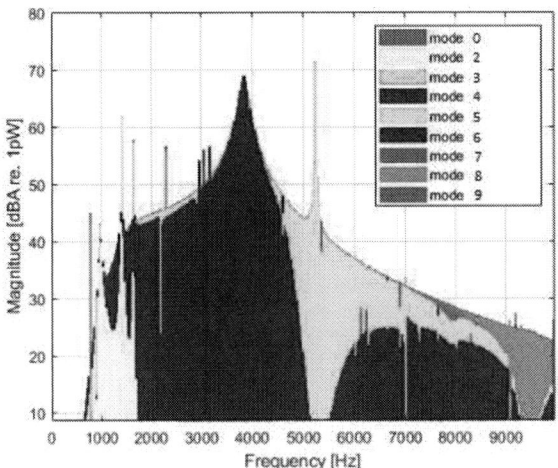

Fig. 7. Contributions of several vibration modes (27 slots on the rotor) to the sound power level

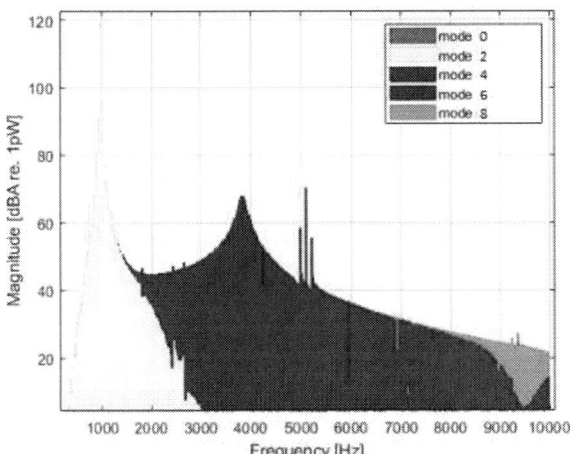

Fig. 8. Contribution of several vibration modes (30 slots on the rotor) to the sound power level

According to (15), for 36 slots on the stator and 28 slots on the rotor, the fourth mode is predominant ($r = 4$), for 27 slots on the rotor, the fifth mode is predominant ($r = 5$), and for 30 slots on the rotor, the second mode is predominant ($r = 2$). Fig. 9 shows the radial stresses which are present in the motor V6AOJ 205-04 with 28 slots on the rotor, loaded with 339 Nm at 1700 min⁻¹, fed with sinusoidal voltage. Two stresses from the mode 4 group with frequencies 142 Hz and 907 Hz and one stress from the mode 8 group with frequency 793 Hz dominate.

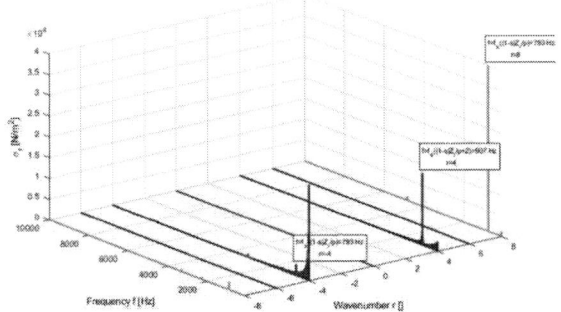

Fig. 9. Noticeable radial stresses and associated frequencies (28 slots)

Fig. 10 shows the radial stresses present in the motor with 27 slots at the same load point. The stress with frequency 142 Hz (mode 4), the stress with frequency 878 Hz (mode 5) and the stress with frequency 765 Hz (mode 9) dominate. Fig. 11 also shows radial stresses, but for the motor with 30 slots on the rotor slots. Here the stresses with frequencies 142 Hz (mode 4), 963 Hz (mode 2) and 850 Hz (mode 6) are dominant.

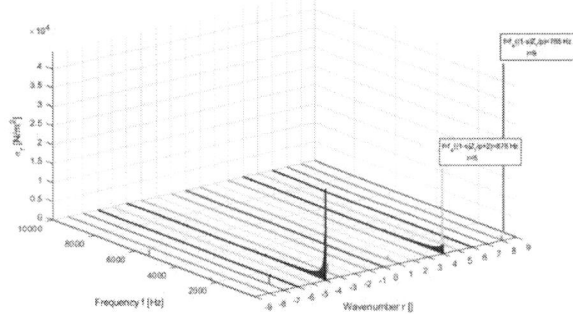

Fig. 10. Noticeable radial stresses and associated frequencies (27 slots)

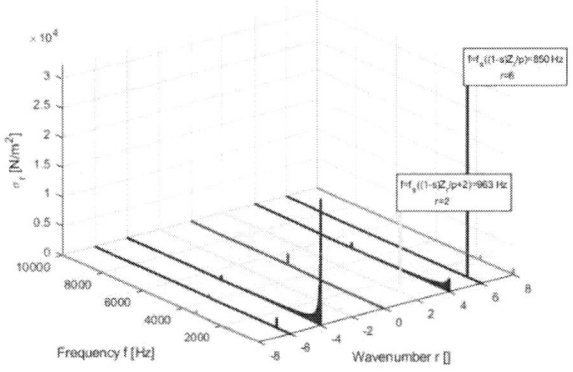

Fig. 11. Noticeable radial stresses and associated frequencies (30 slots)

The comparison of Fig. 6 and Fig. 9 leads to the conclusion that the sound maximum with the frequency around 3800 Hz does not contain the frequencies of the dominant stresses (793 Hz and 907 Hz) shown in Fig. 9. A similar difference is also present when comparing Fig. 7 and Fig. 10, i.e. sound maxima with frequencies around 3800 Hz and 5200 Hz are formed, while the dominant stresses have frequencies 142 Hz, 765 Hz and 878 Hz. Comparison of Fig. 8 and Fig. 11 shows that the sound maximum with frequency around 1000 Hz is close to the peak stress of mode 2 at 963 Hz, which explains why the motor with 30 slots on the rotor is so noisy. The discrepancy between the frequencies at which the sound power is maximized and the frequencies at which the dominant stress occurs is a consequence of the large distance between the natural frequencies of the stator core and the frequencies of the dominant stress.

According to [12], the frequencies of the forces which are exciting stator core vibrations can be calculated using

$$f_r = f_s \left[\frac{k_r N_r}{p} (1 - s) \right] \tag{19}$$

$$f_r = f_s \left[\frac{k_r N_r}{p} (1 - s) \pm 2 \right] \qquad (20)$$

where f_s is the fundamental stator field frequency, k_r is the ordinal number of the air gap magnetic conductivity harmonic due to presence of rotor slots, p is the number of pole pairs, and s is the slip. The dominant harmonic of the magnetic conductivity of the air gap has $k_r = 1$ so for $N_r = 28$, according to (19) and (20), the theoretical vibration frequencies are 793 Hz, 910 Hz and 677 Hz. For $N_r = 27$, the theoretical vibration frequencies are 765 Hz, 881 Hz and 649 Hz. For $N_r = 30$, the theoretical vibration frequencies are 850 Hz, 966 Hz, and 734 Hz. For all slot combinations, equations (19) and (20) give the dominant stress frequencies which are close to the frequencies calculated by MANATEE shown in Figs. 9. to 11.

It is noted that the maximum sound power levels are reached at the modes whose frequencies of the forces which are exciting the stator core vibrations are closest to the natural frequencies of the stator core. This effect is enhanced when the frequency of the dominant stress for the same mode is also relatively close to the natural frequency, which is noticeable for the case of slot numbers 36/30 and vibration mode 2. These frequencies are compared in Table XII. The limited accuracy of the analytical expression (9) for natural frequencies should be considered in this comparison.

The case of 36/30, mode 2, where all three frequencies are relatively close, is highlighted in Table XII. This mode is dominant in producing the high noise level for the 36/30 combination, which exceeds the peak noise levels of all other slot number combinations.

TABLE XII. COMPARISON OF NATURAL STATOR FREQUENCIES, FREQUENCIES OF PEAK SOUND POWER LEVELS, AND FREQUENCIES OF PEAK STRESS FOR ALL THREE SLOT NUMBER COMBINATIONS

Stator/rotor slot number	Natural frequency/mode	Peak sound power level frequency/mode	Peak stress level frequency/mode
36/28	4282 Hz/0 **688 Hz/2** 1784 Hz/3	3800 Hz/4 9400 Hz/8	142 Hz/4 907 Hz/4 793 Hz/8
36/27	3088 Hz/4 4486 Hz/5 5918 Hz/6	3800 Hz/4 5200 Hz/5	142 Hz/4 878 Hz/5 765 Hz/9
36/30	7355 Hz/7 8786 Hz/8 10206 Hz/9	5100 Hz/0 **1000 Hz/2** 3800 Hz/4	142 Hz/4 **963 Hz/2** 850 Hz/6

III. CONCLUSION

The influence of rotor slot number on magnetic noise in a squirrel cage induction motor for traction applications has been analysed in this paper using the software package MANATEE. The natural frequencies of the stator core were calculated analytically for different vibration modes and compared with the frequencies of the peak sound power levels and the frequencies of the peak stress obtained from MANATEE. The results show that the frequencies of the peak sound power levels are more closely related to the natural frequencies than the frequencies of the peak stress due to resonance amplification factor. However, when the stator/rotor slot number combination is selected in which the natural frequency and the frequency of the peak stress are close to each other, as in the case of the stator/rotor slot number combination 36/30 and vibration mode 2, the sound power level near these two frequencies is greatly amplified, resulting in high noise emission.

REFERENCES

[1] Standard IEC 60349-2:2010: Electric traction – Rotating electrical machines for rail and road vehicles – Part 2: Electronic converter – fed alternating current motors, International Electrotechnical Commission, 2010

[2] Koncar - Electrical Industry Inc. Retrieved April 10, 2021, from www.koncar.hr/podrucja-djelovanja/transport/tracnicka-vozila/

[3] K. C. Maliti, Modelling and Analysis of MAGNETIC Noise in Squirrel-Cage Induction Motors, doctoral dissertation, Royal Institute of Technology, Department of Electric Power Engineering, Electrical Machines and Power Electronics, Stockholm, Sweden 2000

[4] S. L. Nau, H. G. G. Mello, "Acoustic Noise in Induction Motors: Causes and Solutions," Proceedings of the 47th IEEE Industry Applications Society Annual Conference on Petroleum and Chemical Industry, pp. 253-263, San Antonio, TX, USA, 2000

[5] J. Le Besnerais, V. Lanfranchi, M. Hecquet and P. Brochet, "Optimal Slot Numbers for Magnetic Noise Reduction in Variable-Speed Induction Motors", IEEE Transactions on Magnetics, Vol. 45, No. 8, pp. 3131-3136, Aug. 2009

[6] J. Le Besnerais. V. Lanfranchi, M. Hecquet, R. Romary, P. Brochet, "Optimal Slot Opening Width for Magnetic Noise Reduction in Induction Motors," IEEE Transactions on Energy Conversion, Vol. 24, No. 4, pp. 869-874, Dec. 2009

[7] M. Donát, "Investigation of the influence of air gap thickness and eccentricity on the noise of the rotating electrical machine," Applied and Computational Mechanics, No. 7, pp. 123-136, 2013

[8] S. L. Nau, "The influence of the skewed rotor slots on the magnetic noise of three-phase induction motors," IET Eight International Conference on Electrical machines and Drives (Conf. Publ. No. 444), pp. 396-399, Cambridge, UK 1997

[9] J. Le Besnerais, Q. Souron, "Effect of Magnetic Wedges on Electromagnetically-induced Acoustic Noise and Vibrations of Electrical Machines," IEEE XXII International Conference on Electrical Machines, pp. 2217-2222, Laussane, Switzerland 2016

[10] J. Pyrhönen, T. Jokinen, V. Hrabovcova, Design of Rotating Electrical Machines, John Wiley & Sons Ltd, 2008

[11] W. R. Finley, "Noise in Induction Motors – Causes and Treatments," IEEE Transactions on Industry Applications, Vol. 27, No. 6, pp. 1204-1213, Nov/Dec 1991

[12] M. Janda, O. Vitek, V. Hajek, Noise of Induction Machines. book chapter, Induction Motors Modelling and Control, IntechOpen 2012

[13] EOMYS Engineering. Retrieved January 15, 2021, from https://eomys.com

Application of a Simplified Inverse Fuzzy Model for an Induction Motor Drive Control

Pavol Fedor
Dept. of El. Engg.
and Mechatronics
Technical University of Košice
Košice, Slovakia
pavol.fedor@tuke.sk

Daniela Perdukova
Dept. of El. Engg.
and Mechatronics
Technical University of Košice
Košice, Slovakia
daniela.perdukova@tuke.sk

Marek Fedor
Dept. of El.
Engg. and Mechatronics
Technical University of Košice
Košice, Slovakia
marek.fedor@tuke.sk

Viliam Fedak
Dept. of El. Engg.
and Mechatronics
Technical University of Košice
Košice, Slovakia
viliam.fedak@tuke.sk

Abstract—**Even, after the adoption of some simplifying assumptions, the three-phase induction motor presents a complex nonlinear dynamic system of the 5th order with multiple inputs. Modern methods of artificial intelligence (fuzzy approach, neural networks) make it possible to find its much simpler and practically usable models based on the description of its important qualitative properties. The presented article deals with the construction of a simplified inverse fuzzy model of an induction motor using an estimation of critical parameters of the Kloss relation, which can be utilized for its control. To build such model, it is necessary to perform only several standard measurements – step responses on the electrical motor. No priori knowledge of motor parameters is required. The obtained simple fuzzy model of the induction motor (IM) is then used for linearization of its torque loop and for design of superior control loops (speed, position). The presented procedure also eliminates the need for transformations of motor quantities, as is usual in vector control.**

Keywords—nonlinear systems control, induction motor, Kloss formula, fuzzy inverse model

I. INTRODUCTION

Nowadays the control of linear systems is elaborated in detail and in practice it is possible to design it analytically for every real system. However, this procedure cannot be applied in the field of nonlinear systems control. Existing control methods are limited to certain selected groups of nonlinear systems due to the invalidity of the superposition principle. This means, there do not exist any universal analytical methods to design their control.

Fuzzy systems present a successful approach to control of many complex nonlinear systems. The fuzzy logic control (FLC) can be considered as one of the most successful applications of the fuzzy systems, which in many cases has become an alternative to the use of conventional control techniques in various areas including: power systems [1-2], mechanical-robotic systems [3-5], automotive systems [6-7] and power electronic systems [8-9]. The fuzzy systems have been implementing satisfactorily also in electrical motors applications such as speed estimation [10] torque ripple minimization [11] and fault identification [12].

The fuzzy logic controller is extensively utilised in real-time IM control using adaptive modelling [13-14]. Also the control based on the inverse fuzzy model is known in the field of FLC [15-16]. This method of the control has many modifications, depending on the specific application [17-18] where the quality of the fuzzy model of the controlled system is also very important [19].

The Slovak Research and Development Agency

Furthermore, FLC can operate in highly linear and nonlinear systems without considering any mathematical model [20-21]. Nevertheless, the accuracy of FLC depends on a suitable design and optimal number of membership functions (MFs), as well as appropriate fuzzy rule generation [22].

In the field of electric drives, the three-phase IM presents a typical representative of a strongly nonlinear system, the analytical description of which, under certain simplifying assumptions, consists of a system of five nonlinear differential equations. The paper shows a methodology of the nonlinear system control design using the linearization by the method of inverse fuzzy model. Based on a general derivation, its application to simple drive control with the IM is presented.

II. MATHEMATICAL MODEL OF AN INDUCTION MOTOR

Several different types of mathematical models of an IM are published in the literature, depending on which motor quantities are chosen as state variables of the model and on the reference system of the selected variables. From the point of view of the control, the IM presents a system of five first-order nonlinear differential equations, which cannot be solved analytically. If the IM stator current and the rotor flux are chosen as system state variables, then its state space mathematical model can be described by the following set equations [10], written down in the matrix form:

$$
\begin{bmatrix} \frac{di_{1x}}{dt} \\ \frac{di_{1y}}{dt} \\ \frac{d\psi_{2x}}{dt} \\ \frac{d\psi_{2y}}{dt} \end{bmatrix} = \begin{bmatrix} -\omega_0 & \omega_1 & -K_{12}\omega_g & -K_{12}\omega_m n_p \\ -\omega_1 & \omega_0 & K_{12}\omega_m n_p & -K_{12}\omega_g \\ M\omega_g & 0 & -\omega_g & \omega_2 \\ 0 & M\omega_g & -\omega_2 & \omega_g \end{bmatrix} \begin{bmatrix} i_{1x} \\ i_{1y} \\ \psi_{2x} \\ \psi_{2y} \end{bmatrix} +
$$
$$
+ \begin{bmatrix} K_{11} & 0 \\ 0 & K_{11} \\ 0 & 0 \\ 0 & 0 \end{bmatrix} \begin{bmatrix} u_{1x} \\ u_{1y} \end{bmatrix} \tag{1}
$$

$$
n_p \frac{M}{L_2} \left(\psi_{2x} i_{1y} - \psi_{2y} i_{1x} \right) - T_L = J \frac{d\omega_m}{dt} \tag{2}
$$

In this model, the motor variables (like the input stator voltage vector $\mathbf{u_1}$, stator current vector $\mathbf{i_1}$ and rotor flux vector $\mathbf{\psi_2}$) are expressed by their components in the rectangular coordinate system $\{x, y\}$, which relatively rotates against the stator by the angular speed ω_1 of the stator rotating magnetic field. The parameters in (1), (2) can

978-1-6654-3236-8/21 $31.00 © 2021 IEEE

be determined directly from the motor parameters according to the following equations:

$$K_{11} = \frac{3}{2}\left(L_{s1} + \frac{L_{s2}L_h}{L_{s2} + L_h}\right)^{-1} \qquad (3)$$

$$K_{12} = -\frac{3}{2}\left(L_{s1} + L_{s2} + \frac{L_{s1}L_{s2}}{L_h}\right)^{-1} \qquad (4)$$

$$\omega_0 = K_{11}\left[R_1 + \left(\frac{M}{L_2}\right)^2 R_2\right] \qquad (5)$$

$$M = \frac{2}{3}L_h \qquad (6)$$

$$\omega_g = \frac{R_2}{L_2} \qquad (7)$$

$$L_2 = \frac{2}{3}(L_{s2} + L_h) \qquad (8)$$

The meaning and used values of IM parameters are given in Tab. 1. The block diagram of the IM corresponding to (1), (2) with the parameters calculated by the relations (3) – (8) is shown in Fig. 1.

The dynamics of the motor variables after its connection to the stator voltage $U_1 = 220$ V with the frequency $\omega_1 = 314$ rad/s is shown in Fig. 2.

TABLE I. ASYNCHRONOUS MOTOR PARAMETERS USED FOR SIMULATION

Symbol	Quantity	Value, dimension
P_N	nominal power	3 kW
U_{1N}	nominal voltage	220 V
I_{1N}	nominal current	6.9 A
T_N	nominal torque	20 Nm
n_N	nominal revolution	1430 rev/min
J	moment of inertia	0,1 kgm^2
R_1	stator phase resistance	1.8 Ω
R_2	rotor phase resistance	1.85 Ω
L_h	main inductance	0.202 H
L_2	inductance defined by equation (8)	0.14 H
$L_{s1} = L_{s2}$	leakage inductance	0.0086 H
M	mutual inductance defined by equation (6)	0.13 H
n_p	number of pole pairs	2
K_{11}	parameter defined by equation (3)	59.35 H^{-1}
K_{12}	parameter defined by equation (4)	-56.93 H^{-1}
ω_0	parameter defined by equation (5)	207.9 s^{-1}
ω_g	rotor winding time constant – equation (7)	13.18 s^{-1}
ω_m	rotor mechanical speed	[rad/s]
ω_1	angular frequency of stator voltage vector	[rad/s]
ω_2	angular slip frequency ($\omega_1 - \omega_m$)	[rad/s]
T_L	load torque	[Nm]
m_1	number of stator phases	3

This relatively complex IM model can be simply replaced by a nonlinear dynamic system of the 1st order, while the nonlinearity of the IM can be described using the so-called Kloss relationship for the motor static torque with the nonlinearity:

$$T = \frac{2\,T_{max}}{\frac{s}{s_{max}} + \frac{s_{max}}{s}} \qquad (9)$$

In this equation, the IM maximum torque T_{max} is:

$$T_{max} = \frac{m_1 U_1^2}{2\,\omega_1\left(R_1 + \sqrt{R_1^2 + X_{sig}^2}\right)} \approx \frac{m_1}{\omega_1}\frac{U_1^2}{2\,X_\delta} =$$

$$= K_m \frac{U_1^2}{\omega_1^2} \qquad (10)$$

$$K_m = \frac{m_1}{2\,L_{s1}} \qquad (11)$$

where:

$$X_\delta = \omega_1 L_\delta \qquad (12)$$

For the slip s and maximum slip s_{max} one applies:

$$s = \frac{\omega_1 - \omega_m}{\omega_1} \qquad (13)$$

$$s_{max} = \pm \frac{R_2'}{\sqrt{R_1^2 + X_\delta^2}} \qquad (14)$$

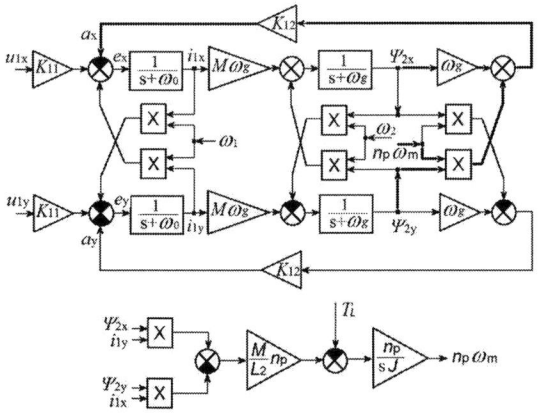

Fig. 1. Block diagram of an IM in the {x,y} coordinate system connected to the stator field

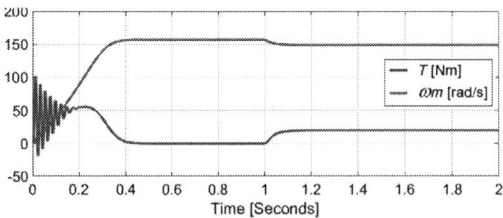

Fig. 2. Time course of the torque and angular speed when connected the IM to $U_1 = 220$ V, $\omega_1 = 314$ rad/s

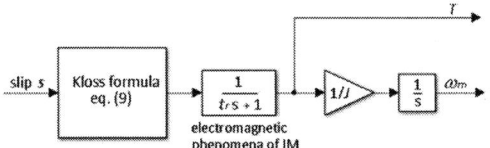

Fig. 3. Block diagram of an induction motor simplified model

When expressing the dynamics of the torque subsystem by the time constant t_r, the block diagram of the simplified model IM is shown in Fig. 3.

The accuracy of this approximation of the IM is given by three parameters of the simplified model: the maximum torque $T_{max}(\omega_1)$ the maximum slip $s_{max}(\omega_1)$ and the substitute time constant of the electromagnetic transients $t_r(\omega_1)$. If the quantitative substitution criterion is chosen as an integral of the root mean square deviation between the simplified and complete IM model, then for the theoretically calculated motor parameters according to the equations (10) to (12) for $\omega_1=314$ rad/s one gets these values of the simplified model: $T_{max} = 58$ Nm, $s_{max} = 0.3185$ and $t_r = 0.01$ s with the value of the criterion function $J = 214$ according to the relation:

$$J = \int \left(T_{real} - T_{simple} \right)^2 dt \tag{15}$$

With optimal parameters of the torque $T_{max} = 58$ Nm, slip $s_{max} = 0.32$ and the time constant $t_r = 0.002$ s the value of the criterion function is $J = 144$. Fig. 4b show the time courses of the actual and approximate torque of the motor for the mentioned cases.

A comparison of the torque courses of the complete and simplified model at steps of the stator frequency ω_l in the vicinity of the nominal speed of the motor is shown in Fig. 5.

It is clear from the figure that the Kloss formula and the simplified block diagram in Fig. 3 approximate the IM well in the vicinity of the working point ω_1, for which the parameters T_{max}, s_{max}, t_r were optimized.

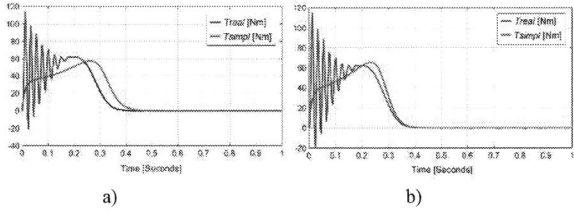

Fig. 4. Optimal approximation of parameters T_{max}, s_{max}, t_r of the simplified AM model: a) calculated parameters; b) optimized parameters

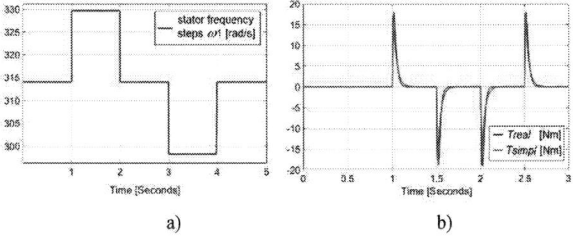

Fig. 5. Fig. 5 a) Stator frequency steps ω_1; b) Comparison of the torque dynamics of the full (T_{real}) and of the simplified (T_{simpl}) model of the IM

III. LINEARIZATION OF THE NONLINEAR TORQUE CHARACTERISTIC OF AN ASYNCHRONOUS MOTOR

The nonlinearity of the IM torque characteristic is given approximately by the Kloss relation (9), which corresponds to the simplified model IM in Fig. 3. The linearization of this static function is generally possible through its inverse function. According to this relation, the torque of the IM is a function of the difference between the angular speed of the stator rotating magnetic field and the mechanical angular speed of the rotor ω_2, where:

$$\omega_2 = \omega_1 - \omega_m = s\,\omega_1 \tag{16}$$

$$\omega_{2max} = s_{max}\,\omega_1 \tag{17}$$

$$T = \frac{2M_{max}\,s_{max}\,\omega_1\,\omega_2}{s_{max}^2\,\omega_1^2 + \omega_2^2} \tag{18}$$

$$T = \frac{2M_{max}\,\omega_{2max}\,\omega_2}{\omega_{2max}^2 + \omega_2^2} \tag{19}$$

Graph if this nonlinear function is shown in Fig. 6a and its inverse function is in Fig. 6b. It is clear from Fig. 5b that the inversion of this particular nonlinearity is not unambiguous, i.e., for example: for the motor torque $T= 50$ Nm the slip angular speed can have two values: $\omega_2 = 48$ rad/s and $\omega_2 = 234$ rad/s.

When linearizing this nonlinear function, therefore it will be necessary to work only in an unambiguous part of the nonlinearity. For this reason, the range of the rotating speed ω_2 is limited to the interval $<0, \omega_{2max}>$. In Fig. 6, the working area defined for ω_2 (and also for the required motor torque at a given speed ω_1) is marked in red.

The analytical inversion of the relation (19) is considerably demanding and it is therefore advantageous to realize the graphical inversion according to Fig. 6b. A clear part of this inversion is shown in Fig. 7.

Figure 8 shows the block diagram of the IM for comparing the linearized simplified and complete IM model.

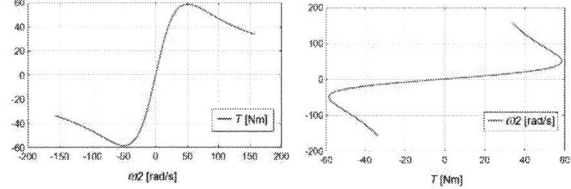

Fig. 6. The static torque characteristic of the induction motor: a) the original function, b) its inversion

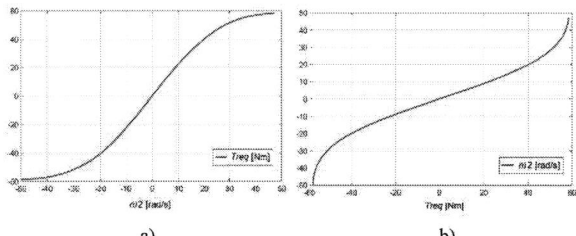

Fig. 7. a) Static torque characteristic of the induction motor and b) its inverse image in the stable part of the Kloss formula

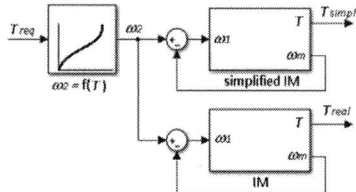

Fig. 8. Linearized torque control loop of the IM

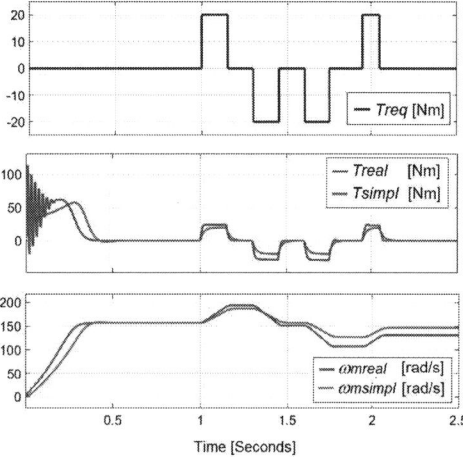

Fig. 9. Comparison of torque dynamics of the complete and simplified IM model with different T_{req} steps

A comparison of the torque dynamics of the complete and simplified model of the IM for different steps of the desired motor torque T_{req} is shown in Fig. 9. The waveforms in this figure show that for the simplified IM model used in the linearization, almost an ideal linear course of the torque control loop corresponding to the 1st order inertia dynamics has been achieved.

IV. FUZZY LINEARIZATION OF IM TORQUE CHARACTERISTIC

As mentioned above, the torque control loop IM can be linearized using an inverse image of its static characteristic. This method has the following disadvantages:

- It is quite difficult to analytically express the inverse image of IM (inverse Kloss formula).
- The inverse Kloss formula for a specific drive depends on the IM parameters, which are known only approximately.
- The basic parameters of the Kloss formula (T_{max}, s_{max}) depend on the instantaneous operating point of the IM (specifically on the value ω_1) according to (10), (12).

In order to avoid these problems, it is advisable to use a procedure in which an optimal invasive image of the IM is obtained directly from measurements on a specific drive. Nowadays the fuzzy systems are considered as a suitable tool for the approximation of nonlinear functions. Chapter III presented the procedure for linearization of the nonlinear characteristic of the IM in the selected operating point (starting the IM, torque steps around the nominal angular speed and constant values T_{max} and s_{max}). If the response of the specific IM has been measured within neighborhood of a selected operating point, then it is possible to obtain a fuzzy image of the valid Kloss formula from this database.

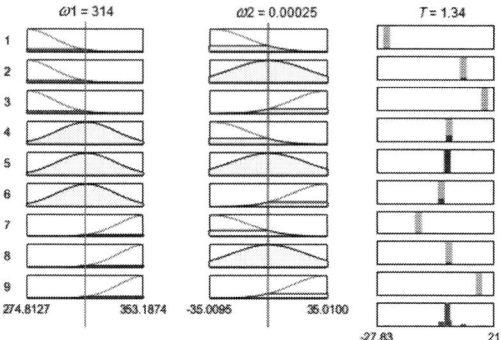

Fig. 10. SUGENO type fuzzy system for approximation of the Kloss formula

The obtained database can be used for searching fuzzy inference structure (FIS) of the IM describing the measured relation between $[\omega_1, \omega_2] \rightarrow T$. Using the measured database, the particular fuzzy model can be designed by standardly known procedures of cluster analysis. For our purpose, subtractive clustering (tool anfis in the MATLAB program), which is a fast and robust data analysis method, was used, having the following parameters: Range of influence = 0.4, Squash factor = 1.25, Accept ratio = 0.4, Reject ratio = 0.01. As a result, the Sugeno type fuzzy system with nine rules is obtained as shown in Fig. 10.

A comparison of an ideal IM torque calculated by the Kloss formula with its fuzzy approximation obtained from the measured database by the fuzzy clustering method is shown in Fig. 11. The waveforms in Fig. 11 confirm the quality of the fuzzy linearization of the IM torque characteristic, where the waveforms of the compared torques almost completely coincide.

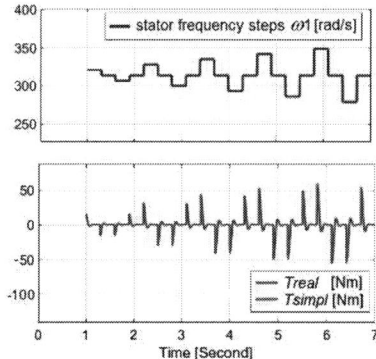

Fig. 11. Comparison of fuzzy approximation of the IM torque according to the Kloss formula with an ideal torque

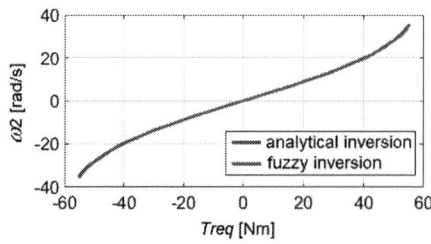

Fig. 12. Comparison of the analytical and fuzzy inversion of the IM nonlinearity

978-1-6654-3236-8/21 $31.00 © 2021 IEEE

Fig. 12 shows that the obtained fuzzy inversion of the nonlinear IM image is almost completely identical to the inversion obtained from the analytical Kloss formula.

The linearization of the IM torque loop using its fuzzy inverse image is shown in Fig. 13.

V. Conclusions

At present, control structures based on the vector control are used for high-quality control of electrical drives with induction motor, where, due to nonlinearities, transformations of IM variables into the selected rotating system is unaivoidable. The quality of the control depends on the accuracy of these transformations, while the accuracy of the transformations depends on the knowledge of the drive parameters. In this way, the task of the IM control is transformed into the task of the control of a similar DC drive with an external excitation.

The presented paper shows an approach to linearization of an IM drive using fuzzy inversion of its fundamental nonlinearity expressed analytically by the Kloss formula. The use of fuzzy approximation of this nonlinearity also eliminates the need for its analytical knowledge and analytical inversion. Such a linearization of the torque loop of the IM drive in principle does not require any transformations and allows to design superior control structures (speed or position control) on basis of known standard methods of linear control.

The use of an inverse fuzzy system of a specific IM also allows its simple application to multi-parameter input values and the adjustment of the nonlinearity of the drive for various operating states and variable motor parameters.

Acknowledgement

This work was supported by the Slovak Research and Development Agency under the Contract no. APVV-19-0210.

References

[1] N. K. Kumar and V. I. Gandhi, "Implementation of fuzzy logic controller in power system applications," J. of Intelligent & Fuzzy Systems, vol. 36, no. 5, pp. 4115-4126, 2019. DOI: 10.3233/JIFS-169971.

[2] M. Gaurkar and Ch. Gowder, "Aapplication of fuzzy logic for power system," IJARIIE, vol. 3, no. 3, 2017. ISSN(O)-2395-43.

[3] J. Bačík J., F. Ďurovský, P. Fedor, and D. Perduková, "Autonomous flying with quadrocopter using fuzzy control and ArUco markers," Intelligent Service Robotics, vol. 10, no. 3, 2017, pp. 185-194. ISSN 1861-2776. DOI: 10.1007/s11370-017-0219-8.

[4] P. Fedor and D. Perduková, "Model based fuzzy control applied to a real nonlinear mechanical system," Iranian Journal of Science and Technology, Trans. of Mechanical Eng., vol. 40, no. 2, pp. 113-124, 2016. ISSN 2228-6187. DOI 10.1007/S40997-016-0005-9.

[5] Ch. H. Chen, Ch. Ch. Wang, Y. T. Wang, and P. T. Wang, "Fuzzy logic controller design for intelligent robots," Mathematical problems in Eng., vol. 2017, Article ID 8984713, 2017. https://doi.org/10.1155/2017/8984713.

[6] V. Ivanov, "A Review of fuzzy methods in automotive engineering applications," European Transportation Research Review, pp. 7-29, 2015.

[7] E. Uzunsoy, "A brief review on fuzzy logic used in vehicle dynamics control," Journal of Innovative Science and Eng. (JISE), vol. 2, no. 1, 2018, pp. 1-7.

[8] I. J. Balaguer, L. Qin, Y. Shuitao, U. Supatti and Z.P. Fang, "Control for grid-connected and intentional islanding operations of distributed power generation," IEEE Transactions on Industrial Electronics, vol. 58, no. 1, pp. 147-157, 2011, ISSN: 0278-0046.

[9] H. R. Chamorro and G. A. Ramos, "Fuzzy Control in Power Electronics Converters for Smart Power Systems," in book: Fuzzy Logic - Controls, Concepts, Theories and Applications. pp. 157-184. 2012. DOI: 10.5772/36311.

[10] P. Brandstetter and M. Kuchar, "Sensorless control of variable speed induction motor drive using RBF neural network," Journal of Applied Logic, vol. 24, pp. 97-108, 2017. DOI: 10.1016/j.jal.2016.11.

[11] B. Kwak, J.H. Um, and J.K. Seok, "Direct active and reactive power control of three-phase inverter for AC motor drives with small DC-link capacitors fed by single-phase diode rectifier," IEEE Trans. Ind. Appl., vol. 55, pp. 3842–3850, 2019.

[12] G. H. Bazan, et al., „Stator fault analysis of three-phase induction motors using information measures and artificial neural networks," Electr. Power Syst. Res., vol. 143, pp. 347–356, 2017.

[13] N. Farah, et al., "A novel self-tuning fuzzy logic controller based induction motor drive system: an experimental approach," IEEE Access, vol. 7, pp. 68172–68184, 2019.

[14] Z. Liu, Z. Zheng and Y. Li, "Enhancing fault-tolerant ability of a nine-phase induction motor drive system using fuzzy logic current controller," IEEE Trans. Energy Convers, vol. 32, pp. 759–769, 2017.

[15] T. Kumbasar, I. Eksin, M. Guzelkaya and E. Yesil, "An inverse controller design method for interval type-2 fuzzy models," Soft Computing - A Fusion of Foundations, Methodologies and Applications, vol. 21, no.10, pp 2665–2686, 2017. https://doi.org/10.1007/s00500-015-1966-0.

[16] R. Babuška, J.M. Sousa and H.B. Verbruggen, "Inverse fuzzy model based predictive control," in Driankov D., Palm R. (eds) Advances in Fuzzy Control. Studies in Fuzziness and Soft Computing, vol 16. Physica, Heidelberg, 1998. https://doi.org/10.1007/978-3-7908-1886-4_6.

[17] M. J. Mahmoodabadi and A. Ziaei, "Inverse dynamics based optimal fuzzy controller for a robot manipulator via particle swarm optimization," Journal of Robotics, vol. 2019, article ID 5052185, 10 p,, 2019. https://doi.org/10.1155/2019/5052185.

[18] S. H. Zareh and A. A. A. Khayyat, „Fuzzy inverse model of magnetorheological dampers for semi-active vibration control of an eleven-degrees of freedom suspension system," J. of System Design and Dynamics, vol. 5. no. 7, pp. 1485-1497, 2011. https://doi.org/10.1299/jsdd.5.1485.

[19] D. Perduková and P. Fedor, „A model-based fuzzy control of an induction motor," Advances in Electrical and Electronic Engineering, VSB-Technical University of Ostrava, vol. 12, no. 5, pp. 635-641, 2015. ISSN 1336-1376. DOI: 10.15598/aeee.v12i5.1229.

[20] S. Rafa, et al., "Implementation of a new fuzzy vector control of induction motor," ISA Trans., vol. 53, pp. 744–754, 2014.

[21] S. Çeven, A. Albayrak and R. Bayır, "Real-time range estimation in electric vehicles using fuzzy logic classifier," Comput. Electr. Eng., vol. 83, pp. 106577, 2020.

[22] M. A. Hannan, J. A. Ali, A. Mohamed, and A. Hussain, "Optimization techniques to enhance the performance of induction motor drives: a review," Renew. Sustain, Energy Rev., vol. 81, pp. 1611–1626, 2017.

978-1-6654-3236-8/21 $31.00 © 2021 IEEE

Index of Authors

Adascalitei, Cristina , 112

Babojelić, Renato, 189
Ban, Željko, 78, 88
Baotić, Mato, 183
Barrett, Michael, 56
Bastovanský, Ronald, 50
Bego, Ozren , 196, 201
Bejvl, Martin, 25
Beng Soh, Chew, 129
Berndl, Sebastian, 43
Blajić, Mislav, 196, 201
Bober, Peter, 238
Bubovich, Alexander, 144
Bulava, Jaroslav, 158
Buzdugan, Mircea, 7

Cao, Shuyu, 129

Despalatović, Marin, 196, 201
Dovdon, Tenuun, 144
Draženović, Karla, 136
Drgona, Peter, 62

Eichler, Jakub, 234
Erica, Dora , 212

Fedak, Viliam, 249
Fedor, Marek, 249
Fedor, Pavol, 249
Ferková, Želmíra, 238
Frivaldsky, Michal, 67

Galkin, Ilya , 144
Geppert, Martin, 19
Glavan, Boris, 37
Gorel, Lukas, 169

Hanić, Ana, 118
Hanić, Zlatko, 37
Hanić, Zlatko , 118
Hargaš, Libor, 158

Hećimović, Ante, 88
Himmelstoss, Felix A, 72
Horváth, Krisztián, 97

Ileš, Šandor, 177, 189

Jakopović, Željko, 136
Jandura, Pavel, 234
Jarek, Grzegorz, 206
Jiang, Hao, 129
Josipović, Goran, 177
Jukić, Filip, 220

Kasprzak, Marcin, 206
Kellner, Jakub, 30
Kierepka, Kamil, 206
Koniar, Dušan, 158
Korotaj, Blaž, 183
Kovačić, Marinko, 37, 118
Kováčik, Michal, 50
Kriletić, Ante , 196, 201
Kristek, Hrvoje, 164
Krčmář, Lukáš , 234
Kutija, Martina, 212, 220
Kyslan, Karol, 102

Makar, Martin, 220
Makarun, Petar, 177
Makyš, Pavol, 169
Malović, Zvonimir, 78
Martis, Claudia, 112
Matuško, Jadranko, 189
Michalak, Jarosław, 206
Milažar, Ivan, 243

Novak, Martin, 228
Novoselnik, Branimir, 183

Pavelek, Miroslav, 13
Pavlić, Ivana, 212
Pelin, Denis, 152
Perdukova, Daniela, 249

Perić, Ante, 136
Peršić, Antonio, 164
Peršić, Vladimir, 164
Petro, Viktor, 102
Pipiska, Michal, 67
Pravica, Luka, 212, 220
Przybyła, Krzysztof, 206

Radu, Martis , 112
Rafajdus, Pavol, 50
Richter, Ales, 234
Rydlo, Pavel, 234

Schröder, Günter, 19
Skorvaga, Jakub, 13
Stepien, Mariusz, 108
Stojanović, Željko, 152

Takács, Kristián, 62
Terzić, Božo, 196, 201

Valouch, Viktor, 25
Vavrúš, Vladimir, 169
Vidlák, Michal, 169
Vilić-Belina, Bruno, 189
Vorobjovs, Maksims , 144
Votzi, Helmut, 72
Vražić, Mario, 37, 164
Vuger, Nikola, 78

Wei, Feng, 129

Zelnik, Richard, 123
Zygmanowski, Marcin A., 206

Šimek, Petr, 25
Šimčák,Marek, 1
Šolc, Ivan, 88
Štefunová, Silvia, 158
Šunde, Viktor, 78, 88, 136
Švec, Marko, 177
Žarko, Damir, 243

IEEE
445 Hoes Lane
Piscataway, NJ 08854-4141

ISBN 978-1-6654-3236-8